T0235599

Communications in Computer and Information Science 1065

Commenced Publication in 2007
Founding and Former Series Editors:
Phoebe Chen, Alfredo Cuzzocrea, Xiaoyong Du, Orhun Kara, Ting Liu,
Krishna M. Sivalingam, Dominik Ślęzak, Takashi Washio, Xiaokang Yang,
and Junsong Yuan

More information about this series at http://www.springer.com/series/7899

Yalin Zheng · Bryan M. Williams ·
Ke Chen (Eds.)

Medical Image Understanding and Analysis

23rd Conference, MIUA 2019
Liverpool, UK, July 24–26, 2019
Proceedings

Springer

Editors
Yalin Zheng ⓘ
University of Liverpool
Liverpool, UK

Bryan M. Williams ⓘ
University of Liverpool
Liverpool, UK

Ke Chen ⓘ
University of Liverpool
Liverpool, UK

ISSN 1865-0929   ISSN 1865-0937 (electronic)
Communications in Computer and Information Science
ISBN 978-3-030-39342-7   ISBN 978-3-030-39343-4 (eBook)
https://doi.org/10.1007/978-3-030-39343-4

This Springer imprint is published by the registered company Springer Nature Switzerland AG
The registered company address is: Gewerbestrasse 11, 6330 Cham, Switzerland

Preface

This volume comprises the proceedings of the 23rd Medical Image Understanding and Analysis (MIUA 2019) Conference, an annual forum organized in the UK for communicating research progress within the community interested in biomedical image analysis. Its goals are the dissemination and discussion of research in medical image processing and analysis to encourage growth, raising the profile of this multidisciplinary field with an ever-increasing real-world applicability. The conference is an excellent opportunity for researchers at all levels to network, generate new ideas, establish new collaborations, learn about and discuss different topics, and listen to speakers of international reputation, as well as presenting their own work in medical image analysis development.

This year's edition was organized by the Centre for Research in Image Analysis (http://liv-cria.co.uk/) and Centre for Mathematical Imaging Techniques (www.liverpool.ac.uk/mathematical-sciences/research/mathematical-imaging-techniques) at the University of Liverpool, in partnership with Springer (www.springer.com), MathWorks (https://uk.mathworks.com/), and Journal of Imaging (www.mdpi.com/journal/jimaging); it was supported by the British Machine Vision Association (BMVA) and OCF (www.ocf.co.uk). The diverse range of topics covered in these proceedings reflect the growth in development and application of medical imaging. The main topics covered in these proceedings are (i) Oncology and Tumour Imaging, (ii) Lesion, Wound, and Ulcer Analysis, (iii) Biostatistics, (iv) Fetal Imaging, (v) Enhancement and Reconstruction, (vi) Diagnosis, Classification, and Treatment, (vii) Vessel and Nerve Analysis, (viii) Image Registration, (ix) Image Segmentation, and (x) Ophthalmology.

The number and level of submissions of this year's edition were excellent. In total, 70 technical papers and 20 abstracts showing clinical and technical applications of image-processing techniques (the latter not considered for inclusion in this volume) were reviewed and revised by an expert team of 93 reviewers. Submissions were received from 20 countries across 4 continents, including the UK (126 authors), China (34), Germany (20), USA (9), Pakistan (8), Italy (7), Denmark (6), Turkey (6), Romania (5), Portugal (4), India (4), Spain (3), New Zealand (3), and one each from Sri Lanka, Thailand, Malaysia, France, Japan, and Iraq. Each of the submissions was reviewed by two to four members of the Program Committee. Based on their ranking and recommendations, 43 of 70 papers were ultimately accepted, 15 were rejected and 12 were invited to present as a poster abstract. We hope you agree that the papers included in this volume demonstrate high quality research and represent a step forward in the medical image analysis field.

We thank all members of the MIUA 2019 Organizing, Program, and Steering Committees and, in particular, all who supported MIUA 2019 by submitting work and attending the conference. We also thank our invited speakers Professor Olaf Ronneberger, Professor Carola-Bibiane Schonlieb, Professor Shaohua K. Zhou, Professor

Sebastien Ourselin, and Professor Simon P. Harding for sharing their success, knowledge, and experiences. We hope you enjoy the proceedings of MIUA 2019.

July 2019

Yalin Zheng
Bryan M. Williams
Ke Chen

Organization

Program Committee Chairs

Yalin Zheng	The University of Liverpool, UK
Ke Chen	The University of Liverpool, UK
Bryan M. Williams	The University of Liverpool, UK

Scientific and Steering Committee

Ke Chen	The University of Liverpool, UK
Víctor González-Castro	University of León, France
Tryphon Lambrou	University of Lincoln, UK
Sasan Mahmoodi	University of Southampton, UK
Stephen McKenna	University of Dundee, UK
Mark Nixon	University of Southampton, UK
Nasir Rajpoot	University of Warwick, UK
Constantino C. Reyes-Aldasoro	City University London, UK
Maria Valdes-Hernandez	The University of Edinburgh, UK
Xianghua Xie	Swansea University, UK
Xujiong Ye	University of Lincoln, UK
Yalin Zheng	The University of Liverpool, UK
Reyer Zwiggelaar	Aberystwyth University, UK

Program Committee

Charith Abhayaratne	The University of Sheffield, UK
Enrique Alegre	Universidad de León, Spain
Paul A. Armitage	The University of Sheffield, UK
Juan Ignacio Arribas	University of Valladolid, Spain
Muhammad Asad	City University London, UK
Lucia Ballerini	The University of Edinburgh, UK
Paul Barber	University College London, UK
Zakaria Belhachmi	Université Haute-Alsace, France
Michael Bennett	University of Southampton, UK
Leonardo Bocchi	University of Florence, Italy
Gunilla Borgefors	Uppsala University, Sweden
Larbi Boubchir	University of Paris 8, France
Joshua Bridge	The University of Liverpool, UK
Gloria Bueno	University of Castilla–La Mancha, Spain
Jorge Novo Buján	Universidade da Coruña, Spain

Maria J. Carreira	University of Santiago de Compostela, Spain
Nathan Cathill	Rochester Institute of Technology, USA
Hammadi Nait Charif	University of Bournemouth, UK
Xu Chen	The University of Liverpool, UK
Yan Chen	Loughborough University, UK
Laurent Cohen	Université Paris-Dauphine, France
Joy Conway	University of Southampton, UK
Tim Cootes	The University of Manchester, UK
Zheng Cui	University of Southampton, UK
María del Milagro Fernández-Carrobles	University of Castilla–La Mancha, Spain
David Alexander Dickie	University of Glasgow, UK
Jin Ming Duan	Imperial College London, UK
Laura Fernández-Robles	University of León, Spain
Samuele Fiorini	University of Genova, Italy
Francesco Fontanella	University of Cassino and Southern Lazio, Italy
Alastair Gale	Loughborough University, UK
Dongxu Gao	The University of Liverpool, UK
María García	University of Valladolid, Spain
Yann Gavet	École nationale supérieure des mines de Saint-Étienne, France
Andrea Giachetti	University of Verona, Italy
Evgin Goceri	Akdeniz University, Turkey
Victor Gonzalez-Castro	University of León, Spain
Juan M. Górriz	University of Granada, Spain
Benjamin Gutierrez Becker	Ludwig Maximilian University, Germany
Matthew Guy	University Hospital Southampton, UK
Jonathon Hare	University of Southampton, UK
Yulia Hicks	University of Cardiff, UK
Krassimira Ivanova	Bulgarian Academy of Sciences, Bulgaria
Andrew King	Kings College London, UK
Frederic Labrosse	Aberystwyth University, UK
Tryphon Lambrou	University of Lincoln, UK
Gabriel Landini	University of Birmingham, UK
Emma Lewis	University Hospital Southampton, UK
Qingde Li	University of Hull, UK
Tom MacGillivray	The University of Edinburgh, UK
Filip Malmberg	Uppsala University, Sweden
Stephen McKenna	University of Dundee, UK
Hongying Meng	Brunel University, UK
Bjoern Menze	Technical University of Munich, Germany
Majid Mirmehdi	University of Bristol, UK
Philip Morrow	Ulster University, UK
Henning Müller	University of Applied Sciences and Arts of Western Switzerland (HES-SO), Switzerland
Susana Munoz-Maniega	The University of Edinburgh, UK

Sponsors

Contents

Diagnosis, Classification and Treatment

Vessel and Nerve Analysis

Posters

Oncology and Tumour Imaging

Tissue Classification to Support Local Active Delineation of Brain Tumors

Albert Comelli[1,2,4]($\boxtimes$) (ID), Alessandro Stefano[2] (ID), Samuel Bignardi[3] (ID),
Claudia Coronnello[1] (ID), Giorgio Russo[2,5] (ID), Maria G. Sabini[5], Massimo Ippolito[6],
and Anthony Yezzi[3]

[1] Fondazione Ri.MED, via Bandiera 11, 90133 Palermo, Italy
acomelli@fondazionerimed.com
[2] Institute of Molecular Bioimaging and Physiology, National Research Council (IBFM-CNR),
Cefalù, Italy
[3] Department of Electrical and Computer Engineering, Georgia Institute of Technology, Atlanta,
GA 30332, USA
[4] Department of Industrial and Digital Innovation (DIID), University of Palermo, Palermo, Italy
[5] Medical Physics Unit, Cannizzaro Hospital, Catania, Italy
[6] Nuclear Medicine Department, Cannizzaro Hospital, Catania, Italy

Abstract. In this paper, we demonstrate how a semi-automatic algorithm we proposed in previous work may be integrated into a protocol which becomes fully automatic for the detection of brain metastases. Such a protocol combines 11C-labeled Methionine PET acquisition with our previous segmentation approach. We show that our algorithm responds especially well to this modality thereby upgrading its status from semi-automatic to fully automatic for the presented application. In this approach, the active contour method is based on the minimization of an energy functional which integrates the information provided by a machine learning algorithm. The rationale behind such a coupling is to introduce in the segmentation the physician knowledge through a component capable of influencing the final outcome toward what would be the segmentation performed by a human operator. In particular, we compare the performance of three different classifiers: Naïve Bayes classification, K-Nearest Neighbor classification, and Discriminant Analysis. A database comprising seventeen patients with brain metastases is considered to assess the performance of the proposed method in the clinical environment.

Regardless of the classifier used, automatically delineated lesions show high agreement with the gold standard ($R^2 = 0.98$). Experimental results show that the proposed protocol is accurate and meets the physician requirements for radiotherapy treatment purpose.

Keywords: Active contour algorithm · Naïve Bayes classification · K-Nearest Neighbor classification · Discriminant Analysis · Segmentation

1 Introduction

Positron Emission Tomography (PET) has the advantage of providing crucial functional information that can be used to evaluate the extent of tumor cell invasion and lead

© Springer Nature Switzerland AG 2020
Y. Zheng et al. (Eds.): MIUA 2019, CCIS 1065, pp. 3–14, 2020.
https://doi.org/10.1007/978-3-030-39343-4_1

to an efficient detection of metastases. In particular, 11C-labeled Methionine (MET) PET conveys complementary information to the anatomical information derived from Magnetic Resonance (MR) imaging [1] or Computerized Tomography (CT) in the brain district, and under favourable conditions, it can even deliver higher performances [2]. As shown by Grosu et al. [3], tumor volumes defined by MR and PET can differ substantially and the integration of the biological tumor volume (BTV) in the radiotherapy treatment planning (RTP) involves a longer median survival than patients in which the RTP is defined using MRI-only. As a final remark, since MET-PET is sensitive to metabolic changes, it can be used for the early detection of the effects produced by radiotherapy. An efficient BTV delineation is crucial not only towards precise RTP [4], but also towards PET quantification accuracy [5].

Despite the fact that manual segmentation is time consuming and operator-dependent (i.e., radiotherapy planning experts tend to draw larger contours than nuclear medicine physicians), the manual approach is still widely adopted in the clinical environment, mostly because of its simplicity. Automatic algorithms do exist, but there is no common agreement on the best choice for a standard PET contouring method [6]. Although the PET high contrast between tumor and healthy tissues can reduce the inter- and intra-observer variability, the BTV delineation is strongly dependent on the segmentation algorithm used [7] and suffers of the typically low resolution of PET images caused by detector crystal size, scanner geometry, and positron range. In addition, image noise is introduced by random coincidences, detector normalization, dead time, scatter, and attenuation.

A well-designed contouring method must meet three basic properties: accuracy, repeatability and efficiency. We refer to "accuracy" as the property of being able to reproduce the segmentation an expert operator would draw, "repeatability" as the property of reproducing the same segmentation starting from different initial segmentations, and "efficiency" as the property of being computationally fast enough to be used in the everyday clinical practice with reduced machine resources. Regarding the latter one, running the algorithm on commercial laptops and in real time would be desirable. These properties may be ensured only by computer-assisted methods supporting clinicians in the planning phase. For this reason, several PET delineation approaches have been proposed and the interested reader may refer to comprehensive reviews, i.e. [8].

In this study, we use an active contour based method which is characterized by flexibility and efficiency in medical image segmentation task meeting the three afore-mentioned basic characteristics (as reported in [9–12]). In particular, we show how a semi-automatic algorithm we introduced in [13–15] is used to design a fully automatic and operator independent protocol to delineate brain tumors. The approach leverages both on the opportunity of using the MET radio-tracer and a local active contour segmentation algorithm, which couples the local optimization guided by the information contained in the PET images with additional information provided by machine learning elements [16–18]. The main idea is to use a properly trained classifier to help the segmentation. Typically, active contour algorithms are formulated as an optimization process where a scalar functional, referred to as "energy" or "objective function", is to be minimized. The objective function depends on the segmentation (i.e. on the shape of a contour), and it is designed in such a way that its minimum corresponds to the best possible segmentation. Indeed, in our algorithm we use an adaptation of the original active contour algorithm [19] to PET imaging and where the mathematical form of the energy functional includes the classification information. In the specific, the classifier is used to label PET tissues as "lesion" (i.e. abnormal tissue), "background" (i.e. normal tissue) or "border-line" (i.e. tissue around the lesion of unclear nature) based on a previous training, which takes into account the portions of the tumor the physician would discard or include [20]. In this way, the performance of the active contour algorithm in the BTV segmentation are enhanced and more similar to a manual segmentation. The idea behind coupling the local region-based active contour with a classifier is to introduce in the segmentation some of the "wisdom" of the physician through a component capable of influencing the final outcome toward what would be the segmentation performed by an operator. In our previous studies [13–15], we proposed a semi-automatic tool to segment oncological lesions and compare it with other state of the art BTV segmentation methods. Results showed that the algorithm outperforms the other ones tested for comparison (T42%, RG [21], FCM clustering [22], RW [23], and original LAC [19]). In this study, we demonstrate how that approach may be integrated into a protocol which becomes fully automatic for the detection of brain metastases. Such a protocol combines 11C-labeled Methionine PET acquisition with our previous segmentation approach. We show that our algorithm responds especially well to this modality thereby upgrading its status from semi-automatic to fully automatic for the presented application. As an additional contribution, we compare the performance of three different classifiers: Naïve Bayes classification, K-Nearest Neighbor classification, and Discriminant Analysis, when these are used to supply information to the LAC method. To assess the performance of the proposed method in clinical environment, seventeen brain tumors have been considered.

2 Materials

2.1 Dataset

Seventeen patients with brain metastases referred to diagnostic MET PET/CT scan have been retrospectively considered. The scan only interested the brain region.

Patients fasted 4 h before the PET examination, and successively were intravenously injected with MET. The PET/CT oncological protocol started 10 min after the injection. Tumor segmentation was performed off-line without influencing the treatment protocol or the patient management. The institutional hospital medical ethics review board approved the study protocol and all patients involved were properly informed released their written consent.

2.2 PET/CT Acquisition Protocol

PET/CT scans were performed using the Discovery 690 PET/CT scanner (General Electric Medical Systems, Milwaukee, WI, USA). The PET protocol includes a SCOUT scan at 40 mA, a CT scan at 140 keV and 150 mA (10 s), and 3D PET scans. The 3D ordered subset expectation maximization algorithm was used to for the PET imaging. Each PET image obtained consists of 256×256 voxels with a grid spacing of 1.17 mm^3 and thickness of 3.27 mm^3. Consequently, the size of each voxel is $1.17 \times 1.17 \times 3.27 \text{ mm}^3$. Thanks to the injected PET radiotracer, tumor appears as hyper-intense region. The CT scan was performed contextually to the PET imaging and used for attenuation correction. Each CT image consists of 512×512 voxels with size $1.36 \times 1.36 \times 3.75 \text{ mm}^3$.

3 Methods

3.1 The Proposed Segmentation

In this section we present a brief overview of the employed segmentation algorithm (Fig. 1). Data from seventeen oncological patients were used partly to train the classifiers, and partly to assess the performances of the delineation algorithm.

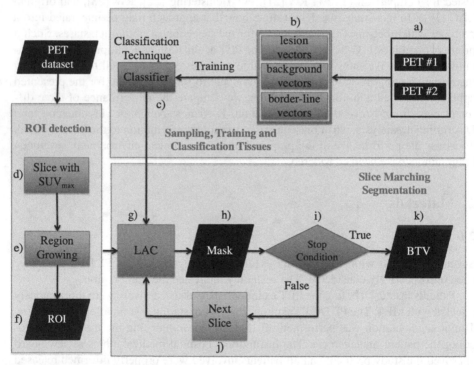

Fig. 1. The segmentation approach [13–15]. (a) part of the dataset is used to train the classifier; (b) samples are obtained for each tissue kind label; (c) training and validation of the classifier; (d) the region containing the lesion is automatically localized; (e) RG is used to identify the user independent mask; (f) once the initial mask is obtained, it is sent to the next logical block of the system; (g) segmentation is performed using the LAC segmentation algorithm through a new energy based on the tissue classification; (h) segmentation mask is propagated to the adjacent slices using a slice-by-slice marching approach; (i) a stop condition is evaluated; (j) segmentation on the next slices is performed until stop condition is false; (k) an operator independent BTV is finally obtained.

Classifiers are used to classify PET voxels according to three labels: tumor, background and border-line. Training and validation were accomplished using a moving window, 3 by 3 voxels, in the PET slices of a data subset. Each window was then reorganized in a 9-element sample vector. The obtained samples were generated and used for training and validation purposes, as detailed in Sect. 3.2. Training and validation steps are required to be performed only once. After, the classifier can be reused on any new dataset. Concerning the practical use of the system on clinical cases, the first step is the automatic identification of the optimal combination of starting ROI and slice containing the tumor to input the subsequent logical steps of the system (Sect. 3.3). Once the ROI is identified, the corresponding mask is feed into the next step of the system, where the segmentation is performed combining a Local region-based Active Contour (LAC) algorithm, appropriately modified to support PET images, and the information derived from the classifiers, which locally drive the active contour (Sect. 3.4). The obtained segmentation is then propagated to the adjacent slices using a slice-by-slice marching approach. Each time convergence criteria are met for a slice, the obtained optimal contour is propagated to the next slice, and evolution is resumed. Starting from the first slice considered, propagation is performed in parallel both upward and downward within the data volume and it is continued until a suitable stopping condition, designed to detect tumor-free slice, is met. Finally, a user independent BTV is obtained.

3.2 Sampling, Training and Performance of Classifiers

In order to normalize the voxel activity and to take into account the functional aspects of the disease, the PET images are pre-processed into SUV images [13]. The classifier (either NB, KNN, or DA) are used to partition PET tissues into three labels: "normal", "abnormal", and "border-line" regions. The goal is to combine each classification with the PET image information which locally drive the LAC delineation. Before integrating the classifier in the LAC method, training and validation phases are required to provide the capability of efficiently classify a newly-encountered tissue into the afore mentioned classes. However, the task requires to be performed only once. Once trained, the classifier is ready to be used on any new case. In order to generate the training input to the classifier, two brain metastases, for a total of 16 PET slices, were used (Fig. 1a). Each PET image containing the lesion is investigated using a moving window of 3×3 voxels in size. The ROI size was empirically determined to provide the best performance on the present dataset. For each new position of the moving window, the selected portion of data was compared with the gold standard. Windows entirely outside or inside the gold standard were labelled as normal or abnormal respectively. Windows comprising no more than three lesion voxels were labeled as border-line tissue. Each window was then reorganized as a 9-element vector (Fig. 1b). The sampling operation produced a total of 834095 samples; 1706 labeled as lesion vectors, 976 border-line, and 831412 labelled as background. The processing time for single slice was: 54.03 s for NB, 40.48 s for KNN, and 26.95 s for DA (iMac computer with a 3.5 GHz Intel Core i7 processor, 16 GB 1600 MHz DDR3 memory, and OS X El Capitan.). This task is performed only once. The 80% of samples was used to train the three classifiers, while the remaining 20% was used to verify the reliability in classifying newly encountered samples (Fig. 1c). The K-Fold cross-validation was integrated in the classifiers to make it reliable and to limit problems such as over-fitting. Once the training step was completed, a validation task was performed to verify the rate of success in classifying new scenarios.

3.3 The Fully Automatic Protocol

By taking advantage of the great sensitivity and specificity of MET in the discrimination between healthy and tumor tissues in the brain district, the proposed protocol identifies the PET slice containing the maximum SUV (SUV_{max}) in the whole PET volume to identify the ROI enclosing the tumor without any operator intervention [24]. Once the current slice with SUV^j_{max} has been identified, an automatic procedure to identify the corresponding ROI starts. The SUV^j_{max} voxel is used as target seed for a rough 2D segmentation based on the region growing (RG) method [25] (Fig. 1e). For each lesion, the obtained ROI represents the output of this preliminary step (Fig. 1f) which is feed into the next logical block of the segmentation algorithm (Fig. 1g) where the actual delineation takes place through the LAC approach. It is worth noting that the RG algorithm is used only to obtain a rough estimate of the tumor contour(s).

3.4 The Modified Local Active Contour Method

In our previous work, the active contour methodology proposed by Lankton et al. [19] was improved. For sake of completeness, we summarize the mathematical development of [21, 22] regarding how the tissue classification, which separates the PET image into lesion, background, and border-line tissues can be integrated in classical LAC in order to further improve the segmentation process. Briefly, the contour energy to be minimized is defined as:

$$E = \oint_C \left(\int_{R_{in}} \chi_l(x, s)(I(x) - u_l(s))^2 dx + \int_{R_{out}} \chi_l(x, s)(I(x) - v_l(s))^2 dx \right) ds \quad (1)$$

- R_{in} denotes the regions inside the curve C
- R_{out} denotes the regions outside the curve C
- s denotes the arc length parameter of C
- χ denotes the characteristic function of the ball of radius l centered around a given curve point $C(s)$
- I denotes the intensity function of the image to be delineated
- $u_l(s)$ and $v_l(s)$ represent the local mean image intensities within the portions of the local neighborhood $\chi_l(x, s)$ inside and outside the curve.

Beyond the identification of the initial ROI, as described in the previous section, the energy (1) has been modified to include a new energy term to separate the PET image into three regions considering tissue classification: $\chi_{lesion}(x)$, $\chi_{border-line}(x)$ and $\chi_{background}(x)$ represent the characteristic functions of the tissue's classification (using KNN or NB or DA) respectively for lesion, background, and border-line tissues.

The first term of the formulated energy functional (1) is essentially a prior term penalizing the overlap between regions which are classified in opposite ways by the contour versus the classifier (no penalty is paid in regions classified as "border-line", for this reason the "$\chi_{\text{border-line}}(x)$" classification is not included in the energy).

To integrate this new prior term and to incorporate SUV in the LAC algorithm, the energy for the PET image segmentation approach is defined as:

$$
E = \oint_C \lambda \left(\int_{R_{in}} \chi_l(x,s)\bar{\text{P}}\text{out}_l(x)dx + \int_{R_{out}} \chi_l(x,s)\bar{\text{P}}\text{in}_l(x)dx \right.
$$
$$
\left. + (1-\lambda) \left(\int_{R_{in}} \chi_l(x,s)(\text{SUV}(x) - u_l(s))^2 dx + \int_{R_{out}} \chi_l(x,s)(\text{SUV}(x) - v_l(s))^2 dx \right) ds \right.
$$
$$
\tag{2}
$$

where the parameter $\lambda \in R^+$ (range between 0 and 1) is chosen subjectively (in our study λ equal to 0.01 provided the best result). $\bar{\text{P}}\text{in}_l(x)$ and $\bar{\text{P}}\text{out}_l(x)$ denote the local mean tissue's classification within the portions of the local neighborhood $\chi_l(x)$ inside and outside the curve respectively (within Ω):

$$
\bar{\text{P}}\text{in}_l(x) = \frac{\int_\Omega \chi_l(x)\chi_{lesion}(x)dx}{\int_\Omega \chi_l(x)dx}, \bar{\text{P}}\text{out}_l(x) = \frac{\int_\Omega \chi_l(x)\chi_{background}(x)dx}{\int_\Omega \chi_l(x)dx} \tag{3}
$$

$u_l(s)$ and $v_l(s)$ denote the local mean SUVs within the portions of the local neighborhood $\chi_l(x,s)$ inside and outside the curve. These neighborhoods are defined by the function χ, the radius parameter l, and the position of the curve C.

Finally, as explained in [13], the process is automatically stopped avoiding the need for any user intervention.

3.5 Framework for Performance Evaluation

A framework for the evaluation of the proposed protocol is presented. The effectiveness of the tissue classification is calculated regarding correct/incorrect classification using sensitivity, specificity, precision, and accuracy scores. Overlap-based and spatial distance-based metrics are considered to determine the accuracy achieved by the proposed computer-assisted segmentation system against the gold-standard (i.e., the manual segmentations performed by three experts are used to define a consolidated reference as described in the next section).

The sensitivity is the number of correctly classified positive samples divided by the number of true positive samples, while the specificity is the number of correctly classified negative samples divided by the number of true negative samples. The precision is related to reproducibility and repeatability and it is defined as the degree to which repeated classifications under unchanged conditions show the same results. The accuracy is defined as the number of correctly classified samples divided by the number of classified samples. Concerning segmentation algorithm performance, the formulations proposed in [26] are used. In particular, mean, and standard deviation of sensitivity, positive predictive value (PPV), dice similarity coefficient (DSC), and Hausdorff distance (HD) were calculated.

3.6 Gold Standard

The ground truth requires exact knowledge of the tumor and the histopathology analysis provides the only valid ground truth for the PET quantitative assessment. Nevertheless, the histopathology analysis is unavailable after the treatment. For this reason, the manual delineations performed by expert clinicians are a commonly accepted substitute for ground truth to assess the clinical effectiveness and feasibility of PET delineation methods [27]. Nevertheless, manual delineation is often influenced by the clinical specialization of the operator. For example, oncologists will, on average, draw smaller BTVs than radio-therapists. For this reason, the segmentations performed by three experts with different expertise (the chief nuclear medicine physician –M.I. author-, the chief radiotherapy physician –M.S. author- and an expert radiotherapy physician –G.R. author-) were used as "ground truths". A simultaneous ground truth estimation tool [28] was employed, and the three segmentations were combined to define a single and consolidated ground truth for each study.

4   Results

4.1 Classifier Validation

The optimal K value of the K-Fold cross-validation integrated in the classifier has been determined as 5 through the trial-and-error method (k range: 5–15, step size of 5) [29, 30]. It corresponds to the highest classification accuracy. The validation results are shown in Table 1.

Table 1. Sensitivity, specificity, precision, and accuracy values for KNN, NB, and DA classifier validations.

	Sensitivity	Specificity	Precision	Accuracy
KNN	97.13%	81.54%	98.88%	95.25%
NB	87.09%	94.94%	99.68%	86.73%
DA	90.00%	85.12%	99.02%	88.31%

4.2 Clinical Testing and Results on Dataset

Seventeen brain lesions were considered. From the initial dataset, two tumors were used in the classifier training. Consequently, the performance of the presented algorithm is investigated in the remaining cases against the ground truth provided by three expert operators (Table 2).

Table 2. Mean sensitivities, PPVs, DSCs and HDs for 15 lesions are reported to assess the differences between the segmentations obtained using the LAC method with KNN, NB and DA classifiers and the "ground truth" provided by the three operators.

	Sensitivity (%)	PPV (%)	DSC (%)	HD (voxels)
KNN	91.54 ± 1.35	85.36 ± 3.72	88.27 ± 1.91	1.18 ± 0.52
NB	89.58 ± 3.40	89.76 ± 3.31	89.58 ± 2.37	1.07 ± 0.61
DA	91.21 ± 1.93	90.78 ± 2.03	90.92 ± 1.35	0.79 ± 0.40

In addition, regardless of the classifier used, automatically segmented BTVs show high agreement with the manually segmented BTVs (determination coefficient $R^2 = 0.98$). Figure 2 reports the comparison between the proposed segmentations and the gold-standards for two patients.

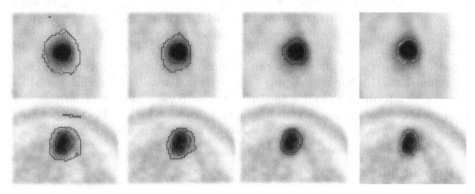

Fig. 2. Two example of brain tumor segmentation for each row using the LAC method coupled with KNN (first column), NB (second column), and DA (third column) classifiers. The proposed segmentations (red contours) and the gold standards (yellow contours) are superimposed. Blue and green contours concern the tissue classification. The region outside the blue boundary represent the "background", while the region inside the green boundaries, and between the curves, represent the "lesion", and the "border-line" region, respectively. In the last column, all retrieved segmentations are superimposed (gold standard in yellow, KNN in magenta, NB in cyan, and DA in light blue). (Color figure online)

5 Discussion

In this study, we described a segmentation protocol which leverages on the properties of MET PET to achieve the fully automatic segmentation of brain cancer. In this context, we used a segmentation algorithm LAC which features an energy function specifically adapted to the PET imaging and which combines the information from the PET data, and feedback from a classifier. The aim of such a protocol is for radiotherapy treatment purpose and for therapy response assessment. Each classifier was purposely and independently trained to label PET tissues into normal, abnormal, and border-line tissues. The training procedure is based on the ground truth obtained using a proper tool [28] starting from the manual segmentations provided by three expert operators. In addition, a smart sampling operation based on a moving window of 3×3 voxels has been implemented. After this preparation step, each classifier was able to label tissues never encountered before and to convey this useful information into the LAC algorithm. As a final remark, a fully automatic stop condition was provided. In this way and by taking advantage of the great sensitivity and specificity of MET PET studies to identify tumors in brain area, the proposed system produces segmentation results which are completely independent by the user.

Performance of the proposed method were obtained by patient studies, for which the ground is impossible to obtain because the treatment alters the tissue morphology. Consequently, although manual delineation may largely differ between different human operators (for example, radiotherapy experts tend to draw larger boundaries than nuclear medicine physicians), with obvious impact on the resulting surrogate of ground truth, the delineation from experts is the only alternative. Here, for performance evaluation purposes, the manual segmentations from three experts was used as a gold-standard. Such manual delineations were used to produce a consolidated reference [28] which was then used to assess the feasibility of the proposed method in the clinical environment. Seventeen patients undergo a PET/CT scan have been considered. Results show that the proposed protocol can be considered clinically feasible using any of the tested classifier, although the DA delivered slightly better results. Automatically segmented BTVs showed high agreement with the gold-standard (R2 = 0.98). Considering the relevance of our study in oncological patient management we will add PET studies to expand our database and, consequently, to improve and further assess the proposed method.

6 Conclusion

We demonstrated how a semi-automatic algorithm we proposed in previous work may be integrated into a protocol which becomes fully automatic for the detection brain metastases. Such a protocol combines 11C-labeled Methionine PET acquisition with our previous segmentation approach. The key features of the protocol include the use of MET radio tracer, the integration of patient-specific functional information obtained by converting PET images to SUV images; a preliminary task to identify an initial, properly localized, operator-independent ROI to be used for the LAC segmentation; the integration of tissue classification (either using KNN, NB, or DA) in the LAC method directly in the formulation of the energy functional to be minimized to enhance the accuracy of the BTV contouring; and a slice-by-slice marching approach with an automatic termination condition. This features make the whole process fully automatic in the context of brain cancer.

Brain metastases were used to assess the performance of the proposed protocol on a statistical basis commonly considered as a reference practice in the PET imaging field. Results showed that this protocol can be used to extract PET parameters for therapy response evaluation purpose and to provide automatic the BTV delineation during radiosurgery treatment planning in the special context of brain cancer.

References

1. Comelli, A., et al.: Automatic multi-seed detection for MR breast image segmentation. In: Battiato, S., Gallo, G., Schettini, R., Stanco, F. (eds.) ICIAP 2017. LNCS, vol. 10484, pp. 706–717. Springer, Cham (2017). https://doi.org/10.1007/978-3-319-68560-1_63
2. Astner, S.T., Dobrei-Ciuchendea, M., Essler, M., et al.: Effect of 11C-Methionine-Positron emission tomography on gross tumor volume delineation in stereotactic radiotherapy of skull base meningiomas. Int. J. Radiat. Oncol. Biol. Phys. **72**, 1161–1167 (2008). https://doi.org/10.1016/j.ijrobp.2008.02.058
3. Grosu, A.L., Weber, W.A., Franz, M., et al.: Reirradiation of recurrent high-grade gliomas using amino acid PET (SPECT)/CT/MRI image fusion to determine gross tumor volume for stereotactic fractionated radiotherapy. Int. J. Radiat. Oncol. Biol. Phys. **63**, 511–519 (2005). https://doi.org/10.1016/j.ijrobp.2005.01.056
4. Borasi, G., Russo, G., Alongi, F., et al.: High-intensity focused ultrasound plus concomitant radiotherapy: a new weapon in oncology? J. Ther. Ultrasound **1**, 6 (2013). https://doi.org/10.1186/2050-5736-1-6
5. Banna, G.L., Anile, G., Russo, G., et al.: Predictive and prognostic value of early disease progression by PET evaluation in advanced non-small cell lung cancer. Oncol. (2017). https://doi.org/10.1159/000448005
6. Angulakshmi, M., Lakshmi Priya, G.G.: Automated brain tumour segmentation techniques— a review. Int. J. Imaging Syst. Technol. (2017). https://doi.org/10.1002/ima.22211
7. Zaidi, H., El Naqa, I.: PET-guided delineation of radiation therapy treatment volumes: a survey of image segmentation techniques. Eur. J. Nucl. Med. Mol. Imaging **37**, 2165–2187 (2010). https://doi.org/10.1007/s00259-010-1423-3
8. Foster, B., Bagci, U., Mansoor, A., et al.: A review on segmentation of positron emission tomography images. Comput. Biol. Med. **50**, 76–96 (2014). https://doi.org/10.1016/j.compbiomed.2014.04.014
9. Khadidos, A., Sanchez, V., Li, C.T.: Weighted level set evolution based on local edge features for medical image segmentation. IEEE Trans. Image Process. (2017). https://doi.org/10.1109/TIP.2017.2666042
10. Göçeri, E.: Fully automated liver segmentation using Sobolev gradient-based level set evolution. Int. J. Numer. Method Biomed. Eng. (2016). https://doi.org/10.1002/cnm.2765
11. Min, H., Jia, W., Zhao, Y., et al.: LATE: a level-set method based on local approximation of taylor expansion for segmenting intensity inhomogeneous images. IEEE Trans. Image Process. (2018). https://doi.org/10.1109/TIP.2018.2848471
12. Goceri, E., Dura, E.: A level set method with Sobolev gradient and Haralick edge detection. In: 4th World Conference on Information Technology (WCIT 2013), vol. 5, pp. 131–140 (2013)
13. Comelli, A., Stefano, A., Russo, G., et al.: A smart and operator independent system to delineate tumours in positron emission tomography scans. Comput. Biol. Med. (2018). https://doi.org/10.1016/J.COMPBIOMED.2018.09.002
14. Comelli, A., Stefano, A., Russo, G., et al.: K-nearest neighbor driving active contours to delineate biological tumor volumes. Eng. Appl. Artif. Intell. **81**, 133–144 (2019). https://doi.org/10.1016/j.engappai.2019.02.005
15. Comelli, A., Stefano, A., Bignardi, S., et al.: Active contour algorithm with discriminant analysis for delineating tumors in positron emission tomography. Artif. Intell. Med. **94**, 67–78 (2019). https://doi.org/10.1016/J.ARTMED.2019.01.002
16. Comelli, A., et al.: A kernel support vector machine based technique for Crohn's disease classification in human patients. In: Barolli, L., Terzo, O. (eds.) CISIS 2017. AISC, vol. 611, pp. 262–273. Springer, Cham (2018). https://doi.org/10.1007/978-3-319-61566-0_25

17. Licari, L., et al.: Use of the KSVM-based system for the definition, validation and identification of the incisional hernia recurrence risk factors. Il Giornale di chirurgia **40**(1), 32–38 (2019)
18. Agnello, L., Comelli, A., Ardizzone, E., Vitabile, S.: Unsupervised tissue classification of brain MR images for voxel-based morphometry analysis. Int. J. Imaging Syst. Technol. (2016). https://doi.org/10.1002/ima.22168
19. Lankton, S., Nain, D., Yezzi, A., Tannenbaum, A.: Hybrid geodesic region-based curve evolutions for image segmentation. In: Proceedings of the SPIE 6510, Medical Imaging 2007: Physics of Medical Imaging, 16 March 2007, p. 65104U (2007). https://doi.org/10.1117/12.709700
20. Comelli, A., Stefano, A., Benfante, V., Russo, G.: Normal and abnormal tissue classification in PET oncological studies. Pattern Recognit. Image Anal. **28**, 121–128 (2018). https://doi.org/10.1134/S1054661818010054
21. Day, E., Betler, J., Parda, D., et al.: A region growing method for tumor volume segmentation on PET images for rectal and anal cancer patients. Med. Phys. **36**, 4349–4358 (2009). https://doi.org/10.1118/1.3213099
22. Belhassen, S., Zaidi, H.: A novel fuzzy C-means algorithm for unsupervised heterogeneous tumor quantification in PET. Med. Phys. **37**, 1309–1324 (2010). https://doi.org/10.1118/1.3301610
23. Stefano, A., Vitabile, S., Russo, G., et al.: An enhanced random walk algorithm for delineation of head and neck cancers in PET studies. Med. Biol. Eng. Comput. **55**, 897–908 (2017). https://doi.org/10.1007/s11517-016-1571-0
24. Stefano, A., Vitabile, S., Russo, G., et al.: A fully automatic method for biological target volume segmentation of brain metastases. Int. J. Imaging Syst. Technol. **26**, 29–37 (2016). https://doi.org/10.1002/ima.22154
25. Stefano, A., et al.: An automatic method for metabolic evaluation of gamma knife treatments. In: Murino, V., Puppo, E. (eds.) ICIAP 2015. LNCS, vol. 9279, pp. 579–589. Springer, Cham (2015). https://doi.org/10.1007/978-3-319-23231-7_52
26. Taha, A.A., Hanbury, A.: Metrics for evaluating 3D medical image segmentation: analysis, selection, and tool. BMC Med. Imaging **15**, 29 (2015). https://doi.org/10.1186/s12880-015-0068-x
27. Hatt, M., Laurent, B., Ouahabi, A., et al.: The first MICCAI challenge on PET tumor segmentation. Med. Image Anal. **44**, 177–195 (2018). https://doi.org/10.1016/j.media.2017.12.007
28. Warfield, S.K., Zou, K.H., Wells, W.M.: Simultaneous truth and performance level estimation (STAPLE): an algorithm for the validation of image segmentation. IEEE Trans. Med. Imaging **23**, 903–921 (2004). https://doi.org/10.1109/TMI.2004.828354
29. Agnello, L., Comelli, A., Vitabile, S.: Feature dimensionality reduction for mammographic report classification. In: Pop, F., Kołodziej, J., Di Martino, B. (eds.) Resource Management for Big Data Platforms. CCN, pp. 311–337. Springer, Cham (2016). https://doi.org/10.1007/978-3-319-44881-7_15
30. Comelli, A., Agnello, L., Vitabile, S.: An ontology-based retrieval system for mammographic reports. In: Proceedings of IEEE Symposium Computers and Communication (2016). https://doi.org/10.1109/ISCC.2015.7405644

Using a Conditional Generative Adversarial Network (cGAN) for Prostate Segmentation

Amelie Grall[1], Azam Hamidinekoo[1(✉)] iD, Paul Malcolm[2] iD,
and Reyer Zwiggelaar[1] iD

[1] Department of Computer Science, Aberystwyth University, Aberystwyth, UK
ameliegrall5@gmail.com, {azh2,rrz}@aber.ac.uk
[2] Department of Radiology, Norwich & Norfolk University Hospital, Norwich, UK
paul.malcolm@nnuh.nhs.uk

Abstract. Prostate cancer is the second most commonly diagnosed cancer among men and currently multi-parametric MRI is a promising imaging technique used for clinical workup of prostate cancer. Accurate detection and localisation of the prostate tissue boundary on various MRI scans can be helpful for obtaining a region of interest for Computer Aided Diagnosis systems. In this paper, we present a fully automated detection and segmentation pipeline using a conditional Generative Adversarial Network (cGAN). We investigated the robustness of the cGAN model against adding Gaussian noise or removing noise from the training data. Based on the detection and segmentation metrics, denoising did not show a significant improvement. However, by including noisy images in the training data, the detection and segmentation performance was improved in each 3D modality, which resulted in comparable to state-of-the-art results.

Keywords: Prostate MRI · Computer Aided Diagnosis · Segmentation · Detection · Generative Adversarial Network

1 Introduction

Prostate cancer is the second most commonly diagnosed cancer among men that has been reported to account for nearly 14% of the total new cancer cases and 6% of the total cancer deaths in 2008 [12]. Incidence rates for prostate cancer are projected to rise by 12% in UK between 2014 and 2035 and it has been estimated that 1 in 8 men will be diagnosed with prostate cancer during their lifetime based on the UK Cancer Research [2]. Commonly used multi-parametric MRI modalities of the prostate include: DWI (Diffusion Weighted Imaging), T1W (T1-weighted), T2W (T2-weighted), ADC (Apparent Diffusion Coefficient), DCE (Dynamic Contrast Enhanced) and MRS (Magnetic Resonance Spectroscopy). T1W and T2W MRI are the most common modalities in radiology and DWI is one of the most recently established modalities, while DCE and MRS are popular

© Springer Nature Switzerland AG 2020
Y. Zheng et al. (Eds.): MIUA 2019, CCIS 1065, pp. 15–25, 2020.
https://doi.org/10.1007/978-3-030-39343-4_2

due to their ability to provide additional information not available in conventional MRI [9]. Computer Aided Diagnosis (CAD) systems have been developed as an alternative to double reading, improving clinicians' accuracy and patient outcomes. These systems are aimed at improving identification of prostate tissue and detection of subtle abnormalities within the prostate gland [7].

For manual delineation of prostate boundaries, traditional machine learning methods have been used and many studies have proposed (semi-)automatic methods, including multi-atlas based methods, model-based methods and graph cut based approaches [7,16,22]. Recently, deep convolutional neural networks (CNNs) have achieved unprecedented results in segmentation by learning a hierarchical representation of the input data, without relying on hand-crafted features [4], some of which performed well [13,19,26] in the "MICCAI PROMISE12" challenge [16].

Among various deep learning based approaches, Generative models [8,11,20] have shown to be one of the most promising models for the segmentation tasks. To the best of our knowledge, no work has been done towards prostate MRI segmentation using generative models.The goal of this work was to develop an automatic detection and segmentation model based on a conditional Generative Adversarial Networks (cGAN) to detect the prostate tissue on various MR modalities (T2W, DWI and ADC). We evaluated the effect of adding or removing noise to/from the training dataset to test the robustness and generalisability of the model from the training to the testing samples.

2   Methodology

The goal of this work was to accurately segment prostate tissue in various MRI modalities. Detecting and accurately localising prostate tissue boundaries is useful for identifying the region of interest (RoI) in CAD systems to be used for radiotherapy or tracking disease progression. We also investigated whether the performance could be improved by adding/removing noise to/from the training data in the learning process.

2.1   Datasets

Data from 40 patients with biopsy-proven prostate cancer was collected, which included ADC, DWI and T2W MR imaging from the Department of Radiology at the Norfolk and Norwich University Hospital, Norwich, UK. All images were obtained with a 3 T magnet GE Medical Systems, using a phased array pelvic coil, with a field of view equal to $100 \times 100 \, \text{mm}^2$, 6 mm slice thickness, acquiring 512×512 images, while covering the whole prostate and seminal vesicles in both transverse and coronal planes. For each case, there were between 16 to 32 axial images. All images were manually annotated by an expert radiologist to indicate the precise location of the prostate on a slice-by-slice basis. This dataset was randomly split into training and testing sets as 75% and 25% of the whole database, based on cases, ensuring that there was no case overlap between the

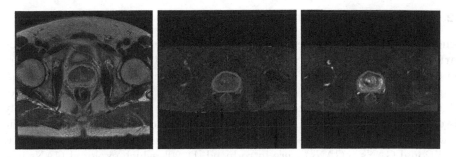

Fig. 1. Samples images for various considered modalities with the prostate tissue being annotated. from left to right: T2W, ADC and DWI.

splits. Figure 1 contains examples of raw images of each modality along with the annotated prostate boundaries.

We aimed to investigate the consequences of additional noise and noise removal in the training process for the detection and segmentation of the prostate tissue. We created additional datasets from the raw images, where Gaussian noise with different σ values (0.5, 1, 2) was added to all modalities (DWI, ADC, T2W), resulting in three noisy datasets (representing low, medium, high level noisy datasets) for each modality.

Over the past years, although the resolution, signal-to-noise ratio and speed of magnetic resonance imaging technology have been increased, MR images are still affected by artifacts and noise. MR images suffer significantly from Rician noise: for low signal intensities (SNR < 2), the noise follows a Rice distribution; however, according to [3], for SNR larger than 2, the noise distribution is nearly Gaussian. There are a wide range of MRI de-noising algorithms; and among them, the Non-Local Means (NLM) filter has been shown to be able to achieve state-of-the-art de-noising performance [3]. The NLM filter averages across voxels that capture similar structures instead of recovering the true intensity value of a voxel by averaging the intensity values of neighboring voxels, as the Gaussian smoothing filter does. Accordingly, in a patch-based approach, the NLM avoids blurring structural details [3]. We have used the NLM filter to create datasets of de-noised images from the raw images. In our de-noising implementation, the required parameters in the NLM model (the search radius (M) and the neighbourhood radius (d)) were empirically adjusted as $M = 5$ and $d = 2$. In order to investigate the power of de-noising with regards to the value of h_i that controls the attenuation of the exponential function (see [4] for more details), two other datasets were also created with different values of h_i. For one dataset, the value of h_i was obtained based on [5] (denoted as h_i^1). In the second dataset, h_i was multiplied by 3 (denoted as h_i^2) to create further blurred images to enhance the effects of de-noising.

2.2   Model Architecture

Generative models have been studied as a fundamental approach in machine learning to reconstruct or generate natural images [10]. These models, capture the underlying structure in data by defining flexible probability distributions. Image generation through a Generative Adversarial Network (GAN) was proposed by Goodfellow *et al.* [8], which estimated a loss function aimed to classify the output image as real or fake via a discriminative model. At the same time, the network trained a generative model to minimize the loss function. Subsequently, conditional GANs (cGANs) were introduced [11,20], which explored GANs in a conditional setting, i.e. applying a specific condition on the input image to train a conditional generative model.

Prostate tissue detection and segmentation on MRI scans in 3D can be defined as translating each 2D MR scan into the corresponding semantic label map representing the desired tissue in the image. In the architecture of the utilised cGAN, a U-Net-based structure [11,23] was used as the generator (G) and a convolutional PatchGAN classifier [11] was used as the discriminator (D). The discriminator penalised structures at image patch scale, which was determined to be 70×70 due to the optimal patch size reported by [11]. The aim of the generator was to minimise the overall loss function (G^*) in Eq. 1 with the definition of $\mathcal{L}_{cGAN}$ in Eq. 2,

$$G^* = arg \, min_G \, max_D \, \mathcal{L}_{cGAN}(G, D) + \lambda \mathcal{L}_1 \qquad (1)$$

where

$$\mathcal{L}_{cGAN}(G, D) = E_{x,y}[log D(x, y)] + E_{x,z}[log(1 - D(x, G(x, z)))] \qquad (2)$$

The generator was trained to generate mask images (y) from paired prostate images z, with the aim of being similar to the mask images of the real observed images from the training data repository (x) in an $L1$ sense. This was formulated as G: x,y $\rightarrow$ y. Meanwhile, the adversarially trained discriminator attempted to detect the generator's fake images as much as possible. To achieve this, D was updated by maximising $[log D(x, y) + log(1 - D(x, G(x, z)))]$ and G was updated by maximising $[log D(x, G(x, z)) + L_1(y, G(x, z))]$. The schematic of this model is shown in Fig. 2. This model converged when G recovered the training sample distribution and D was equal to 0.5 everywhere (for the mathematical proof, see [8].

2.3   Training and Post-processing

We combined raw images with de-noised images and images with added noise to create various datasets as demonstrated in Fig. 3. For each targeted MRI modality (DWI, T2W and ADC), we trained a cGAN model on a specific dataset and tested the detection and segmentation performance on the raw test images. The model was trained via a stochastic gradient descent solver with a learning rate of 0.00004 and batch-Size of 16 based on the parameter tuning approach to select

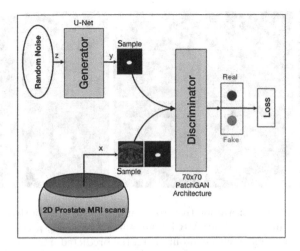

Fig. 2. Schematic of the used cGAN model.

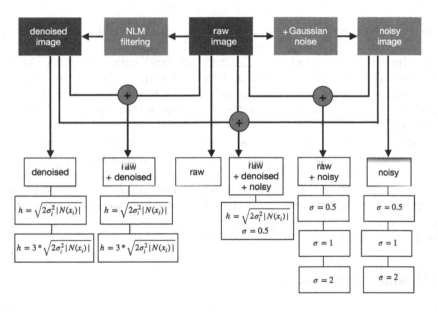

Fig. 3. Generated datasets including raw, noisy and de-noised samples. In our experiments, three levels of noise were added to the raw images and accordingly three noisy datasets were generated. Besides, two de-noising models were applied to the raw images. Overall, 12 datasets were created for training purpose for each targeted MRI modality (T2W, DWI and ADC).

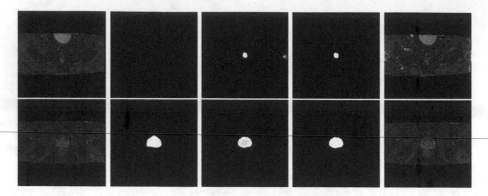

Fig. 4. Two example cases: a poor (top row) and appropriate (bottom row) segmentation example. From left to right: MR image, annotated mask, predicted mask, mask after post processing, image with contours from the predicted mask (red) and contours from the mask after post processing (green). Black mask image in *(b)* represents that no prostate tissue was annotated on the slice. (Color figure online)

the most appropriate parameters. The network was trained till convergence with the total number of iterations equal to 50. All network training was performed within the Torch framework using pix2pix software[1] with a NVIDIA GeForce GTX 1080 GPU on Intel Core i7-4790 Processor with an Ubuntu 16.04 operating system. Additional programming was done in Python and the code can be accessed at https://github.com/AzmHmd/Analysing-MRI-Prostate-scans-with-cGANs.git.

In post-processing, objects close to the image border were removed and only the objects that were located between 1/4 and 3/4 of the height, and 1/4 and 3/4 of the width were kept (based on the empirical observation of all the images). Subsequently, using connected component analysis, the largest object was extracted. In addition, the top and bottom slices of the prostate were disregarded. Figure 4 presents an appropriate and a poor detection and segmentation output with respect to the average model performance.

3   Results and Discussion

The evaluation of the detection and segmentation performances were done using common metrics [16]: (1) the Dice Similarity Coefficient (DSC), (2) the 95% Hausdorff distance (HD), (3) Jaccard Index (JI), (4) Precision, (5) Recall, (6) Accuracy and (7) F.Score for each image of the 10 unseen cases in the test set. Table 1 reports the quantitative results in terms of mean and standard deviation of the three segmentation metrics for each dataset and for each specific modality. Based on our quantitative results, across each modality, the best results were obtained for the dataset combining raw and noisy images. Adding a low

---

[1] https://phillipi.github.io/pix2pix.

Table 1. Quantitative results of the segmentation/detection for all datasets and modalities.

Modality	Dataset	DSC	JI	HD_{95}	F.Score
ADC	Raw	0.665 ± 0.187	0.526 ± 0.202	12.327 ± 6.393	0.682
	Noisy ($\sigma = 0.5$)	$\mathbf{0.734 \pm 0.180}$	$\mathbf{0.608 \pm 0.208}$	$\mathbf{10.279 \pm 6.277}$	0.775
	Raw + Noisy ($\sigma = 0.5$)	$\mathbf{0.732 \pm 0.153}$	$\mathbf{0.599 \pm 0.176}$	$\mathbf{11.550 \pm 5.437}$	0.700
	Noisy ($\sigma = 1$)	0.691 ± 0.229	0.569 ± 0.238	12.184 ± 10.517	0.721
	Raw + Noisy ($\sigma = 1$)	0.707 ± 0.155	0.567 ± 0.171	11.647 ± 4.556	0.642
	Noisy ($\sigma = 2$)	0.579 ± 0.204	0.435 ± 0.196	14.248 ± 8.235	0.678
	Raw + Noisy ($\sigma = 2$)	0.654 ± 0.204	0.518 ± 0.218	13.515 ± 6.791	0.756
	Denoised ($h_i 1$)	0.715 ± 0.186	0.587 ± 0.211	12.010 ± 6.791	0.693
	Raw + Denoised ($h_i 1$)	0.709 ± 0.164	0.573 ± 0.183	$\mathbf{10.967 \pm 5.733}$	0.698
	Denoised ($h_i 2$)	0.648 ± 0.178	0.503 ± 0.184	12.514 ± 7.253	$\mathbf{0.789}$
	Raw + Denoised ($h_i 2$)	0.719 ± 0.197	0.593 ± 0.211	13.050 ± 5.313	0.716
	Raw + Denoised ($h_i 1$) + Noisy ($\sigma = 0.5$)	0.660 ± 0.184	0.518 ± 0.191	12.287 ± 7.063	0.693
DWI	Raw	0.744 ± 0.254	0.644 ± 0.259	9.821 ± 9.895	0.664
	Noisy ($\sigma = 0.5$)	$\mathbf{0.788 \pm 0.156}$	$\mathbf{0.675 \pm 0.191}$	$\mathbf{8.656 \pm 8.910}$	0.656
	Raw + Noisy ($\sigma = 0.5$)	$\mathbf{0.787 \pm 0.146}$	$\mathbf{0.670 \pm 0.177}$	11.744 ± 5.223	0.735
	Noisy ($\sigma = 1$)	0.772 ± 0.222	0.669 ± 0.224	10.178 ± 13.406	0.680
	Raw + Noisy ($\sigma = 1$)	0.629 ± 0.233	0.498 ± 0.237	12.943 ± 7.100	$\mathbf{0.754}$
	Noisy ($\sigma = 2$)	0.718 ± 0.201	0.593 ± 0.216	9.578 ± 6.663	0.700
	Raw + Noisy ($\sigma = 2$)	0.100 ± 0.102	0.056 ± 0.064	22.476 ± 8.905	0.466
	Denoised ($h_i 1$)	0.709 ± 0.226	0.589 ± 0.234	10.516 ± 8.405	0.696
	Raw + Denoised ($h_i 1$)	0.649 ± 0.202	0.512 ± 0.209	12.354 ± 7.731	0.731
	Denoised ($h_i 2$)	0.719 ± 0.174	0.586 ± 0.186	12.670 ± 9.480	0.646
	Raw + Denoised ($h_i 2$)	0.758 ± 0.185	0.642 ± 0.214	10.328 ± 10.409	0.678
	Raw + Denoised ($h_i 1$) + Noisy ($\sigma = 0.5$)	0.782 ± 0.172	0.670 ± 0.206	11.979 ± 5.471	0.656
T2W	Raw	0.668 ± 0.160	0.521 ± 0.165	11.819 ± 7.036	0.750
	Noisy ($\sigma = 0.5$)	0.670 ± 0.174	0.526 ± 0.174	12.591 ± 10.661	0.733
	Raw + Noisy ($\sigma = 0.5$)	0.729 ± 0.144	0.592 ± 0.164	11.981 ± 11.293	0.777
	Noisy ($\sigma = 1$)	0.496 ± 0.163	0.345 ± 0.147	15.553 ± 9.139	0.730
	Raw + Noisy ($\sigma = 1$)	$\mathbf{0.735 \pm 0.161}$	$\mathbf{0.603 \pm 0.174}$	$\mathbf{9.599 \pm 4.373}$	$\mathbf{0.812}$
	Noisy ($\sigma = 2$)	$\mathbf{0.742 \pm 0.118}$	$\mathbf{0.603 \pm 0.138}$	$\mathbf{9.875 \pm 4.533}$	0.763
	Raw + Noisy ($\sigma = 2$)	0.721 ± 0.154	0.584 ± 0.170	10.193 ± 4.601	0.761
	Denoised ($h_i 1$)	0.707 ± 0.176	0.572 ± 0.192	12.603 ± 14.093	0.809
	Raw + Denoised ($h_i 1$)	0.706 ± 0.191	0.574 ± 0.196	11.376 ± 8.836	0.790
	Denoised ($h_i 2$)	0.692 ± 0.203	0.559 ± 0.198	13.365 ± 15.227	0.768
	Raw + Denoised ($h_i 2$)	0.712 ± 0.178	0.577 ± 0.182	15.636 ± 19.430	0.746
	Raw + Denoised ($h_i 1$) + Noisy ($\sigma = 0.5$)	0.722 ± 0.171	0.589 ± 0.183	10.015 ± 4.608	0.793

noise level ($\sigma = 0.5$) to images and combining the original (raw) and noisy images could enable the network to learn better by improving the regularisation of the training procedure. We defined the model trained on the raw data as the "base-learning". In the ADC and DWI modalities, training the network on images with low added noise ($\sigma = 0.5$) improved the results compared to the base-learning, while by adding more noise ($\sigma = 1$, $\sigma = 2$) to the raw images, the results did not show improvement. For example, in the ADC modality, the DSC mean for the raw dataset was 0.665 and improved to the value of 0.734 by adding a low noise level, but the performance did not change significantly or even decreased to the values of 0.691 and 0.589 for $\sigma = 1$ and $\sigma = 2$, respectively. The T2W modality provided similar outcome. By increasing the noise to $\sigma = 1$ in this modality, the dataset composed of only noisy images showed the worst results, with a DSC mean value of 0.496, but increasing the noise ($\sigma = 2$), the best results were achieved and the DSC mean reached the maximum value of 0.742. Despite differences between results from datasets composed of different noise levels, datasets combining raw and noisy images achieved the best results (with DSC vales of 0.73, 0.78 and 0.74 for the ADC, DWI and T2W modalities, respectively, with the smallest spread of values from the average) and the outcome was consistent between the different evaluation metrics for analysing the segmentation performance. It was observed that the dataset composed of raw, de-noised and noisy images did not show improvements compared to the combination of raw and noisy images, although this dataset clearly incorporated most samples. Feeding a learning system with sharped, blurred and noisy samples - resulting from different image effects- was expected to improve the outcome but the results showed that not necessarily the number of training samples affects the learning performance but also the number of discriminative samples is important i.e. the difference between the raw and de-noised data was not significant and so the obtained results did not differ much but the nature of raw and noisy images was slightly different, which led to improvement in learning. Furthermore, the effect of noise on various modalities was different. Considering noisy samples, T2W images were affected by higher levels of noise, while the ADC or DWI images were affected by lower noise levels, considering that the additional noise could affect the surrounding structures of the prostate. Applying a de-noising method to the raw images did not improve the results significantly and datasets composed of only de-noised images or both de-noised and raw images provided the lowest results compared to datasests containing noisy images. We have also tested the robustness of the model towards blurred images by multiplying the parameter h by 3 (indicated by h_i^2 in the results tables) and the same results were obtained for de-noised images and blurred images, as well as the combination of them with the raw images. Thus, we concluded that for the model, they looked similar and no additional learning was achieved with more blurred images used, which demonstrated the robustness of the model behaviour towards such modifications in the training set that could be regarded as poor image quality. Finally, in terms of detection performance, considering numerical values shown in Table 1, the best results (in term of the F.Score) from the combination of raw plus noisy images was confirmed, which was in agreement with our previous observations. Indeed, the F.Score reached an average score of 0.812 and 0.754 for the T2W and DWI modalities, using raw plus noisy images, respectively, while the best score for the ADC modality for the detection was achieved with denoised images. Table 2 summarises and compares the segmentation performance of various methods using the dataset released in the "MICCAI PROMISE12" challenge. In this challenge, multi-center and multi-vendor training data was released to train the models. To be mentioned, we have used a different dataset (as explained in Sect. 2.1) for training our proposed model and for testing. With these considerations and using the same evaluation metrics, comparative results were obtained on the explained testing set. This work was the first use of cGANs for prostate MRI analysis. The proposed pipeline can be helpful in health-care appellations, due to its optimal performance, end-to-end learning scheme with integrated feature learning, capability of handling complex and multi-modality data plus simplicity of implementation.

Table 2. Comparison of state-of-the-art segmentation approaches for prostate T2W images.

Reference	Methodology	HD	DSC
[25]	Active appearance models	5.54 ± 1.74	0.89 ± 0.03
[1]	Region-specific hierarchical segmentation using discriminative learning	6.04 ± 1.67	0.87 ± 0.03
[18]	Interactive segmentation tool called Smart Paint	7.34 ± 3.08	0.85 ± 0.08
[27]	Convex optimization with star shape prior	7.15 ± 2.08	0.84 ± 0.04
[17]	3D active appearance models	6.72 ± 1.42	0.83 ± 0.06
[14]	Probabilistic active shape model	11.08 ± 5.85	0.77 ± 0.10
[6]	Local appearance atlases and patch-based voxel weighting	5.89 ± 2.59	0.76 ± 0.26
[15]	Multi-atlas approach	7.95 ± 3.21	0.80 ± 0.07
[21]	Zooming process with robust registration and atlas selection	7.07 ± 1.64	0.65 ± 0.34
[24]	Deformable landmark-free active appearance models	8.48 ± 0.59	0.70 ± 0.10
cGAN model	Conditional generative adversarial model	9.59 ± 4.37	0.73 ± 0.16

4   Conclusion

Among many deep learning based models, Generative Adversarial Networks have been among the successful supervised deep learning models, proposed as a general purpose solution for image-to-image translation treated under specific conditional settings. Our contribution in this paper was to present the simple but effective cGAN model to analyse prostate tissue in 3D MRI sequences. Feeding the model with a suitable input, it was able to compute parameters layer by layer in the generator and the discriminator parts and estimate a final output that corresponded to prostate boundaries in different scans of various MRI modalities. The objective of training was to minimise the difference between the prediction of the network and the expected output (manual annotated boundaries by an expert radiologist). Our quantitative results indicate the robustness and generalisability of the model to image quality along with the effectiveness of including noise in the training samples for the task of detection and segmentation.

5 Future Work

In future, we extend the segmentation task to the related prostate structures or substructures and subsequently put more focus on cancer localization and lesion identification considering the discrimination between transitional, central and peripheral zones (TZ, CZ, PZ), the neurovascular bundles or the seminal vesicles. Tumor staging is another perspective of focus that we could investigate when we acquire more case studies to be sufficient to train a deep network. Investigating other data augmentation approaches is another perspective that can be considered in future work.

Acknowledgments. The authors would like to acknowledge Dr. Alun Jones and Sandy Spence for their support and maintenance of the GPU and the computer systems used for this research.

References

1. Birkbeck, N., Zhang, J., Requardt, M., Kiefer, B., Gall, P., Kevin Zhou, S.: Region-specific hierarchical segmentation of MR prostate using discriminative learning. In: MICCAI Grand Challenge: Prostate MR Image Segmentation, vol. 2012 (2012)
2. Cancer-Research-UK: Prostate cancer statistics (2018). http://www.cancerresearchuk.org/health-professional/cancer-statistics/statistics-by-cancer-type/prostate-cancer
3. Chen, G., Zhang, P., Wu, Y., Shen, D., Yap, P.T.: Denoising magnetic resonance images using collaborative non-local means. Neurocomputing **177**, 215–227 (2016)
4. Cheng, R., et al.: Active appearance model and deep learning for more accurate prostate segmentation on MRI. In: Medical Imaging 2016: Image Processing, vol. 9784, p. 97842I. International Society for Optics and Photonics (2016)
5. Coupé, P., Yger, P., Prima, S., Hellier, P., Kervrann, C., Barillot, C.: An optimized blockwise nonlocal means denoising filter for 3-D magnetic resonance images. IEEE Trans. Med. Imaging **27**(4), 425–441 (2008)
6. Gao, Q., Rueckert, D., Edwards, P.: An automatic multi-atlas based prostate segmentation using local appearance-specific atlases and patch-based voxel weighting. In: MICCAI Grand Challenge: Prostate MR Image Segmentation, vol. 2012a (2012)
7. Ghose, S., et al.: A survey of prostate segmentation methodologies in ultrasound, magnetic resonance and computed tomography images. Comput. Methods Programs Biomed. **108**(1), 262–287 (2012)
8. Goodfellow, I., et al.: Generative adversarial nets. In: Advances in Neural Information Processing Systems, pp. 2672–2680 (2014)
9. Hegde, J.V., et al.: Multiparametric MRI of prostate cancer: an update on state-of-the-art techniques and their performance in detecting and localizing prostate cancer. J. Magn. Reson. Imaging **37**(5), 1035–1054 (2013)
10. Hinton, G.E., Osindero, S., Teh, Y.W.: A fast learning algorithm for deep belief nets. Neural Comput. **18**(7), 1527–1554 (2006)
11. Isola, P., Zhu, J.Y., Zhou, T., Efros, A.A.: Image-to-image translation with conditional adversarial networks. arXiv preprint (2017)
12. Jemal, A., Bray, F., Center, M.M., Ferlay, J., Ward, E., Forman, D.: Global cancer statistics. CA Cancer J. Clin. **61**(2), 69–90 (2011)

13. Karimi, D., Samei, G., Shao, Y., Salcudean, T.: A novel deep learning-based method for prostate segmentation in T2-weighted magnetic resonance imaging (2017)
14. Kirschner, M., Jung, F., Wesarg, S.: Automatic prostate segmentation in MR images with a probabilistic active shape model. In: MICCAI Grand Challenge: Prostate MR Image Segmentation, vol. 2012 (2012)
15. Litjens, G., Karssemeijer, N., Huisman, H.: A multi-atlas approach for prostate segmentation in MR images. In: MICCAI Grand Challenge: Prostate MR Image Segmentation, vol. 2012 (2012)
16. Litjens, G., et al.: Evaluation of prostate segmentation algorithms for MRI: the PROMISE12 challenge. Med. Image Anal. **18**(2), 359–373 (2014)
17. Maan, B., van der Heijden, F.: Prostate MR image segmentation using 3D active appearance models. In: MICCAI Grand Challenge: Prostate MR Image Segmentation, vol. 2012 (2012)
18. Malmberg, F., Strand, R., Kullberg, J., Nordenskjöld, R., Bengtsson, E.: Smart paint a new interactive segmentation method applied to MR prostate segmentation. In: MICCAI Grand Challenge: Prostate MR Image Segmentation, vol. 2012 (2012)
19. Milletari, F., Navab, N., Ahmadi, S.A.: V-net: fully convolutional neural networks for volumetric medical image segmentation. In: Fourth International Conference on 3D Vision (3DV), pp. 565–571. IEEE (2016)
20. Mirza, M., Osindero, S.: Conditional generative adversarial nets. arXiv preprint arXiv:1411.1784 (2014)
21. Ou, Y., Doshi, J., Erus, G., Davatzikos, C.: Multi-atlas segmentation of the prostate: a zooming process with robust registration and atlas selection. In: MICCAI Grand Challenge: Prostate MR Image Segmentation, vol. 7, pp. 1–7 (2012)
22. Pasquier, D., Lacornerie, T., Vermandel, M., Rousseau, J., Lartigau, E., Betrouni, N.: Automatic segmentation of pelvic structures from magnetic resonance images for prostate cancer radiotherapy. Int. J. Radiat. Oncol. Biol. Phys. **68**(2), 592–600 (2007)
23. Ronneberger, O., Fischer, P., Brox, T.: U-net: convolutional networks for biomedical image segmentation. In: Navab, N., Hornegger, J., Wells, W.M., Frangi, A.F. (eds.) MICCAI 2015. LNCS, vol. 9351, pp. 234–241. Springer, Cham (2015). https://doi.org/10.1007/978-3-319-24574-4_28
24. Toth, R., Madabhushi, A.: Deformable landmark-free active appearance models: application to segmentation of multi-institutional prostate MRI data. In: MICCAI Grand Challenge: Prostate MR Image Segmentation, vol. 2012 (2012)
25. Vincent, G., Guillard, G., Bowes, M.: Fully automatic segmentation of the prostate using active appearance models. In: MICCAI Grand Challenge: Prostate MR Image Segmentation, vol. 2012 (2012)
26. Yu, L., Yang, X., Chen, H., Qin, J., Heng, P.A.: Volumetric convnets with mixed residual connections for automated prostate segmentation from 3D MR images. In: AAAI, pp. 66–72 (2017)
27. Yuan, J., Qiu, W., Ukwatta, E., Rajchl, M., Sun, Y., Fenster, A.: An efficient convex optimization approach to 3D prostate MRI segmentation with generic star shape prior. In: MICCAI Grand Challenge: Prostate MR Image Segmentation, vol. 7512, pp. 82–89 (2012)

A Novel Application of Multifractal Features for Detection of Microcalcifications in Digital Mammograms

Haipeng Li[1]([⊠]) [iD], Ramakrishnan Mukundan[1]([⊠]) [iD], and Shelley Boyd[2]([⊠]) [iD]

[1] Department of Computer Science and Software Engineering, University of Canterbury, Christchurch, New Zealand
haipeng.li@pg.canterbury.ac.nz, mukundan@canterbury.ac.nz
[2] Canterbury Breastcare, St. George's Medical Center, Christchurch, New Zealand
shelley.boyd@pacificradiology.com

Abstract. This paper presents a novel image processing algorithm for automated microcalcifications (MCs) detection in digital mammograms. In order to improve the detection accuracy and reduce false positive (FP) numbers, two scales of sub-images are considered for detecting varied sized MCs, and different processing algorithms are used on them. The main contributions of this research work include: use of multifractal analysis based methods to analyze mammograms and describe MCs texture features; development of adaptive α values selection rules for better highlighting MCs patterns in mammograms; application of an effective SVM classifier to predict the existence of tiny MC spots. A full-field digital mammography (FFDM) dataset INbreast is used to test our proposed method, and experimental results demonstrate that our detection algorithm outperforms other reported methods, reaching a higher sensitivity (80.6%) and reducing FP numbers to lower levels.

Keywords: Multifractal analysis · SVM · Mammogram · Microcalcifications detection · Image enhancement

1 Introduction

In accordance with WHO's statistics [1], over 1.5 million women are diagnosed with breast cancer each year which makes it the most frequent cancer among women. The greatest number of cancer-related deaths among women are caused by breast cancer. In 2015, 570,000 women died from breast cancer – that is approximately 15% of all cancer deaths among women. In New Zealand also, breast cancer is the most common cancer among women, with approximately nine women diagnosed every day [2]. There has been no effective or proven method to prevent breast cancer, as its precise cause remains unknown. Early detection of breast cancer therefore plays an important role in reducing the risks and associated morbidity and mortality. Mammography, which uses low-energy X-rays to identify abnormalities in the breast, is the only widely accepted imaging method used for routine breast cancer screening. However, reading and interpreting suspicious

© Springer Nature Switzerland AG 2020
Y. Zheng et al. (Eds.): MIUA 2019, CCIS 1065, pp. 26–37, 2020.
https://doi.org/10.1007/978-3-030-39343-4_3

regions in mammograms is a repetitive task for radiologists, leading to a 10%–30% rate of undetected lesions [3]. To decrease this rate, computer-aided detection and diagnosis (CAD) systems have been developed to assist radiologists in the interpretation of the medical images [4–6].

The MCs are small deposits of calcium in the breast, and they can be a primary indicator of breast cancer. However, accurate detection of MCs in mammograms can be very challenging and difficult. Breasts contain variable quantities of glandular, fatty, and connective tissues, and if there are a large number of glandular tissues, the mammograms could be very bright, making small MCs poorly visible. In many countries, including New Zealand, radiologists still analyze and interpret mammograms using manual operations and visual observations, which cannot guarantee constantly identical criteria among a large number of patients. Many parts of the world utilize double reading, where at least two radiologists interpret each image, to reduce this variability. This is a time consuming and expensive measure which may not be feasible in areas with staffing or budgetary shortages.

Several methods for automatically detecting MCs have been reported in literature. Simple nonlinear functions to modify contourlet coefficients and spectral approaches were used in [5, 6] for enhancing MCs. These methods worked well with screen-film mammogramgs (SFM) but were not tested on digital mammograms. A method using morphological transformations and watershed segmentation was proposed to segment MCs in [7]. A possibilistic fuzzy c-means (PFCM) clustering algorithm and weighted support vector machine (WSVM) were integrated in [8] for detecting MCs.

However, problems such like high FP numbers and varied FP rate levels in different experimental datasets are prominent and make those algorithms difficult to be applied directly. A novel MCs detection method was developed in this research work by enhancing local image texture patterns based on multifractal analysis methods and designing an SVM classifier. Multifractal features have been used as texture descriptors in applications in the field of medical image analysis [9, 10]. Although few research work have considered using multifractal methods to analyze microcalcifications in mammograms [11, 12], they all used a SFM dataset in experiments and the proposed methods were not totally automatic. To the best of our knowledge, there is no other research work reported before, extracting texture features from α-images for automatic MCs detection based on digital mammogram dataset.

This paper is organized as follows: The next section gives a description of the dataset used and a brief introduction of the processing pipeline. Section 3 introduces multifractal measures and their applications in image enhancement. Section 4 shows the main steps of MCs detection using our proposed method, and Sect. 5 presents experimental results and comparative analysis. Conclusions and some future work directions are given in Sect. 6.

2   Materials and Methods

2.1   Microcalcifications Detection in Mammograms

MCs' size, shape and local contrast vary differently in each mammogram, which leads to the difficulty of accurate detection and frequent occurrence of FPs. In accordance with opinions of radiologists, individual MCs in size of 0.1 to 0.5 mm should be observed and considered in mammograms, and if there are over 3 MC spots within 1 cm^2 area, we need to further consider clusters of MCs and their benign or malignant category.

For example, if the image resolution of mammograms is 50 μm per pixel, then a MC spot with size 0.1 mm corresponds to only two pixels size, and 0.5 mm corresponds to ten pixels. Detecting tiny spots precisely from mammograms is a challenging task, as there is not only MCs possessing high local contrast, but also other components, such like gland or fibro tissues, which could be recognized as MCs incorrectly by detection algorithms.

Obviously, the detecting accuracy of individual MCs affects the grouping of MC clusters and their classification. High FPs will contribute to more MC clusters and may lead to wrong classifications.

2.2 Dataset

INbreast [13] is a FFDM database which contains 115 cases and 410 images including bilateral mediolateral oblique (MLO) and craniocaudal (CC) views. Over half of the images in this database contain calcifications, which reflects the real population, where calcifications are the most common finding in mammography. In addition, this dataset offers carefully associated ground truth (GT) annotations. The ground truth for MCs contain their image locations specified in pixel coordinates. Image resolution in digital dataset is much higher than that in other SFM datasets [14], which is close to clinical applications and more suitable for testing the proposed method. We therefore gave importance to this digital mammogram dataset, INbreast, in our experiments.

2.3 Processing Stages

A processing pipeline containing the main steps in the proposed method is shown in Fig. 1. As a necessary pre-processing step, breast region is segmented firstly from each original mammogram. Some research work focused on precise breast region segmentation in mammograms due to artifacts or low contrast along the breast skin line in images from SFM databases [15, 16]. In our experiments, the pixel intensity based fractal dimension is computed and all pixels in a mammogram are grouped into 10 layers by using their fractal dimension values. Then, breast region contours can be identified clearly and corresponding mask images are generated. In order to avoid too high FP numbers, a linear structure operator is imported to differentiate some line structures (gland or fibro tissues) from MCs as mentioned in Sect. 2.1. More details of the proposed method are given in next two sections.

3 Multifractal Analysis and Image Texture Enhancement

3.1 Multifractal Measures

There are four commonly used intensity measures in multifractal analysis: Maximum measure, Inverse-minimum measure, Summation measure and Iso measure [17].

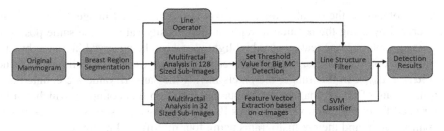

Fig. 1. Processing stages for detecting MCs in mammograms.

The function of a multifractal measure is denoted as $\mu_w(p)$, where p is the central pixel within a square window of size w. Let $g(k, l)$ represents the intensity value of the pixel at the position of (k, l) inside the window, and Ω denotes the set of all neighbourhood pixels of p in the window. Then the four measures can be formulated respectively as follows:

$$Maximum : \mu_w(p) = max_{(k,l \in \Omega)} g(k, l) \tag{1}$$

$$Inverse - minimum : \mu_w(p) = 1 - min_{(k,l) \in \Omega} g(k, l) \tag{2}$$

$$Summation : \mu_w(p) = \sum_{(k,l) \in \Omega} g(k, l) \tag{3}$$

$$Iso : \mu_w(p) = \#\{(k, l) | g(p) \cong g(k, l), (k, l) \in \Omega\} \tag{4}$$

where, $\#$ is the number of pixels. In our experiments, pixel intensities in images are firstly normalized into the range of [0, 1] by dividing the maximum grey level value when considering Maximum and Inverse-minimum measures. Such normalization brings better image enhancing results due to the amplifying effect of the logarithmic function when computing the Hölder exponent (Sect. 3.2) in the domain of (0, 1).

3.2 Hölder Exponent and α-Image

The local singularity coefficient, also known as the Hölder exponent [18] or α-value, is used to depict the pointwise singularity of the object and the multifractal property quantitatively. Hölder exponent reflects the local behaviour of a function $\mu_w(p)$, which can be calculated by different intensity based measures introduced in Sect. 3.1. The variation of the intensity measure with respect to w can be characterized as follows:

$$\mu_w(p) = Cw^{\alpha_p}, \ w = 2i + 1, \ i = 0, 1, 2, \ldots, d \tag{5}$$

$$\log(\mu_w(p)) = \alpha_p \log(w) + log(C) \tag{6}$$

where, C is an arbitrary constant and d is the total number of windows used in the computation of α_p. The value of α_p can be estimated from the slope of the linear regression line in a log-log plot where $log(\mu_w(p))$ is plotted against $log(w)$.

After computing the α value for each pixel in the processed image, α-images can be generated by using the α value to replace the intensity value in the same position. In α-images, some image features are highlighted and can be observed in more obvious patterns than original images. All the α values in an α-image constitute a limited α value range [α_{min}, α_{max}]. In this range, the pixels having the same α value are counted and an α-histogram of the α-image can be obtained based on this counting, which can be further used for extracting texture features [9, 10, 19]. Figure 2 shows some α-images of a mammogram and their α-histograms using four multifractal measures.

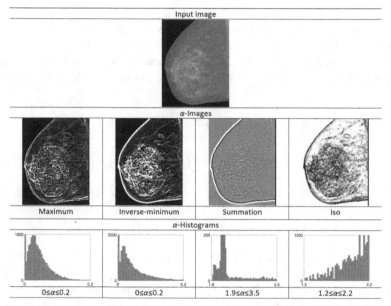

Fig. 2. An original mammogram and its α-images calculated using Maximum, Inverse-minimum, Summation and Iso measures and their corresponding α-histograms.

3.3 Image Texture Enhancement

For better highlighting some image features which are not apparently clear in the intensity image and its α-image, the α-range [α_{min}, α_{max}] could be further subdivided into some subintervals within narrower α-ranges. In each subinterval, only the pixels possessing similar α values are retained, thus effectively enhancing partial features in the image. An image only containing the pixels in one subinterval is called an α-slice. In our experiments, we find that some α-slices under the deliberate subdivision of the α-range enhance and highlight partial texture features significantly in mammograms, which could be used to identify ROI and extract relevant features. In this research work, such characteristics are used to detect MCs from background tissues. As seen in Fig. 3, the selection of a narrow range of α values helps us to highlight MCs while eliminating other neighbourhood pixels containing different α values.

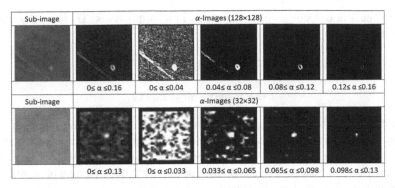

Fig. 3. Sub-images and their α-images (Inv-min measure) in different α value ranges.

3.4 α-Value Range Selection

Among different mammograms or even in two local regions from one image, the pixel intensity and the local tissue density could vary, leading to various α-value ranges corresponding to α-images. Therefore, it is often difficult to select one fixed α-value range in which α-slice describes MC features best. In our experiments, we found that in a higher α-value sub-range MCs textures usually could be highlighted better than that in a lower sub-range, as illustrated in Fig. 3. This can be ascribed to the much clearer local contrast of MCs, which generates higher α value. In our proposed method, adaptive α-value range selection rules are designed for 128 and 32 sized sub-images separately as follows.

For one sub-image, its α-image and a whole α value range $[\alpha_{min}, \alpha_{max}]$ are generated as discussed above, then a specific percentage (P) of total points with the highest α values in this sub-image are retained for selecting a suitable α-value range, which means that there exist an α threshold value denoted by α_t, and it satisfies:

$$\alpha_{min} < \alpha_t < \alpha_{max} \tag{7}$$

$$\sum_{i=t}^{max} n(\alpha_i) = P \times total\ pixel\ numbers \tag{8}$$

where, $n(\alpha_i)$ denotes the number of points possessing the α_i value in this α-image. After testing different values of P (from 0.01 to 0.1) in our experiments, we found that when $P = 0.04$, α-image in range of $[\alpha_t, \alpha_{max}]$ shows better MCs patterns in 128 sized sub-images, and when $P = 0.1$, $[\alpha_t, \alpha_{max}]$ is more suitable for highlighting MCs in 32 sized sub-images.

4 Microcalcifications Detection

4.1 Linear Structure Operator

Tissue regions having a nearly linear structure and high local contrast could potentially be misclassified as MCs. Therefore, a line operator algorithm is used in the proposed method for identifying such tissue structures. Research work in [20] demonstrated its effectiveness in detecting linear structures in mammograms. Traditionally, the line operator algorithm is used as follows [21].

$$S(x) = L(x) - N(x) \tag{9}$$

$$L(x) = \max_{\theta_i \in [0,\pi)} L_{\theta_i}(x) \tag{10}$$

where, x denotes each pixel's location in mammograms; $S(x)$ is the line strength signal; $N(x)$ denotes the average local background intensity around x; and $L_{\theta_i}(x)$ is the average grey-level in the orientation of θ_i, in our experiments, θ_i uses 12 equally-spaced orientation angles. Here, a linear structure is defined as a straight line in length of 31 pixels and with a specific angle of θ_i, keeping the pixel x in its middle point. A 5×5 square window with x in the center is considered as the local background area when computing $N(x)$. Comparing to the line operator applied in other research work, our proposed method does not use pixel intensity values to calculate $L_{\theta_i}(x)$ and $N(x)$ but uses each pixel's α value to measure $S(x)$.

4.2 MC Detection Based on Size

Due to the heterogeneous feature of MCs' size, it will be difficult to detect separate MC spots by using identical rules as their diameters range from 0.1 mm to over 1 mm. By considering this fact, two scales of sub-images, 128×128 and 32×32 pixels slide windows, are proposed to detect varied sized MCs.

Big MC Detection. Specifically, a threshold value T_{area} for defining the size of big MCs is needed, and $T_{area} = 25$ pixels is assigned in our experiments, aiming at recognizing potential MC spots with the area over 25 pixels or the diameter over 5-pixel length in 128 sized sub-images. As MCs' shapes are irregular as well, for example, they are not limited to round shapes but could also be present in oval, rod, stellate or aciform shapes. Therefore, there is no morphological detection rules designed for discerning MCs in this method; instead, a line operator (Sect. 4.1) is considered to avoid incorrectly recognizing some gland or fibro tissues in rod or line shapes as potential MCs. When there are too much overlapping points (T_{over}) between a detected MC spot and a linear structure, the current spot will not be considered as a MC.

Small MC Detection. For small MC spots with the area below 25 pixels or diameter under 5 pixels, 32 sized patches are divided from the breast region in each original mammogram and are used to detect the existence of small MCs. Because some very tiny MCs only occupy 2 to 4 pixels area in a whole mammographic image and do not possess as high image contrast as other big MCs or gland tissues, it is almost impossible to detect such small MC spots by using global texture features. However, after narrowing slide window size to 32 pixels and performing the multifractal based image texture enhancing scheme (Sect. 3.3), those tiny MC spots could be highlighted significantly in this local region by calculating the α-image.

An SVM classifier based on α-values and intensities is designed and trained with the aim of detecting tiny MCs in each patch. The feature vector X used in this classifier consists of six elements: $X(p) = [\alpha_1, \alpha_2, \alpha_3, i_1, i_2, i_3]$, where p is the currently considered pixel in a patch, and $\alpha_1, \alpha_2, \alpha_3$ are average α-values calculated from neighborhood areas of size 3×3 pixels, 5×5 pixels and 7×7 pixels respectively around p, and i_1, i_2, i_3 are computed in the same way by using pixel intensities instead of α-values.

Training Set of the SVM Classifier. A large number of image patches of size 32×32 pixles containing breast region are cropped from mammograms and used as training set for the proposed classifier. For example, one mammogram in INbreast dataset contains 3328×4084 pixels, and its breast region usually occupies about half of its total image region after segmenting operations; therefore, about 2000 to 3000 sub-images could be extracted from it. Finally, we selected 10 mammograms with pixel-level MC ground truth information from INbreast dataset to prepare training samples. Those MCs spots with 1 to 5 ground truth pixel-labels connecting together are picked out from each sub-image and are used to calculate their feature vectors $X(p_h) = [\alpha_1, \alpha_2, \alpha_3, i_1, i_2, i_3]$, $h = 1, 2 \ldots K_1$. These feature vectors are marked with a class label '1' manually indicating 'MC category'. And the other points without MC ground truth labels in sub-images are used to generate feature vectors $X'(p_j)$ $(j = 1, 2.. K_{-1})$ to form a 'normal category' part in the training set with a manual class label '-1'.

For this SVM classifier, a linear kernel function is used and the training set contains totally 150 training samples, with 50 (i.e. $K_1 = 50$) of them belonging to MC category and remaining 100 (i.e. $K_{-1} = 100$) samples are in the normal category.

5 Experimental Results and Analysis

First, breast region segmentation mentioned in Sect. 2.3 is done, then the linear operator and two scales slide window policy both with 50% overlapping areas are executed respectively in breast region areas.

After detecting MCs in 128 and 32 sized sub-images, their result images I_{MC1} and I_{MC2} are combined to output a final detection image $I_{MC} = I_{MC1} + I_{MC2}$. In order to audit and analyze experimental results, we define some rules for counting true positive (TP) and FP. In I_{MC1}, if one detected MC region overlaps the GT contour or its inside area, this detected region is counted as one TP. Otherwise, this region is regarded as one FP. In I_{MC2}, due to some GT for tiny MCs only label one pixel in the mammogram, we regulate that if one detected point in MCs is less than three pixels distance from a GT point, then this point is counted as one TP. Otherwise, this point is sent to the FP group. If SN denotes the total pixels number in the breast region of one mammogram and GT_n denotes ground truth number of negative points, then $GT_n = SN - GT_p$, where GT_p is the number of MCs offered by INbreast dataset. Figure 4 illustrates the main processing steps in our experiment and one MC detection result image I_{MC} in a local region.

$$FN = GT_p - TP, \ TN = GT_n - FP \quad (11)$$

$$Sensitivity = \frac{TP}{TP + FN}, \quad Specificity = \frac{TN}{TN + FP} \quad (12)$$

From the result image I_{MC}, we find that FP number is still very high and majority of them are caused by I_{MC2} (detected small MCs), as some spots with slightly high

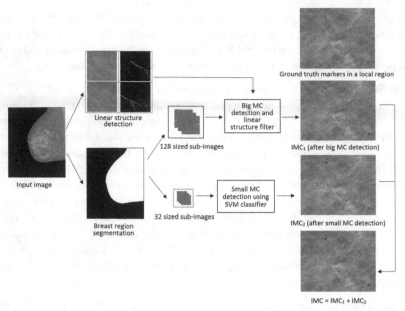

Fig. 4. Main processing steps in our experiments and an example of MC detection in a local region.

contrast are recognized as MCs, indicating that the designed SVM classifier is very sensitive towards local high contrasts. Therefore, a threshold value T is set for filtering those spots and reducing FP numbers. Here, T means that in I_{MC2} there should be at least T detected points around the target MC point and meantime these points satisfy the judging rule mentioned above. In our experiments, five values (1, 2, 3, 4, and 5) are assigned to T respectively.

In INbreast dataset, 50 mammograms with pixel-level MC ground truth information are selected randomly and tested using our method. These mammograms have been noted "MCs" by radiologists and classified into different Breast Imaging Reporting and Data System (BI-RADS) [22] categories. Confusion matrices with different T values are given in Table 1, showing the performance of the proposed method.

Table 1. Confusion matrices with different T values.

		Predicted T = 1		Predicted T = 2		Predicted T = 3		Predicted T = 4		Predicted T = 5	
		Positive	Negative	Positive	Negative	Positive	Negative	Positive	Negative	Positive	Negative
Actual	Positive	27.2	4.7	25.7	6.2	24.1	7.7	22	9.8	19.8	12.1
	Negative	164	$>10^6$	89.7	$>10^6$	53.1	$>10^6$	32.9	$>10^6$	22.9	$>10^6$
Sensitivity		0.8527		0.8056		0.7579		0.6918		0.6207	
Specificity		0.9998		0.9999		0.9999		1.0000		1.0000	

There are some other methods proposed for detecting MCs in the literature [23, 24], including Bayesian surprise method, mathematical morphology and outlier detection, and they are tested using the same dataset. Therefore, our methods are analyzed and compared with the reported results in [23]. Results comparison in Table 2 and the free response receiver operating characteristic (FROC) curve analysis in Fig. 5 show that our detection result outperforms other methods. Bayesian surprise demonstrated a better performance than other methods with a higher sensitivity (60.3%) in [23], but the average FP number (108) indicated that a further improvement is needed. While by using our method, 80.6% of sensitivity is achieved, which is much higher than the reported methods and the average FP number (90) is lower than Bayesian surprise method and mathematical morphology. When the average FP number is reduced to 53 by setting $T = 3$, the sensitivity (75.8%) still is the highest among these considered approaches.

Table 2. Results comparison between schemes reported in [23] and our method.

Method		Sensitivity (%)	FP (Average number per image)
Outlier detection		45.8	60
Mathematical morphology		40.3	225
Bayesian surprise		60.3	108
Our proposed method with different threshold values	$T = 2$	80.6	90
	$T = 3$	75.8	53
	$T = 4$	69.2	33
	$T = 5$	62.1	23

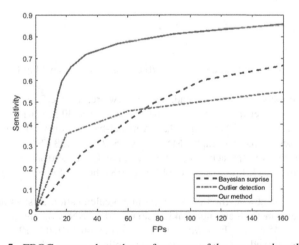

Fig. 5. FROC curves show the performance of the proposed method.

Although the proposed method displayed better results than others did, the FP number is still high for radiologists and improving work should be considered in the future work. Some possible solutions and research directions are discussed in the next section.

6 Conclusions

This paper has introduced a novel MCs detection method in digital mammograms based on image multifractal analysis and an SVM classifier. A breast region segmentation scheme and a linear operator algorithm based on α values are discussed and used in the pre-processing step. To the best of our knowledge, this is the first time an approach based on image multifractal analysis and feature vectors extracted from α-images are used with the aim of automatically identifying MCs in each whole mammographic image. The usefulness of multifractal based feature descriptors and their applications in image texture enhancing algorithms have been demonstrated through experimental results. The proposed algorithm integrates a texture enhancing process in two scales of sub-images for detecting MCs in different sizes, then a linear structure filter was developed and an SVM classifier based on selected α-values and pixel intensity values was trained for generating accurate detecting results. Experimental results show that the proposed method works well in digital mammograms and outperforms other MCs detection algorithms.

However, the high FP number still is a big challenge to improve the detection accuracy to a higher level, and a rising recall rate caused by FP is unacceptable for radiologists and their patients in the clinical screening mammography. Some other image texture descriptors, such like gray-level co-occurrence matrix (GLCM), local binary patterns (LBP), could be considered to be integrated into this approach. In addition, high breast density areas contributed by fibroglandular tissues in breast regions should be paid more attention to, as MCs detection usually is more difficult and higher FP numbers are generated in such regions.

References

1. WHO. http://www.who.int/cancer/detection/breastcancer/en/index3.html. Accessed 11 Dec 2018
2. Breast Cancer Foundation New Zealand. https://www.breastcancerfoundation.org.nz/breast-awareness/breast-cancer-facts/breast-cancer-in-nz. Accessed 30 Jan 2019
3. Sampat, M.P., Bovik, A.C., Whitman, G.J.: A model-based framework for the detection of speculated masses on mammography. Med. Phys. **35**, 2110–2123 (2008)
4. Singh, B., Kaur, M.: An approach for classification of malignant and benign microcalcification clusters. Sadhana-Acad. Proc. Eng. Sci. **43**(3) (2018). https://doi.org/10.1007/s12046-018-0805-2
5. Guo, Y.N., et al.: A new method of detecting micro-calcification clusters in mammograms using contourlet transform and non-linking simplified PCNN. Comput. Methods Programs Biomed. **130**, 31–45 (2016). https://doi.org/10.1016/j.cmpb.2016.02.019
6. Mehdi, M.Z., Ben Ayed, N.G., Masmoudi, A.D., Sellami, D., Abid, R.: An efficient micro-calcifications detection based on dual spatial/spectral processing. Multimedia Tools Appl. **76**(11), 13047–13065 (2017). https://doi.org/10.1007/s11042-016-3703-9

7. Ciecholewski, M.: Microcalcification segmentation from mammograms: a morphological approach. J. Digit. Imaging **30**(2), 172–184 (2017). https://doi.org/10.1007/s10278-016-9923-8
8. Liu, X.M., Mei, M., Liu, J., Hu, W.: Microcalcification detection in full-field digital mammograms with PFCM clustering and weighted SVM-based method. Eurasip J. Adv. Signal Process. 1–13 (2015). https://doi.org/10.1186/s13634-015-0249-3
9. Ibrahim, M., Mukundan, R.: Multifractal techniques for emphysema classification in lung tissue images. In: 3rd International Conference on Future Bioengineering (ICFB), pp. 115–119 (2014)
10. Ibrahim, M., Mukundan, R.: Cascaded techniques for improving emphysema classification in CT images. Artif. Intell. Res. **4**(2), 112–118 (2015). https://doi.org/10.5430/air.v4n2p112
11. Sahli, I.S., Bettaieb, H.A., Ben Abdallah, A., Bhouri, I., Bedoui, M.H.: Detection and segmentation of microcalcifications in digital mammograms using multifractal analysis. In: 5th International Conference on Image Processing, Theory, Tools and Applications 2015, pp. 180–184 (2015)
12. Stojic, T., Rejin, I., Rejin, B.: Adaptation of multifractal analysis to segmentation of microcalcifications in digital mammograms. Phys. A-Stat. Mech. Appl. **367**, 494–508 (2006). https://doi.org/10.1016/j.physa.2005.11.030
13. Moreira, I.C., Amaral, I., Domingues, I., Cardoso, A., Cardoso, M.J., Cardoso, J.S.: INbreast: toward a full-field digital mammographic database. Acad. Radiol. **19**(2), 236–248 (2012). https://doi.org/10.1016/j.acra.2011.09.01414
14. Suckling, J., et al.: Mammographic Image Analysis Society (MIAS) database v1.21 [Dataset] (2015). https://www.repository.cam.ac.uk/handle/1810/250394
15. Shi, P., Zhong, J., Rampun, A., Wang, H.: A hierarchical pipeline for breast boundary segmentation and calcification detection in mammograms. Comput. Biol. Med. **96**, 178–188 (2018). https://doi.org/10.1016/j.compbiomed.2018.03.011
16. Rampun, A., Morrow, P.J., Scotney, B.W., Winder, J.: Fully automated breast boundary and pectoral muscle segmentation in mammograms. Artif. Intell. Med. **79**, 28–41 (2017). https://doi.org/10.1016/j.artmed.2017.06.001
17. Braverman, B., Tambasco, M.: Scale-specific multifractal medical image analysis. Comput. Math. Methods Med. (2013). https://doi.org/10.1155/2013/262931
18. Falconer, K.: Random Fractals. Fractal Geometry: Mathematical Foundations and Applications, 2nd edn. Wiley, Chichester (2005)
19. Reljin, I., Reljin, B., Pavlovic, I., Rakocevic, I.: Multifractal analysis of gray-scale images. In: MELECON 2000: Information Technology and Electrotechnology for the Mediterranean Countries, vol. 1–3, Proceedings, pp. 490–493 (2000)
20. Wang, J., Yang, Y.Y., Nishikawa, R.M.: Reduction of false positive detection in clustered microcalcifications. In: 2013 20th IEEE International Conference on Image Processing (ICIP 2013), pp. 1433–1437 (2013)
21. Zwiggelaar, R., Astley, S.M., Boggis, C.R.M., Taylor, C.J.: Linear structures in mammographic images: detection and classification. IEEE Trans. Med. Imaging **23**(9), 1077–1086 (2004). https://doi.org/10.1109/Tmi.2004.828675
22. Sickles, E.A., D'Orsi, C.J., Bassett, L.W., et al.: ACR BI-RADS® mammography. In: ACR BI-RADS® Atlas, Breast Imaging Reporting and Data System. American College of Radiology, Reston (2013)
23. Domingues, I., Cardoso, J.S.: Using Bayesian surprise to detect calcifications in mammogram images. In: 2014 36th Annual International Conference of the IEEE Engineering in Medicine and Biology Society, pp. 1091–1094 (2014)
24. Zhang, E.H., Wang, F., Li, Y.C., Bai, X.N.: Automatic detection of microcalcifications using mathematical morphology and a support vector machine. Bio-Med. Mater. Eng. **24**(1), 53–59 (2014). https://doi.org/10.3233/Bme-130783

Wilms' Tumor in Childhood: Can Pattern Recognition Help for Classification?

Sabine Müller[1,2(⊠)], Joachim Weickert[2], and Norbert Graf[1]

[1] Department of Pediatric Oncology and Hematology,
Saarland University Medical Center, Homburg, Germany
`graf@uks.eu`
[2] Mathematical Image Analysis Group, Saarland University,
Campus E1.7, Saarbrücken, Germany
`{smueller,weickert}@mia.uni-saarland.de`

Abstract. Wilms' tumor or nephroblastoma is a kidney tumor and the most common renal malignancy in childhood. Clinicians assume that these tumors develop from embryonic renal precursor cells - sometimes via nephrogenic rests or nephroblastomatosis. In Europe, chemotherapy is carried out prior to surgery, which downstages the tumor. This results in various pathological subtypes with differences in their prognosis and treatment.

First, we demonstrate that the classical distinction between nephroblastoma and its precursor lesion is error prone with an accuracy of 0.824. We tackle this issue with appropriate texture features and improve the classification accuracy to 0.932.

Second, we are the first to predict the development of nephroblastoma under chemotherapy. We use a bag of visual model and show that visual clues are present that help to approximate the developing subtype.

Last but not least, we provide our data set of 54 kidneys with nephroblastomatosis in conjunction with 148 Wilms' tumors.

1 Introduction

Wilms' tumor, or nephroblastoma, accounts for 5% of all cancers in childhood and constitutes the most frequent malignant kidney tumor in children and juveniles [16]. About 75% of all patients are younger than five years - with a peak between two and three years [5,11]. Nephroblastoma is a solid tumor, consisting mainly of three types of tissue: blastema, epithelium and stroma [21]. In Europe, diagnosis and therapy follow the guidelines of the International Society of Pediatric Oncology (SIOP) [6,10]. One of the most important characteristics of this therapy protocol is a preoperative chemotherapy. During this therapy, the tumor tissue changes, and a total of nine different subtypes can develop [6]. Depending on this and the local stage, the patient is categorized into one of the three risk groups (low-, intermediate-, or high-risk patients) and further therapy is adapted accordingly. Of course, it would be of decisive importance for therapy and treatment planning to determine the corresponding subtype as early as possible. It is currently not known how this can be achieved.

© Springer Nature Switzerland AG 2020
Y. Zheng et al. (Eds.): MIUA 2019, CCIS 1065, pp. 38–47, 2020.
https://doi.org/10.1007/978-3-030-39343-4_4

However, there are very few research results in this direction so far: To the best our knowledge, there is only the recent work of Hötker et al. [9], where they show that diffusion-weighted MRI might be helpful in making this distinction. Unfortunately, diffusion-weighted MR images are not yet recorded as standard. Due to a relatively low incidence of this disease, it is also difficult to sensitise the clinical staff in this direction. On the other hand, a T_2 sequence is part of the therapy protocol and always recorded - even if there are no parameter specifications. We can show that even this standard sequence might be sufficient to predict subtype tendencies.

In about 40% of all children with nephroblastoma, so-called nephrogenic rests can be detected. Since these only occur in 0.6% of all childhood autopsies, they are considered a premalignant lesion of Wilms' tumors [2]. The diffuse or multifocal appearance of nephrogenic rests is called nephroblastomatosis [13,15]. Despite the histological similarity, nephroblastomatosis does not seem to have any invasive or metastatic tendencies. In order to adapt the therapy accordingly and not to expose children to an unnecessary medical burden on the one hand and to maximize their chances of survival on the other, it is necessary to distinguish nephroblastoma and its precursor nephroblastomatosis at the beginning of treatment. Its visual appearance has been described as homogenous and small abdominal mass [4,18]. However, all existing publications describe the visual appearance on usually very small data sets [7,18]. So far, it has never been validated statistically to what extent the described features are sufficient for classification. Thus, we review this current clinical practise. For this purpose, we have created a data set and evaluate whether the assumed properties can solve the classification problem between these two entities. In addition, we propose further properties that dramatically simplify the problem.

In summary our main contributions are:

- We demonstrate that the assumptions about nephroblastomatosis are mostly correct, but not sufficient to ensure a reliable classification. We solve this problem by including more texture features in the classification procedure.
- We are the first to show that T_2 imaging can be used to predict tumor development under chemotherapy in advance. We extract a variety of features and create a collection of visual properties from each image. We use this visual vocabulary to create a histogram of the relative frequency of each pattern in an image of a given subtype. We then use this information for subtype determination.
- We provide a data set with images of nephroblastomatosis and nephroblastoma from a total of 202 different patients.[1]

---

[1] The data set can be accessed at www.mia.uni-saarland.de/nephroblastomatosis.

2   Materials and Methods

Nephroblastoma is the most common kidney tumor in childhood, although it is always difficult to collect a sufficient amount of data from children and adolescents. This problem is partially solved within large-scale multi-center studies on Wilms' tumor [17, 22].

2.1   Data Sets

In recent years, the SIOP studies have collected clinical and imaging data from more than 1000 patients, possible through networking of many hospitals. Unfortunately, this has also caused a major problem: The MR images were taken on devices from different manufacturers with different magnetic field strengths over several years. In addition, there are no uniform parameter sets and the individual sequences (of the same type) can vary dramatically.

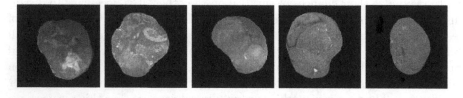

Fig. 1. Exemplary images from our data set. From left to right: epithelial dominant, stromal dominant, blastemal dominant, regressive, nephroblastomatosis.

We made sure that the main parameter settings of the T_2 sequences included in our data set are as similar as possible - this has drastically reduced the amount of imaging data available. Nevertheless, we have compiled a data set of 202 patients, see Table 1. All data sets are T_2-weighted images (axial 2D acquisition) with 3.4 mm to 9.6 mm slice thickness and inslice-sampling ranging from 0.3 mm to 1.8 mm.

In a first step, we cubically resampled all images to a grid size of one in x and y direction, but refrained from resampling in z direction as the interpolation error would be too high. Then, we linearly rescaled image intensities for simplicity to the interval $[0, 1]$. In the end, a human expert with years of experience in the field of nephroblastoma annotated the tumor regions using the method of Müller et al. [14]. We mask everything except the tumor areas and embed them in a square shaped image; see Fig. 1.

Research Ethics of the Study. All images were received as part of the SIOP 2001 prospective clinical trial. This trial received Ethical Approval from 'Ärztekammer des Saarlandes', Germany, No.: 248/13. Informed consent was given by parents or legal guardians of all enrolled children with nephroblastoma. In addition, all DICOM files were anonymized before analysis.

Table 1. Detailed information about our data set.

Patient characteristics		
Age	Range (month)	1–153
	Average	34.3
Gender	Female	50.9%
	Male	49.1%
Metastasis (Wilms' Tumor)		22 (14.86%)
Tumor characteristics		
Nephroblastoma subtypes	Diffuse anaplastic	3
	Blastemal	18
	Regressive	50
	Mixed	29
	Stromal	28
	Epithelial	17
	Necrotic	3
	Total	148
Nephroblastomatosis		54
Total		202

2.2 Features

First, we like to investigate and improve the currently clinically applied distinction between nephroblastoma and nephroblastomatosis. In order to imitate the clinically used properties as accurately as possible, we apply texture features and evaluate their significance.

Next, we evaluate if it is possible to predict the development of a nephroblastoma under chemotherapy based on standard T_2 sequences. Also in this case we like to know if the overall structure of a tumor layer already contains information about the subtype. For this purpose we use a Bag of Visual Words model.

Texture Features. Haralick et al. [8] established the basic assumption that gray-level co-occurence matrices contain all available textural information of an image. These second order Haralick texture features are extensively used in recent years in the area of medical image analysis to diagnose and differentiate cancer [20, 23, 24].

The basis of co-occurrence characteristics is the second-order conditional probability density function of an given image. Here, the elements of the co-occurrence matrix for the structure of interest represent the number of times intensity levels occur in neighboring pixels. Several features can be extracted from this matrix, e.g. contrast, homogeneity, entropy, autocorrelation [8, 19]. We use these features to distinguish nephroblastomatosis and Wilms' tumors.

Bag of Visual Words Model. The basic idea of a bag of visual words model is to represent an image as a set of local visual features. For this purpose we calculate the SURF features [1] of each 8th tumor pixel for a patch of size 7×7. The patches of the training images are then clustered with k-means [12] where cluster centroids are visual dictionary vocabularies. This allows us to determine a frequency histogram of the features in each training and test image. We use this information to train a bagged random forest classifier with 300 decision trees [3].

3   Experiments

We use our data set consisting of nephroblastomatoses and Wilms' tumors to perform several experiments. First, we want to know how accurate the clinical assumption is that nephroblastomatosis and Wilms' tumor can be distinguished by size and homogeneity. Then, we analyze the effectiveness of texture features and incorporate them to improve our classification results. In the second part of our experiments, we address the problem of subtype classification of nephroblastoma. We want to evaluate whether there is a possibility of estimating the development of the tumor under chemotherapy. All parameters in our experiments are empirically determined.

3.1   Nephroblastoma vs. Nephroblastomatosis

We first validate the general assumption that nephroblastoma can be distinguished from their predecessors by homogeneity and size. Subsequently, we show how this distinction can be significantly improved.

For this purpose we randomly select 54 out of our 148 Wilms' tumors. We then subdivide these into 27 test and training data sets again by chance. We proceed analogously with nephroblastomatosis data sets. Since the diffuse anaplastic and necrotic subtypes are under-represented, we made sure that they occur exclusively in the test-sets. From each of our data sets we draw the middle slice of the annotated tumor region and train a random forest classifier to distinguish these two classes (nephroblastomatosis and Wilms' tumor) with 3-fold cross validation. We repeat this procedure 5 times and calculate the average accuracy at the end.

Verifying Clinical Assumptions. In clinical practice, homogeneity and size of an abdominal tumor are generally used to make a distinction between nephroblastoma and nephroblastomatosis [15]. In order to validate this approach, we calculated these two feature for all data sets and used it for classification, see Table 2. The average accuracy of 0.824 indicates that homogeneity and size are valuable properties to distinguish a nephroblastoma from its precursor lesion. Nevertheless, it seems not sufficient to build clinical decisions on. Thus, we add more visual texture properties to the classification procedure [8, 19].

Table 2. Evaluation of clinical assumptions for classification: Nephroblastoma versus Nephroblastomatosis.

	Predicted	
	Nephroblastoma	Nephroblastomatosis
Nephroblastoma	0.833 ± 0.079	0.167 ± 0.079
Nephroblastomatosis	0.185 ± 0.067	0.815 ± 0.067

Feature Selection with Random Forests. Haralick et al. [8] and Soh and Tsatsoulis [19] suggested a number of additional texture features. In a first step we calculate all of these 23 features and train a bagged random forest classifier with 300 ensemble learners. This also gives us the opportunity to evaluate the influence of each feature on the final classification. It turned out that the following nine features are decisive: size, information measure correlation 1 and 2, cluster prominence, sum entropy, dissimilarity, maximum probability, energy and autocorrelation. Surprisingly, the feature of homogencity is not important when the above information is given. We evaluated these features as previously on five randomly selected data sets and 3-fold cross validation. It turns out that this additional information dramatically improves the classification performance to an accuracy of 0.932, see Table 3.

Table 3. Classification result with appropriate feature selection: Nephroblastoma versus Nephroblastomatosis.

	Predicted	
	Nephroblastoma	Nephroblastomatosis
Nephroblastoma	0.926 ± 0.064	0.074 ± 0.064
Nephroblastomatosis	0.063 ± 0.027	0.937 ± 0.027

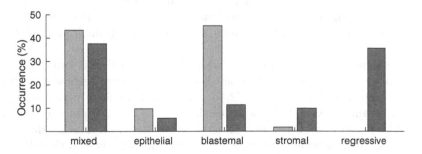

Fig. 2. Subtype distribution without (red) and with (blue) pre-operative chemotherapy. (Color figure online)

3.2   Subtype Determination

A Wilms' tumor consists of the tissue types stroma, epithelium, and blastema
[21]. Depending on the chosen therapy strategy, the subtypes are distributed
differently, see Fig. 2. In Europe, the key concept in therapy planning is a preoperative chemotherapy. This aims to shrink the tumor but also to make it
more resistant to ruptures [6]. During this phase of therapy, various subtypes
emerge, some of which differ dramatically in their prognosis. In the following
we consider the standard group of intermediate risk patients. This consists of
mainly regressive, epithelial dominant, stromal dominant, and mixed (none of
the tissue types predominates) tumors. Since the blastemal dominant type has
the worst prognosis, we also include it. Unfortunately, it is not yet possible to
predict which of the subtypes develops during chemotherapy. Clinicians assume
- based on subtype distributions before and after chemotherapy - that mainly
blastemal tissue is destroyed during this phase of therapy, see Fig. 2. However,
there is currently no possibility to determine the histological components without
a biopsy, exclusively based on imaging data.

We evaluate how far we can get in subtype determination with simple but
standard T_2 sequences. Since this problem is much more complex than the distinction between nephroblastoma and nephroblastomatosis, we need more data.
Therefore, we select one slice from each annotated tumor from the lower third
of the annotation, one from the upper third and the middle slice. In this way we
generate a total of 54 images of a blastemal dominant tumor, 150 of a regressive tumor, 87 of a mixed tumor, 84 of a stromal dominant tumor and 51 of an
epithelial dominant tumor.

Depending on the classification problem, we always take as many images as
there are in the smaller class and divide them randomly into training and test
sets. In this way we ensure that the results are not aimed at the frequency of the
images but only at the discrimination. Then we calculate the visual vocabulary
for each data set to generate a bag of visual words. With this information we then
train a random forest with 300 ensemble learners and 3-fold cross validation. We
repeat this process 5 times, analogous to the differentiation of nephroblastomatosis, including the newly generated training and test set. Here, we optimize
the size of the vocabulary on the training set and select a value from the interval
$[10, 100]$.

We compare all selected subtypes with all others in Fig. 3. Our results are
strictly above the chance level (dashed line) while average accuracy of regressive is 0.70, epithelial dominant 0.72, stromal dominant 0.66, mixed 0.67, and
blastemal dominant 0.64. This indicates that we are on the right way and that
it should be possible to distinguish these subtypes based on imaging data.

There are also several cases where our classification is surprisingly accurate. The accuracy of the distinction between regressive and epithelial dominant subtypes is 0.80. This leads to the following conclusions: 1. Tumors that
are epithelial dominant prior to chemotherapy are less likely to regress than
those that are rich in stroma or blastemal tissue. This coincides with subtype

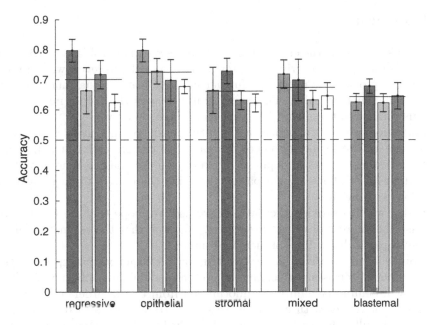

Fig. 3. Evaluation results showing mean and standard deviation of between-class classification accuracy for regressive, epithelial, stromal, mixed and blastemal subtypes. Mean performance is indicated with black lines. The dashed line marks the chance level. red: regressive, blue: epithelial, yellow: stromal, green: mixed, white: blastemal. (Color figure online)

distributions before and after chemotherapy. 2. Epithelial areas can be distinguished from other types of tissue by visual features.

Furthermore, the differentiation between regressive and mixed subtypes is relatively accurate with 0.73. This allows conclusions similar to those of the epithelial type. In addition, the epithelial dominant subtype is also well distinguishable from the stromal dominant one, i.e. classification accuracy of 0.7. We also tried to use neural networks to solve our classification problem. Unfortunately it turned out that we do not have a sufficient amount of data to re-train enough layers of a pretrained network. Therefore, all our attempts with neural networks showed low performance.

We ensured that the main parameter settings of images included in our data set are as similar as possible. However, several parameters differ dramatically in many cases. Since these cannot be compensated, the data is unfortunately not completely comparable and a considerable parameter noise is present. We firmly believe that the classification would improve significantly if this kind of noise in the data were lower. We therefore hope that in the near future a standardization of MRI sequences will be established in the medical area.

4 Conclusions

We demonstrated that the classical distinction between nephroblastomatosis and nephroblastoma is not as trivial as previously assumed. However, we were able to solve this problem by proposing further intuitive features that make the distinction much more reliable. This significantly reduces the risk of misdiagnosis and thus minimizes the medical burden on affected children.

In addition, we are the first to address the considered unsolvable problem of subtype determination prior to chemotherapy. We can show that it is basically possible to estimate this development. Even though the imaging is not standardized and therefore shows a high parameter noise, there are still visual features that allow a distinction.

Finally, we also provide the data set we use. We hope that we will be able to arouse the interest of other researchers. We hope that the estimation of the subtype in particular will be of increased interest.

In our current research, we are working on the exact visual representation of the individual classes, especially the epithelial dominant subtype. We hope that we will be able to gain more information from this in order to identify at least individual types with certainty. Most importantly, we are convinced that this research will enhance the chances of survival of the affected children. If it is possible to detect especially blastemal dominant tumors (after chemotherapy) early, the therapy can be adapted much earlier such that the recovery process of the child can be improved.

Acknowledgements. J. Weickert has received funding from the European Research Council (ERC) under the European Union's Horizon 2020 research and innovation programme (grant agreement no. 741215, ERC Advanced Grant INCOVID).

References

1. Bay, H., Tuytelaars, T., Van Gool, L.: SURF: speeded up robust features. In: Leonardis, A., Bischof, H., Pinz, A. (eds.) ECCV 2006. LNCS, vol. 3951, pp. 404–417. Springer, Heidelberg (2006). https://doi.org/10.1007/11744023_32
2. Beckwith, J.B., Kiviat, N.B., Bonadio, J.F.: Nephrogenic rests, nephroblastomatosis, and the pathogenesis of Wilms' tumor. Pediatr. Pathol. **10**(1–2), 1–36 (1990)
3. Breiman, L.: Random forests. Mach. Learn. **45**(1), 5–32 (2001)
4. Cox, S.G., Kilborn, T., Pillay, K., Davidson, A., Millar, A.J.: Magnetic resonance imaging versus histopathology in Wilms tumor and nephroblastomatosis: 3 examples of noncorrelation. J. Pediatr. Hematol. Oncol. **36**(2), e81–e84 (2014)
5. Davidoff, A.M.: Wilms' tumor. Curr. Opin. Pediatr. **21**(3), 357–364 (2009)
6. Graf, N., Tournade, M.F., de Kraker, J.: The role of preoperative chemotherapy in the management of Wilms' tumor: the SIOP studies. Urol. Clin. North Am. **27**(3), 443–454 (2000)
7. Gylys-Morin, V., Hoffer, F., Kozakewich, H., Shamberger, R.: Wilms' tumor and nephroblastomatosis: imaging characteristics at gadolinium-enhanced MR imaging. Radiology **188**(2), 517–521 (1993)

8. Haralick, R.M., Shanmugam, K., et al.: Textural features for image classification. IEEE Trans. Syst. Man Cybern. **6**, 610–621 (1973)
9. Hötker, A.M., et al.: Diffusion-weighted MRI in the assessment of nephroblastoma: results of a multi-center trial (2018, Submitted)
10. Kaste, S.C., et al.: Wilms' tumour: prognostic factors, staging, therapy and late effects. Pediatr. Radiol. **38**(1), 2–17 (2008)
11. Kim, S., Chung, D.H.: Pediatric solid malignancies: neuroblastoma and Wilms' tumor. Surg. Clin. North Am. **86**(2), 469–487 (2006)
12. Lloyd, S.: Least squares quantization in PCM. IEEE Trans. Inf. Theory **28**(2), 129–137 (1982)
13. Lonergan, G.J., Martinez-Leon, M.I., Agrons, G.A., Montemarano, H., Suarez, E.S.: Nephrogenic rests, nephroblastomatosis, and associated lesions of the kidney. Radiographics **18**(4), 947–968 (1998)
14. Müller, S., Ochs, P., Weickert, J., Graf, N.: Robust interactive multi-label segmentation with an advanced edge detector. In: Rosenhahn, B., Andres, B. (eds.) GCPR 2016. LNCS, vol. 9796, pp. 117–128. Springer, Cham (2016). https://doi.org/10.1007/978-3-319-45886-1_10
15. Owens, C.M., Brisse, H.J., Olsen, Ø.E., Begent, J., Smets, A.M.: Bilateral disease and new trends in Wilms tumour. Pediatr. Radiol. **38**(1), 30–39 (2008)
16. Pastore, G., Znaor, A., Spreafico, F., Graf, N., Pritchard-Jones, K., Steliarova-Foucher, E.: Malignant renal tumours incidence and survival in European children (1978–1997): report from the Automated Childhood Cancer Information System project. Eur. J. Cancer **42**(13), 2103–2114 (2006)
17. Reinhard, H., et al.: Outcome of relapses of nephroblastoma in patients registered in the SIOP/GPOH trials and studies. Oncol. Rep. **20**(2), 463–467 (2008)
18. Rohrschneider, W.K., Weirich, A., Rieden, K., Darge, K., Tröger, J., Graf, N.: US, CT and MR imaging characteristics of nephroblastomatosis. Pediatr. Radiol. **28**(6), 435–443 (1998)
19. Soh, L.K., Tsatsoulis, C.: Texture analysis of SAR sea ice imagery using gray level co-occurrence matrices. IEEE Trans. Geosci. Remote Sens. **37**, 780–795 (1999). CSE Journal Articles p. 47
20. Soomro, M.H., et al.: Haralick's texture analysis applied to colorectal T2-weighted MRI: a preliminary study of significance for cancer evolution. In: Proceedings of 13th International Conference on Biomedical Engineering, pp. 16–19. IEEE (2017)
21. Vujanić, G.M., Sandstedt, B.: The pathology of Wilms' tumour (nephroblastoma): the International Society of Paediatric Oncology approach. J. Clin. Pathol. **63**(2), 102–109 (2010)
22. Vujanić, G.M., et al.: Revised International Society of Paediatric Oncology (SIOP) working classification of renal tumors of childhood. Med. Pediatr. Oncol. **38**(2), 79–82 (2002)
23. Wibmer, A., et al.: Haralick texture analysis of prostate MRI: utility for differentiating non-cancerous prostate from prostate cancer and differentiating prostate cancers with different gleason scores. Eur. Radiol. **25**(10), 2840–2850 (2015)
24. Zayed, N., Elnemr, H.A.: Statistical analysis of Haralick texture features to discriminate lung abnormalities. J. Biomed. Imaging **2015**, 12 (2015)

Multi-scale Tree-Based Topological Modelling and Classification of Micro-calcifications

Zobia Suhail[1](✉)(iD) and Reyer Zwiggelaar[2](✉)(iD)

[1] Punjab University College of Information Technology,
University of the Punjab, Lahore, Pakistan
zobia.suhail@pucit.edu.pk
[2] Department of Computer Science, Aberystwyth University,
Aberystwyth, Wales, U.K.
rrz@aber.ac.uk

Abstract. We propose a novel method for the classification of benign and malignant micro-calcifications using a multi-scale tree-based modelling approach. By taking the connectivity between individual calcifications into account, micro-calcification trees were build at multiple scales along with the extraction of several tree related micro-calcification features. We interlinked the distribution aspect of calcifications with the tree structures at each scale. Classification results show an accuracy of 85% for MIAS and 79% for DDSM, which is in-line with the state-of-the-art methods.

Keywords: Micro-calcification · Classification · Computer Aided Diagnosis · Modelling · Benign · Malignant

1 Literature Review

Recent advances in image processing, pattern recognition and machine learning techniques have been incorporated in Computer Aided Diagnosis (CAD) system development [10]. CAD systems have been developed for diagnosis of micro-calcifications in order to assist radiologists to improve their diagnosis [3–5,8]. The use of CAD systems in clinical practice could increase the sensitivity about 10% compared to diagnosis without a CAD system [21]. Various aspects of micro-calcifications have been studied in the past including morphology, shape, texture, distribution and intensity for the classification of benign and malignant micro-calcifications [5,8]. Some of the developed approaches focused on the indivisual micro-calcification features [2,18,25] whereas others used global features from clustered micro-calcifications [4,24,26,28].

Existing methods have used topological modelling of micro-calcifications in order to extract cluster-level features to be used for classification [4,26]. Chen et al. [4] presented a method for the classification of benign and malignant micro-calcification based on the connectivity and topology of micro-calcification clusters. The method used dilation at multiple scales in addition to building graph like structures at each scale. They extracted eight graph-based features at each scale which were then aggregated as a feature vector to classify between benign and malignant microcalcifications. They

© Springer Nature Switzerland AG 2020
Y. Zheng et al. (Eds.): MIUA 2019, CCIS 1065, pp. 48–58, 2020.
https://doi.org/10.1007/978-3-030-39343-4_5

used the segmented images in the binary format from three different dataset (MIAS, DDSM and a non public database). Alternative work regarding topological modeling of micro-calcification clusters has been done by [26], where they introduced meretopological barcodes for the classification of benign and malignant micro-calcifications. They applied morphological operations on the segmented micro-calcification images over multiple scales. The morphological operations they used were the implementation of RCC8D (8 Region Connected Calculus) and their evaluation was based on image patches from MIAS and DDSM dataset containing micro-calcifications. Both methods used segmented micro-calcifications for experimental evaluation.

A lot of work has been presented in the area of selecting global or local features (intensity, texture, etc.) [1,24] from Region of Interest (RoI) containing microcalcifications to classify them as benign or malignant. Getting inspired by the good classification results achieved from topological modelling (85%) [4] and keeping in mind the need to study different forms of the topology for micro-calcification clusters, we proposed a novel method that extracted tree-based features at multiple scales. Current work is an extension of the work done in [29] that involved extracting the fixed-scale tree-based features to classify micro-calcifications as benign or malignant. In recent years, deep learning based classification showed improved performance in the classification accuracy particularly for medical images. Although deep learning based methods have been widely used with different architectures [6,9,20], a more traditional NaiveBayes classification gives promising results.

2   Approach: Scale-Invariant Modelling of Micro-calcification

In this section, we propose a novel method for the classification of benign and malignant micro-calcifications using a multi-scale tree-based modelling approach. By taking the connectivity between individual calcifications into account, micro-calcification trees were build at multiple scales along with the extraction of several tree related microcalcification features. We interlinked the distribution aspect of calcifications with the tree structures at each scale.

2.1   Dataset

DDSM Dataset. We used segmented RoIs extracted from the DDSM [13] database. The dataset has been used by other researchers for the performance evaluation of their algorithm [4,26]. The RoIs for the DDSM dataset are of variable sizes (average size of image patches is 482×450 pixels), but the proposed algorithm is invariant to the image size. There were 149 benign cases and 139 malignant cases in the dataset. The RoIs were probability images regarding calcification presence [22]. For evaluation of our algorithm, we used a subset of these 288 RoIs which were all classified as diffuse/scattered micro-calcification clusters according to the BIRADS standard [17]. Only those RoIs were used that did not contain mass. The BIRADS classification for the DDSM database has been provided by expert radiologist and is provided as part of the dataset.

Fig. 1. Original mammographic data from DDSM dataset with segmented benign (top row) and malignant (bottom row) micro-calcifications RoIs.

This subset contains 129 RoIs, of which 71 were malignant and 58 benign. Some example RoIs from the used database can be found in Fig. 1, where the 2nd and 4th columns are representing the annotations/segmentations.

MIAS Dataset. Like DDSM, the MIAS dataset used for this work contains segmented micro-calcifications [27]. Images in the dataset are 512×512 RoIs and digitized to 50μm/pixel with the linear optical density in the range of 0–3.2. The dataset consists of 20 RoIs (all are biopsy proven). Eleven RoIs are benign and 9 are malignant. The dataset provides annotation where the central part of calcification have been marked by value 1 whereas the boundaries are marked by value 2. Some example RoIs from the MIAS database are shown in Fig. 2, where the left image is showing micro-calcifications that include the boundary (D_1), whereas the right image is showing the micro-calcifications without the boundary regions (D_2). We used both RoIs (with and without the boundary pixels), to highlight the overall effect of using boundary pixels on the classification's accuracy. The overall evaluation of using these two variation of dataset reveals an important conclusion that the boundary area of micro-calcification is also important for characterizing them as either benign or malignant.

2.2   Method

The proposed method used multi-scale tree-based modelling for micro-calcification cluster classification. Unlike similar approaches presented in the literature [4, 26] which used multiple dilation operations (with different structuring elements) as scales to show the connectivity/relationship between the calcifications, the original structure of micro-calcification do not change at multiple scales instead we defined scales as the distance between pixels.

Fig. 2. Sample images from the MIAS database: The left image shows micro-calcifications that included boundary pixels, whereas the right image shows micro-calcifications without the boundary pixels included.

Input. The input to the scale-invariant approach were the datasets mentioned in Sect. 2.1. For the DDSM dataset the binarized form of the probability images were used, where only pixels with a probability higher than 0.27 were used:

$$\forall_{i,j} P(x_i, y_j) = \begin{cases} 1, & \text{if } P(x_i, y_j) \geq 0.27 \\ 0, & \text{otherwise} \end{cases}$$

The particular value of 0.27 as a threshold probability was based on work presented in [29].

Whereas for MIAS dataset, two variations of the dataset were used, where D_1 contained the boundary as well as the central part of the calcification and D_2 contained only the central area of the calcification. For the MIAS dataset, D_1 was defined as:

$$\forall_{i,j} P(x_i, y_j) = \begin{cases} 255, & \text{if } P(x_i, y_j) \in 1, 2 \\ 0, & \text{otherwise} \end{cases}$$

Secondly, for D_2, we take only the central part of the calcifications:

$$\forall_{i,j} P(x_i, y_j) = \begin{cases} 255, & \text{if } P(x_i, y_j) = 1 \\ 0, & \text{otherwise} \end{cases}$$

Reducing RoIs to 1 Pixel/Calcification and Generating Leaf Nodes. We reduced all individual micro-calcifications to a single point. For that, we extract the region-based properties of the binary image after labeling it. We use a module 'measure' for both labeling the image and getting regional properties, that was provided with the Scikit-image image processing tool-kit available in Python (version 2.7.0). The extracted properties from the labeled binary image included a detailed description for each connected component (area, bounding box, coordinates, etc.) in addition to the centroid of each region as a tuple (x,y). We retained only the pixels corresponding to the centroid position of each region and discard others. In this way we reduced all regions in the image to a single pixel, that was representing the central position of each connected-component (Fig. 3).

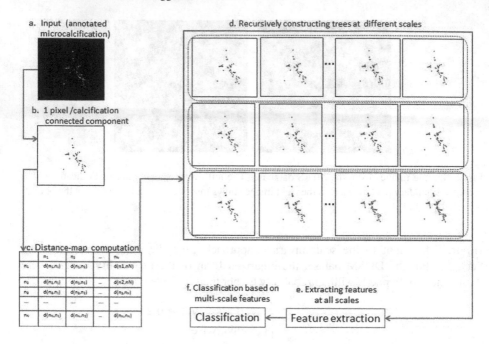

Fig. 3. Process for the multi-scale tree based modelling and classification of micro-calcification clusters in mammograms at different scales. In d. from top to bottom the scales are 16, 22 and 39.

These isolated pixels were converted into a node data structure [11], which was used as the basis for a multi-scale tree-based modelling approach, for which we used the node structure as defined in Table 1. Apart from the left and right child, we added node id's (unique for each node) and a connected-components list (that represents the connected nodes) as application specific components to the traditional binary node structure. At the initial stage each pixel was represented as a connected-component where left and right child had been assigned NULL values.

Table 1. Node structure

Node-Id	Representing Unique ID for each node
Left child	Representing left child of the node
Right child	Representing right child of the node
Connected-components	List of connected pixels P(x,y)

Distance-Map Computation. After creating node structures, we computed the distance between all leaf nodes to define their connectivity. A Euclidean distance was used to represent the distance between leaf nodes N_i and N_j (using Eq. 1). This distance map had $n \times n$ dimensions with n leaf nodes.

$$D_{(N_i,N_j)} = \sqrt{(x_{N_j} - x_{N_i})^2 + (y_{N_j} - y_{N_i})^2} \tag{1}$$

Tree-Generation at Multiple Scales. Recursively, we build a tree-structure at a particular scale 's' that took a list of leaf nodes as input. At each step the function searched for the nodes having distance $\leq$'s' and if it found such a pair of nodes, it merged the nodes by assigning both the nodes a new parent. At the same time the connected-component list of newly created parent was populated by the pixels of both the child nodes. In this way the recursive function continued until it found no node beyond a particular scale 's' to be merged. The resulting representation for the RoIs at a particular scale 's' was a set of trees where some trees had height ≤ 1 and some had height ≥ 2. We assign a label to each tree as: if height of the tree was ≤ 1 (i.e. tree contained 1 or at-most a group of 2 micro-calcifications), the label is benign and if the tree height was ≥ 2, it was given as malignant label.

Algorithm 1 elaborates the procedure of constructing binary trees recursively at a particular scale 's': This process of tree generation used scales 1, 2, 3 ... 79, where an increasing number defines larger connected regions.

Algorithm 1. Trees construction at a particular scale 's' from leaf nodes

---

Input: List of nodes (initially represented as leaf nodes)
   do recursively connect leaf nodes by:
      1. finding the closest pair of nodes (having distance $\leq s$);
      2. removing this pair from the list of nodes;
      3. merge these nodes together as a binary tree where the nodes are now represented as leaf nodes of a binary tree, and also appeared as connected-components for the root of binary tree;
      4. adding the root of the tree to the list of nodes;
   while no further pairs of nodes below distance 's' are found
return: list of nodes, in which closest nodes have been merged together as trees

---

Multiscale Tree-Based Micro-calcification Feature Extractions. The next step was to extract features from the trees that have been generated at multiple scales. These features were used as descriptors for benign and malignant RoIs. We took the following features for all the trees:

– *no. of benign trees* (as defined in the previous section labels had been assigned for all the constructed trees, this was a count of trees having a benign label).
– *no. of malignant trees* (count of constructed trees that have a malignant label).
– *max. tree height* (maximum height for all the constructed trees).
– *min. tree height* (minimum height for all the constructed trees).

- *no. of leaf nodes* belonging to the benign trees (as each tree is composed of leaf nodes, all the leaf nodes representing the pixels that are included in that tree, this feature will count all the leaf nodes from all the constructed trees whose labels were benign).
- *no. of leaf nodes* belonging to the malignant trees (count the leaf nodes for all the constructed trees whose labels were malignant).

As we use 79 scales, the length of feature vector is equal to 474.

2.3   Results and Discussion

After feature extraction, the next step was classification. By using the NaiveBayes classifier, classification accuracy for D_1 was 85% whereas for D_2 the accuracy was 80%. The Weka machine learning tool [12] was used for the performance evaluation (for both MIAS and DDSM datasets). In addition, a 10-Fold Cross Validation (10-FCV) scheme was used for evalating the results. The confusion matrix for the classification results for the MIAS dataset (D_1 and D_2) can be found in Tables 2(A) and (B), respectively.

Figure 4 shows examples of the proposed algorithm at different scales (1, 16 and 39) for a malignant and benign RoI. As can be seen from Fig. 4 (top row) the malignant RoI started creating more dense trees at lower scales as compared to the benign RoI (bottom row). The final results for the benign and malignant RoIs showed different tree structures (Fig. 4 (last column)), where the malignant RoI formed denser trees compared to the benign RoI.

Table 2. Classification results for MIAS dataset.

(a) Based on the boundary and central area of the calcifications.

	Benign	Malignant
Benign	10	1
Malignant	2	7

(b) Based on only the central area of calcifications.

	Benign	Malignant
Benign	10	1
Malignant	3	6

Results for the DDSM dataset have been evaluated using multiple machine learning classifiers. As explained in Sect. 2.1, for DDSM there were in total 288 RoIs (149 benign and 139 malignant). The same features were selected for the DDSM dataset as for the MIAS dataset for evaluating the classification performance. Overall classification accuracy of 79% was achieved using the J48 classifier with bootstrap aggregation. The J48 classifier is an implementation of the C4.5 algorithm [23]. In J48, a decision tree was build that was then used for classification. For bootstrap aggregation, the bag size was set to 70 and number of iterations was set to 20. The remaining parameters were set at the default values as provided by Weka. Like MIAS, 10-FCV was used for evaluating the results for the DDSM datset. The classification results for the DDSM dataset can be found in Table 3.

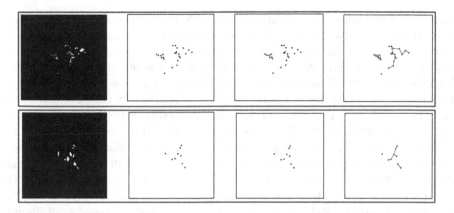

Fig. 4. Results of constructed trees for benign and malignant RoIs: first column showing the sample RoIs (top row: malignant, bottom row: benign), second column shows constructed trees at scale equal to 1; third column shows the trees constructed at scale 16, whereas the final column shows trees at scale 39.

The results for the features extracted at multiple scales for tree based modeling showed improved results for the entire DDSM dataset compared to the fixed scale approach [29] (where 55% classification accuracy was achieved for the full DDSM dataset and 55% and 60% accuracy was achieved for both variants of the MIAS dataset, respectively). The results for the DDSM dataset (as shown in Table 3) were more consistent: for 149 benign RoIs, 122 were reported as benign and for 139 malignant RoIs, 106 were reported as malignant. For the MIAS dataset, the results for dataset D_1 (Table 2(A)) seems more consistent compare to dataset D_2 (Table 2(B)). This showed that the boundary information of the calcifications was also important for the discrimination of benign and malignant micro calcifications.

Table 3. Classification results for DDSM dataset.

	Benign	Malignant
Benign	122	27
Malignant	33	106

3   Comparison

In this section, we compare the current approach with existing approaches developed for the classification of micro-calcification using topological modeling. Ma et al. [18] proposed a method for classifying benign and malignant micro-calcifications using roughness of the individual micro-calcification. They reported overall 80% of classification accuracy for 183 cases from DDSM dataset. Work proposed by Shen et al. [25], used measures of compactness, moments and Fourier descriptors as shape features as a measure of roughness of the contours of calcifications for classifying the benign and malignant micro-calcifications. They reported 100% classification accuracy on 18 mammographic cases, however the dataset used was unspecified. Topological modelling has proven to provide more accurate results in terms of classifying micro-calcification [4,26,28]. Recently, Suhail et al. [28] proposed a multiscale morphological approach for the classification of micro-calcifications. Several class extractors were used, where each class extractor computed the probability of pixels belong to a certain class. They achieved overall 90% accuracy for the MIAS and 79% accuracy for the DDSM dataset. Chen et al. [4] proposed a topological modelling where a graphical model was used to extract the features representing connectivity of calcification at multiple scales. Overall 86% and 95% classification accuracy was achieved for the DDSM and MIAS dataset, respectively. Similar work was presenetd by Strange et al. [26], where the mereotopological barcode was introduced combining concepts of discrete mereotopology and computational topology. They reported classification accuracy of 80% and 95% for the DDSM and MIAS datasets respectively. The proposed approach achieved good classification accuracy (79% for the DDSM dataset and 85% for the MIAS). In addition, the tree based features can be visualized at each scale to observe the cluster connectivity of micro-calcifications.

4   Future Work

The effect of using complete binary trees at each scale 's' will be studied in the future to study the effect of overall computation using a multi-scale approach. The current approach of classifying micro-calcification using multi-scale tree-based features can be extended to other application areas like surveillance and traffic control systems [15,16,19].

5   Conclusion

Identification of benign and malignant calcification is an important part of the breast cancer diagnostic process contributing towards correct treatment [7,14]. We have introduced a novel methods for the identification of benign and malignant calcifications based on a tree representation of the distribution of micro-calcifications and a distance metric, where modelling of micro-calcifications has been presented by using a scale-invariant approach. After building tree-like structures at multiple scales, feature-sets were extracted at all scales and used as descriptors for benign and malignant micro-calcifications' classification. Feature-set representing the distribution of calcification at each scale were defined by the distance between the individual calcifications. The method showed good classification results (79% for the DDSM dataset and 85% for the MIAS), comparable to other state-of-the-art approaches developed for the classification of benign and malignant micro-calcifications.

References

1. Chan, H.P., et al.: Computerized classification of malignant and benign microcalcifications on mammograms: texture analysis using an artificial neural network. Phys. Med. Biol. **42**(3), 549 (1997)

2. Chan, H., et al.: Computerized analysis of mammographic microcalcifications in morphological and texture feature spaces. Med. Phys. **25**(10), 2007–2019 (1998)
3. Chen, Z., Denton, E., Zwiggelaar, R.: Modelling breast tissue in mammograms for mammographic risk assessment. In: Medical Image Understanding and Analysis (MIUA), pp. 37–42 (2011)
4. Chen, Z., Strange, H., Oliver, A., Denton, E., Boggis, C., Zwiggelaar, R.: Topological modeling and classification of mammographic microcalcification clusters. IEEE Trans. Biomed. Eng. **62**(4), 1203–1214 (2015)
5. Cheng, H., Cai, X., Chen, X., Hu, L., Lou, X.: Computer-aided detection and classification of microcalcifications in mammograms: a survey. Pattern Recogn. **36**(12), 2967–2991 (2003)
6. Goceri, E., Goceri, N.: Deep learning in medical image analysis: recent advances and future trends. In: International Conferences Computer Graphics, Visualization, Computer Vision and Image Processing (2017)
7. Elmore, J., Barton, M., Moceri, V., Polk, S., Arena, P., Fletcher, S.: Ten-year risk of false positive screening mammograms and clinical breast examinations. N. Engl. J. Med. **338**(16), 1089–1096 (1998)
8. Elter, M., Horsch, A.: CADx of mammographic masses and clustered microcalcifications: a review. Med. Phys. **36**(6), 2052–2068 (2009)
9. Goceri, E., Gooya, A.: On the importance of batch size for deep learning. In: Conference on Mathematics (ICOMATH2018), An Istanbul Meeting for World Mathematicians, Minisymposium on Approximation Theory and Minisymposium on Math Education, Istanbul, Turkey (2018)
10. Goceri, E., Songul, C.: Biomedical information technology: image based computer aided diagnosis systems. In: International Conference on Advanced Technologies, Antalya, Turkey, p. 132 (2018)
11. Goodrich, M.T., Tamassia, R.: Data Structures and Algorithms in Java. Wiley, New York (2008)
12. Hall, M., Frank, E., Holmes, G., Pfahringer, B., Reutemann, P., Witten, I.: The Weka data mining software: an update. ACM SIGKDD Explor. Newsl. **11**(1), 10–18 (2009)
13. Heath, M., Bowyer, K., Kopans, D., Moore, R., Kegelmeyer, W.: The digital database for screening mammography. In: Proceedings of the 5th International Workshop on Digital Mammography, pp. 212–218 (2000)
14. Howard, J.: Using mammography for cancer control: an unrealized potential. CA Cancer J. Clin. **37**(1), 33–48 (1987)
15. Hussain, N., Yatim, H., Hussain, N., Yan, J., Haron, F.: CDES: a pixel-based crowd density estimation system for Masjid Al-Haram. Saf. Sci. **49**(6), 824–833 (2011)
16. Joshi, M., Mishra, D.: Review of traffic density analysis techniques. Int. J. Adv. Res. Comput. Commun. Eng. **4**(7) (2015)
17. Lazarus, E., Mainiero, M., Schepps, B., Koelliker, S., Livingston, L.: BI-RADS lexicon for US and mammography: interobserver variability and positive predictive value. Radiology **239**(2), 385–391 (2006)
18. Ma, Y., Tay, P., Adams, R., Zhang, J.: A novel shape feature to classify microcalcifications. In: 17th IEEE International Conference on Image Processing (ICIP), pp. 2265–2268. IEEE (2010)
19. Marana, A., Velastin, S., Costa, L., Lotufo, R.: Automatic estimation of crowd density using texture. Saf. Sci. **28**(3), 165–175 (1998)
20. Mohsen, H., El-Dahshan, E., El-Horbaty, E., Salem, A.: Classification using deep learning neural networks for brain tumors. Future Comput. Inform. J. **3**(1), 68–71 (2018)
21. Nishikawa, R.: Detection of Microcalcifications. CRC Press, Boca Raton (2002)
22. Oliver, A., et al.: Automatic microcalcification and cluster detection for digital and digitised mammograms. Knowl. Based Syst. **28**, 68–75 (2012)

23. Patil, T., Sherekar, S., et al.: Performance analysis of naive bayes and J48 classification algorithm for data classification. Int. J. Comput. Sci. Appl. **6**(2), 256–261 (2013)
24. Ren, J.: ANN vs. SVM: which one performs better in classification of MCCs in mammogram imaging. Knowl. Based Syst. **26**, 144–153 (2012)
25. Shen, L., Rangayyan, R., Desautels, J.: Application of shape analysis to mammographic calcifications. IEEE Trans. Med. Imaging **13**(2), 263–274 (1994)
26. Strange, H., Chen, Z., Denton, E., Zwiggelaar, R.: Modelling mammographic microcalcification clusters using persistent mereotopology. Pattern Recogn. Lett. **47**, 157–163 (2014)
27. Suckling, J., et al.: The mammographic image analysis society digital mammogram database. In: Exerpta Medica. International Congress Series, vol. 1069, pp. 375–378 (1994)
28. Suhail, Z., Denton, E.R., Zwiggelaar, R.: Multi-scale morphological feature extraction for the classification of micro-calcifications. In: 14th International Workshop on Breast Imaging (IWBI 2018), vol. 10718, p. 1071812. International Society for Optics and Photonics (2018)
29. Suhail, Z., Denton, E.R., Zwiggelaar, R.: Tree-based modelling for the classification of mammographic benign and malignant micro-calcification clusters. Multimedia Tools Appl. **77**(5), 6135–6148 (2018)

Lesion, Wound and Ulcer Analysis

Automated Mobile Image Acquisition of Skin Wounds Using Real-Time Deep Neural Networks

José Faria, João Almeida, Maria João M. Vasconcelos⓪, and Luís Rosado(✉)⓪

Fraunhofer Portugal AICOS, Rua Alfredo Allen 455/461, 4200-135 Porto, Portugal
luis.rosado@fraunhofer.pt

Abstract. Periodic image acquisition plays an important role in the monitoring of different skin wounds. With a visual history, health professionals have a clear register of the wound's state at different evolution stages, which allows a better overview of the healing progress and efficiency of the therapeutics being applied. However, image quality and adequacy has to be ensured for proper clinical analysis, being its utility greatly reduced if the image is not properly focused or the wound is partially occluded. This paper presents a new methodology for automated image acquisition of skin wounds via mobile devices. The main differentiation factor is the combination of two different approaches to ensure simultaneous image quality and adequacy: real-time image focus validation; and real-time skin wound detection using Deep Neural Networks (DNN). A dataset of 457 images manually validated by a specialist was used, being the best performance achieved by a SSDLite MobileNetV2 model with mean average precision of 86.46% using 5-fold cross-validation, memory usage of 43 MB, and inference speeds of 23 ms and 119 ms for desktop and smartphone usage, respectively. Additionally, a mobile application was developed and validated through usability tests with eleven nurses, attesting the potential of using real-time DNN approaches to effectively support skin wound monitoring procedures.

Keywords: Skin wounds · Mobile health · Object detection · Deep learning · Mobile devices

1 Introduction

Nowadays, chronic wounds are considered a worldwide problem and are one of the major health issues prevailing in Europe. The annual incidence estimate for acute and chronic wounds stands at 4 million in the region and the wound management market is expected to register a compound annual growth rate of 3.6% during 2018–2023 [1]. Moreover, the numbers noted previously have a tendency to increase due to the raise of life expectancy, the consequent population aging and the fact that older people have higher risk for chronic wounds given that wounds heal at a slower rate and incidences of diseases also increases [6].

© Springer Nature Switzerland AG 2020
Y. Zheng et al. (Eds.): MIUA 2019, CCIS 1065, pp. 61–73, 2020.
https://doi.org/10.1007/978-3-030-39343-4_6

Wound healing is a complex, dynamic and lengthy process that involves several factors like skin condition and the presence of other pathologies. The monitoring of skin wounds healing can be improved and its cost reduced by using mobile health (m-Health), a rising digital health sector that provides healthcare support, delivery and intervention via mobile technologies such as smartphones. The usage of m-Health for skin wound monitoring opens a new range of unexplored possibilities, such as: improve the image acquisition process by embedding automated quality and adequacy validation; enable remote monitoring of patients at home through frequent sharing of skin wound pictures; or even enable double-check and requests of second opinions between healthcare professionals.

This work presents a new approach for automated image acquisition of skin wounds, by simultaneously merging automated image quality and adequacy control. Particularly, an image focus validation approach was developed to perform real-time image quality control. From the adequacy perspective, it was assumed that only images with a detected skin wound would be suitable for clinical monitoring, so different deep learning algorithms were studied and tested for that purpose. With this work, we aim to simplify the image acquisition process of skin wounds via mobile device, and consequently facilitate monitoring procedures.

This paper is structured as follow: Sect. 1 presents the motivation and objectives of this work; Sect. 2 summarizes the related work and applications found on the literature; Sect. 3 describes the system, the algorithms and mobile application developed; in Sect. 4, the results and discussion are presented; and finally conclusions and future work are drawn in Sect. 5.

2   Related Work

Several m-Health solutions are already commercially available to share and monitor skin wounds information, such as PointClickCare,[1] +WoundDesk[2] and MOWA - Mobile Wound Analyzer[3]. In general, these applications take advantage of mobile devices features like the embedded camera sensor and portability to help healthcare professionals and patients in the collection of data (pictures and textual information) for skin wounds monitoring. Regarding image processing analysis, some of these applications already provide some automation in terms of wound size and color, which can either be autonomous or semi-autonomous (e.g. a reference marker is generally used to estimate wound size). However, to the best of our knowledge, none of those solutions perform automated quality assessment of the picture taken. Additionally, the image acquisition seems to be always manual i.e. the user has to tap the smartphone screen to perform image capture that can be a significant source of motion blur in close-up photos.

Regarding image processing and analysis approaches, two different tasks can be considered relevant for skin wound images: object detection and segmentation. Object detection consists on detecting the wound localization by returning

---

[1] https://pointclickcare.com/.
[2] https://wounddesk.com/.
[3] https://www.healthpath.it/mowa.html/.

a bounding-box, while segmentation consists in clearly defining the boundaries of the wound and returning a mask of the area of interest. In recent years, Deep Neural Networks (DNN) approaches have been proposed on the literature to tackle both tasks. In terms of segmentation, one of the first approaches was proposed by Wang et al., 2015 [17], where a Convolutional Neural Network (CNN) was used to segment skin wounds area, coupled to a Support Vector Machine (SVM) classifier to detect wound infection and estimate healing time. More recently, Liu et al. [10] improved the previously defined model by adding data augmentation and post-processing features, being tested and compared different backbone networks like the MobileNetV1 [4]. In 2018, Li et al. [7] created a composite model that combined traditional methods for pre-processing and post-processing with DNN to improve the overall result. In terms of object detection, Goyal et al., 2018 [3] recently proposed an algorithm to detect and locate diabetic foot ulcers in real-time. The trained final model was a Faster R-CNN [12] with Inception-V2 [15] that detects the wound with a mean average precision of 91.8%. Still, this approach only detects diabetic foot ulcers, being necessary a more generic approach to support the monitoring of different skin wound types.

In summary, the previously referred DNN approaches already report very promising results in terms of skin wound image analysis. Segmentation methodologies are clearly much more addressed in the literature than object detection, probably due to the more complex and challenging nature of the task. However, most of the reported segmentation approaches are unsuitable to support image acquisition in real-time on mobile devices due to: (i) required processing power; and (ii) the used DNN models are not currently compatible with mobile operative systems. Nevertheless, it should be noted that object detection tasks requires significantly less processing power, and the retrieved bounding-box is enough to support real-time image acquisition. Thus, the search for new approaches for skin wound automated localization, merged with simultaneous image quality validation and suitable for mobile environments, is an area that still needs further research and development.

3 System Overview

The proposed system allows the automated mobile image acquisition of skin wounds and is comprised by an image acquisition methodology and a mobile application. For each frame obtained from the camera preview, the image acquisition methodology starts by checking the image quality through an image focus validation approach, followed by the adequacy control where the skin wound detection is performed (Fig. 1). After guaranteeing the quality and adequacy on a certain number of consecutive frames, an image of the skin wound is automatically acquired without additional user interaction. Also, a mobile application was designed, developed and optimized through usability testing, to enable the intuitive interaction between the developed methodology and the user.

3.1 Image Focus Validation

Most mobile devices manufacturers already incorporate the autofocus functionality that is usually designed according to the specific characteristics of the embedded camera. However, current API of Android OS only allow developers to force the autofocus, not providing reliable methods neither to assess autofocus failed attempts nor focus check on acquired images. This is particularly relevant for close-up photos (e.g. small skin wounds), where several autofocus attempts might be needed until the desired focus distance is achieved by the user. Therefore, in the presented approach we include an extra validation layer to evaluate the focus of camera preview frames that can be executed in real-time and is suited for different mobile devices models. In particular, the developed algorithm starts by generating the image I_{Gray} by converting each camera preview frame from the RGB colorspace to grayscale. A new image I_{Blur} is then generated by applying a median blur filter to I_{Gray} with kernel size $kernel_{Size}$ that is calculated according to the following equation:

$$kernel_{Size} = \begin{cases} \dfrac{min(I_{Gray}^{Width}, I_{Gray}^{Height})}{125}, & \text{if } \dfrac{min(I_{Gray}^{Width}, I_{Gray}^{Height})}{125} = odd \\ \dfrac{min(I_{Gray}^{Width}, I_{Gray}^{Height})}{125} + 1, \text{otherwise} \end{cases} \quad (1)$$

The selected metric for focus assessment was the Tenenbaum gradient [16], being separately calculated for I_{Gray} and I_{Blur} using the formula:

$$TENG = G_x(i,j)^2 + G_y(i,j)^2, \quad (2)$$

where G_x and G_y are the horizontal and vertical gradients computed by convolving the focus region image with the Sobel operators. The selection of this focus metric was based on a previous work that compared a with range of focus metrics and reported its remarkable discriminative power of the Tenenbaum gradient [13]. It should be noted that the magnitude of the absolute $TENG$ value greatly depends on the specific characteristics of each frame (e.g. texture, edges,

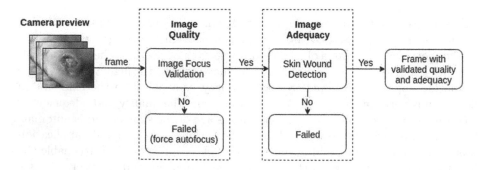

Fig. 1. Frame validation diagram of the image acquisition methodology.

etc.). So in order to achieve an adaptive approach that effectively generalizes for different image characteristics, the difference between the mean Tenenbaum gradient values $TENG^{MEAN}$ of I_{Gray} and I_{Blur} was used:

$$TENG^{DIFF} = TENG^{MEAN}_{I_{Gray}} - TENG^{MEAN}_{I_{Blur}}. \qquad (3)$$

The rationale behind using an artificially blurred image is based on the fact that a well-focused I_{Gray} will present a much higher $TENG^{DIFF}$, when compared with an I_{Gray} with motion and/or out-of-focus blur. After extensive testing with different mobile devices models on distinct skin wound types, the threshold $TENG^{DIFF} > 300$ was empirically chosen to consider the focus of camera preview frame properly validated.

3.2 Skin Wound Detection

Dataset: To the best of our knowledge, there is no publicly available image database that includes skin wound images with respective localization ground truth. Therefore, a database of skin wound images was collected with the collaboration of three healthcare institutions. The images were acquired by different healthcare professionals during one year, under uncontrolled lighting environment and complex background, and using Android devices, such as Samsung S8 and One Plus 1. After the removal of duplicate and low quality images, the dataset contained a total of 166 images. Considering that this volume of data might be insufficient for CNN's training, 255 extra images were selected from the public medetec dataset [11] to increase the dataset to a total of 457 images. Padding and resize operations were latter applied in order to have final images with 512×512 pixels of resolution. Additionally, the pre-annotation of the open wound localization was firstly performed by the authors (inflamed area was not considered), being the bounding-boxes following adjusted and/or validated by an healthcare professional with several years of experience in this clinical field. Unlike previous works [3], our dataset contains different types of skin wounds (e.g. traumatic, ulcers, excoriations, etc.), which can be depicted in Fig. 2.

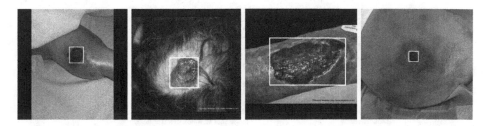

Fig. 2. Illustrative examples of skin wound images and respective ground truth localization on the used dataset.

Real-Time DNN for Skin Wound Detection: In this section we detail our methodology in terms of DNN architecture, Transfer Learning, Data Augmentation and Performance Metrics.

(i) **Architecture**: In order to select the most suitable DNN architecture, we took into account our restrictions and requirements: mobile devices are limited in terms of processing power and memory, but we need an approach that runs in real-time while delivering accurate results. For that purpose, we used a recently proposed architecture that outperformed state-of-art real-time detectors on COCO dataset, both in terms of accuracy and model complexity [14]. In particular, this approach merges the Single Shot Multi-Box Detector (SSD) [9] and the MobileNetV2 [14] architectures (see Fig. 3). The SSD consists on a single feed-forward end-to-end CNN that directly predicts classes and anchors offsets, without needing a second stage per-proposal classification operation. The SSD can be divided in two main parts: (i) the feature extractor, where high level features are extracted; and (ii) the detection generator, which uses the features extracted to detect and classify the different objects. In this work, we used the MobileNetV2 as the feature extractor component and a lightweight CNN that uses bottleneck depth-separable convolution with residuals as the basic building block. This approach allows to build smaller models with a reasonable amount of accuracy, that are simultaneously tailored for mobile and resource constrained environments.

(ii) **Transfer Learning**: In order to deal with the reduced dimension of our dataset, we used transfer learning, a technique that allows to improve the generalization capability of the model by embedding knowledge obtained from the training of different datasets. The transfer learning techniques can be divided in partial and full transfer learning: the former uses information from only a few convolutional layers, while the latter uses all the layers already trained. In the present work, we used full transfer learning from a model previously trained with the MS-COCO, a dataset that consists on more than 80000 images with 90 different classes [8].

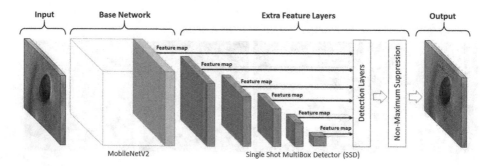

Fig. 3. High level diagram of the SSD MobileNetV2 meta-architecture.

(iii) **Performance Metrics**: Four metrics were used to evaluate the performance of the different models: *Inference Precision, Speed, Memory,* and *Model Size.* The *Inference Precision* is calculated via two similar, but distinct metrics: (a) the Mean Average Precision (mAP) used in the Pascal VOC challenge (*Pascal mAP*) [2]; and (b) the mAP usually applied to COCO dataset (*COCO mAP*) [8]. Both metrics start by calculating the intersection over union (*IoU*) between the ground truth and the prediction. Being BB_{GT} and BB_P the bounding boxes of the ground truth and prediction, the *IoU* is given by:

$$IoU = \frac{Area(BB_{GT} \cap BB_P)}{Area(BB_{GT} \cup BB_P)}. \tag{4}$$

After obtaining the *IoU* for each prediction, the *Precision* is calculated:

$$Precision = \frac{TruePositives}{TruePositives + FalsePositives}. \tag{5}$$

The *Pascal mAP* considers that every prediction with *IoU* greater than 0.5 is a true positive, and the respective *Precision* is calculated. On other hand, the *COCO mAP* is the mean value of all *Precisions* calculated for different *IoU* values, ranging from 0.5 to 0.95, with incremental intervals of 0.05 units.

3.3   Mobile Application

The mobile application aims to support healthcare professionals, such as nurses, to effectively support skin wound monitoring procedures. With the purpose of helping the image acquisition process and the construction of our database, a mobile application was developed for the Android OS, with a minimum supported version of API 23, minimum resolution display 1920 × 1080 pixels and 8 MP camera. The application allows the automatic and manual acquisition of skin wounds in an easy and intuitive way, as explained next. Usability tests (detailed in Sect. 4.2) were initially performed in a first version of the application (see Fig. 4(a)), which led to a second version of the application (see Fig. 4(b) and (c)).

The user is firstly guided to center the skin wound inside the square (see Fig. 4(a)). When the camera preview frames are considered focused through our image focus validation approach, the left icon on the top of the screen becomes green and skin wound detection algorithm starts running. Once the skin wound is detected, the right icon also becomes green and the image is automatically acquired. In case the developed automated image acquisition presents unexpected behaviours, and to guarantee that the user is always able to acquire an image, it is possible to change to manual capture mode by clicking on the bottom left button of the screen. In the manual acquisition mode, both the image focus validation and user acquisition guidance using the central square are still performed, but it is required to click in the camera button to trigger image capture (see Fig. 4(c)).

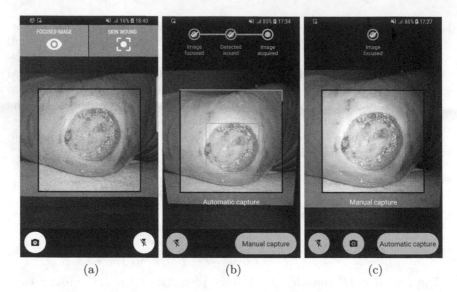

(a)   (b)   (c)

Fig. 4. Application screenshots of: (a) first version of the automatic mode; (b) and (c): second version of the automatic and manual mode, respectively.

4   Results and Discussion

4.1   DNN Performance for Skin Wound Detection

Experiments: Our models were trained using the Tensorflow Object Detection API [5] on a desktop with the following specifications: Intel core i7-7700K 4.20 GHz and Nvidia GeForce Titan X with 12 GB RAM. The experiments were performed on the described dataset using 5-fold cross-validation. In order to ensure that the whole dataset is evaluated, we randomly split the dataset in 5 sets with 91 images each (20%). For each fold, the remaining 366 images are split into 70% for training (320 images) and 10% for validation (46 images). To evaluate the proposed approach, a systematic analysis of 5 different network hyperparameters and configurations were performed on the selected architecture: (i) Data augmentation; (ii) Image size; (iii) Learning rate; (iv) L2 regularizer; and (v) SDD vs SDDLite frameworks comparison. For data augmentation we used image rotations, flips and resizes, as well as small brightness and contrast adjustments. It is worth noting that color changes were limited to a maximum of 10% variation after visual validation to ensure transformation adequacy.

Being the real-time execution in mobile devices one major requirement of our application, three image sizes were tested (128×128, 256×256 and 512×512) to assess the trade-off between inference precision and speed. For the considered hyperparameters, we used four learning rates of 0.001, 0.004, 0.0001 and 0.0004, as well as four L2 regularizer weights of 0.004, 0.0004 and 0.00004. The last configuration tested was the usage of the SSD or SSDLite frameworks: SSDLite is an adaptation of the SSD where all the regular convolutions are replaced by depthwise separable convolutions in the prediction layers, creating a friendlier architecture for mobile devices [14].

Results: In terms of inference precision, the best performance was achieved by a SSDLite MobileNetV2 model with the following configurations: RMS_Prop optimizer with learning rate of 0.004 and decay factor of 0.95; Momentum optimizer value of 0.9, with a decay of 0.9 and epsilon of 1.0; L2 regularizer weight of 0.00004 with truncated normal distribution, standard deviation of 0.03 and mean of 0.0; and batch normalization with decay 0.9997 and epsilon of 0.001. The model was trained with a batch size of 6 and maximum of 100000 epochs, with early stopping if no gain in the loss value is verified after 10000 epochs. The different evaluation metrics for the best performer model can be consulted in Table 1, where results are detailed for different image sizes.

Table 1. Performance metrics for the best performer model (on desktop).

Image size (px)	Pascal mAP (%)	COCO mAP (%)	Speed (ms)	Memory (MB)	Model size (MB)
512 × 512	86.98	63.29	85.47	168.22	12.9
256 × 256	86.46	59.22	22.99	43.05	12.9
128 × 128	73.76	40.57	9.81	10.66	12.9

As we can see in the Table 1, the usage of 512 × 512 images led to the best inference precision, but it is by far the most resource-demanding model. In comparison, the model trained with 256 × 256 images presents a slight reduction in the precision, but the inference time and memory usage drops by 73.1% and 74.4%, respectively. As a side note, the model presented a constant size of 12.9 MB, which is suitable for deploy on mobile devices.

Since the inference speeds previously discussed were obtained on a desktop, the same models were following deployed on three mobile devices to assess its applicability for real-time usage, namely: (i) Samsung S8 (12 MP camera, 4 GB RAM, Exynos 8895 Octa-core (4 × 2.3 GHz, 4 × 1.7 GHz) CPU and Mali-G71 MP20); (ii) Google Pixel 2 (12 MP camera, 4 GB RAM, Qualcomm Snapdragon 835 Octa-core (4 × 2.35 GHz Kryo, 4 × 1.9 GHz Kryo) CPU and Adreno 540 GPU); and (iii) Nexus 5 (8 MP camera, 2 GB RAM, Qualcomm Snapdragon 800 Quad-core 2.3 GHz Krait 400 CPU and Adreno 330 GPU). In particular, two high-end devices were selected (Samsung S8 and Google Pixel 2) in order to assess the performance of similar mobile devices, but from different manufacturers. In turn, the inclusion of a low-end device (Nexus 5) aims to evaluate the impact of significantly lower processing power on inference time. It should be noted that Tensorflow models can currently be executed in mobile devices via two different approaches: Tensorflow Mobile (TFmobile) and Tensorflow Lite (TFlite). The TFlite is a recent evolution of TFmobile, and in most cases allows to create lighter and faster models. However, only supports a limited set of operators, but our architecture is fully compatible with TFlite supported operators. The results for the best performer model for the different mobile devices are depicted on Table 2.

Table 2. Inference time of the best performer model for different mobile devices.

Image size (px)	TFmobile inference speed (ms)			TFlite inference speed (ms)		
	S8	Pixel 2	Nexus 5	S8	Pixel 2	Nexus 5
512 × 512	787.67	604.39	4168.80	423.09	506.95	923.207
256 × 256	212.05	179.09	1053.33	119.07	144.87	288.74
128 × 128	98.96	73.41	430.10	51.19	49.83	105.12

As showed in Table 2, TFlite presents faster inference speeds, which makes it the technology of choice for the deploy of our model on mobile platforms. Additionally, these results also confirmed a critical issue on the 512 × 512 model: despite delivering the highest inference precision, its memory usage and inference speed are significantly worse, making it unsuitable for real-time usage on mobile devices. In turn, the two remaining models showed much better inference speeds and memory usage, but each one has its own advantages and drawbacks. In terms of inference precision, the 256 × 256 model clearly outperforms the 128 × 128 model, particularly in terms of the *COCO mAP*. However, the 128 × 128 model presents by far the best inference speed and smaller memory usage. In fact, the inference speeds of the 256 × 256 model for the Nexus 5 is clearly not suitable for real-time usage, so we envision the usage of the 256 × 256 model for high-end devices and the 128 × 128 model for low-end devices.

4.2   User Interface Evaluation

The user interface of the mobile application was evaluated through the assistance of usability tests. The primary objective of these tests was to identify whether nurses could take photos of skin ulcers using the mobile application developed. For that, formative usability tests were conducted, not only to detect the problems in the usability of the application but also to understand why it happened. Iterative tests were performed in order to test the solutions implemented and find new usability problems that might not be detected previously. The test was performed using a Samsung S8 running the mobile application and an image of a pressure ulcer printed on paper. The session was audio and sound recorded and informed consents for the tests were obtained previously.

The usability test sessions took place in two healthcare institutions, with eleven nurses with ages between 30 and 34 years old. Three iterations were performed: the first with four participants, the second with four participants and the third with five participants. The eleven participants were not all distinct, since two of them participated in two sessions. Participants were required to complete a total of seven small tasks, such as acquiring the image in the automatic mode or to test acquiring in the manual mode. Task success, assistance and errors were collected in order to evaluate user performance and efficiency.

In the first iteration session the main usability problem identified by the participants concerned with the automatic acquisition mode of the application. The feedback provided was unanimously that the information showed on top of the screen ("focused and wound") was not clear or easy to understand (see Fig. 4(a)). One should notice that there was no explanation how the participant should use the application or what the symbols meant before the session. Also in the first session none of the participants were able to change for the manual acquisition mode by themselves. Based on the reported findings, design changes were performed as observed in Fig. 4(b). In the second session the issues identified previously did not persist and the participants were able to successfully complete every tasks. However, this session took place in the same institution and some of the participants were already acquainted with the application. Due to this fact, the authors decided to perform another session with participants that have never used the application. It was concluded that the problems found on the first iteration did not occur in this last session and new problems did not emerged.

5   Conclusions and Future Work

In this work, a new methodology for automated image acquisition of skin wounds via mobile devices is presented. Particularly, we propose a combination of two different approaches to ensure simultaneous image quality and adequacy, by using real-time image focus validation and real-time DNN for skin wound detection, respectively. Given the lack of freely available image datasets, a dataset of 457 images with different types of skin wounds was created, with ground truth localization properly validated by a specialist. In our experiments we performed a systematic analysis of the different network hyperparameters and configurations for the selected architectures, being the best performance achieved by a SSDLite MobileNetV2 model with mean average precision of 86.46%, memory usage of 43 MB, and inference speeds of 23 ms and 119 ms for desktop and smartphone usage, respectively. Additionally, a mobile application was developed and properly validated through usability tests with eleven nurses, which attests the potential of using real-time DNN approaches to effectively support skin wound monitoring.

As future work, we are focused in complementing our dataset with more annotated images, in order to increase the robustness of our solution. Additionally, we also aim to obtain a more detailed annotation of those images to explore new segmentation and classification tasks, namely the segmentation of the inflamed area that usually surrounds the open wound, as well as the automatic classification of the skin wound type (e.g. traumatic, ulcers, excoriations, etc.).

Acknowledgments. This work was done under the scope of MpDS - "Medical Pre-diagnostic System" project, identified as POCI-01-0247-FEDER-024086, according to Portugal 2020 is co-funded by the European Structural Investment Funds from European Union, framed in the COMPETE 2020.

References

1. Europe wound management market - segmented by product, wound healing therapy and geography - growth, trends, and forecast (2018–2023). Technical report, Mordor Intelligence, April 2018
2. Everingham, M., Van Gool, L., Williams, C.K.I., Winn, J., Zisserman, A.: The PASCAL visual object classes (VOC) challenge. Int. J. Comput. Vis. **88**(2), 303–338 (2010)
3. Goyal, M., Reeves, N., Rajbhandari, S., Yap, M.H.: Robust methods for real-time diabetic foot ulcer detection and localization on mobile devices. IEEE J. Biomed. Health Inf. **23**, 1730–1741 (2018)
4. Howard, A.G., et al.: Mobilenets: efficient convolutional neural networks for mobile vision applications. CoRR abs/1704.04861 (2017)
5. Huang, J., et al.: Speed/accuracy trade-offs for modern convolutional object detectors. In: 2017 IEEE Conference on Computer Vision and Pattern Recognition (CVPR), pp. 3296–3297 (2018)
6. Järbrink, K., et al.: The humanistic and economic burden of chronic wounds: a protocol for a systematic review. Syst. Rev. **6**(1), 15 (2017)
7. Li, F., Wang, C., Liu, X., Peng, Y., Jin, S.: Corrigendum to "a composite model of wound segmentation based on traditional methods and deep neural networks". Comput. Intell. Neurosci. **2018**, 1 (2018)
8. Lin, T.-Y., et al.: Microsoft COCO: common objects in context. In: Fleet, D., Pajdla, T., Schiele, B., Tuytelaars, T. (eds.) ECCV 2014. LNCS, vol. 8693, pp. 740–755. Springer, Cham (2014). https://doi.org/10.1007/978-3-319-10602-1_48
9. Liu, W., et al.: SSD: single shot multibox detector. In: Leibe, B., Matas, J., Sebe, N., Welling, M. (eds.) ECCV 2016. LNCS, vol. 9905, pp. 21–37. Springer, Cham (2016). https://doi.org/10.1007/978-3-319-46448-0_2
10. Liu, X., Wang, C., Li, F., Zhao, X., Zhu, E., Peng, Y.: A framework of wound segmentation based on deep convolutional networks. In: 2017 10th International Congress on Image and Signal Processing, BioMedical Engineering and Informatics (CISP-BMEI), pp. 1–7 (2017)
11. Medetec: Wound database (2014). http://www.medetec.co.uk/files/medetec-image-databases.html
12. Ren, S., He, K., Girshick, R., Sun, J.: Faster R-CNN: towards real-time object detection with region proposal networks. IEEE Trans. Pattern Anal. Mach. Intell. **39**(6), 1137–1149 (2017)
13. Rosado, L., et al.: μSmartScope: towards a fully automated 3D-printed smartphone microscope with motorized stage. In: Peixoto, N., Silveira, M., Ali, H.H., Maciel, C., van den Broek, E.L. (eds.) BIOSTEC 2017. CCIS, vol. 881, pp. 19–44. Springer, Cham (2018). https://doi.org/10.1007/978-3-319-94806-5_2
14. Sandler, M., Howard, A., Zhu, M., Zhmoginov, A., Chen, L.: MobileNetV2: inverted residuals and linear bottlenecks. In: 2018 IEEE/CVF Conference on Computer Vision and Pattern Recognition, pp. 4510–4520 (2018)

15. Szegedy, C., Vanhoucke, V., Ioffe, S., Shlens, J., Wojna, Z.: Rethinking the inception architecture for computer vision. In: The IEEE Conference on Computer Vision and Pattern Recognition (CVPR), pp. 2818–2826, June 2016
16. Tenenbaum, J.M.: Accommodation in computer vision. Technical report, Stanford University California Department of Computer Science (1970)
17. Wang, C., et al.: A unified framework for automatic wound segmentation and analysis with deep convolutional neural networks. In: 2015 37th Annual International Conference of the IEEE Engineering in Medicine and Biology Society, pp. 2415–2418 (2015)

Hyperspectral Imaging Combined with Machine Learning Classifiers for Diabetic Leg Ulcer Assessment – A Case Study

Mihaela A. Calin[1]($\boxtimes$) (iD), Sorin V. Parasca[2,3] (iD), Dragos Manea[1] (iD),
and Roxana Savastru[1] (iD)

[1] National Institute of Research and Development for Optoelectronics – INOE 2000,
Magurele, Romania
mantonina_calin@yahoo.com

[2] Carol Davila University of Medicine and Pharmacy Bucharest, Bucharest, Romania

[3] Emergency Clinical Hospital for Plastic, Reconstructive Surgery and Burns,
Bucharest, Romania

Abstract. Diabetic ulcers of the foot are a serious complication of diabetes with huge impact on the patient's life. Assessing which ulcer will heal spontaneously is of paramount importance. Hyperspectral imaging has been used lately to complete this task, since it has the ability to extract data about the wound itself and surrounding tissues. The classification of hyperspectral data remains, however, one of the most popular subjects in the hyperspectral imaging field. In the last decades, a large number of unsupervised and supervised methods have been proposed in response to this hyperspectral data classification problem. The aim of this study was to identify a suitable classification method that could differentiate as accurately as possible between normal and pathological biological tissues in a hyperspectral image with applications to the diabetic foot. The performance of four different machine learning approaches including minimum distance technique (MD), spectral angle mapper (SAM), spectral information divergence (SID) and support vector machine (SVM) were investigated and compared by analyzing their confusion matrices. The classifications outcome analysis revealed that the SVM approach has outperformed the MD, SAM, and SID approaches. The overall accuracy and Kappa coefficient for SVM were 95.54% and 0.9404, whereas for the other three approaches (MD, SAM and SID) these statistical parameters were 69.43%/0.6031, 79.77%/0.7349 and 72.41%/0.6464, respectively. In conclusion, the SVM could effectively classify and improve the characterization of diabetic foot ulcer using hyperspectral image by generating the most reliable ratio among various types of tissue depicted in the final maps with possible prognostic value.

Keywords: Minimum distance technique · Spectral angle mapper · Spectral information divergence · Support vector machine · Classification map · Accuracy

1 Introduction

Diabetic ulcers of the foot are a serious complication of diabetes with huge impact on the patient's life. Assessing which ulcer will heal spontaneously is of paramount importance.

© Springer Nature Switzerland AG 2020
Y. Zheng et al. (Eds.): MIUA 2019, CCIS 1065, pp. 74–85, 2020.
https://doi.org/10.1007/978-3-030-39343-4_7

Hyperspectral imaging (HSI) has been used lately to complete this task, since it has the ability to extract data about the wound itself and surrounding tissues.

The HSI method, originally developed by Goetz [1] for remote sensing applications, has been explored over the past two decades for various medical applications. Promising results have been reported so far in the detection of cancer, diabetic foot ulcer, peripheral vascular disease, or for the assessment of blood oxygenation levels of tissue during surgery [2]. The method consists of collecting a series of images in many adjacent narrow spectral bands and reconstruction of reflectance spectrum for every pixel of the image [1]. The set of images (typically hundreds of images) is a three-dimensional hypercube, known also as spectral cube. The hypercube includes two spatial dimensions (x and y) and one spectral dimension (wavelength). Therefore, from the hypercube data analysis, the information about an object can be extracted from spectral images in correlation with spatial information for identification, detection, classification and mapping purposes of the investigated scene. The abundant spatial and spectral information available provides great opportunities to efficiently characterize objects in the scene, but the large amount of data requires advanced analysis and classification methods. Although many unsupervised or supervised methods have been proposed in recent decades as an answer to this hyperspectral data classification problem in remote sensing area [3], their development and application in the medical field are still in their infancy. Only a few classification methods have been successfully tested for disease diagnosis and image-guided surgery purposes, including: (1) support vector machines (SVMs) methods applied for tongue diagnosis [4], gastric cancer detection [5], prostate cancer detection [6, 7], head and neck tumor detection [8], brain cancer detection [9, 10] and mapping skin burn [11]; (2) artificial neural networks (ANN) was used in brain tumor detection [10], melanoma detection [12], and kidney stone types characterization [13]; (3) spectral angle mapper (SAM) method with applications in mapping skin burn [11], chronic skin ulcers diagnosis [14], melanoma detection [15], retinal vasculature characterization [16]; (4) spectral information divergence (SID) method used for pathological white blood cells segmentation [17] and retinal vasculature characterization [16]; and (5) maximum likelihood classification (MLC) used for colon cancer diagnosis [18] and melanoma detection [15].

Most applications of these classification methods in hyperspectral medical imaging have been focused on cancer detection, but their results in predicting the same cancerous tissue attributes have proven to be variable [15]. Therefore, for a precise diagnosis, choosing the most appropriate classification method for each individual medical purpose is a particularly important task.

In this study, the performance of four different supervised machine learning approaches i.e. minimum distance technique (MD), spectral angle mapper (SAM), spectral information divergence (SID), and support vector machine (SVM) were investigated and compared to identify the most appropriate classification method that could differentiate as accurately as possible between normal and pathological biological tissues in hyperspectral images of diabetic foot ulcers.

The main objectives of the study were focused on: (1) establishing the experimental conditions for hyperspectral images acquisition; (2) improving the hyperspectral image quality using appropriate preprocessing and processing methods; (3) implementing the MD, SAM, SID, and SVM classifications method on hyperspectral images for tissue mapping and (4) assessing the performance of the classifiers in diabetic foot ulcer characterisation using a confusion matrix.

2 Materials and Methods

2.1 Patient

The study was performed on a 42 years-old male patient with diabetic ulceration of the leg. The wound was cleansed and any trace of povidone iodine (a brown antiseptic solution) was washed away. Informed consent was obtained from the patient prior to participation in this study. Approval for image acquisition was obtained from the Ethics Committee of the Emergency Clinical Hospital for Plastic, Reconstructive Surgery and Burns, Bucharest.

2.2 Hyperspectral Image Acquisition, Calibration and Processing

A visible/near-infrared (VNIR) pushroom hyperspectral imaging systems was used to acquire hyperspectral images of the wound and surrounding skin in the spectral range between 400 and 800 nm (Fig. 1). The HSI system consists of five components: (1) an imaging spectrograph (ImSpector V8E, Specim, Oulu, Finland) equipped with a 19 deg field-of-view Xenoplan1.4/17 lens (Schneider, Bad Kreuznach, Germany) which allows the acquisition of 205 spectral bands at a spectral resolution of 1.95 nm; (2) a charge-coupled device (DX4 camera, Kappa, Gleichen, Germany); (3) an illumination unit consisting of two 300 W halogen lamps (OSRAM, Munich, Germany) equipped with diffusion filters (Kaiser Fototechnik GmbH & Co. KG, Buchen, Germany) fixed 50 cm above the foot, on both sides and at an angle of 45° from the investigated area; (4) a computer for acquisition, processing and analysis of hyperspectral data, and (5) a tripod (Manfrotto 055XDB, Cassola, Italy) with mobile tripod head (Manfrotto MVH502AH). The hyperspectral data acquisition was controlled by SpectralDAQ software (Specim, Oulu, Finland) and the processing and analysis of data were performed with ENVI v.5.1 software (Exelis Visual Information Solutions, Boulder, Colorado, USA).

Before any other processing and analysis step, a correction of the acquired hyperspectral images is required to eliminate the inherent spatial non-uniformity of the artificial light intensity on the scene and the dark current in the CCD camera [19]. Therefore, the hyperspectral images were calibrated with white and dark reference images using the following expression:

$$I_C = \frac{I_O - I_D}{I_W - I_D}$$

where I_C is the calibrated hyperspectral image of the patient, I_O is the original hyperspectral image of the patient, I_D is the dark reference image obtained by completely covering the system lens and I_W is the white reference image of a white PTFE reference tile (model WS-2, Avantes, Apeldoorn, Netherlands) with approximately 98% reflectance in spectral range 350–1800 nm.

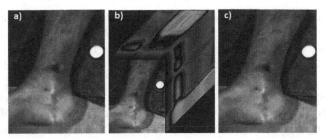

Fig. 1. The hyperspectral image of diabetic leg ulcer. (a) original HSI image (each pixel in the image has an intensity value expressed by a digital number (DN)); (b) 3D hypercube; (c) calibrated HSI image (each pixel in the image has a normalized intensity value expressed by relative reflectance (R))

The information contained in the calibrated HSI images is often accompanied by the noise added to the signal by the pushbroom hyperspectral imaging system, which limits the potential of different data analysis tasks such as classification or may even make them ineffective. Therefore, data noise reduction is mandatory as a data processing step before any data analysis to produce accurate results. To accomplish this data processing task, minimum noise fraction (MNF) transform as modified from Green et al. [20] was applied to calibrated HSI images (Fig. 2).

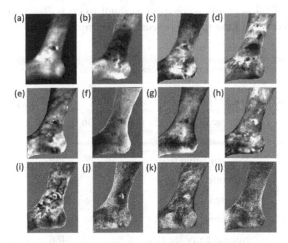

Fig. 2. The first twelve minimum noise fraction (MNF) images of hyperspectral image data from the patient's leg: (a) MNF image 1 (EV: 437.5037); (b) MNF image 2 (EV: 176.1214); (c) MNF image 3 (EV: 52.8996); (d) MNF image 4 (EV: 34.1270); (e) MNF image 5 (EV: 32.0423); (f) MNF image 6 (EV: 15.5353); (g) MNF image 7 (EV: 10.8718); (h) MNF image 8 (EV: 9.6374); (i) MNF image 9 (EV: 4.7406); (j) MNF image 10 (EV: 2.2279); (k) MNF image 11 (EV: 1.9908); and (l) MNF image 12 (EV: 1.7530)

MNF is a linear transformation that includes a forward minimum noise fraction and an inverse MNF transform. The forward MNF transform reorders the hyperspectral data cube according to the signal-to-noise ratio in two parts: one associated with large eigenvalues and coherent eigenimages, and a second with near-unity eigenvalues and noise-dominated images [21]. In our case, a subset of 10 MNF images associated with large eigenvalues (EV > 2) were identified as containing more signal than noise and the remaining MNF images with small eigenvalues (EV < 2) were found as dominated by noise.

The inverse MNF transform was than applied on these noise-reduced images to obtain less noisy images (with the same number of bands as the calibrated HIS images) in the spectral range (400–800) nm. These reduced-noise images were considered as a basis for further data classification.

2.3 Hyperspectral Image Classification

Four classification methods i.e. minimum distance technique (MD), spectral angle mapper (SAM), spectral information divergence (SID) and support vector machine (SVM) were compared regarding their ability to accurately differentiate between normal and pathological biological tissues in a hyperspectral image of diabetic leg ulcer.

In performing the image classification, two steps were taken. The first step consisted of defining the training regions for each main tissue class identified in hyperspectral image under the assistance of clinicians (chronically glabrous skin, new epithelial tissue, granulating wound, cyanotic skin, hyperemic skin, less hyperemic glabrous skin, normal hairy leg skin, exposed bone, superficial abrasion) using ROI tool of the ENVI software (Table 1). A total of 9904 pixels were used as input to each classifier (approximately 32% of the total pixels) and the rest were used for testing.

Table 1. Training dataset: tissue class names and number of samples

Class no.	Tissue type	No. of samples
1	Chronically ischemic glabrous skin	4180
2	Granulating wound	235
3	New epithelial tissue	253
4	Cyanotic skin	478
5	Hyperemic skin	1830
6	Less hyperemic glabrous skin	1070
7	Normal hairy leg skin	1488
8	Abrasion	274
9	Exposed bone	96
TOTAL		9904

In the second step, the MD, SAM, SID, and SVM algorithms were run on the leg hyperspectral image using as input data the training dataset representative of tissue classes defined in the previous step as follows:

- **Minimum distance technique (MD)**

The MD is a supervised approach that uses the mean vectors of each class and calculates the Euclidean distance $D_j(x)$ from each unknown pixel (x) to the mean vector of each training class (m_j), according to the following equation [22]:

$$D_j(x) = \|x - m_j\| \quad \text{for } j = 1, 2, \dots, M$$

where:

$m_j = $ mean vector of a training class, defined as:

$$m_j = \frac{1}{N_j} \sum_{x \in \omega_j} x \quad \text{for } j = 1, 2, \dots, M$$

$N_j = $ number of training vectors from class ω_j.
$M = $ number of spectral bands

The distance is calculated for every pixel in the image and each pixel is classified in the class of the nearest mean unless a standard deviation or distance threshold is specified.

- **Spectral Angle Mapper (SAM)**

The SAM is a classification method that evaluates the spectral similarity by calculating the angle between each spectrum in the image (pixel) and the reference spectra ("training classes") extracted from the same image, considering each spectrum as vector in a space with dimensionality equal to the number of spectral bands [23]. The angle (α) between two spectral vectors, image spectrum (X_i) and reference spectrum (X_r) is defined by the following equation:

$$\alpha = cos^{-1} \frac{X_i \cdot X_r}{|X_i||X_r|}$$

Spectral angle (α) is calculated for each image spectrum and each reference spectrum selected for hyperspectral image analysis. The smaller the spectral angle, the more the image spectrum is more similar to a given reference spectrum.

- **Spectral information divergence (SID)**

Spectral divergence information (SID) is a spectral classification method that uses a divergence measure to compare the spectral similarity between a given pixel spectrum and a reference spectrum [24]. In this method, each spectrum is treated as a random variable and SID measure is defined as follows:

$$SID(x, y) = D(x\|y) + D(y\|x)$$

where: $x = (x_1, x_2, \ldots\ldots x_N)$ and $y = (y_1, y_2, \ldots\ldots y_N)$ refer to the two spectral vectors that can come from ASCII files or spectral libraries, or can be extracted directly from an image (as ROI average spectra) and $D(x\|y)$ and $D(y\|x)$ are the relative entropy of y with respect to x defined as:

$$D(x\|y) = \sum_{i=1}^{N} p_i log\left(\frac{p_i}{q_i}\right)$$

In the equations above, N refers to the number of spectral bands and $p_i = x_i/\sum_{j=1}^{N} x_j$, $q_i = y_i/\sum_{j=1}^{N} y_j$ are the probability measure for spectral vectors x and y.

Therefore, SID measures the spectral similarity between two pixels by using the relative entropy to account for the spectral information contained in each pixel. The lower the SID value, the more likely the pixels are similar.

- **Support vector machine (SVM)**

SVM is the supervised learning algorithm most commonly used in the medical hyperspectral data classification. The SVM approach consists in finding the optimal hyperplane that maximizes the distance between the nearest points of each class (named support vectors) and the hyperplane [25, 26]. Given a labeled training data set (x_i, y_i); $i = 1, 2, \ldots\ldots, l$ where $x_i \in R^n$ and $y \in \{1, -1\}^l$, and a nonlinear mapping $\Phi(\cdot)$, usually to a higher dimensional (Hilbert) space, $\Phi: R^l \rightarrow H$, the SVM require the solution of the following optimization problem [27]:

$$min_{w,b,\xi} \frac{1}{2} w^T w + C \sum_{i=1}^{l} \xi_i$$

$$\text{subject to constrains}: y_i\left(w^T \emptyset(x_i) + b\right) \geq 1 - \xi_i \ \forall i = 1, 2, \ldots\ldots, l$$
$$\xi_i \geq 0 \qquad\qquad \forall i = 1, 2, \ldots\ldots, l$$

where: w and b are the hyperplane parameters; C is a regularization parameter and ξ is the slack variable that allows to accommodate permitted errors appropriately. Given that the training algorithm depends only on the data obtained through the dot products in H, i.e. the functions of the formula $\Phi(x_i) \cdot \Phi(x_j)$, a kernel function $K(x_i, x_j) = \Phi(x_i) \cdot \Phi(x_j)$, can be defined and a nonlinear SVM can be constructed using only the kernel function without considering the explicit mapping function Φ [25]. Commonly used kernel functions in hyperspectral imaging are linear, polynomial, radial basis function or sigmoid kernels [28]. The classification accuracy of nonlinear SVM methods strongly depends on the kernel type and parameters setting [29].

In this study the SVM with radial basis kernel function (RBF) as implemented in ENVI v.5.1 software was used to classify different tissue type in the hyperspectral image of the diabetic leg using the pairwise classification method. The default parameters of this classifier (gamma in kernel function = 0.005, penalty parameter = 100.000, pyramid level = 0 and classification probability threshold = 0.00) were used for leg hyperspectral image classification.

All four methods were used to classify the leg hyperspectral image to identify which one is most effective in characterizing the ulcers. Their classification accuracies were compared by analyzing the confusion matrices. The overall accuracy (OA), Kappa coefficient (K), producer's accuracy (PA) and the user's accuracy (UA) were computed for each classifier.

3   Results

The classification maps generated by all four classifiers MD, SAM, SID, and SVM from the leg hyperspectral image are shown in Fig. 3.

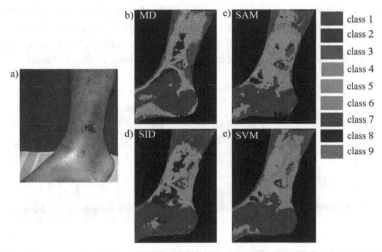

Fig. 3. Classification maps obtained by implementing the MD, SAM, SID and SVM classifiers on the hyperspectral image of the diabetic leg. (a) digital image; (b) MD classification; (c) SAM classification; (d) SID classification; (e) SVM classification

From visual inspection, class 1 seems the best classified one by all methods, maybe due to its uniformity and large number of pixels. Class 3 shows errors on all maps (less with SVM), explained probably by the thinnest of skin in the shightely pigmented area above the ulcer interpreted as newly formed epithelium. Class 9 (exposed bone) is well classified while class 8 is more extended for MD, SAM, SID than clinical control. Classes 5 and 6 show multiple variations, may be due to their dispigmentation.

The performances of the classifiers were evaluated using the confusion matrix and the related accuracy statistics are illustrated in Fig. 4.

When comparing the statistical indicators of each classifier, the overall accuracy and Kappa coefficient for SVM were higher than those of the MD, SAM and SID classifiers. The overall accuracy and Kappa coefficient for SVM were 95.54% and 0.9404, whereas for the other three approaches (MD, SAM and SID) these statistical parameters were 69.43%/0.6031, 79.77%/0.7349 and 72.41%/0.6464, respectively. The individual class accuracy also indicated that the SVM approach is the best in the discrimination of different normal and pathological tissue classes. Producer's and user's accuracy for SVM ranged between (47.43–100)% and (55.05–99.98)% respectively. The lowest performance was detected for MD, for which the producer's and user's accuracy varied between (29.92–92.82)% and (20.41–99.97)% respectively. The producer's and user's accuracy of the classification of classes 2, 3 and 4 (granulating wound, new epithelial tissue, cyanotic skin) was reported to be always lower than 48.62% and 60.81% for MD, SAM, and SID, respectively, but reaching values of 87.87% and 74.47% respectively for SVM. For these classes the low producer's accuracy reported by MD, SAM, and SID suggests that not all collected training samples for each class were also found to belong in the same class. For instance, for class 2, SAM reported that only 90 pixels, out of a total of 235 pixels, belong to this class, the rest being classified as class 3 (53 pixels), class 4 (3 pixels), class 6 (27 pixels), class 7 (6 pixels), and class 9 (53 pixels). Better classification results were obtained for SVM that reported a smaller number of misclassified pixels (110 pixels) for class 2. For the same three classes, the low user's accuracy reported by MD, SAM, and SID suggests that only few pixels classified as class 2, class 3 or class 4 can be expected to actually belong to these classes.

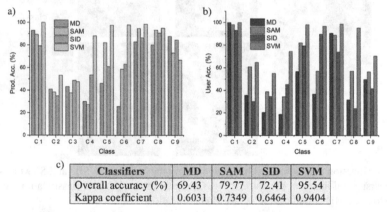

Fig. 4. Statistical indicators for the MD, SAM, SID, and SVM classifiers implemented on the leg hyperspectral image. (a) producer's accuracy for nine tissue classes; (b) user's accuracy for nine tissue classes; (c) overall accuracy and Kappa coefficient for all classifiers.

All the classifiers performed well for classes with a high number of training input pixels (classes 1, 7, 8) in both user's and producer's accuracy as well as for class 9 representing a limited area of exposed bone in the central part of the ulcer (maybe due to its uniformity and distinct reflectance spectrum). Differences between MD, SAM and SID are hardly noticeable in this respect. Low values were recorded for classes 2, 3, 4 and occasionally 5 and 6, which are the ones related with the ulcer per se. SVM showed by far better figures for both user's and producer's accuracy, even for classes in which the other classifiers behaved poorly.

Overall, all tissue classes were clearly differentiated in all classification methods investigated herein, but with poor producer's as well as user's accuracy for classes 2, 3 and 4. A special position has SVM, which seems to be the best classifying method for this type of image.

4 Discussion and Conclusion

For many applications of hyperspectral imaging in the medical field an important goal is to discriminate as accurately as possible between the pathological tissue areas and normal tissue areas or to identify the type of pathology using pattern recognition. This requires a method to classify each pixel in the image into a particular class of tissue. Various methods for classification of hyperspectral data have been developed, especially for applications in the Earth observation area, and an enormous amount of literature is now available [30]. Some of these classification methods could also be applied to classify medical hyperspectral data.

In this study, the problem of the classification of a leg hyperspectral image using four of the most popular classification methods, minimum distance technique (MD), spectral angle mapper (SAM), spectral information divergence (SID) and support vector machine (SVM), has been addressed. The main objective was to search for an optimal classifier for diabetic ulcers and surrounding skin. The results of our study showed that SVM is by far the best classification method for this type of images. It displayed the best statistical accuracy tested with the confusion matrix at all parameters (overall accuracy, Kappa coefficient, user's and producer's accuracy), and gave the best results in classifying pixels related with the ulcer per se, which constitutes the main advantage when considering future clinical applications.

The results reported here are generally in agreement with results reported by other authors who implemented the same or other classifiers on medical hyperspectral images. For example, Fei et al. [6] shown that the hyperspectral imaging and least squares support vector machines (LS-SVMs) classifier were able to reliably detect prostate tumors in an animal model. Ibraheem [15] performed a comparison of different classifiers, maximum likelihood (ML), spectral angle mapper (SAM) and K-means algorithms, for melanoma detection and reported true-positive results of 88.28%, 81.83% and 79% for ML, SAM, and K-means, respectively. In a previous study [11], the authors compared spectral angle mapper and support vector machine for mapping skin burn using hyperspectral imaging and shown that the overall classification accuracy of support vector machine classifier exceeded (91.94%) that of the spectral angle mapper classifier (84.13%).

The major limitations of this study are that it used only one case and the low number of training pixels (mainly in the area strictly related to the ulcer). However, even under such conditions, it is worth noting that the SVM is the best classifier for medical hyperspectral images of chronic diabetic ulcers.

In conclusion, these preliminary data revealed that SVM had better performance compared to the MD, SAM and SID approaches and could effectively classify and improve the characterization of diabetic foot ulcer using hyperspectral image by generating the most reliable ratio among various types of tissue depicted in the final maps with possible prognostic value.

Acknowledgments. This work was financed by the Romanian Ministry of Research and Innovation by means of the Program No. 19PFE/17.10.2018.

References

1. Goetz, A.F.H., Vane, G., Solomon, J.E., Rock, B.N.: Imaging spectrometry for earth remote sensing. Science **228**(4704), 1147–1153 (1985)
2. Calin, M.A., Parasca, S.V., Savastru, D., Manea, D.: Hyperspectral imaging in the medical field: present and future. Appl. Spectrosc. Rev. **49**(6), 435–447 (2014). https://doi.org/10.1080/05704928.2013.838678
3. Chutia, D., Bhattacharyya, D.K., Sarma, K.K., Kalita, R., Sudhakar, S.: Hyperspectral remote sensing classifications: a perspective survey. Trans. GIS **20**(4), 463–490 (2016)
4. Liu, Z., Zhang, D., Yan, J.-Q., Li, Q.-L., Tang, Q.-L.: Classification of hyperspectral medical tongue images for tongue diagnosis. Comput. Med. Imaging Graph. **31**, 672–678 (2007)
5. Akbari, H., Uto, K., Kosugi, Y., Kojima, K., Tanak, N.: Cancer detection using infrared hyperspectral Imaging. Cancer Sci. **102**(4), 852–857 (2011). https://doi.org/10.1111/j.1349-7006.2011.01849.x
6. Fei, B., Akbari, H., Halig, L.V.: Hyperspectral imaging and spectral-spatial classification for cancer detection. In: 5th International Conference on BioMedical Engineering and Informatics 2012, Chongqing, vol. 1, pp. 62–64. IEEE (2012). https://doi.org/10.1109/bmei.2012.6513047
7. Akbari, H., et al.: Hyperspectral imaging and quantitative analysis for prostate cancer detection. J. Biomed. Opt. **17**(7), 076005 (2012)
8. Lu, G., Halig, L., Wang, D., Qin, X., Chen, Z.G., Fei, B.: Spectral-spatial classification for noninvasive cancer detection using hyperspectral imaging. J. Biomed. Opt. **19**(10), 106004 (2014)
9. Fabelo, H., Ortega, S., Ravi, D., Kiran, B.R., et al.: Spatio-spectral classification of hyperspectral images for brain cancer detection during surgical operations. PLoS ONE **13**(3), e0193721 (2018). https://doi.org/10.1371/journal.pone.0193721
10. Ortega, S., Fabelo, H., Camacho, R., De Laluz Plaza, M., Callicó, G.M., Sarmiento, R.: Detecting brain tumor in pathological slides using hyperspectral imaging. Biomed. Opt. Express **9**(2), 818 (2018). https://doi.org/10.1364/BOE.9.000818
11. Calin, M.A., Parasca, S.V., Manea, D.: Comparison of spectral angle mapper and support vector machine classification methods for mapping skin burn using hyperspectral imaging. In: Fournier, C., Georges, M.P., Popescu, G. (eds.) Unconventional Optical Imaging, SPIE Photonics Europe 2018, SPIE, Strasbourg, vol. 10677, p. 106773P (2018). https://doi.org/10.1117/12.2319267
12. Aboras, M., Amasha, H., Ibraheem, I.: Early detection of melanoma using multispectral imaging and artificial intelligence techniques. Am. J. Biomed. Life Sci. **3**(2–3), 29–33 (2015). https://doi.org/10.11648/j.ajbls.s.2015030203.16. Special Issue: Spectral Imaging for Medical Diagnosis "Modern Tool for Molecular Imaging"
13. Blanco, F., López-Mesas, M., Serranti, S., Bonifazi, G., Havel, J., Valiente, M.: Hyperspectral imaging based method for fast characterization of kidney stone types. J. Biomed. Opt. **17**(7), 076027 (2012)
14. Denstedt, M., Pukstad, B.S., Paluchowski, L.A., Hernandez-Palacios, J.E., Randeberg, L.L.: Hyperspectral imaging as a diagnostic tool for chronic skin ulcers. In: Photonic Therapeutics and Diagnostics IX, SPIE BiOS 2013, SPIE, San Francisco, vol. 8565, p. 85650N (2013)
15. Ibaheem, I.: maximum likelihood and spectral angle mapper and K-means algorithms used to detection of melanoma. Am. J. Biomed. Life Sci. **3**(2–3), 8–15 (2015). https://doi.org/10.11648/j.ajbls.s.2015030203.12
16. Kashani, A.H., Wong, M., Koulisis, N., Chang, C.-I., Martin, G., Humayun, M.S.: Hyperspectral imaging of retinal microvascular anatomy. J. Biomed. Eng. Inf. **2**(1), 139–150 (2016)

17. Guan, Y., Li, Q., Liu, H., Zhu, Z., Wang, Y.: Pathological leucocyte segmentation algorithm based on hyperspectral imaging technique. Opt. Eng. **51**(5), 053202 (2012)
18. Lall, M., Deal, J.: Classification of normal and lesional colon tissue using fluorescence excitation-scanning hyperspectral imaging as a method for early diagnosis of colon cancer. In: The National Conference on Undergraduate Research (NCUR) Proceedings, University of Memphis, Memphis, pp. 1063–1073 (2017)
19. Polder, G., Gerie, W.A.M., Van, D.H.: Calibration and characterization of imaging spectrographs. Near Infrared Spectrosc. **11**, 193–210 (2003)
20. Green, A.A., Berman, M., Switzer, P., Craig, M.D.: A transformation for ordering multispectral data in terms of image quality with implications for noise removal. IEEE Trans. Geosci. Remote Sens. **26**(1), 65–74 (1988)
21. Kruse, F.A., Richardson, L.L., Ambrosia, V.G.. Techniques developed for geologic analysis of hyperspectral data applied to near-shore hyperspectral ocean data. In: ERIM 4th International Conference on Remote Sensing for Marine and Coastal Environments 1997, Environmental Research Institute of Michigan (ERIM), Orlando, vol. I, pp I-233–I-246 (1997)
22. Minimum distance classification. https://semiautomaticclassificationmanual-v4.readthedocs. io/en/latest/remote_sensing.html#classification-algorithms. Accessed 11 Feb 2019
23. Kruse, F.A., Lefkoff, A.B., Boardoman, J.W.: The spectral image processing system (SIPS) - interactive visualization and analysis of imaging spectrometer data. Remote Sens. Environ. **44**(2–3), 145–163 (1993)
24. Chang, C.-I.: An information-theoretic approach to spectral variability, similarity, and discrimination for hyperspectral image. IEEE Trans. Inf. Theory **46**(5), 1927–1932 (2000). https://doi.org/10.1109/18.857802
25. Burges, C.J.C.: A tutorial on support vector machines for pattern recognition. Data Min. Knowl. Discovery **2**, 121–167 (1998). Fayyad, U. (ed.) Kluwer Academic. https://www.slideshare.net/Tommy96/a-tutorial-on-support-vector-machines-for-pattern-recognition
26. Vapnick, V.N.: Statistical Learning Theory. Wiley, Hoboken (1998)
27. Hsu, C.W., Chang, C.C., Lin, C.J.: A practical guide to support vector classification. National Taiwan University. http://www.csie.ntu.edu.tw/~cjlin/papers/guide/guide.pdf
28. Mercier, G., Lennon, M.: Support vector machines for hyperspectral image classification with spectral-based kernels. In: IEEE International, Geoscience and Remote Sensing Symposium. Proceedings, Toulouse, pp. 288–290. IEEE (2003). https://doi.org/10.1109/igarss.2003.1293752
29. Pal, M., Mather, P.M.: Assessment of the effectiveness of support vector machines for hyperspectral data. Future Gener. Comput. Syst. **20**(7), 1215–1225 (2004)
30. Tuia, D., Volpi, M., Copa, L., Kanevski, M., Munoz-Mari, J.: A survey of active learning algorithms for supervised remote sensing image classification. IEEE J. Sel. Top. Sign. Proces. **5**(3), 606–617 (2011). https://doi.org/10.1109/JSTSP.2011.2139193

Classification of Ten Skin Lesion Classes: Hierarchical KNN *versus* Deep Net

Robert B. Fisher[1](✉) ⓘ, Jonathan Rees[1], and Antoine Bertrand[2]

[1] University of Edinburgh, Edinburgh, Scotland
rbf@inf.ed.ac.uk
[2] INP Grenoble, Grenoble, France

Abstract. This paper investigates the visual classification of the 10 skin lesions most commonly encountered in a clinical setting (including melanoma (MEL) and melanocytic nevi (ML)), unlike the majority of previous research that focuses solely on melanoma *versus* melanocytic nevi classification. Two families of architectures are explored: (1) semi-learned hierarchical classifiers and (2) deep net classifiers. Although many applications have benefited by switching to a deep net architecture, here there is little accuracy benefit: hierarchical KNN classifier 78.1%, flat deep net 78.7% and refined hierarchical deep net 80.1% (all 5 fold cross-validated). The classifiers have comparable or higher accuracy than the five previous research results that have used the Edinburgh DER-MOFIT 10 lesion class dataset. More importantly, from a clinical perspective, the proposed hierarchical KNN approach produces: (1) 99.5% separation of melanoma from melanocytic nevi (76 MEL & 331 ML samples), (2) 100% separation of melanoma from seborrheic keratosis (SK) (76 MEL & 256 SK samples), and (3) 90.6% separation of basal cell carcinoma (BCC) plus squamous cell carcinoma (SCC) from seborrheic keratosis (SK) (327 BCC/SCC & 256 SK samples). Moreover, combining classes BCC/SCC & ML/SK to give a modified 8 class hierarchical KNN classifier gives a considerably improved 87.1% accuracy. On the other hand, the deepnet binary cancer/non-cancer classifier had better performance (0.913) than the KNN classifier (0.874). In conclusion, there is not much difference between the two families of approaches, and that performance is approaching clinically useful rates.

Keywords: Skin cancer · Melanoma · RGB image analysis

1 Introduction

The incidence of most types of skin cancer is rising in fair skinned people. The causes of the increase are not certain, but it is hypothesized that increased ultraviolet exposure and increasing population ages are the main causes. Irrespective of the cause, skin cancer rates are increasing, as is awareness of skin cancer. This increased awareness has also led to increased reporting rates. In addition to the health risks associated with cancer, a second consequence is the increasing

© Springer Nature Switzerland AG 2020
Y. Zheng et al. (Eds.): MIUA 2019, CCIS 1065, pp. 86–98, 2020.
https://doi.org/10.1007/978-3-030-39343-4_8

medical cost: more people are visiting their primary medical care practitioner with suspicious lesions, and are then forwarded onto dermatology specialists. As many, perhaps a majority, of the referrals are for normal, but unusual looking, lesions, this leads to a considerable expense. Eliminating these unnecessary referrals is a good goal, along with improving outcomes.

A second issue is that there are different types of skin cancer, in part arising from different cell types in the skin. Most people are familiar with melanoma, a dangerous cancer, but it is considerably less common than, for example, basal cell carcinoma. Because of the rarity of melanoma, a primary care practitioner might only encounter one of these every 5–10 years, leading to the risk of over or under referring people onto a specialist. Hence, it is good to have tools that can help discriminate between different skin cancer types.

A third issue is the priority of referrals, which might be routine or urgent. Melanoma and squamous cell carcinoma metastasize and are capable of spreading quickly, and thus need to be treated urgently. Other types of skin cancer grow more slowly, and may not even need treatment. Moreover, there are many normal lesion types that may look unusual at times. This also motivates the need for discrimination between the types of cancer and other lesions.

This motivates the research presented here, which classifies the 10 lesion types most commonly encountered by a general practice doctor. Because of the healthcare costs arising from false positives, and the health and potentially life costs of incorrect decisions, even small improvements in performance can result in considerable cost reduction and health increases.

The research presented here uses standard RGB camera data. There is another research stream based on dermoscopy [6], which is a device typically using contact or polarized light. This device can give better results than RGB image data, but has been typically limited to melanoma *versus* melanocytic nevus (mole) discrimination. Here, we focus on 10 types of lesion instead of 2, so use RGB images.

This paper presents two approaches to recognizing the 10 classes. The first more traditional approach is based on a combination of generic and hand-crafted features, feature selection from a large pool of potential features, and a hierarchical decision tree using a K-nearest neighbor classifier. A second deepnet approach is also presented for comparison. The cross-validated hierarchical 10 class accuracy (78.1%) and the cross-validated deepnet with refinement accuracy (80.1%) are comparable to the best previous performances. **The key contributions of the paper are: (1) a hierarchical decision tree structure and associated features best suited for discrimination at each level, (2) a deepnet architecture with BCC/SCC and ML/SK refinement that has 2% better performance, (3) improved classification accuracy (87.1% for 8 merged classes as compared to 78.1% for 10 classes), (4) a malignant melanoma *versus* benign nevi classification accuracy of 99.5%, (5) a malignant melanoma *versus* seborrheic keratosis classification accuracy of 100%, and (6) a clinically relevant BCC/SCC *versus* seborrheic keratosis classification accuracy of 90.6%.**

2   Background

There is a long history of research into automated diagnosis of skin cancer, in part because skin cancer is the most common cancer [1]. A second factor is the lesions appear on the skin, thus making them amenable to visual analysis. The most commonly investigated issues are (1) the discrimination between melanoma (the most serious form of skin cancer) and melanocytic nevi (the most-commonly confused benign lesion) and (2) the segmentation of the boundary between normal and lesion skin (typically because the boundary shape is one factor commonly used for human and machine diagnosis). The most commonly used imaging modalities are color and dermoscopy (a contact sensor) images. A general review of this research can be found in [11, 13, 14].

A recent breakthrough is the Stanford deep neural network [5], trained using over 129K clinical images (including those used here), and covering over 2000 skin diseases. Their experiments considered four situations: (1) classification of a lesion into one of 3 classes (malignant, benign and non-neoplastic (which is not considered in the work presented here)), (2) refined classification into 9 classes (5 of which correspond to the classes considered here), (3) keratinocyte carcinomas (classes BCC and SCC here) *versus* benign seborrheic keratoses (class SK here) and (4) malignant melanoma (class (MEL) *versus* melanocytic nevi (class ML). In case 1, their deep net achieved 0.721 accuracy as compared to the dermatologist accuracy of approximately 0.658. In case 2, the deep net achieved 0.554 accuracy as compared to the dermatologist accuracy of 0.542. In cases 3 and 4, accuracy values are not explicitly presented, but from the sensitivity/specificity curves, one can estimate approximately 0.92 accuracy for the deepnet approach, with the dermatologists performing somewhat worse.

One important issue raised in [11] is the absence of quality public benchmark datasets, especially covering more than the classification of melanoma *versus* melanocytic nevi. From 2013, the Edinburgh Dermofit Image Library [2] (1300 lesions in 10 classes, validated by two clinical dermatologists and one pathologist - more details in Sect. 3) has been available. It was part of the training data for the research of Esteva *et al.* described above, and is the core dataset for the research results described below.

The first result was by Ballerini *et al.* [2] which investigated the automated classification of 5 lesion classes (AK, BCC, ML, SCC, SK - see Sect. 2 for labels), and achieved 3-fold cross-validated accuracy of 0.939 on malignant *versus* benign, and 0.743 over 960 lesions from the 5 classes. A 2 level hierarchical classifier was used. Following that work, Di Leo [4] extended the lesion analysis to cover all 10 lesion classes in the dataset of 1300 lesions. That research resulted in an accuracy of 0.885 on malignant *versus* benign and 0.67 over all lesions from the 10 classes.

More recently, Kawahara *et al.* [9] developed a deep net to classify the 10 lesion classes over the same 1300 images. The algorithm used a logistic regression classifier applied to a set of features extracted from the final convolutional layers of a pre-trained deep network. Multiple sizes of image were input to enhance scale invariance, and each image was normalized relative to its mean RGB value to enhance invariance to skin tone. As well as substantially improving the 10 lesion accuracy, the proposed method did not require segmented lesions (which was required by the approach proposed here). Follow-on research by Kawahara and Hamarneh [10] developed a dual tract deep network with the image at 2 resolutions through the two paths, which were then combined at the end. Using auxiliary loss functions and data augmentation, the resulting 10-class perfor-mance was 0.795, which was stated as an improvement on the methods of [9], although the new methodology used less training data and so the initial baseline was lower. A key benefit of the multi-scale approach was the ability to exploit image properties that appear at both scales. As with the original research, the proposed method did not require segmented lesions.

3   Edinburgh DERMOFIT Dataset

The dataset used in the experiments presented in this paper was the Edinburgh DERMOFIT Dataset[1] [2]. The images were acquired using a Canon EOS 350D SLR camera. Lighting was controlled using a ring flash and all images were cap-tured at the same distance (approximately 50 cm) with a pixel resolution of about 0.03 mm. Image sizes are typically 400×400 centered on the cropped lesions plus about an equal width and height of normal skin. The images used here are all RGB, although the dataset also contains some registered depth images. The ground truth used for the experiments is based on agreed classifications by two dermatologists and a pathologist. This dataset has been used by other groups [4,5,9,10], as·discussed above.

The dataset contains 1300 lesions from the 10 classes of lesions most com-monly presented to consultants. The first 5 classes are cancerous or pre-cancerous: actinic keratosis (AK): 45 examples, basal cell carcinoma (BCC): 239, squamous cell carcinoma (SCC): 88, intraepithelial carcinoma (IEC): 78, and melanoma (MEL): 76. The other five classes are benign, but commonly encountered: melanocytic nevus/mole (ML): 331 examples, seborrheic kerato-sis (SK): 257, pyogenic granuloma (PYO): 24, haemangioma (VASC): 96, and dermatofibroma (DF): 65.

4   Hierarchical Classifier Methodology

The process uses RGB images, from which a set of 2500+ features are extracted. The key steps in the feature extraction are: (1) specular highlight removal, (2) lesion segmentation, (3) feature extraction, (4) feature selection, (5) hierarchical decision tree classification. Because some lesions had specular regions, the com-bination of the ring-flash and camera locations results in specular highlights. These were identified ([2], Section 5.2) using thresholds on the saturation and intensity. Highlight pixels were not used in the feature calculations.

Lesion segmentation used a binary region-based active contour approach, using statistics based on the lesion and normal skin regions. Morphological open-ing was applied afterwards to slightly improve the boundaries. Details can be found in [12]. This produced a segmentation mask which covers the lesion.

---

[1] homepages.inf.ed.ac.uk/rbf/DERMOFIT/datasets.htm.

4.1   Feature Calculation

The majority of the 2551 features are calculated from generalized co-occurrence texture matrices. More details of the features are given in Section 4.2 of [2]. The texture features are described by "XY FUNC DIST QUANT", where X,Y $\in$ { R, G, B }, { L, a, b } or { H, S, V } gives the co-occurring color channels from the lesion, DIST is the co-occurring pixel separation, (5, 10, ... 30), QUANT is the number of gray levels $\in$ {64, 128, 256}. The co-occurrence matrices are computed at 4 orientations and then averaged. From each matrix, 12 summary scalar features are extracted, including FUNC $\in$ { Contrast, Cluster-Shade, Correlation, Energy, MaxProbability, Variance}, as described in [7]. As well as using these features directly, the difference (l-s: lesion-normal skin) and ratio (l/s: lesion/normal skin) features were computed.

After these features were calculated, a z-normalization process was applied, where the mean and standard deviation were estimated from the inner 90% of the values. Features above or below the 95th or 5th percentile were truncated. Some of the top features selected (see next section) for use by the decision tree are listed in Table 1.

Because there is much correlation between the color channels, and the different feature scales and quantizations, a feature reduction method was applied. The feature calculation process described above resulted in 17079 features. These features were cross-correlated over the 1300 lesions. The features were then sequentially examined. Any feature whose absolute correlation was greater than 0.99 with a previously selected feature was removed. This reduced the potential feature set to 2489. Some additional lesion-specific features (for each of R, G, B) were added to give 2551 features (also normalized):

- Ratio of mean lesion periphery to lesion center color
- Ratio of mean lesion color to non-lesion color
- Std dev of lesion color
- Ratio of lesion color std dev to non-lesion color std dev
- Ratio of mean lesion color to lesion color std dev
- Six gray-level moment invariants
- Given a unit sum normalized histogram of lesion pixel intensities H_l and normal skin pixel intensities H_n, use features $mean(H_l.-H_n)$, $std(H_l.-H_n)$, $mean(H_l./H_n)$, $std(H_l./H_n)$, $H_l.-H_n$ and $H_l./H_n$. The latter 2 features are histograms and the Bhattacharyya distance is used.

4.2   Feature and Parameter Selection

From the 2551 initial features, greedy Forward Sequential Feature Selection (en.wikipedia... .org/wiki/Feature_selection) is used to select an effective subset of features for each of the 9 tests shown in Fig. 1. This stage results in 2–7 features selected for each test. Table 2 lists the number of features selected and the top 2 for each of the tests. The full set of features used can be

Table 1. Overview of top selected features. See text body for details.

68	GB Correlation d10 L64	222	HH ClusterShade d5 L128
245	HH Energy d5 L256	267	HH Correlation d30 L256
272	aa Contrast d5 L64	523	ns SS Correlation d15 L64
622	ns bb Homogeneity d10 L64	630	ns bb Correlation d5 L128
832	l-s SS Dissimilarity d5 L64	834	l-s HH Variance d5 L64
950	l-s aa Autocorrelation d5 L64	959	l-s La Variance d5 L64
1467	l/s HH ClusterShade d25 L256	1536	l/s La Variance d5 L64
1556	l/s aa Contrast d15 L64	1824	so sigma Im G s3 n4
2303	M-m lo sigma Re R s1 G s1	2422	sigma Im G s1
2433	l-s mu Re G s1	2441	l/s mu Re G s1
2503	G mean(l)/std(l)	2526	R std(hist(l)./ hist(s))

seen at: `homepages.inf.ed.ac.uk/rbf/`... `DERMOFIT/SCusedfeatures.htm`. We tersely describe in Table 1 the top features of the tests. The tests use a K-Nearest Neighbor classifier with a Euclidean distance measure $\sum_r (x_r - n_r)^2$ where x_r is the rth property of the test sample and n_r is from a neighbor sample.

Also included in the parameter optimization stage was the selection of the optimal value K value to use in the K-Nearest Neighbor algorithm. The K values reported for the different tests reported in Table 2 were found by considering odd values of K from 3 to 19. The best performing K was selected over multiple cross-validated trials, but generally there was only 1–3% variation in the results for K in 7–19. Performance evaluation, and parameter and feature selection used 5-fold cross validation (using the Matlab `cvpartition` function), with 1 of the 5 subsets as an independent test set. The splits kept the lesion classes balanced.

4.3 Hierarchical Decision Tree

The lesion classification uses a hierarchical decision tree, where a different K-NN is trained for each decision node. Other classifiers (*e.g.* Random Forest or multi-class SVM) or decision node algorithms (*e.g.* SVM) could be investigated, but we chose a K-NN because of the intuition that there was much in-class variety and so data-driven classification might perform better. Several varieties of deepnet based classification are presented in the next section.

The choice of branching in the tree is motivated partly by clinical needs (*i.e.* cancer *versus* non-cancer) and partly by classifier performance (*i.e.* the lower levels are chosen based on experimental exploration of performance). Figure 1 shows the selected decision tree. Exploration (based on experience rather than a full systematic search) of different cancer subtree structures showed that the PYO/VASC branch was most effectively isolated first, with then a two way split between IEC, MEL and DF *versus* the rest. These two initial decisions could be performed almost perfectly, thus reducing the decision task to smaller decisions (and without propogating errors).

While a general classification is valuable, from a clinical perspective several more focussed binary decisions are important, namely melanoma (MEL) *versus* melanocytic nevi (ML), melanoma (MEL) *versus* seborrheic keratosis (SK), and basal cell carcinoma (BCC) plus squamous cell carcinoma (SCC) *versus* seborrheic keratosis (SK). We implemented these binary decisions each with a single test, after again doing feature selection, also as reported in Table 2.

Table 2. Top 2 features for each of the key K-NN decisions in the decision tree and cross-validated performance on ground-truthed data.

Test	K	Num of features used	Feat 1	Feat 2	Accuracy
PYO/VASC vs rest	11	5	2503	834	0.974
MEL/IEC/DF vs rest	13	2	2433	2441	1.000
MEL vs IEC/DF	19	3	959	832	0.831
AK/BCC/SCC vs ML/SK	17	9	222	1536	0.916
AK vs BCC/SCC	17	4	2422	272	0.876
PYO vs VASC	15	3	523	622	0.852
IEC vs DF	15	7	630	1467	0.888
BCC vs SCC	15	7	245	1556	0.814
ML vs SK	11	7	950	267	0.850
MEL vs ML	9	2	2303	1824	0.995
MEL vs SK	3	1	2433		1.000
BCC/SCC vs SK	11	5	68	2526	0.906

5   Decision Tree Experiment Results

Evaluation of the classification performance using the decision tree presented above used leave-one out cross-validation. The confusion matrix in Table 4 summarizes the performance for the detailed classification results over the 10 classes. The mean accuracy over all lesions (micro-average - averaging over all lesions) was 0.781 and the accuracy over all classes (macro-average - averaging over the performance for each class) was 0.705. Mean sensitivity is 0.705 and mean specificity is 0.972 (when averaging the sensitivities and specificities of each class over all 10 classes). A comparison of the results with previous researchers is seen in Table 3.

The new 10 class results are comparable those of [9,10] and considerably better than the others. Combining classes BCC/SCC & ML/SK to give an 8 class decision produces considerably better results. "Kawahara and Hamarneh [10] repeated the experiments from Kawahara *et al.* [9], but changed the experimental setup to use half the training images, and omitted data augmentation in order to focus on the effect of including multi-resolution images." (private communication from authors).

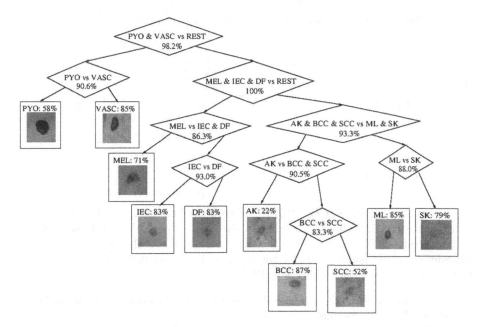

Fig. 1. Decision tree for lesion classification, also giving example images of the 10 lesion types. The numbers given at the decision boxes are the test data accuracies over the relevant classes (*i.e.* ignoring other classes that also go down that tree path). The numbers at the leaves of the tree are the final test accuracies over the whole dataset.

The confusion matrix for the final classification is shown in Table 4. There are three significant observations: (1) The AK lesion class, which has the worst performance, is mainly confused with the BCC and SCC classes. Visual inspection of the AK lesions shows that many of the lesions look a little like BCC and SCC lesions. (2) Many of the misclassifications are between the ML and SK classes. Neither of these are cancerous, so confusion between the classes has no real consequences. (3) Many of the other misclassifications are between the BCC and SCC classes. Both are cancers, but SCC needs more urgent treatment.

Merging BCC/SCC and ML/SK into 2 classes (and using the same tree) improves the classification rate to 0.871 (*i.e.* over an 8 class decision problem.).

To compare with the third experiment of Esteva *et al.* [5] (keratinocyte carcinomas (classes BCC and SCC here) *versus* benign seborrheic keratoses (class SK here)), we explored a KNN classifier with features selected from the same pool of features. Best accuracy of 0.906 was achieved with K = 9 and top features 168 and 186 over 5-fold cross-validation using the set of 327 BCC + SCC and 257 SK lesions. This is comparable to the accuracy (0.92) estimated from Esteva *et al.*'s sensitivity/specificity curves. To compare with the fourth experiment of Esteva *et al.* [5] (malignant melanoma (class (MEL) *versus* benign nevi (class ML)), we again explored a KNN classifier with features selected from the same pool of features. Accuracy of 0.995 (compared to their 0.92) was achieved with K = 9 and

Table 3. 10 class performance comparison with previous research.

Paper	Lesion (Micro) Accuracy	Class (Macro) Accuracy
Ballerini *et al.* [2]	0.743*	0.592
Di Leo *et al.* [4]	0.67	-
Esteva *et al.* [5]	>0.554+	-
Kawahara *et al.* [9]	0.818^x	-
Kawahara *et al.* [10]	0.795	-
New algorithm	0.781	0.705
New algorithm (BCC/SCC & ML/SK merged)	**0.871**	**0.736**

*: The results of Ballerini *et al.* consider only 5 classes (AK, BCC, ML, SCC, SK).
+: The results of Esteva *et al.* cover 9 classes, 5 of which roughly correspond to those considered in this paper. x: 0.818 was reported in [9] but 0.795 was reported and compared to in the later publication [10].

Table 4. Confusion Matrix: Row label is true, Column label is classification. Micro-average = 0.781. Macro-average accuracy = 0.705.

	AK	BC	ML	SC	SK	ME	DF	VA	PY	IE	TOT	ACC
AK	10	20	1	11	3	0	0	0	0	0	45	0.22
BCC	2	208	7	17	5	0	0	0	0	0	239	0.87
ML	2	10	280	0	39	0	0	0	0	0	331	0.85
SCC	0	34	0	46	8	0	0	0	0	0	88	0.52
SK	2	21	27	5	202	0	0	0	0	0	257	0.79
MEL	0	0	0	0	0	54	7	2	0	13	76	0.71
DF	0	0	0	0	0	4	54	3	0	4	65	0.83
VASC	0	0	0	0	0	6	1	82	3	5	97	0.85
PYO	0	0	0	0	0	2	0	7	14	1	24	0.58
IEC	0	0	0	0	0	5	5	1	2	65	78	0.83

only 2 features (2303 and 1824) over 5-fold cross-validation using the set of 331 ML and 76 MEL lesions. This compares to an accuracy of 0.92 estimated from Esteva *et al.*'s sensitivy/specificity curves [5]. Similarly, the clinically important discrimination between malignant melanoma (MEL) *versus* seborrheic keratosis (SK) (256 SK + 76 MEL samples) in this dataset gave perfect 5-fold cross-validated classification using K = 3 and 1 feature (2433). This high performance suggests that the dataset should be enlarged; however, melanoma is actually a rather uncommon cancer compared to, for example, BCC.

6   Deep Net Classifier Methodology

Given the general success of deepnets in classification image tasks, we investigated [3] three variations of a classifier based on the Resnet-50 architecture [8] pretrained on the ImageNet dataset and then tuned on the skin lesion samples in the same 5-fold cross-validation manner. The variations were:

1. A standard deepnet with 10 output classes, where the class with the highest activation level is selected.
2. A hierarchy of classifiers with the same structure as the decision tree presented above, except where each classification node is replaced with a Resnet-50 classifier.
3. A standard deepnet with 10 output classes, with a refinement stage. If the top two activation levels for a lesion from the standard deepnet were either {BCC, SCC} or {ML, SK}, then the lesion went to an additional binary Resnet-50 trained to discriminate the two classes.

Preprocessing of the segmented images: (1) produced standard 224 * 224 images, and (2) rescaled the RGB values of the whole image to give the background normal skin a standard value (computed by mean across the whole dataset). Data augmentation was by flipping, translation and rotation as there was no preferred orientation in the lesion images. Further augmentations such as color transformations, cropping and affine deformations as proposed by [15] could be added, which improved their melanoma classification accuracy by about 1% on the ISIC Challenge 2017 dataset. Training performance optimization used a grid search over the network hyperparameters. As deepnets are known to train differently even with the same data, the main result is an average over multiple (7) trainings. This is configuration (1) below.

Several other configurations were investigated using the same general deepnet: (2) The decision tree structure from Sect. 5, except each decision node is replaced by a 2 class deep net. (2x) For comparison, we list the performance of the decision tree from Sect. 5. (3) If the deepnet selected one of BCC, SCC, ML, or SK and the activation level was less than 0.88, then the lesion was re-classified using a 2 class BCC/SCC or ML/SK deep net. (4) A classifier for 8 classes, where the cancerous classes SCC/BCC were merged, and the benign classes ML/SK were merged. (4x) For comparison, the performance of the 8 class decision tree from Sect. 5. (5) A deepnet producing only a two-class cancer/not cancer decision (5x) A two-class cancer/not cancer decision using the same KNN (K = 17, 10 features) methods used in the decision tree. The resulting performances are shown in Table 5.

The results show that there is not much difference in cross-validated performance between the basic decision tree (0.781) and basic deepnet (0.787). Given the variability of deepnet training and cross-validation, there is probably no statistical significance between these. Interestingly, the reproduction of the decision tree with deepnets replacing the KNN classifiers produced distinctly worse performance (0.742 *vs* 0.781). It is unclear why. It is clear that applying the refinement based on easily confused classes gives better 10-class results

Table 5. Summary of deepnet results and comparisons with KNN classifier.

Case	Algorithm	Accuracy
1	Flat Resnet-50	0.787 ± 1.0
2	Decision tree with Resnet-50 nodes	0.742
2x	10 class decision tree from Sect. 5	0.781
3	Flat Resnet-50 with BCC/SCC and ML/SK refinement	0.801
4	8 class deep net with BCC/SCC and ML/SK merged	0.855 ± 0.4
4x	8 class decision tree from Sect. 5	0.871
5	Resnet-50 2 class cancer vs non-cancer	0.913 ± 0.71
5x	KNN based 2 class cancer vs non-cancer	0.874 ± 0.01

(0.801 *vs* 0.787), but still not as good as combining the BCC/SCC and ML/SK (0.855 *vs* 0.801). The 8-class deepnet performed worse than the 8-class decision tree (0.855 *vs* 0.871). Possibly the binary cancer/non-cancer classifier performed better when using the deepnet (0.913 *vs* 0.874).

7   Discussion

The approaches presented in this paper have achieved good performance on the 10 lesion type classification task and even better performance on the modified 8 class and binary problems. These lesions are those most commonly encountered in a clinical context, so there is a clear potential for medical use. Of interest is the fact that this was achieved using traditional features (and is thus more 'explainable') as well as using a deep learning algorithm.

Although the demonstrated performance is good, and generally better than dermatologist performance [5], there are still limitations. In particular, the BCC *versus* SCC, and ML *versus* SK portions of the classification tree have the lowest performance. Another poorly performing decision is AK *versus* other cancers (although the performance rate looks good, the large imbalance between the classes masks the poor AK identification - a class with few samples). We hypothesize that part of the difficulties arise because of the small dataset size, particularly for classes AK, DF, IEC, and PYO. When the K-NN classifier is used, it is necessary to have enough samples to have a good set of neighbors. This can also affect the other classes, because some lesion types may have difference appearance subclasses (*e.g.* BCC).

Another complication related to the dataset size is the correctness of the ground-truth for the lesion classes. Although 2 clinical dermatologists and a pathologist concur on the lesion diagnosis for the 1300 lesions in the dataset, generating that consistent dataset also showed up many differences of opinion about the diagnoses. It is probable that there are some lesions that were incorrectly labeled by all three professionals (the real world is not tidy), although, since one of the three was a pathologist, the dataset is probably reasonably correctly labeled. Again, having additional correctly labeled samples would help overcome outlying mis-labeled samples, given the nearest neighbor classifier structure.

Another limitation arises from the fact that the images in this dataset are all acquired under carefully controlled lighting and capture conditions. By contrast, the images used by Esteva *et al.* [5] came from many sources, which is probably one of the reasons their performance is so much lower. In order to make the solution found here be practically usable, it must work with different cameras and under different lighting conditions. This is a direction for future research.

Acknowledgments. The research was part of the DERMOFIT project (homepages. inf.ed.ac.uk/rbf/DERMOFIT) funded originally by the Wellcome Trust (Grant No: 083928/Z/07/Z). Much of the ground-truthing, feature extraction and image analysis was done by colleagues B. Aldridge, L. Ballerini, C. Di Leo, and X. Li as reported previously.

References

1. American Cancer Society. Cancer Facts & Figures (2016)
2. Ballerini, L., Fisher, R.B., Aldridge, R.B., Rees, J.: A color and texture based hierarchical K-NN approach to the classification of non-melanoma skin lesions. In: Celebi, M.E., Schaefer, G. (eds.) Color Medical Image Analysis. Lecture Notes in Computer Vision and Biomechanics, vol. 6. Springer, Dordrecht (2013). https:// doi.org/10.1007/978-94-007-5389-1_4
3. Bertrand, A.: Classification of skin lesions images using deep nets. Intern report, INP Grenoble (2018)
4. Di Leo, C., Bevilacqua, V., Ballerini, L., Fisher, R., Aldridge, B., Rees, J.: Hierarchical classification of ten skin lesion classes. In: Proceedings of Medical Image Analysis Workshop, Dundee (2015)
5. Esteva, A., et al.: Dermatologist-level classification of skin cancer with deep neural networks. Nature **542**, 115–118 (2017)
6. Ferris, L.K., et al.: Computer-aided classification of melanocytic lesions using dermoscopic images. J. Am. Acad. Dermatol. **73**(5), 769–776 (2015)
7. Haralick, R.M., Shanmungam, K., Dinstein, I.: Textural features for image classification. IEEE Trans. Syst. Man Cybern. B Cybern. **3**(6), 610–621 (1973)
8. He, K., Zhang, X., Ren, S., Sun, J.: Deep residual learning for image recognition. In: Proceedings of CVPR (2016)
9. Kawahara, J., BenTaieb, A., Hamarneh, G.: Deep features to classify skin lesions. In: Proceedings of IEEE 13th International Symposium on Biomedical Imaging (ISBI) (2016)
10. Kawahara, J., Hamarneh, G.: Multi-resolution-tract CNN with hybrid pretrained and skin-lesion trained layers. In: Wang, L., Adeli, E., Wang, Q., Shi, Y., Suk, H.-I. (eds.) MLMI 2016. LNCS, vol. 10019, pp. 164–171. Springer, Cham (2016). https://doi.org/10.1007/978-3-319-47157-0_20
11. Korotkov, K., Garcia, R.: Computerized analysis of pigmented skin lesions: a review. Artif. Intell. Med. **56**(2), 69–90 (2012)
12. Li, X., Aldridge, B., Ballerini, L., Fisher, R., Rees, J.: Depth data improves skin lesion segmentation. In: Yang, G.-Z., Hawkes, D., Rueckert, D., Noble, A., Taylor, C. (eds.) MICCAI 2009. LNCS, vol. 5762, pp. 1100–1107. Springer, Heidelberg (2009). https://doi.org/10.1007/978-3-642-04271-3_133

13. Maglogiannis, I., Doukas, C.N.: Overview of advanced computer vision systems for skin lesions characterization. IEEE Trans. Inf. Technol. Biomed. **13**(5), 721–733 (2009)
14. Masood, A., Al-Jumaily, A.A.: Computer aided diagnostic support system for skin cancer: a review of techniques and algorithms. Int. J. Biomed. Imaging **2013**, 22 (2013)
15. Perez, F., Vasconcelos, C., Avila, S., Valle, E.: Data augmentation for skin lesion analysis. In: Stoyanov, D., et al. (eds.) CARE/CLIP/OR 2.0/ISIC -2018. LNCS, vol. 11041, pp. 303–311. Springer, Cham (2018). https://doi.org/10.1007/978-3-030-01201-4_33

Biostatistics

Multilevel Models of Age-Related Changes in Facial Shape in Adolescents

Damian J. J. Farnell$^{(\boxtimes)}$, Jennifer Galloway, Alexei I. Zhurov, and Stephen Richmond

School of Dentistry, Cardiff University, Heath Park, Cardiff CF14 4XY, Wales

{FarnellD,GallowayJL,ZhurovAI,RichmondS}@cardiff.ac.uk

Abstract. Here we study the effects of age on facial shape in adolescents by using a method called multilevel principal components analysis (mPCA). An associated multilevel multivariate probability distribution is derived and expressions for the (conditional) probability of age-group membership are presented. This formalism is explored via Monte Carlo (MC) simulated data in the first dataset; where age is taken to increase the overall scale of a three-dimensional facial shape represented by 21 landmark points and all other "subjective" variations are related to the width of the face. Eigenvalue plots make sense and modes of variation correctly identify these two main factors at appropriate levels of the mPCA model. Component scores for both single-level PCA and mPCA show a strong trend with age. Conditional probabilities are shown to predict membership by age group and the Pearson correlation coefficient between actual and predicted group membership is $r = 0.99$. The effects of outliers added to the MC training data are reduced by the use of robust covariance matrix estimation and robust averaging of matrices. These methods are applied to another dataset containing 12 GPA-scaled (3D) landmark points for 195 shapes from 27 white, male schoolchildren aged 11 to 16 years old. 21% of variation in the shapes for this dataset was accounted for by age. Mode 1 at level 1 (age) via mPCA appears to capture an increase in face height with age, which is consistent with reported pubertal changes in children. Component scores for both single-level PCA and mPCA again show a distinct trend with age. Conditional probabilities are again shown to reflect membership by age group and the Pearson correlation coefficient is given by $r = 0.63$ in this case. These analyses are an excellent first test of the ability of multilevel statistical methods to model age-related changes in facial shape in adolescents.

Keywords: Multilevel principal components analysis · Multivariate probability distributions · Facial shape · Age-related changes in adolescents

1 Introduction

The importance of modeling the effects of groupings or covariates in shape or image data is becoming increasingly recognized, e.g., a bootstrapped response-based imputation modeling (BRIM) of facial shape [1], a linear mixed model of optic disk shape [2], or variational auto-encoders more generally (see, e.g., [3–5]). Multilevel principal components analysis (mPCA) has also been shown [6–10] to provide an efficient method of

© Springer Nature Switzerland AG 2020
Y. Zheng et al. (Eds.): MIUA 2019, CCIS 1065, pp. 101–113, 2020.
https://doi.org/10.1007/978-3-030-39343-4_9

modeling shape and image texture in such cases. Previous calculations using the mPCA approach have focused on: facial shape for a population of subjects that demonstrated groupings by ethnicity and sex [7, 8], image texture for two expressions (neutral and smiling) [9, 10], and time-series shape data tracked through all phases of a smile [10]. Here we consider how age-related changes in facial shape can be modelled by multilevel statistical approaches for Monte Carlo (MC) simulated data and for real data by using a model that is illustrated schematically in Fig. 1.

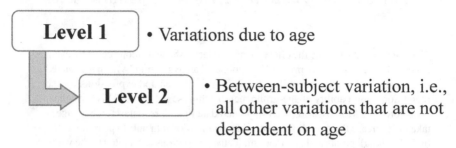

Fig. 1. Multilevel model of the effects of age on facial shape.

2 Methods

2.1 Mathematical Formalism

3D landmark points are represented by a vector z for each shape. Single-level PCA is carried out by finding the mean shape vector μ over all shapes and a covariance matrix

$$\Sigma_{k_1,k_2} = \frac{1}{N-1} \sum_{i=1}^{N} (z_{ik_1} - \mu_{ik_1})(z_{ik_2} - \mu_{ik_2}). \tag{1}$$

k_1 and k_2 indicate elements of this covariance matrix and i refers to a given subject. The eigenvalues λ_l and (orthonormal) eigenvectors u_l of this matrix are found readily. For PCA, one ranks all of the eigenvalues into descending order and one retains the first l_1 components in the model. The shape z is modeled by

$$z^{model} = \mu + \sum_{l=1}^{l_1} a_l u_l, \tag{2}$$

The coefficients $\{a_l\}$ (also referred to as "component scores" here) are found readily by using a scalar product with respect to the set of orthonormal eigenvectors, i.e., $a_l = u_l \cdot (z - \bar{z})$, for a fit of the model to a new shape vector z. The component score a_l is standardized by dividing by the square root of the eigenvalue λ_l.

Level 1   **Level 2**   **Data**

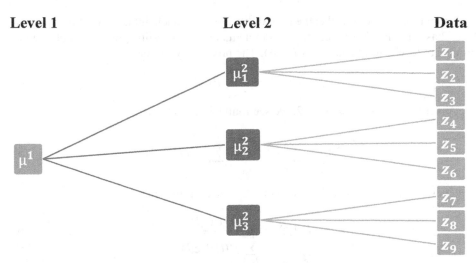

Fig. 2. Multilevel model represented as a tree. Shapes μ_l^2 at level 2 are average shapes over all shape data z in a given group l (e.g., 3 shapes per group are shown above). The shape μ^1 at level 1 is the average shape over all of the shape data μ_l^2 at level 2 (e.g., 3 groups at level 2 are shown above).

Multilevel PCA (mPCA) allows us to isolate the effects of various influences on shape at different levels of the model. This allows us to adjust for each subjects' individual facial shape in order to obtain a clearer picture of those changes due to the primary factor, i.e., age here. The covariance matrix at level 2 is formed with respect to all subjects in each age group l and then these covariance matrices are averaged over all age groups to give the level 2 covariance matrix, Σ_2. The average shape for group l at level 2 is denoted by μ_l^2. By contrast, the covariance matrix at level 1, Σ_1, is formed with respect to the shapes μ_l^2 at each age group at level 2. The overall "grand mean" shape at level 1 is denoted by μ^1. These relationships for the multilevel model are illustrated as a tree diagram in Fig. 2. mPCA uses PCA with respect to the covariance matrices at the two levels separately. The l-th eigenvalue at level 1 is denoted by λ_l^1, with associated eigenvector u_l^1, whereas the l-th eigenvalue at level 2 is denoted by λ_l^2, with associated eigenvector u_l^2. We rank all of the eigenvalues into descending order at each level of the model separately, and then we retain the first l_1 and l_2 eigenvectors of largest magnitude at the two levels, respectively. The shape z is modeled by

$$z^{model} = \mu^1 + \sum_{l=1}^{l_1} a_l^1 u_l^1 + \sum_{l=1}^{l_2} a_l^2 u_l^2, \tag{3}$$

where μ^1 is the "grand mean" at level 1, as described above. The coefficients $\{a_l^1\}$ and $\{a_l^2\}$ (again referred to as "component scores" here) are determined for any new shape z by using a global optimization procedure in MATLAB R2017 with respect to an appropriate cost function [6–10]. The mPCA component scores a_l^1 and a_l^2 may again be standardized by dividing by the square roots of λ_l^1 and λ_l^2, respectively.

From Fig. 2, we define that the probability along a branch linking level 1 to group l at level 2 as $P(l)$. Furthermore, we may define that the probability along a branch linking group l at level 2 to the data z as $P(z|l)$. The probability of both is therefore,

$$P(z, l) = P(l)P(z|l). \tag{4}$$

Assuming m groups at level 2, we see immediately also that

$$P(z) = \sum_{l=1}^{m} P(z, l) = \sum_{l=1}^{m} P(l)P(z|l). \tag{5}$$

These results lead on to Bayes theorem, which implies that

$$P(l|z) = \frac{P(l)P(z|l)}{\sum\limits_{l=1}^{m} P(l)P(z|l)}. \tag{6}$$

Here we shall use a multivariate normal distribution at level 2, which is given by

$$P(z|l) = N(z|\mu_l^2, \Sigma_2). \tag{7}$$

For small numbers of groups m at level 2, one might set $P(l)$ to be constant. In this case, the conditional probability that a given shape z belongs to group l is given by,

$$P(l|z) = \frac{N(z|\mu_l^2, \Sigma_2)}{\sum\limits_{l=1}^{m} N(z|\mu_l^2, \Sigma_2)}. \tag{8}$$

For larger numbers of groups m at level 2, it might be more appropriate to model $P(l)$ as a multivariate normal distribution also, by using

$$P(l) = N(\mu_l^2|\mu^1, \Sigma_1). \tag{9}$$

The conditional probability that a given shape z belongs to group l is now given by,

$$P(l|z) = \frac{N(\mu_l^2|\mu^1, \Sigma_1)N(z|\mu_l^2, \Sigma_2)}{\sum\limits_{l=1}^{m} N(\mu_l^2|\mu^1, \Sigma_1)N(z|\mu_l^2, \Sigma_2)}. \tag{10}$$

The extension of this approach to three or more levels is straightforward.

2.2   Image Capture, Preprocessing, and Subject Characteristics

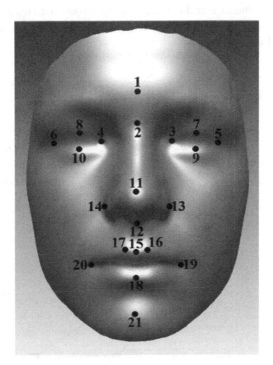

Fig. 3. Illustration of the 21 landmark points for dataset 1.

MC simulations were used initially to explore age-related changes on facial shape in dataset 1. A template facial shape containing 21 landmarks points in three dimensions was constructed firstly, as shown in Fig. 3. The effects of age were simulated by applying a scale factor that grew linearly with age (in arbitrary units) to all points in this template equally. All other "subjective" variation was included by altering the width of the face for all subjects randomly (irrespective of age). A small amount of normally distributed random error was added to all shapes additionally. However, there were essentially just two factors affecting facial shape in dataset 1. We expect the overall change in scale to be reflected at level 1 (age) of the multilevel model shown in Fig. 1 and changes in the width to be reflected at level 2 (all other sources of variation). All shapes were centered on the origin and the average scale across all shapes was set to be equal to 1. Note that 30 age groups were used here with 300 subjects per group in forming the original model and 100 per group for a separate testing dataset. The effects of outliers in the training set for single-level PCA and mPCA were explored by carrying out additional calculations with an extra 5% of the data containing strong outliers. Such shapes were outlying in terms of facial width and overall scale. Dataset 2 contained real data of 195 shapes from 27 white, male subjects (aged 11 to 16) selected from two large comprehensive schools in the South Wales Valleys area in Rhonda Cynon Taf. Those with craniofacial anomalies were excluded. Ethical approval was obtained from the director of education, head teachers, school committees, and the relevant ethics committees of Bro Taf. Written informed consent was obtained before obtaining the 3D laser scans. 12 landmark points along the centerline of the face and between the eyes were then used to describe facial shape. All shapes were GPA transformed and the average scale across all shapes was again set to be equal to 1.

3   Results

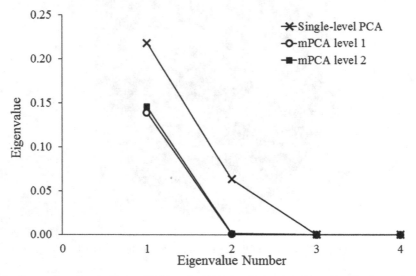

Fig. 4. Eigenvalues for single-level PCA and mPCA level 1 (age) and level 2 (all other variations) for dataset 1.

Eigenvalues for single-level PCA and mPCA are shown in Fig. 4 for dataset 1. These results for mPCA demonstrated a single non-zero eigenvalue for the level 1 (age) and a single large eigenvalue for the level 2 (all other variations), as expected. Results for the eigenvalues for single-level PCA are of comparable magnitude to those results of mPCA, as one would expect, and they follow a very similar pattern.

Modes of variation of shape for dataset 1 are presented in Fig. 5. The first mode at level 1 (age) via mPCA and mode 1 via single-level PCA both capture increases in overall size of the face, as required. The first mode at level 2 (all other variations) for mPCA clearly corresponds to changes in the width of the face, as required. However, mode 2 via single-level PCA clearly mixes the effects of overall changes in size and also width of the face. Such "mixing" is a limitation of single-level PCA.

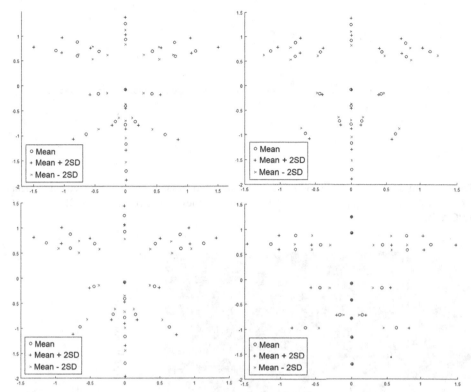

Fig. 5. Modes of shape variation in the frontal plane only for dataset 1: (upper left) = mode 1 via single-level PCA; (upper right) = mode 2 via single-level PCA; (lower left) = mode 1 at level 1 (age) via mPCA; (lower right) = mode 1 at level 2 (all other variations) via mPCA. (Landmark points are illustrated in Fig. 3).

Results for the standardized component 'scores' for mPCA for shape are shown in Fig. 6. Component 1 for level 1 (age) mPCA demonstrates differences due to age clearly because the centroids are strongly separated. Indeed, there is a clear progression of these centroids with age. By contrast, component 1 for level 2 (all other variations except age) via mPCA does not seem to reflect changes due to age very strongly (not shown here). Results for both components 1 and 2 via single-level PCA also demonstrate a clear trend with age.

Results for the (conditional) probabilities of group membership of Eq. (8) are shown as a heat map in Fig. 7 for dataset 1. (Note that very similar results are seen by using Eq. (10) for a multivariate normal distribution for $P(l)$ and so these results are not presented here.) A strong trend in the maximal probabilities is observed in Fig. 7 that clearly reflects the groupings by age. Age-group membership for each shape was predicted by choosing the group for which the conditional probability was highest. The Pearson correlation coefficient of actual versus predicted age group (from 1 to 30) is given by $r = 0.99$.

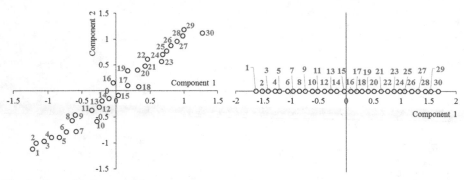

Fig. 6. Centroids for each of the 30 age groups (indicated by labels) of standardized component scores with respect to shape for dataset 1 for (left) single-level PCA (modes 1 and 2) and (right) mPCA for mode 1 at level 1 (age).

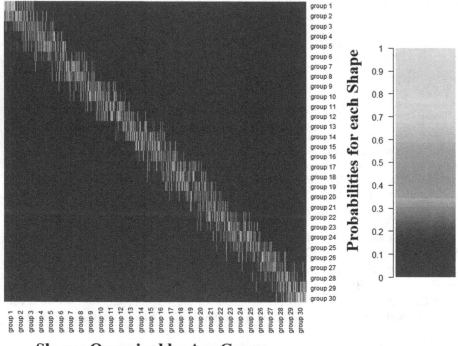

Fig. 7. Heat map of (conditional) probabilities of group membership of Eq. (8) for the 3000 test shapes used in dataset 1 (30 age groups and 100 shapes per group). High probabilities lie "along the diagonal," which reflects the groupings by age correctly.

The effects of adding outlying shapes to the training set were to increase the magnitude of eigenvalues and to add "random scatter" to points in the major modes of variation for both single-level PCA and mPCA. Model fits for the test set were seen to demonstrate a progression with age with that was less clear than in Fig. 6 due to this source of additional error. Furthermore, the overall scale of the (standardized) component scores was increased and conditional probabilities of Eqs. (8) and (10) became less efficient at predicting group membership, e.g., Pearson's r was reduced. Robust covariance matrix estimation and robust (median) averaging of covariance matrices was found to reduce the effects of outliers in these initial studies.

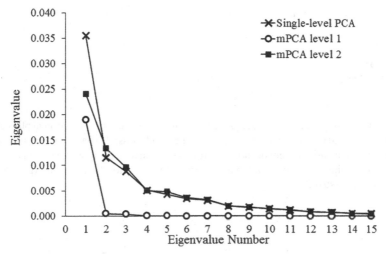

Fig. 8. Eigenvalues for single-level PCA and mPCA level 1 (age) and level 2 (all other variations) for dataset 2.

Eigenvalues for single-level PCA and mPCA are shown in Fig. 8 for dataset 2. The results for mPCA demonstrate a single large non-zero eigenvalue for the level 1 (age) only, which is presumably due to the small number of landmark points and/or the small number of groups at this level. However, level 2 (all other variations) does now have many large non-zero eigenvalues, which is reasonable for "real data." Results for the eigenvalues via single-level PCA are again of comparable magnitude to those results of mPCA and they follow a very similar pattern. mPCA calculations suggest that age contributed approximately 21% of the total variation for this 3D shape dataset.

Mode 1 at level 1 (age) via mPCA is shown in Fig. 9 for dataset 2. This mode represents an overall increase in face height and a decrease in the distance between the endocanthion and pronasale. Broadly, one might interpret this as an elongation in facial shape (possibly also a flatter face), which is consistent with the growth of children [11, 12]. Subtle differences are observed only between modes 1 at levels 1 (age) and 2 (all other variations) via mPCA, although we believe that these differences would become more apparent with increased number of landmark points. Mode 1 via single-level PCA is similar to both of these modes via mPCA. Mode 2 via single-level PCA and mode 2 at level 2 via mPCA are similar; both modes appear to relate to shape changes relating to the eyes and prominence of the chin. However, all modes are difficult to resolve with so few landmark points, and future studies of age-related changes in facial shape in adolescents will include more such landmark points.

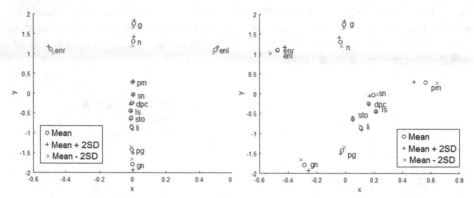

Fig. 9. Mode 1 of shape variation at level 1 (age) via mPCA for dataset 2: (left) frontal plane; (right) sagittal plane. Overall changes in shape indicate a longer and (possibly) flatter face with increasing age. (Glabella (g), nasion (n), endocanthion left (enl), endocanthion right (enr), pronasale (prn), subnasale (sn), labiale superius (ls), labiale inferius (li), pogonion (pg), gnathia (gn), philtrum (dpc), and stomion (sto)).

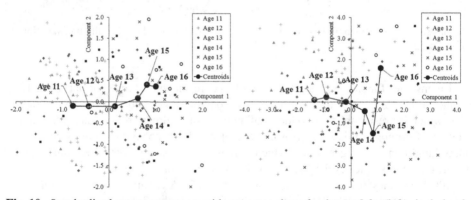

Fig. 10. Standardized component scores with respect to shape for dataset 2 for (left) single-level PCA and (right) mPCA at level 1 (age). A strong progression of the group centroids with increasing age is observed in both diagrams.

Results for the standardized component 'scores' via mPCA are shown in Fig. 10. Component 1 for level 1 (age) via mPCA demonstrates differences due to age clearly because the centroids are strongly separated. Indeed, a clear progression of these scores with age is again seen via mPCA at level 1. Component 2 for level 1 (age) shows a possible difference between ages 15 and 16, although this is probably due to random error because the sample size for age 16 was quite small. Component scores for level 2 (all other variations) for mPCA again do not seem to reflect changes due to age very strongly (not shown here). This is an encouraging result given that we found only very subtle differences between modes 1 at levels 1 and 2 via mPCA. A clear trend with age is also seen in Fig. 10 for both components 1 and 2 via single-level PCA.

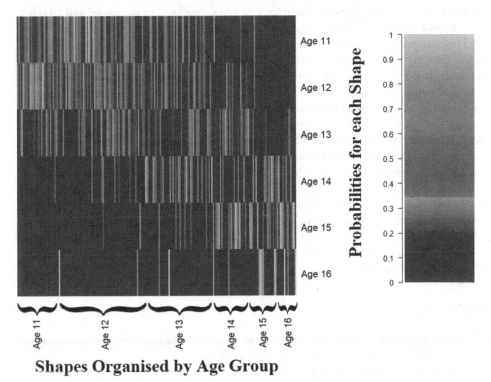

Shapes Organised by Age Group

Fig. 11. Heat map of (conditional) probabilities of group membership of Eq. (8) for the 195 shapes in dataset 2 (27 white, male subjects, aged 11 to 16 years old) using "miss-one-out" testing. Higher probabilities lie (broadly) "along the diagonal," which reflects the groupings by age.

Results for the probabilities of group membership of Eq. (8) are shown as a heat map in Fig. 11 for dataset 2. (Note that very similar results are again seen by using Eq. (10) for a multivariate normal distribution for $P(l)$ and so these results are not presented here.) "Miss-one-out" testing was used here, i.e., the model in each case was formed from all shape data except for the shape being tested. A trend is again observed in the maximal probabilities in Fig. 11 that reflects the groupings by age, although this trend is not quite as clear as for the MC-simulated data. We expect that this trend will become clearer with increased number of landmark points. Again, age-group membership for each shape was predicted by choosing the group for which the conditional probability was highest. The Pearson correlation coefficient of actual versus predicted group membership (from 11 to 16 years old) is given by $r = 0.63$.

4 Conclusions

The effect of age on 3D facial shape data has been explored in this article. The formalism for mPCA has been described and it was seen that mPCA allows us to model variations at different levels of structure in the data, i.e., age at one level of the model and all other variations at another level. Two datasets were considered, namely, MC-simulated data of 21 landmark points and real data for 195 shapes of 12 landmark points for 27 white, male subjects aged 11 to 16 years old. Eigenvalues appeared to make sense for both datasets. In particular, examination of these eigenvalues suggested that age contributed approximately 21% to the total variation in the shapes for the real data in dataset 2. Modes of variation also appeared to make sense for both datasets. Evidence of clustering by age group was seen in the component scores for both the simulated data and also the real data. An initial exploration of the associated multivariate probability distribution for such multilevel architectures was presented. Conditional probabilities were used to predict group membership. Results for the predicted and actual group memberships were positively correlated and the Pearson correlation coefficients were $r = 0.99$ and $r = 0.63$ for the MC-simulated and the real data, respectively. These results are an encouraging initial exploration of the use of multilevel statistical methods to explore and understand pubertal and age-related changes in facial shape. The interested reader is referred, e.g., to Refs. [11–15] for more information about age-related facial shape changes in adolescents.

References

1. Claes, P., Hill, H., Shriver, M.D.: Toward DNA-based facial composites: preliminary results and validation. Forensic Sci. Int. Genet. **13**, 208–216 (2014)
2. MacCormick, I.J., et al.: Accurate, fast, data efficient and interpretable glaucoma diagnosis with automated spatial analysis of the whole cup to disc profile. PLoS ONE **10**, e0209409 (2019)
3. Doersch, C.: Tutorial on variational autoencoders (2016). arXiv:1606.05908
4. Pu, Y., et al.: Variational autoencoder for deep learning of images, labels and captions. Advances in Neural Information Processing Systems, pp. 2352–2360. Curran Associates Inc., Red Hook, USA (2016)
5. Wetzel, S.J.: Unsupervised learning of phase transitions: from principal component analysis to variational autoencoders. Phys. Rev. E **96**, 022140 (2017)
6. Lecron, F., Boisvert, J., Benjelloun, M., Labelle, H., Mahmoudi, S.: Multilevel statistical shape models: a new framework for modeling hierarchical structures. In: 9th IEEE International Symposium on Biomedical Imaging (ISBI), pp. 1284–1287 (2012)
7. Farnell, D.J.J., Popat, H., Richmond, S.: Multilevel principal component analysis (mPCA) in shape analysis: a feasibility study in medical and dental imaging. Comput. Methods Programs Biomed. **129**, 149–159 (2016)
8. Farnell, D.J.J., Galloway, J., Zhurov, A., Richmond, S., Perttiniemi, P., Katic, V.: Initial results of multilevel principal components analysis of facial shape. In: Valdés Hernández, M., González-Castro, V. (eds.) MIUA 2017. CCIS, vol. 723, pp. 674–685. Springer, Cham (2017). https://doi.org/10.1007/978-3-319-60964-5_59

9. Farnell, D.J.J., Galloway, J., Zhurov, A., Richmond, S., Pirttiniemi, P., Lähdesmäki, R.: What's in a Smile? Initial results of multilevel principal components analysis of facial shape and image texture. In: Nixon, M., Mahmoodi, S., Zwiggelaar, R. (eds.) MIUA 2018. CCIS, vol. 894, pp. 177–188. Springer, Cham (2018). https://doi.org/10.1007/978-3-319-95921-4_18
10. Farnell, D.J.J., et al.: What's in a Smile? Initial analyses of dynamic changes in facial shape and appearance. J. Imaging **5**, 2 (2019)
11. Lin, A.J., Lai, S., Cheng, F.: Growth simulation of facial/head model from childhood to adulthood. Comput. Aided Des. Appl. **7**, 777–786 (2010)
12. Bhatia, S.N., Leighton, B.C.: Manual of Facial Growth: A Computer Analysis of Longitudinal Cephalometric Growth Data. Oxford University Press, Oxford (1993)
13. Bishara, S.E.: Facial and dental changes in adolescents and their clinical implications. Angle Orthod. **70**(6), 471–483 (2000)
14. Kau, C.H., Richmond, S.: Three-dimensional analysis of facial morphology surface changes in untreated children from 12 to 14 years of age. Am. J. Orthod. Dentofac. Orthop. **134**(6), 751–760 (2008)
15. West, K.S., McNamara Jr., J.A.: Changes in the craniofacial complex from adolescence to midadulthood: a cephalometric study. Am. J. Orthod. Dentofac. Orthop. **115**(5), 521–532 (1999)

Spatial Modelling of Retinal Thickness in Images from Patients with Diabetic Macular Oedema

Wenyue Zhu[1](✉), Jae Yee Ku[1,2], Yalin Zheng[1,2], Paul Knox[1,2],
Simon P. Harding[1,2], Ruwanthi Kolamunnage-Dona[3],
and Gabriela Czanner[1,4]

[1] Department of Eye and Vision Science, Institute of Ageing and Chronic Disease,
University of Liverpool, Liverpool L7 9TX, UK
{Wenyue.Zhu,yzheng}@liverpool.ac.uk
[2] St Paul's Eye Unit, Royal Liverpool University Hospital, Liverpool L7 8XP, UK
[3] Department of Biostatistics, University of Liverpool, Liverpool L69 3GL, UK
[4] Department of Applied Mathematics, Liverpool John Moores University,
Liverpool L3 3AF, UK
G.Czanner@ljmu.ac.uk

Abstract. For the diagnosis and monitoring of retinal diseases, the spatial context of retinal thickness is highly relevant but often underutilised. Despite the data being spatially collected, current approaches are not spatial: they involve analysing each location separately, or they analyse all image sectors together but they ignore the possible spatial correlations such as linear models, and multivariate analysis of variance (MANOVA). We propose spatial statistical inference framework for retinal images, which is based on a linear mixed effect model and which models the spatial topography via fixed effect and spatial error structures. We compare our method with MANOVA in analysis of spatial retinal thickness data from a prospective observational study, the Early Detection of Diabetic Macular Oedema (EDDMO) study involving 89 eyes with maculopathy and 168 eyes without maculopathy from 149 diabetic participants. Heidelberg Optical Coherence Tomography (OCT) is used to measure retinal thickness. MANOVA analysis suggests that the overall retinal thickness of eyes with maculopathy are not significantly different from the eyes with no maculopathy ($p = 0.11$), while our spatial framework can detect the difference between the two disease groups ($p = 0.02$). We also evaluated our spatial statistical model framework on simulated data whereby we illustrate how spatial correlations can affect the inferences about fixed effects. Our model addresses the need of correct adjustment for spatial correlations in ophthalmic images and to improve the precision of association in clinical studies. This model can be potentially extended into disease monitoring and prognosis in other diseases or imaging technologies.

Keywords: Spatial modelling · Correlated data · Simulation · Retinal imaging · Diabetic Macular Oedema

© Springer Nature Switzerland AG 2020
Y. Zheng et al. (Eds.): MIUA 2019, CCIS 1065, pp. 114–126, 2020.
https://doi.org/10.1007/978-3-030-39343-4_10

1   Introduction

Diabetic Macular Oedema (DMO) is a consequence of diabetes that involves retinal thickness changes in the area of the retina called the macula. Although the macula is only approximately 5 mm in diameter, the densely pack photoreceptors in the macula give rise to our central high acuity and colour vision. A healthy macula plays an essential role for activities such as reading, recognizing faces and driving. Macular disease can cause loss of central vision; DMO is the most common cause of vision loss among people with diabetic retinopathy.

DMO is caused by an accumulation of fluid (oedema) in the macula thought to be secondary to vascular leakage. It has been proposed that macular thickness is associated with visual loss [9]. For measuring retinal thickness, OCT is now widely used for the diagnosis and monitoring of DMO as it is able to produce high-resolution cross-sectional images of the retina [12].

The macula is often divided into nine subfields as initially described by the Early Treatment of Diabetic Retinopathy Study (ETDRS) research group [11]. These subfields comprise three concentric circles/rings with radii of 500, 1500 and 3000 μm subdivided into four regions (superior, temporal, inferior and nasal) as shown in Fig. 1. These subfields are named by their location as the central (CS), superior inner (SI), temporal inner (TI), nasal inner (NI), superior outer (SO), temporal outer (TO), inferior outer (IO) and nasal outer (NO). OCT measurements provide retinal thickness measurements for each of these nine subfields.

The simplest approach to analyse retinal thickness in these nine subfields is to analyse them separately in nine separate analyses. Sometimes only the measurement of the central subfield is used and the other measurements are discarded. However, this ignore the spatial context of measurements. If spatial dependency between the measurements of different subfields is not analysed properly, it can affect the precision of estimates and lead to inaccurate results in statistical tests.

A more complex approach is to properly spatially analyse the data [6]. Spatial statistical model takes into account the spatial correlations [4] thus in our data provides a means of incorporating spatial information from retinal thickness measurements in different subfields into statistical analyses. It may provide information of value in discriminating between disease states and for detection of retinal disease [7]. It has already been applied widely in other medical imaging contexts such as functional neuroimaging and cardiac imaging, where spatial correlations are also captured in the model. For example, Bowman et al. constructed a spatial statistical model for cardiac imaging from single photon emission computed tomography (SPECT) [2], and Bernal-Rusiel et al. explored the spatial structures in Magnetic Resonance Image (MRI) data in patients with Alzheimer's disease [1]. However, the application of spatial statistics to ophthalmic images has not yet been extensively studied.

Another concern in the analysis of ophthalmic images is the unit of analysis issue. Often, the correlation between two eyes from the same individual is ignored. Treating the eyes as independent introduces spuriously small standard errors. Although there is a continuing concern regarding this problem and methods [15, 16] are available for adjusting the correlation between the two eyes, the majority of studies do not take this into account when data from both eyes are available. This methodological problem has not improved much over the past two decades [17].

In this paper, we aimed to present a new statistical spatial inference framework for retinal images and to study the effect of the spatial correlations on the analysis of the spatial data. This framework is based on a linear mixed effect model with a spatial (Gaussian, autoregressive-1, exponential and spherical) error structure for the analysis of OCT imaging data, where correlation between eyes from the same patient and demographic data is adjusted in the model. We conducted a simulation study to validate our model and study the benefits of using a spatial modelling framework when spatial correlations exist.

The organization of the rest of the paper is as follows. The image dataset and the statistical modelling framework are presented in Sect. 2. In Sect. 3, we present results from the real data set. Simulation setting and simulation results are presented in Sect. 4. Discussion of our work and the conclusion are presented in Sects. 5 and 6.

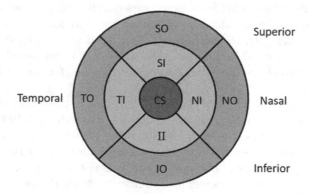

Fig. 1. Early Treatment of Diabetic Retinopathy (ETDRS) gird centred on the fovea with the radii of the central circle being 500 μm, inner circle 1500 μm and outer circle 3000 μm and the nine sectors (also called subfields).

2   Methods

2.1   Image Dataset

To illustrate the use of our proposed model, we apply it to retinal thickness measurements from the Early Detection of Diabetic Macular Oedema (EDDMO) study which is a prospective observational study conducted in the Royal Liverpool University Hospital. This study was performed in accordance with the ethical standards laid down in the Declaration of Helsinki, with local ethical and governance approval.

There were 150 participants with diabetes at their baseline visit in the study. Self-reported ethnic background revealed that approximately 90% of the participants were Caucasians. Participants with diabetes and co-existing pathologies (intracranial lesions n = 1, and ocular pathologies n = 4) were excluded from the analysis. A small number of participants did not have data collected from both eyes. All participants were examined by an ophthalmologist with a slit lamp. All the participants with diabetes had a dilated fundoscopy examination. Based on examination findings, eyes of participants with diabetes were categorized into two groups, either no maculopathy (M0) or maculopathy (M1). A summary of the dataset stratified by clinical diagnosis based on slit lamp examination is shown in Table 1. Retinal thickness measurements for both eyes were obtained using Heidelberg Spectralis OCT. Although the measurements of both foveal centre point thickness and central subfield mean thickness are available using OCT, central subfield mean thickness is more commonly used in clinical research when tracking center-involving DMO [3]. Therefore only central subfield mean thickness (CS) is used in the statistical analysis in this paper.

Table 1. Number of eyes for the analysis of overall thickness of retina

	M0	M1	Total
Left eyes	91	40	131
Right eyes	77	49	126
Total	168	89	257

2.2   Statistical Model

Our spatial statistical model has the general form described in Eq. (1), which is based on a linear mixed effect model with two-level of nested random effects.

In the model, Y_{ij} is the response vector for ith individual in the nested level of grouping, X_{ij} is the fixed effect vector (e.g. age, age, sex, glycated haemoglobin (HbA1c), axial length) associated with *beta*, b_i is the first level of random effects (e.g. individual level random effect) associated with Z_i, and u_{ij} is the second level of random effects (e.g. eye level random effect nested within each individual) associated with D_{ij}.

$$Y_{ij} = X_{ij}\beta + Z_i b_i + D_{ij} u_{ij} + \epsilon_{ij}, \qquad i = 1, ..., m; j = 1, ..., n_i$$

(1)

$$b_i \sim N(0, G_1), \qquad u_{ij} \sim N(0, G_2), \qquad \epsilon_{ij} \sim N(0, \Sigma_s)$$

where the first level random effect b_i is independent of the second level random effect u_{ij}, and ϵ_{ij} are within group error representing spatial correlation in the images which is assumed to be independent of random effect.

2.3 Spatial Dependency

The covariance matrix $\boldsymbol{\Sigma}_s$ for ϵ_{ij} can be decomposed to $\boldsymbol{\Sigma}_s = \sigma_s^2 \boldsymbol{\Psi}_{ij}$ where $\boldsymbol{\Psi}_{ij}$ is a positive-definite matrix which can be decomposed to $\boldsymbol{\Psi}_{ij} = \boldsymbol{\Lambda}_{ij} \boldsymbol{C}_{ij} \boldsymbol{\Lambda}_{ij}$. $\boldsymbol{\Lambda}_{ij}$ is a diagonal matrix and $\boldsymbol{C}_{ij}$ is correlation matrix with parameter γ. In our model, $\boldsymbol{\Lambda}_i$ is a identity matrix and it is easy to write that $cor(\epsilon_{ijk}, \epsilon_{ijk'}) = [\boldsymbol{C}_{ij}]_{kk'}$.

The spatial dependency $cor(\epsilon_{ijk}, \epsilon_{ijk'})$ is modelled with four different structures where the correlation is modelled as autoregressive-1 model, Gaussian model, exponential model and spherical model where γ can take the value of $\gamma_a, \gamma_g, \gamma_e, \gamma_s$ respectively.

For the lag autoregressive model, the correlation function decreases in absolute value exponentially with lag δ ($\delta = 1, 2, ...$) model has the form of

$$s_{kk'} = \gamma_a^\delta, \tag{2}$$

For spatial structured correlation, if $d_{kk'}$ is denoted as the Euclidean distance between location k and k'. The Gaussian correlation has the form of,

$$s_{kk'} = exp(-\gamma_g d_{kk'}^2), \tag{3}$$

the exponential model has the form of

$$s_{kk'} = exp(-\gamma_g d_{kk'}), \tag{4}$$

and the spherical model has the form of

$$s_{kk'} = 1 - 1/2(3\gamma_s d_{kk'} - \gamma_s^3 d_{kk'}^3). \tag{5}$$

2.4 Statistical Inference

In a traditional mixed effect model defined by Laird and Ware [5], the maximum likelihood estimator $\hat{\beta}$ for fixed effect β is as follows,

$$\hat{\beta} = (X' \Sigma X)^{-1} X' \Sigma^{-1} (Y - Z\hat{b}) \tag{6}$$

with the prediction $\hat{b}$ for random effect is obtained as follows,

$$\hat{b} = (Z' \Sigma_0^{-1} Z + G_0^{-1})^{-1} Z' \Sigma_0^{-1} (Y - X\beta_0) \tag{7}$$

where $Y|b \sim N(X\beta + Zb, \Sigma)$, $b \sim N(0, G)$ and Σ_0, G_0 β_0 are given values in EM algorithm during iterations. And the maximum likelihood estimator $\hat{\theta}$ for the precision estimates which define the parameters in the overall covariance matrix can be optimized through EM iterations or Newton-Raphson iterations [5].

If we describe a two level random effect model ignoring the spatial dependency, which assume $\boldsymbol{\Psi}_{ij}$ in model (1) as a identity matrix. We will have

$$\boldsymbol{Y}_{ij}^* = \boldsymbol{X}_{ij}^* \beta + \boldsymbol{Z}_i^* b_i + \boldsymbol{D}_{ij}^* u_{ij} + \epsilon_{ij}^*, \qquad i = 1, ..., m; j = 1, ..., n_i \tag{8}$$

$$\epsilon_{ij}^* \sim N(0, \sigma_s^2 I)$$

We define θ_1 as parameter in matrix $\boldsymbol{G}_1$ and θ_2 as parameter in matrix $\boldsymbol{G}_2$. The likelihood function for model (2) can be written as

$$L(\beta, \theta, \sigma_s^2 | y^*) =$$

$$\prod_{i=1}^{m} \int \prod_{j=1}^{n_i} \left[\int p(y_{ij}^* | b_i, u_{ij}, \beta, \theta_s^2) p(u_{ij} | \theta_2, \sigma_s^2) du_{ij} \right] p(b_i | \theta_1, \sigma_s^2) \tag{9}$$

where $p(\cdot)$ is the probability density function. Using the same idea in [5], we have the profiled log-likelihood for $l(\theta_1, \theta_2 | y)$. And the maximum likelihood estimator $\hat{\theta} = (\hat{\theta}_1, \hat{\theta}_2)$ can also be obtained via EM iterations or Newton-Raphson iterations.

If we describe a two level random effect model with the spatial dependency as shown in model (1), we use a linear transformation for y_{ij} with $y_{ij} = (\Psi_{ij}^{-1/2}) y_{ij}^*$. Then the likelihood function for model (1) can be written as

$$L(\beta, \theta, \sigma_s^2, \gamma | y) = \prod_{i=1}^{m} \prod_{j=1}^{n_i} p(y_{ij} | \beta, \theta, \sigma_s^2, \gamma)$$

$$= \prod_{i=1}^{m} \prod_{j=1}^{n_i} p(y_{ij}^* | \beta, \theta, \sigma_s^2) |\Psi_{ij}^{-1/2}| \tag{10}$$

$$= L(\beta, \theta, \sigma_s^2 | y^*) \prod_{i=1}^{m} \prod_{j=1}^{n_i} |\Psi_{ij}^{-1/2}|$$

Then Eq. (10) can be linked with Eq. (9), leading to a solution for all unknown parameters in model (1).

2.5   Statistical Analysis

In our application for the analysis of the results from the EDDMO study, the model used is described as follows,

$$y_{ijk} = x_{ijk}\beta + b_i + u_{ij} + \epsilon_{ijk}, \quad i = 1, .., m; j = 1, ..., n_i; k = 1, .., 9 \tag{11}$$

$$b_i \sim N(0, \sigma_1^2), \quad u_{ij} \sim N(0, \sigma_1^2), \quad \epsilon_{ij} \sim N(0, \sigma_s^2 \Psi_{ij})$$

where β is a vector, x_{ijk} are covariates for ith participant from j eye in sector k, b_i denotes the random effect for participant i, u_{ij} denote the random effect for j eye in participant i, m is the number of participant and $\max n_i = 2$. After statistical analyses, uncorrelated covariates such as sex, duration of diabetes and duration of diabetes were deleted from the model. In the final model, the fixed effect β and the predictor variable x_{ijk} used are follows,

$$x_{ijk}\beta = \beta_0 + \beta_1 * Age_i + \beta_2 * Diagnosis_{ij}$$
$$+ \beta_{3(k)} * Sector_{ijk} + \beta_{4(k)} * Sector_{ijk} * Diagnosis_{ij} \tag{12}$$

where Age_i is a continuous variable which represent the age for ith participant; $Diagnosis_{ij}$ is a categorical variable represent the diagnosis for j eye from ith participant, which include diabetic eye without maculopathy (baseline) and diabetic eye with maculopathy; $Sector_{ijk}$ is a categorical variable from 1 to 9 which represent the 9 sectors in ETDRS grid with central subfield (CS) thickness as a baseline; $Sector_{ijk} * Diagnosis_{ij}$ represent interaction term between sector and diagnosis with CS*Healthy as baseline.

The mixed effect model with two levels of random effects was fitted using the nlme-R package [10] and the spatial dependency Ψ_{ij} was fitted with structures as described in Sect. 2.3. Missing observations were tested whether they were missing at random and then handled using multiple imputation method in mice-R package [13]. Likelihood ratio test and information criterion (Akaike Information Criterion, AIC; Bayesian information criterion, BIC) were used to compare the models and to find the best model for the inference.

3　Results

To get the visual insight into the data, we made pairwise visualisations of mean profiles of retinal thickness measurements for all nine sectors at patients' baseline visits as shown in Fig. 2. This shows a large within group variability and suggests a pattern for the mean profiles of retinal thickness over the nine sectors. The mean retinal thickness profile of maculopathy group was consistently higher than the no-maculopathy group as shown in Fig. 2.

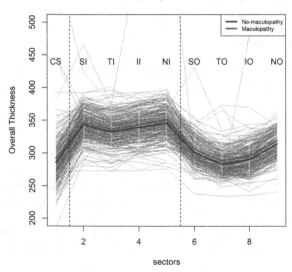

Fig. 2. Pairwise visualizations for mean profiles of retinal overall thickness over 9 sectors at baseline visit

Univariate MANOVA was performed and demonstrated that maculopathy vs no-maculopathy eyes were not different in retinal thickness over the nine sectors ($p = 0.11 > 0.05$). Then we considered the correlations between the two eyes and the spatial correlation between the nine sectors in statistical analyses using our model described in Sect. 2, which also allows heteroscedasticity between groups. We investigated different spatial dependency structures described in Sect. 2.3; an exponential correlation structure was the most informative with the lowest AIC and BIC.

With the two levels of random effects model and an exponential correlation structure, we can detect the difference in the main effect of diagnosis between the maculopathy and no-maculopathy groups ($p = 0.0218 < 0.05$, Table 2). The effect size between groups is 4.4982 with standard error equals 1.9543. Moreover, compared with other correlation structures mentioned in Sect. 2.3, an exponential correlation structure gives the best model with the lowest AIC and BIC. However, we did not detect a shape effect, which is measured as the interaction term between diagnosis and sector mathematically, between maculopathy group and no-maculopathy group ($p = 0.9715$). In the final model as described in Eq. (11), estimators for age, main effect for diagnosis, variance component estimators for random effect and residuals, and heteroscedicity range are shown in Table 2.

Table 2. Estimator for age, main effect, variance component estimators for random effect and residuals, and heteroscedicity range in the final model

		Estimate (Std. Err)	p value
Estimator for age	$\hat{\beta}_1$	-0.2681 (0.0962)	0.0061*
Estimator for main effect	$\hat{\beta}_2$	4.4982 (1.9543)	0.0218*
Variance component	$\hat{\sigma}_1$	15.0307	
	$\hat{\sigma}_2$	7.9328	
$\hat{\Sigma}_s$	$\hat{\sigma}_s$	22.5541	
	$\hat{\gamma}_e$	0.6620	
Heteroscedicity range among diagnosis group	Maculopathy	1	
	No-maculopathy	0.5790	

*statistically significant with $p < 0.05$

We also found a negative correlation between age and the mean retinal thickness profile ($p = 0.0011$). And we further used a likelihood ratio test to confirm the significance of the eye within patient random effect (u_{ij}) in the model ($p = 0.0005$).

4   Simulation

We carried out a simulation study to check the performance of our final model and to investigate the importance of incorporating spatial correlation in the statistical imaging analyses. We simplified the two-level of nested random effect (11) into a one-level random effect model with spatial exponential correlation. Covariates were chosen based on statistical analyses results from the EDDMO study, including nine locations from the ETDRS grid and a negatively continuous correlated risk factor (e.g. age in EDDMO). Only two disease groups (e.g. maculopathy group versus no-maculopathy group) were considered in the simulation as we are interested the effect of the spatial correlation and the power of our model rather than the clinical outcomes. The aim of our simulation study was to establish how well the spatial approach is able to estimate the risk factor

Table 3. Simulation studies: scenario 1: no correlation between simulated outcomes

Non-spatial approach using linear regression			Spatial approach using linear mixed effect without correlation		
	Risk factor	Diagnosis for main effect		Risk factor	Diagnosis for main effect
True	−0.3	6.1	True	−0.3	6.1
n = 200	β_1	β_2	n = 200	β_1	β_2
Estimates	−0.2997	6.1013	Estimates	−0.2999	6.1015
SE	0.0012	0.1549	SE	0.0013	0.1548
SD	0.0013	0.1449	SD	0.0013	0.1448
CP	95.5%	96.5%	CP	98.5%	95.5%

SE, mean of standard error estimates; SD, Monte Carlo standard deviation of the estimates across the simulated data; CP, coverage probability for the estimates.

Table 4. Simulation studies: scenario 2: moderate exponential spatial correlation between simulated outcomes ($\gamma_e = 0.5$)

Non-spatial approach using linear regression			Spatial approach using linear mixed effect without correlation		
	Risk factor	Diagnosis for main effect		Risk factor	Diagnosis for main effect
True	−0.3	6.1	True	−0.3	6.1
n = 200	β_1	β_2	n = 200	β_1	β_2
Estimates	−0.3001	6.0590	Estimates	−0.3000	6.0585
SE	0.0056	0.6927	SE	0.0083	0.6927
SD	0.0077	0.6563	SD	0.0077	0.6569
CP	84.0%	96.0%	CP	96.6%	95.9%

SE, mean of standard error estimates; SD, Monte Carlo standard deviation of the estimates across the simulated data; CP, coverage probability for the estimates.

and to test the difference between the diagnosis group in terms of the main effect and the shape effect.

Sample size were chosen as $n = 200$ participants with one eye per individual where 70% of the eyes does not have maculopathy and 30% of the eyes have maculopathy. In order to investigate how the spatial correlation can change the statistical inferences, we set three simulation scenarios in this section, one without correlation, one with a moderate ($\gamma_e = 0.5$) and the other with a high correlation ($\gamma_e = 0.1$) structure between different locations. All the simulation results in this section are based on 1000 Monte Carlo replications. The simulation results including the true parameter values, sample size, Monte Carlo standard deviation, the mean of standard error estimates, the coverage probabilities for the estimates and the power to detect the shape effect are reported in Tables 3, 4, 5 and 6.

In scenario 1 where no correlation exists between simulated outcomes, our spatial approach performed the same as the non-spatial approach in terms of the estimates when the sample size equals 200 (Table 3). The estimates were practically unbiased, the Monte Carlo standard deviation agreed with the mean of standard error estimates, and the coverage probability was around 95%, which is reasonable.

Table 5. Simulation studies: scenario 3: high exponential spatial correlation between simulated outcomes ($\gamma_e = 0.1$)

Non-spatial approach using linear regression			Spatial approach using linear mixed effect without correlation		
	Risk factor	Diagnosis for main effect		Risk factor	Diagnosis for main effect
True	−0.3	6.1	True	−0.3	6.1
$n = 200$	β_1	β_2	$n = 200$	β_1	β_2
Estimates	−0.3000	6.1180	Estimates	−0.3002	6.1120
SE	0.0056	0.6918	SE	0.0138	0.6927
SD	0.0170	0.7043	SD	0.0141	0.7033
CP	**57.2%**	95.1%	CP	94.9%	95.1%

SE, mean of standard error estimates; SD, Monte Carlo standard deviation of the estimates across the simulated data; CP, coverage probability for the estimates.

Table 6. Power of detecting the difference in shape between groups in spatial approach and non-spatial approach ($p < 0.01$)

Detection of shape effect $p < 0.01$	Non-spatial approach using linear regression	Spatial approach using linear mixed effect
No correlation	100%	100%
Moderate correlation	88.1%	95.3%
High correlation	88.9%	100%

For comparison, Tables 4 and 5 present the results based on moderate exponential correlation where $\gamma_e = 0.5$, and a high exponential correlation where $\gamma_e = 0.1$. In Table 4, we can see a lower coverage probability in the non-spatial approach compared with our spatial approach. When higher spatial correlations exist, the coverage probability for the estimates of β_1 is much worse (Table 5). Using our spatial approach, the estimates of the parameters were practically unbiased with a reasonable coverage probability both in a moderate correlation setting and a high correlation setting. As expected, when there is no correlation between the simulated data as reported in Table 6, the spatial approach and the non-spatial approach were the same in detecting the shape effect (i.e. the interaction term). However, our spatial approach performed much better in the other two correlation settings.

5   Discussion

Rather than analysing only the retinal thickness in central subfield (a Welch's test with winsorized variances was performed for retinal thickness between groups in central subfield, $p = 0.38 > 0.05$), we proposed here a nested linear mixed effect model to analyse spatially related data, generated from a study of DMO. We showed that this approach is capable of incorporating spatial correlations in images and the correlation between two eyes from the same patient, and we found differences in mean retinal thickness between no-maculopathy versus maculopathy group. We also showed an exponential spatial correlation between sectors provides the best model. Simulations demonstrated that our spatial approach is able to provide more accurate inference on the risk factor and has the ability to detect the main effect and shape effect between diagnostic groups. A further interesting study would be early detection of diabetic retinopathy to discriminate between healthy eyes from eyes with retinopathy but without DMO, which would be useful for clinicians in planning early treatment for patients.

Our approach can be applied to further investigate the spatial context of other features [8] in images with other retinal diseases such as diabetic retinopathy and central vascular occlusion with the aim of developing a flexible anistropic spatial dependency structures which can be adapted to other medical images. Moreover, it would be useful to predict the disease occurrence and the time of occurrence by extending our spatial modelling into spatio-temporal modelling which incorporates longitudinal images [14].

6   Conclusion

Spatially collected data from retinal images presents both important opportunities and challenges for understanding, detecting and diagnosing eye disease. We extend the standard analytic approach into spatial methods that adjust for spatial correlations between the image sectors and correlation between eyes from the same patients. Our simulation results confirmed the advantage of the spatial modelling to provide more powerful statistical inference: power increases from

88.1% to 95.3%, 88.9% to 100% for moderate or high spatial correlations. In the future, the spatial approach has the ability to extend the model into prediction or prognosis (i.e. the predictive modelling) and develop personal clinical management and monitoring tool.

Acknowledgement. Wenyue Zhu would like to acknowledge the PhD funding from Institute of Ageing and Chronic Disease and Institute of Translational Medicine at University of Liverpool and the Royal Liverpool University Hospital.

References

1. Bernal-Rusiel, J.L., et al.: Spatiotemporal linear mixed effects modeling for the mass-univariate analysis of longitudinal neuroimage data. Neuroimage **81**, 358–370 (2013)
2. Bowman, F.D., Waller, L.A.: Modelling of cardiac imaging data with spatial correlation. Stat. Med. **23**(6), 965–985 (2004)
3. BuAbbud, J.C., Al-latayfeh, M.M., Sun, J.K.: Optical coherence tomography imaging for diabetic retinopathy and macular edema. Curr. Diab. Rep. **10**(4), 264–269 (2010)
4. Cressie, N.: Statistics for spatial data. Terra Nova **4**(5), 613–617 (1992)
5. Laird, N.M., Ware, J.H., et al.: Random-effects models for longitudinal data. Biometrics **38**(4), 963–974 (1982)
6. Lindquist, M.A., et al.: The statistical analysis of fMRI data. Stat. Sci. **23**(4), 439–464 (2008)
7. MacCormick, I.J., et al.: Accurate, fast, data efficient and interpretable glaucoma diagnosis with automated spatial analysis of the whole cup to disc profile. PloS one **14**(1), e0209409 (2019)
8. MacCormick, I.J., et al.: Spatial statistical modelling of capillary non-perfusion in the retina. Sci. Rep. **7**(1), 16792 (2017)
9. Nussenblatt, R.B., Kaufman, S.C., Palestine, A.G., Davis, M.D., Ferris, F.L.: Macular thickening and visual acuity: measurement in patients with cystoid macular edema. Ophthalmology **94**(9), 1134–1139 (1987)
10. Pinheiro, J., Bates, D., DebRoy, S., Sarkar, D., R Core Team: nlme: Linear and Nonlinear Mixed Effects Models (2018). https://CRAN.R-project.org/package=nlme. R package version 3.1-137
11. Early Treatment Diabetic Retinopathy Study Research Group: Grading diabetic retinopathy from stereoscopic color fundus photographs-an extension of the modified Airlie House classification: ETDRS report no. 10. Ophthalmology **98**(5), 786–806 (1991)
12. Hee, M.R., et al.: Quantitative assessment of macular edema with optical coherence tomography. Arch. Ophthalmol. **113**(8), 1019–1029 (1995)
13. Van Buuren, S., Groothuis-Oudshoorn, K.: MICE: multivariate imputation by chained equations in R. J. Stat. Softw. **45**(3), 1–67 (2011). https://www.jstatsoft.org/v45/i03/
14. Vogl, W.D., Waldstein, S.M., Gerendas, B.S., Schmidt-Erfurth, U., Langs, G.: Predicting macular edema recurrence from spatio-temporal signatures in optical coherence tomography images. IEEE Trans. Med. Imaging **36**(9), 1773–1783 (2017)
15. Ying, G.S., Maguire, M.G., Glynn, R., Rosner, B.: Tutorial on biostatistics: linear regression analysis of continuous correlated eye data. Ophthalmic Epidemiol. **24**(2), 130–140 (2017)

16. Ying, G.S., Maguire, M.G., Glynn, R., Rosner, B.: Tutorial on biostatistics: statistical analysis for correlated binary eye data. Ophthalmic Epidemiol. **25**(1), 1–12 (2018)
17. Zhang, H.G., Ying, G.S.: Statistical approaches in published ophthalmic clinical science papers: a comparison to statistical practice two decades ago. Br. J. Ophthalmol. **102**(9), 1188–1191 (2018)

Fetal Imaging

Incremental Learning of Fetal Heart Anatomies Using Interpretable Saliency Maps

Arijit Patra[✉] and J. Alison Noble

Institute of Biomedical Engineering, University of Oxford, Oxford, UK
arijit.patra@eng.ox.ac.uk

Abstract. While medical image analysis has seen extensive use of deep neural networks, learning over multiple tasks is a challenge for connectionist networks due to tendencies of degradation in performance over old tasks while adapting to novel tasks. It is pertinent that adaptations to new data distributions over time are tractable with automated analysis methods as medical imaging data acquisition is typically not a static problem. So, one needs to ensure that a continual learning paradigm be ensured in machine learning methods deployed for medical imaging. To explore interpretable lifelong learning for deep neural networks in medical imaging, we introduce a perspective of understanding forgetting in neural networks used in ultrasound image analysis through the notions of attention and saliency. Concretely, we propose quantification of forgetting as a decline in the quality of class specific saliency maps after each subsequent task schedule. We also introduce a knowledge transfer from past tasks to present by a saliency guided retention of past exemplars which improve the ability to retain past knowledge while optimizing parameters for current tasks. Experiments on a clinical fetal echocardiography dataset demonstrate a state-of-the-art performance for our protocols.

Keywords: Saliency · Interpretability · Continual learning

1 Introduction

Medical image analysis pipelines have made extensive use of deep neural networks in recent years with state-of-the-art performances on several tasks. In diagnostic ultrasound, the availability of trained sonographers and capital equipment continues to be scarce. For congenital heart disease diagnosis in particular, the challenges become even more pronounced with the actual identification and processing of relevant markers in sonography scans being made difficult through the presence of speckle, enhancements and artefacts over a small region of interest. As with other applications of deep networks in medical image analysis [1], the retention of performance on already learnt information while adapting to new data distributions has been a challenge. Often, a requirement for deep networks is the availability of large labeled datasets. In medical imaging tasks however,

© Springer Nature Switzerland AG 2020
Y. Zheng et al. (Eds.): MIUA 2019, CCIS 1065, pp. 129–141, 2020.
https://doi.org/10.1007/978-3-030-39343-4_11

data is often not abundant or legally available. There exist intra-patient variations, physiological differences, different acquisition methods and so on. Not all necessary data may be available initially but accumulated over time, and be used to establish overall diagnosis. Incremental learning systems are those that leverage accumulated knowledge gained over past tasks to optimize adaptation to new tasks. Such optimization may not always ensure the traversal through the parameter space in a manner suitable for old tasks. This causes degradation in the performance of old tasks while adapting to new ones, and a balance is desired between stability of old knowledge and plasticity to absorb new information.

Literature Review. The loss of learnt features from prior tasks on retraining for new tasks leading to a diminished performance on old tasks is a phenomenon called 'catastrophic forgetting' [3]. Many methods have been proposed to build a lifelong learner. These are broadly classified [4] into (i) architectural additions to add new parameters for new data distributions, such as Progressive Networks [11] where new parameter sets get initialised for new tasks with a hierarchical conditional structure imposed in the latter case and the memory footprint is of the order of the number of parameters added (ii) memory and rehearsal based methods where some exemplars from the past are retained for replay (rehearsal [16], or are derived by generative models in pseudo-rehearsal, for replay while learning on new tasks. Examples include iCaRL [5], end-to-end lifelong learning [6] etc. In these, certain informative exemplars from the past are retained and used as representative of past knowledge on future learning sessions (iii) regularization strategies, which include methods to enforce preservation of learnt logits in parts of the network like in distillation strategies in Learning without Forgetting [7], or estimate parameter importance and assign penalties on them to ensure minimal deviation from learnt values over future tasks like in Elastic Weight Consolidation [8] and Synaptic Intelligence based continual learning [9]. In medical imaging, incremental addition of new data has been sporadically addressed, notably in [2,12], despite clinical systems often acquiring data in non-deterministic phases. While there have been efforts apart from transfer learning [10] to resolve the paucity of labeled data, these have concentrated on augmentation, multitask learning [1], and so on. In the domain of interpretability of medical images, there has been a focus on utilizing attention mechanisms to understand decisions of machine learning models, such as attention mechanisms for interpretation in ultrasound images in fetal standard plane classification [13], pancreas segmentation [14] and so on. Utilizing the notions of interpretability in a continual setting or for enabling learning in incremental sessions is yet to be studied in literature and we introduce notions of class saliency and explainability for assessing and influencing continual learning mechanisms.

Contributions. Our contributions are (a) a novel method to avoid catastrophic forgetting in medical image analysis (b) quantifying model forgetting and incremental performance via saliency map quality evolution over multiple learning sessions (c) saliency quality in individual sessions to choose informative exemplars for

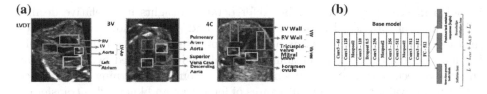

Fig. 1. (a) Fetal cardiac anatomical classes (b) Classification and knowledge distillation scheme for our model.

class-wise rehearsal over successive learning sessions. Usage of saliency map quality for evaluating incremental learning performance and saliency maps to select exemplars to retain for replay and distillation based knowledge regularization are new to computer vision and medical imaging to our knowledge. 'Incremental learning' and 'continual learning' are used interchangeably in literature.

2 Methodology

The aim of the study is to introduce the usage of interpretability as a building block for incremental learning in clinical imaging using fetal cardiac anatomy classification as a proof-of-concept. Our classes of interest are anatomical structures apparent in standard fetal cardiac planes (four-chamber or 4C, three-vessels or 3V and left ventricular outflow tract or LVOT view). Our problem deals with a class-incremental setting for detecting fetal cardiac structures- Ventricle Walls (VW), Foramen ovale (FO), valves (mitral, tricuspid) (4C view structures), left atrium (LA), right ventricles (RV), Aorta (LV and aorta are seen as a continuous cavity and labeled as LV-Ao hereafter) and right atrium (LVOT structures), Pulmonary Artery (PA), Aorta, Superior Vena Cava (SVC) (3V view). These structures are considered for study because of their relevance in assessment of congenital heart disease [17]. Out of these, structures are learnt in sets of 3, first as base categories in the initial task, followed by incremental task sessions. The remaining 3 are shown to the base-trained model in incremental stages in our class incremental learning experiments. (VW, Valves, FO) and (RV, PA, Aorta) and (LV-Ao, SVC, LA) are then the class groups introduced in successive task sessions. This simulates a setting where the algorithm needs to adapt to new data distributions in the absence of a majority of exemplars from past distributions.

We propose a dual utilization of saliency to implement this continual learning setting. First, we define novel quality metrics for class averaged attention maps that also quantify the ability of a model to learn continually. CNNs learn hierarchical features that are aggregated towards a low-dimensional representation and the inability of the model to retain knowledge is manifested in this hierarchy as well. Since attention maps point out the most relevant pixels in an input image towards the classification decision made on it by a model, a digression of focus from these pixels is indicative of degradation of past knowledge. Thus, attention map quality can be used to quantify not only the overall decline in performance over old tasks, but offer detailed insights into relative decline at the level of individual classes in the task, and also for individual instances in a class (which can be used as a measure of some examples being especially 'difficult to retain'). This attention based analysis is motivated by the fact that there has been no standard agreement on how continual learning performance ought to be evaluated. Present measures of forgetting and knowledge transfer do not allow a granular assessment of learning processes or distinctions on the basis of 'difficulty' of an instance, nor do they allow a scope for explainability of the continual learning process. Creating attention maps by finding class activations allows for feature level explainability of model decision on every learning cycle.

2.1   Saliency Map Quality

At the conclusion of a given task of N classes, the validation set of each class is passed through the model which is then subjected to class activation mappings (CAM) to obtain attention maps using the GradCAM approach [15]. Note that the specific method to obtain attention/saliency maps here is for demonstration only and any suitable method may replace it. We consider only the maps resulting from correct predictions because the explanations are generated even otherwise but is suboptimal for further inference. Each instance's map represents the understanding of the model for the decision taken on that instance. Averaged over the validation set of a class in a task, this average saliency map represents the average explanation of the model for the decisions of classification. In the absence of a ground truth, estimating a quality metric for the explanation based saliency map is non-trivial. We try to assess the extent of forgetting by tracking the difference between the activations obtained just after the class or task has been learnt and that after a few other tasks are learnt subsequently. This can be performed both for individual instances of classes and by considering classwise average saliency maps. Past literature has explored saliency map quality in terms of being able to mimic human gaze fixations or as a weak supervision for segmentation or detection tasks, in which ground truth signals were enforced, even if weakly. That apart, saliency maps were evaluated by [16] in context of their attempts at designing explainable deep models. They interpret the efficiency of saliency maps as a reasonable self-sufficient unit for positive classification of the base image. Then, the smaller the region that could give a confident classification in the map the 'better' the saliency. Mathematically, this was expressed as a log of the ratio of the area and the class probability if that area was fed to the classifier alone. This quantity called the SSR (Smallest Sufficient Region) is expressed in [16] as $|log(a)\text{-}log(p)|$, where a is the minimum area that gives a class probability for the correct class as p. This method assumes that the concentration of informative pixels is a good indicator of attentive features, and is unsuitable for cases where features of interest are distributed spatially in a non-contiguous manner (say a fetal cardiac valve motion detection, lesions in x-rays for lung cancer classification etc.). As such, considering our dataset of fetal heart ultrasound, attentive regions may be distributed over a spatial region and the non-contiguous informative regions can be adequately quantified only through metrics designed for multiple salient region estimation and cannot be adequately expressed by SSR. We extend from the SSR concept, and instead of thinking in terms of concentration of information consider the extent of regions of informative content. To do so, we consider a grid of fixed uniform regions on the input image. Each grid region is taken as a small rectangular space whose information content is evaluated, the size of these small regions is fixed as a hyperparameter (we consider 224×224 image inputs, and 16×16 grid regions by optimizing for computational cost and accuracy). Each of these grids is evaluated by the trained model after a task for their prediction probabilities by themselves. Then for each grid region, a quantity $|log(A_g) - log(p)|$ can be used, with $log(A_g)$ being constant for fixed size grid regions, to estimate the contribution of the region to the overall saliency map. The smaller this quantity, the more informative this region is. A threshold can be imposed and all n grid regions contributing can be summed up to express the Overall Saliency Quality (OSQ) as:

$$\sum_{k=1}^{n} |\log\left(A_g^k\right) - \log\left(p\right)| \tag{1}$$

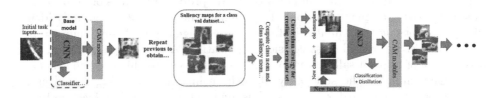

Fig. 2. Schematic of saliency curriculum based exemplar retention - After training for a session, CAM modules compute attention maps from instances for OSQ calculations and selecting retention examples. Figures for representation purposes only

This threshold depends on the desired class probability for the correct class (this expression would be valid even for incorrect classifications, but we choose regions only with correct predictions). Thus, a quality measure can be derived for each saliency map at all stages of tasks. The expression above essentially gives an 'absolute' measure of saliency map quality. Retention of exemplars for efficient continual learning was done by random selection, by nearest class mean and so on. These methods do not consider actual performance while making the retention decision but derive from input distributions if not selecting randomly. While in ideal cases, the class mean will reflect exemplars with high quality saliency maps (as one would trivially expect that the average class representation has the most volume of data and hence is more influential on the learning curve for these classes), it is not always true in case of classes with significant diversity and multiple sub-clusters of exemplars. In the latter, selecting exemplars by methods like herding [5], the need for retaining exemplars close to the class means is not optimal for retaining the diversity of class information, and the diversity of informative representations needs to be accounted for, which can be informed by saliency based retention strategies (Fig. 2).

2.2 Saliency Driven Continual Curriculum

We attempt to use the attention maps of past representations to help actively preserve knowledge and at the same time improve generalization to future classes. This is similar to transfer of knowledge from *Task N* to *Task (N+1)* through the attention maps of the former being used to condition the latter. After each task schedule we generate class activation maps using [9] for the validation data per class and estimate the quality measure defined in the first part. Following our need for retaining prior knowledge while moving to the next task, we consider a selective retention strategy. In order to account for fixed memory allocations per class, a fixed number of exemplars may be retained. We try to establish the most informative of instances through an optimal map quality curriculum over available instances in a class. This relies on studies of explainable representation learning that if a classification decision were established through an empirical risk minimization objective, a majority of instances for that category would have low-dimensional feature representations in a close vicinity and away from hyperplanes separating different clusters [3]. We propose two ways of preparing a saliency based exemplar retention: (1) Assuming the presence of definitely identifiable salient cues in available instances, an average class saliency map can be understood as an overall class decision explanation. A relative proximity at the pixel space of a given saliency map of a given instance to this average saliency map would indicate the suitability of such an instance for being stored as a representative exemplar for its class. This approach is termed the Average Representative Distance Selection (ARDS). (2) Another alternative is to pre-select a set of most representative exemplars from a validation dataset of the class, in terms of the class confidence probabilities, and use the normalized cumulative distance from their saliency maps to every other instance's saliency map. This is termed Distributed Exemplar Distance Selection (DEDS), and is primarily useful for cases where the salient regions of interest have a non-trivial spatial variation within the same class exemplar set. ARDS is suited when strong cues are localised and class prototype saliency maps are useful. DEDS is useful in cases of a diversity of cues not similarly located on all exemplars or when a significant shift is caused by affine transformations. ARDS is computationally more efficient as a prototype-to-exemplar distance is computed in a single step. Actual choices between the two would depend on data characteristics. In both cases, the saliency map is treated as a probability distribution and similarity is assessed by KL divergence between reference saliency maps and instance maps,

$$D_{KL}[e, d] = \int (e(x) - d(x))\log(e(x)/d(x))dx \qquad (2)$$

considering that *e(x)* and *d(x)* are the respective saliency maps being compared. As the computation over the pixel space is discretized, the integral form is replaced with a discrete summation,

$$D_{KL}[e,d] = \sum_{x=1}^{N_{pixels}} (e(x) - d(x))\log(e(x)/d(x)) \tag{3}$$

As for choosing instances for the ARDS or DEDS calculations, it would be superfluous to compute for entire training sets. We choose 100 exemplars from each class's training set (based on their confidence probabilities in final epoch) for this process. The retention examples are chosen from training sets as they are replayed in future sessions and hence validation set examples cannot be used except as saliency benchmarks, i.e, while the benchmarks for average saliency or representative exemplar sets are derived from validation sets, these can't be retained as validation data can't be used in any part of training. In both cases, 30 instances are finally chosen after a grid search over integral number of samples between 10 and 50 (with the upper bound governed by memory constraints and lower bound on performance thresholds), to be retained in memory for future rehearsal. Choosing higher numbers of exemplars was found to lead to minimal performance gains (a detailed study of these trade-offs is kept for future work).

3   Model and Objective Functions

For our architecture, we implement a convolutional network with 8 convolutional layers, interspersed alternately with maxpooling and a dropout of 0.5 (Fig. 1(b)). This is followed by a 512-way fully-connected layer before a softmax classification stage. The focus of the work is not to achieve possible state-of-the-art classification accuracies on the tasks and datasets studied, but to investigate catastrophic forgetting in learning incrementally. Thus, the base network used is significantly simplified to keep the order of magnitude of the number of parameters within that of other continual learning approaches reported in literature [4] and enable a fair benchmarking. For a loss function, we implement a dual-objective of minimizing a shift on learnt representations in the form of prior logits in the final model layers using a knowledge distillation approach [17], and performing a cross-entropy classification on the current task classes. A quadratic regularization is additionally imposed with a correction factor that is set as a hyperparameter by grid search. The overall objective function is:

$$L = L_{cur} + L_{KD} + L_r \tag{4}$$

where L is the total objective composed of the softmax cross entropy as the current loss, the knowledge distillation loss on past logits, and the regularization term for the previously trained parameters.

$$L_{cur}(X_n, Y_n) = -\frac{1}{|N_n|}\sum_{i=1}^{N_n}\sum_{j=1}^{J_n} y_n^{ij}.\log(p_n^{ij}) \tag{5}$$

where N_n is the number of examples in a batch, J_n is the number of classes, $y_n{}^{ij}$ is the one-hot encoded label for an instance, and $p_n{}^{ij}$ is the softmax prediction. For the distillation terms, the original labels not being available, $y_o{}^{ij}$ is computed with new and retained examples and compared to stored logits for the old examples $p_o{}^{ij}$, giving a loss term:

$$L_{KD}(X_n, Y_o) = -\frac{1}{|N_n|} \sum_{i=1}^{N_n} \sum_{j=1}^{J_o} y_o^{ij'} \cdot \log(p_o^{ij'}) \tag{6}$$

where $y_o{}^{ij'} = \frac{(y_o{}^{ij})^{1/\lambda}}{\sum_j (y_o{}^{ij})^{1/\lambda}}$ and $p_o{}^{ij'} = \frac{(p_o{}^{ij})^{1/\lambda}}{\sum_j (p_o{}^{ij})^{1/\lambda}}$, where the distillation temperature λ is set at 2.0 over hyperparameter search on values from 1 to 10.

The parameter regularization is imposed for already trained weights for the past classes and penalises the shift for current adaptation through a Frobenius norm over the parameters as $L_r = \mu \sum_j ||w_o - w'||^2$. μ is set to 0.4 by grid search. The idea here is that while the retained exemplars from the past tasks are seen with the present data, the process of optimization should update parameters in a manner compatible with past prediction features. This ensures that parameters are adapted to present distributions without drastic shifts that adversely affect their ability to arrive at the optimal representations for previously seen examples. A distillation framework is implemented here as such a loss term in conjunction with a cross-entropy objective can enforce a regularization on representations from the past, which is not achieved by a simple parametric regularization.

Data. For fetal echocardiography data, we consider a clinically annotated dataset of 91 fetal heart videos from 12 subjects of 2–10 s with 25–76 fps. Obtaining 39556 frames of different standard planes and background, we crop out relevant anatomical structures in patches of 100×100 from the frames, leading to a total of 13456 instances of 4 C view structures (Ventricle Walls, Valves, Foramen ovule), 7644 instances of 3V view structures (Pulmonary Artery, Superior Vena Cava, Descending Aorta) and 6480 LVOT structures (Right Atrium, Left ventricle-Aorta continuum and Right Ventricle). A rotational augmentation scheme is applied with angular rotations of 10° considering the rotational symmetry of the actual acquisition process. Instances sourced from 10 subjects are used for training sets and the rest for validation.

Training. For initial training, the number of base classes (N) are taken as 3. In the (N + 1)th task (N > 1) following the creation of exemplar sets of the immediate past task, the training process is started for the (N + 1)th task and so on (this goes on for 3 sessions in total in our case as we deal with a total of 9 classes of sub-anatomies). Batches are created between old and new data, and to further improve performance a distillation based regularization with representative logits of the past tasks is used along with the cross-entropy loss for the present task and exemplars. The training stage essentially involves a base training with the first set of classes, carried over 50 epochs with a learning rate of 0.001. This is followed by a class activation mapping stage with the recently trained model, and an averaged saliency map calculation per class. This model is now fine-tuned for the group of classes for the next task with a joint distillation and cross-entropy loss over past logits and new labels for 50 epochs. The process continues over the remaining task sessions. CAM stages are carried out only after entire task session is completed and not in-between epochs. In those models where past exemplars are retained and rehearsed, these are interspersed with the batches during fine-tuning over the new class sets.

Baselines. For baseline comparisons, the network model described is adapted with the protocol in iCaRL [5], where the representation learning stages are followed by a storage of class-specific exemplars using a herding algorithm [5]. This is implemented in our datasets by computing average prototype representations through the penultimate fully-connected layers for the classes seen till the previous task. A multi-class adaptation of Learning without Forgetting (LwF.MC) [7] is attempted with our network and objective function without the weight regularization term, and logits for distillation are retained. For the end-to-end learning (E2EL) [8] method adapted to our network, a fixed memory version is followed to be a more accurate benchmark to our own fixed memory per class assumptions. The representative memory fine-tuning protocol in E2EL is implemented for baselines with the same training configurations as the initial training, except that the learning rate is reduced to 1/10th the initial value. A progressive distillation and retrospection scheme (PDR) [9] is implemented with replicated versions of our network serving as the teacher network for the distillation and the retrospection phases, with the exemplars generating 'retrospection' logits for regularization while progressively learning on new data which are presented as a second set of logits which have been learned separately in another replication of the base network. In these implementations, storage of prior exemplars follows the same protocols as used in the original implementations.

4   Results and Discussion

We consider configurations of our approach in terms of the exemplar retention method used and adopt OSQ in tandem: (1) map quality with OSQ and ARDS, (2) map quality with OSQ + DEDS. Variants of our approach without storing rehearsal exemplars are also considered in terms of training a vanilla CNN with similar architecture as the base CNN used in other approaches. This is same as using a simple CNN baseline with transfer learning over task sessions. Another version CNN (TL) also functions without retention strategies but with convolutional layers frozen while further task finetuning (this would reflect in a much higher difference between initial task performance and subsequent task values). The reported performances include the average accuracies at incremental levels, with and without using the salient retention scheme in step 2, benchmarked with our adaptations of methods in iCaRL [5], LwF.MC [7], E2EL [18] and Progressive distillation (PDR) [19]. Using these benchmarks indirectly also allow comparison between different exemplar retention strategies, such as with naive and herding based methods explored in iCaRL and LwF. Also reported is the average saliency map quality in each stage. This OSQ metric is a proxy for the level of forgetting over multiple stages, and a difference between consecutive stages in the OSQ represents the decline in the model's ability to seek out most salient image regions over classes. Also, the OSQ is an indicator of stagewise model interpretability since accurate model explanations rely directly on the quality of saliency maps for medical images.

The reported OSQ over learning sessions is the saliency quality averaged over all validation exemplars available for previously seen classes. We report the average value for tasks so far, since we want to look at broad trends in the overall saliency to assess the overall ability to retain knowledge. For future extensions, it is straight-forward to obtain these values at both class and instance level and only requires them to be input to the trained model and class activation maps processed before the OSQ calculation. There is a difference in accuracies for baseline methods compared to original implementations due to our using the same base network for all baselines and models for uniform assessments (e.g. iCarL originally used embeddings from 32 layer ResNet on CIFAR 100 but we use the iCaRL baseline on our data and our base network). A trend is established where the inclusion of exemplars based on a saliency driven approach is seen to have a marked improvement on mitigation of forgetting, based on the OSQ metrics introduced here, and also on the validation accuracies averaged on previously seen classes for the task sessions considered (the past accuracy % in Table 1 for a task stage refers to validation accuracy of validation sets from previously seen classes, and the present accuracy % is the validation accuracy obtained on the present validation set). The saliency quality variation roughly corresponds to the past accuracy percentages demonstrating the efficacy of using map quality as a metric for evaluation of continual learning algorithms in medical image analysis. The methods that consider the retention of exemplars from the past overall show not only a better performance with respect to past task accuracies, but also demonstrate considerably higher current learning performances. This implies that a feature importance based identification of informative examples for classes of physiological markers not only improves the ability to better rehearse on past data during future learning stages but also transfers salient representative knowledge leading to better initialization of the parameters for improving performances on the present as well. Here, the diversity of the examples that can constitute a single class type representing a physiological region requires that multiple salient features can be used to explain the final optimization decisions and a diversity of informative exemplars need to be chosen for optimal forward propagation of knowledge while learning on future increments of tasks. The proposed pipeline ensures an inherent continual explainability of the decisions and how they shift over new data arrivals (Fig. 3).

Conclusion. In this proof-of-concept for saliency aware continual learning paradigms, we presented metrics for assessment of continual learning in terms of saliency allowing for instance and class level understanding of the basis for prediction and a shift in the learning of the same. We also utilised the saliency from the past task as a selected representative for prior tasks during subsequent learning and developed a joint curriculum for creating such sets of exemplars.

Table 1. Evolution of model performance over task sessions. Past accuracy refers to validation accuracy on past session classes (or past tasks). Map quality (MQ) is reported for present task session (leftmost column for each task session head) and for previous session classes.

Method	Task 1		Task 2				Task 3			
	MQ	Task 1 acc. %	MQ (T2)	MQ (T1)	Task 2 acc. %	Past acc. %	MQ (T3)	MQ (T2)	Task 3 acc. %	Past acc. %
Ours (OSQ+ADRS)	0.933	0.812	0.942	0.915	0.845	0.704	0.913	0.876	0.632	0.568
Ours (OSQ+DEDS)	0.946	0.812	0.938	0.923	0.863	0.691	0.887	0.843	0.636	0.593
OSQ+std. CNN	0.871	0.811	0.840	0.631	0.778	0.592	0.852	0.802	0.661	0.455
OSQ+CNN (TL)	0.827	0.813	0.822	0.612	0.702	0.511	0.772	0.647	0.560	0.322
iCaRL	0.811	0.775	0.831	0.622	0.713	0.616	0.834	0.674	0.658	0.321
LwF.MC	0.773	0.762	0.767	0.668	0.732	0.529	0.822	0.731	0.621	0.301
E2EL	0.818	0.793	0.792	0.703	0.742	0.554	0.785	0.706	0.603	0.295
PDR	0.842	0.802	0.837	0.711	0.759	0.514	0.791	0.721	0.581	0.342

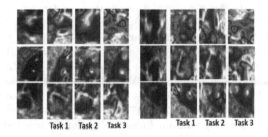

Task 1   Task 2   Task 3   Task 1   Task 2   Task 3

Fig. 3. Map quality variation over successive tasks sequentially, for class examples seen originally in task session 1. Saliency maps for these task 1 validation instances show a shift in attentive features over subsequent task sessions (areas in red are of higher salience) where models learn other classes. Quantification of this shift with the OSQ is treated as a proxy metric for forgetting here. (Color figure online)

Our method makes the continual learning process interpretable to a degree, and thereby ensures that the forgetting and retention characteristics of models are explainable. Given the foundations laid here, multiple future directions are possible starting with an exploration of classwise characteristics in terms of forgetting and retention performances and the trends of decline in map quality over intra-class variations. This is likely to be a natural follow-up of the ideas proposed here. Another immediate direction would be to study other strategies for saliency-curriculum driven exemplar retention. While we used a fixed number of exemplars for retention and rehearsal over future tasks, other approaches like a variable retention based on class difficulty are possible. Future directions can also include expanding to different tasks and datasets, using the saliency based exemplar scheme with other lifelong learning methods, using generative replay for estimating past saliency maps and images without need to retain exemplars, and so on. While we have demonstrated on approaches on distillation-based preservation, and using class activation derived saliency maps, the concept is generally applicable with any other continual learning pipeline, and can use other methods of estimating saliency maps, with different base architectures or objectives.

References

1. Patra, A., Huang, W., Noble, J.A.: Learning spatio-temporal aggregation for fetal heart analysis in ultrasound video. In: Cardoso, M.J., et al. (eds.) DLMIA/ML-CDS -2017. LNCS, vol. 10553, pp. 276–284. Springer, Cham (2017). https://doi.org/10.1007/978-3-319-67558-9_32
2. Ozdemir, F., Fuernstahl, P., Goksel, O.: Learn the new, keep the old: extending pretrained models with new anatomy and images. In: Frangi, A.F., Schnabel, J.A., Davatzikos, C., Alberola-López, C., Fichtinger, G. (eds.) MICCAI 2018. LNCS, vol. 11073, pp. 361–369. Springer, Cham (2018). https://doi.org/10.1007/978-3-030-00937-3_42
3. Goodfellow, I., et al.: An Empirical Investigation of Catastrophic Forgetting in Gradient-Based Neural Networks (2015). https://arxiv.org/abs/1312.6211
4. Parisi, G.I., Kemker, R., Part, J.L., Kanan, C., Wermter, S.: Continual Lifelong Learning with Neural Networks: A Review. https://arxiv.org/abs/1802.07569
5. Rebuffi, S.A., Kolesnikov, A., Sperl, G., Lampert, C.H.: iCaRL: incremental classifier and representation learning. In: CVPR 2017 (2017)
6. Aljundi, R., Chakravarty, P., Tuytelaars, T.: Expert gate: lifelong learning with a network of experts. In: ICCV 2017 (2017)
7. Li, Z., Hoiem, D.: Learning without forgetting. IEEE Trans. Pattern Anal. Mach. Intell. **40**(12), 2935–2947 (2017)
8. Kirkpatrick, J., et al.: Overcoming catastrophic forgetting in neural networks. Proc. Nat. Acad. Sci. **114**(13), 3521–3526 (2017)
9. Zenke, F., et al.: Continual learning through synaptic intelligence. In: ICML 2017 (2017)
10. Pan, S.J., Qiang, Y.: A survey on transfer learning. IEEE Trans. Knowl. Data Eng. **22**(10), 1345–1359 (2010)
11. Rusu, A.A., et al.: Progressive neural networks, arxiv:1606.04671 (2016)
12. Kim, H.-E., Kim, S., Lee, J.: Keep and learn: continual learning by constraining the latent space for knowledge preservation in neural networks. In: Frangi, A.F., Schnabel, J.A., Davatzikos, C., Alberola-López, C., Fichtinger, G. (eds.) MICCAI 2018. LNCS, vol. 11070, pp. 520–528. Springer, Cham (2018). https://doi.org/10.1007/978-3-030-00928-1_59
13. Schlemper, J., et al.: Attention Gated Networks: Learning to Leverage Salient Regions in Medical Images, arXiv preprint arXiv:1808.08114
14. Oktay, O., et al.: Attention U-Net: Learning Where to Look for the Pancreas. arXiv preprint arXiv:1804.03999
15. Selvaraju, R.R., et al.: Grad-CAM: visual explanations from deep networks via gradient-based localization. In: ICCV 2017 (2017)

16. Dabkowski, P., Gal, Y.: Real time image saliency for black box classifiers. In: Advances in Neural Information Processing Systems (2018)
17. Hinton, G., Vinyals, O., Dean, J.: Distilling the knowledge in a neural network. arXiv preprint arXiv:1503.02531 (2015)
18. Castro, F.M., Marín-Jiménez, M.J., Guil, N., Schmid, C., Alahari, K.: End-to-end incremental learning. In: Ferrari, V., Hebert, M., Sminchisescu, C., Weiss, Y. (eds.) ECCV 2018. LNCS, vol. 11216, pp. 241–257. Springer, Cham (2018). https://doi.org/10.1007/978-3-030-01258-8_15
19. Hou, S., Pan, X., Loy, C.C., Wang, Z., Lin, D.: Lifelong learning via progressive distillation and retrospection. In: Ferrari, V., Hebert, M., Sminchisescu, C., Weiss, Y. (eds.) ECCV 2018. LNCS, vol. 11207, pp. 452–467. Springer, Cham (2018). https://doi.org/10.1007/978-3-030-01219-9_27

Improving Fetal Head Contour Detection by Object Localisation with Deep Learning

Baidaa Al-Bander[1(✉)], Theiab Alzahrani[2], Saeed Alzahrani[2],
Bryan M. Williams[3], and Yalin Zheng[3]

[1] Department of Computer Engineering, University of Diyala, Baqubah, Diyala, Iraq
baidaa.q@gmail.com
[2] Department of Electrical Engineering and Electronics, University of Liverpool,
Liverpool, UK
{pstalzah,s.g.a.alzahrani}@liverpool.ac.uk
[3] Department of Eye and Vision Science, University of Liverpool, Liverpool, UK
{bryan,yalin.zheng}@liverpool.ac.uk

Abstract. Ultrasound-based fetal head biometrics measurement is a key indicator in monitoring the conditions of fetuses. Since manual measurement of relevant anatomical structures of fetal head is time-consuming and subject to inter-observer variability, there has been strong interest in finding automated, robust, accurate and reliable method. In this paper, we propose a deep learning-based method to segment fetal head from ultrasound images. The proposed method formulates the detection of fetal head boundary as a combined object localisation and segmentation problem based on deep learning model. Incorporating an object localisation in a framework developed for segmentation purpose aims to improve the segmentation accuracy achieved by fully convolutional network. Finally, ellipse is fitted on the contour of the segmented fetal head using least-squares ellipse fitting method. The proposed model is trained on 999 2-dimensional ultrasound images and tested on 335 images achieving Dice coefficient of 97.73 ± 1.32. The experimental results demonstrate that the proposed deep learning method is promising in automatic fetal head detection and segmentation.

Keywords: Fetal ultrasound · Object detection and segmentation · Deep learning · CNN · FCN

1 Introduction

Ultrasound imaging (US) is the primary modality used in daily clinical practice for assessing the fetus condition such as detecting of possible abnormalities and estimating of weight and gestational age (GA) [1]. Fetal biometrics from ultrasound used in routine practice include occipital-frontal diameter (OFD), femur length (FL), biparietal diameter (BPD), crown-rump length, abdominal

© Springer Nature Switzerland AG 2020
Y. Zheng et al. (Eds.): MIUA 2019, CCIS 1065, pp. 142–150, 2020.
https://doi.org/10.1007/978-3-030-39343-4_12

circumference, and head circumference (HC) [2,3]. Fetal head-related measurements including BPD and HC are usually used for estimating the gestational age and fetal weight between 13 and 25 weeks of gestation [4–6]. The 2-dimensional fetal US scan is characterised by its non-invasive nature, real time capturing, wide availability and low cost. However, the US manual examination is highly dependent on the training, experience and skills of sonographer due to the image artefacts and poor signal to noise ratio [7].

Manual investigation of US images is also a time-consuming process and therefore developing automatic US image analysis methods is a significant task. Automated fetal head boundary detection is often performed as a prerequisite step for accurate biometric measurements. The automated fetal head contour detection from US images can be basically fulfilled by developing effective segmentation algorithms which is able to extract the segmented head structure. A number of fetal head segmentation methods have been developed over the past few years with varying degrees of success, including parametric deformable models [8], Hough transform-based methods [9], active contour models [10], and machine learning [11–13]. However, the presence of noise and shadow, intensity inhomogeneity, and lack of contrast in US images make the traditional segmentation methods are not sufficient or have a limited success on fetal head detection. It is a strong need to develop more accurate segmentation algorithm which is able to tackle the presented fetal head detection problems in US images.

Recently, deep convolutional neural networks (CNNs) has revolutionised the field of computer vision achieving great success in many medical image analysis tasks including image segmentation, detection and classification. In terms of segmentation accuracy, fully convolutional network (FCN) [14] has dominated the field of segmentation. FCN has demonstrated improved results in automatic fetal head detection and segmentation in [15–17]. However, FCN has some challenges which need to be tackled. The challenges are represented by being expensive to acquire pixel level labels for network training and having difficulties with imbalanced data samples which lead to a biased representation learned by the network.

In this paper, we propose deep learning based method to segment fetal head in ultrasound. The proposed method aims to improve the segmentation accuracy by incorporating object localisation mechanism in segmentation framework achieved by merging Faster R-CNN [18] with FCN [14]. This incorporation allows to leverage object detection labels to help with the learning of network, alleviating the need for large scale pixel level labels. The rest of this paper is organised as follows. In Sect. 2, the materials and proposed method are described. Results of the proposed method are presented and discussed in Sect. 3. Finally, the work is concluded in Sect. 4.

2   Materials and Methods

2.1   Materials

A publicly available dataset has been used in the training and evaluation of the proposed method [19]. The ultrasound images were captured from 551 pregnant women who received screening exam after (12–20) weeks of gestation. The dataset includes 1334 2D ultrasound images of the standard plane and the corresponding manual annotation of the head circumference, which was made by an experienced sonographer. The data is randomly spilt into a training set of 999 images and a test set of 335 images. The size of each ultrasound image is 800 × 540 pixels with a pixel size ranging from 0.052 to 0.326 mm.

2.2   Methods

The framework of our proposed method can be divided into two stages: (i) fetal head segmentation by adapting Mask R-CNN (Regional Convolutional Neural Network) [20], and (ii) fetal head ellipse fitting using least-squares ellipse fitting method. Mask R-CNN [20] which was originally developed for object instance segmentation combined both localisation and segmentation in one architecture has been adapted to detect fetal head boundary. The proposed fetal head segmentation method comprises four major parts:

1. The feature extraction module is the first step of our method. The feature extraction module is a standard convolutional neural network consisting of convolutional and pooling layers. This module serves as a backbone feature extractor for the segmentation task. Ultrasound images and their masks are resized into 512 × 512 × 3 and passed through the backbone network. We use Resnet101 architecture [21] as a backbone network. Instead of training the model from scratch, transfer learning is exploited by initialising the first 50 layers of the model with pre-trained Resnet50 weights from ImageNet competition. The resulted feature map becomes the input for Faster R-CNN.
2. The object localisation represented by fetal head is achieved using Faster R-CNN which is well-known deep learning based object detection model [18]. It is adopted to generate and predict a number of bounding boxes producing multiple ROIs. The object localisation in Faster R-CNN is achieved by Region Proposal Network (RPN). The RPN scans over the backbone feature maps resulted from ResNet101 producing candidate region proposals/ anchors. The candidate region proposals/anchors are examined by a classifier and regressor to check the occurrence of foreground regions. Two outputs are generated by RPN for each anchor which are anchor class (foreground or background) and bounding box adjustment to refine the anchor box location. Then, the top anchors/candidate bounding boxes which are likely to contain objects are picked. The location and size of the candidate bounding boxes (ROIs) are further refined to encapsulate the object. After that, the final selected proposals (regions of interest) are passed to the next stage.

3. The dimensions of candidate bounding boxes (ROIs) generated by the RPN are adjusted by applying ROIAlign to have same dimensions as they have different sizes. ROIAlign technique samples the feature map at different points and then apply a bilinear interpolation to produce a fixed size feature maps. These feature maps are fed into a classifier to make decision whether the ROI is positive or negative region.
4. The positive regions (fetal head region) selected by the ROI classifier is passed into the mask branch in Mask R-CNN which is known as mask network. The mask network is fully convolutional neural network (FCN) that generates masks on the localised ROI. The output of this stage is the segmented region of fetal head.

The model is trained and weights are tuned using Adam optimiser for 75 epochs with adaptive learning rates $(10^{-4} - 10^{-6})$ and momentums of 0.9. Due to small training data set, heavily image augmentation is applied during training by randomly cropping of images to $256 \times 256 \times 3$, randomly rotate the images in the range of $(-15, 15)°$, random rotation 90 or $-90°$, and random scaling of image in the range (0.5, 2.0). The network is trained under multi-task cross-entropy loss function combining the loss of classification, localisation and segmentation mask: $L = L_{cls} + L_{bbox} + L_{mask}$, where L_{cls} and L_{bbox} are class and bounding box loss of Faster R-CNN, respectively, and L_{mask} is mask loss of FCN.

Finally, an ellipse is fitted to the predicted segmentation contours of fetal head using least-squares fitting method to mimic the measurement procedure used by the trained sonographers.

3   Results and Discussion

All of the experiments were run on an HP Z440 with NVIDIA GTX TITAN X 12 GB GPU card, an Intel Xeon E5 3.50 GHz and 16 GB RAM. Keras built on the top of Tensorflow has been used to implement the proposed system. The performance of the proposed method for segmenting the fetal head when compared with the ground truth was evaluated using many evaluation metrics such as Dice coefficient, mean absolute difference, mean difference, and mean Hausdorff distance which measures the largest minimal distance between two boundaries. The measurements can be defined as follows:

$$Dice(A, B) = \frac{2|A.B|}{|A| + |B|} \tag{1}$$

$$MeanAbsoluteDifference(A, B) = \frac{\sum |A - B|}{N \times M} \tag{2}$$

$$MeanDifference(A, B) = \frac{\sum A - B}{N \times M} \tag{3}$$

$$HausdorffDistance(A, B) = max(h(A, B), h(B, A)) \tag{4}$$

where
$$h(A, B) = max_{a \in A} min_{b \in B} \parallel a - b \parallel$$

A, B are the ground truth mask and resulted segmentation map from the proposed method, respectively. a, b are two sets of points from A and B, respectively, which represent the points on fetal head contour. N, M represent the dimensions of ground truth or predicted mask.

Figures 1 and 2 show some example of segmentation results. Figure 1 presents image examples as resulted from trained model used for validation without ellipse fitting. Figure 2 shows the image examples where ellipse is fitted and overlaid the test images.

The proposed system was evaluated on 355 US images achieving Dice coefficient of 97.73 ± 1.32, mean absolute difference (mm) of 2.33 ± 2.21, mean difference (mm) of 1.49 ± 2.85, and mean Hausdorff distance (mm) of 1.39 ± 0.82. The obtained results are comparable and often outperform the existing automated fetal head segmentation methods. Our model achieves higher performance than most recent work carried out by Heuvel et al. [19] who tested their method on the same 355 test images reporting Dice coefficient of 97.10 ± 2.73, mean absolute difference (mm) of 2.83 ± 3.16, mean difference (mm) of 0.56 ± 4.21, and mean Hausdorff distance (mm) of 1.83 ± 1.60.

Although Wu et al. [15] reported slightly better Dice coefficient of 98.4, yet, they reported boundary distance of 2.05 which is higher error than the boundary distance reported by our method. Furthermore, they tested their method on only 236 fetal head images and their results are affected by a refinement stage which is combing FCN with auto-context scheme. Sinclair et al. [16] reported comparable Dice coefficient of 98, however, they trained their model on large training set of 1948 images (double of our training data) and tested only on 100 images. Moreover, we obtain higher Dice coefficient than Li et al. [13] who achieved 96.66 ± 3.15 on 145 test images.

4   Conclusions

In this paper, an automated method to segment fetal head in ultrasound images has been presented. The developed method, which is based on merging Faster R-CNN and FCN, has proved to be efficient in fetal head boundary detection. Incorporating object localisation with segmentation has been proved to be comparable or superior to current approaches in extracting the fetal head measurements from the US data. The proposed system has been evaluated on a fairly large and independent dataset which included US images of all trimesters. The obtained results demonstrated that the proposed deep learning method is promising in segmenting anatomical structure of fetal head efficiently and accurately. The proposed object localisation-segmentation framework is generic and will be easily extended and developed to other ultrasound image segmentation tasks.

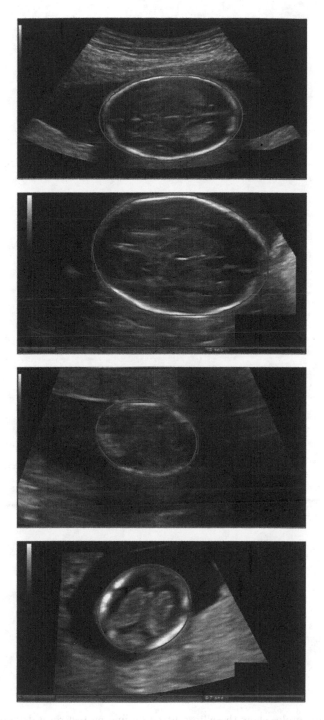

Fig. 1. Results of our model on four randomly images. Blue colour: without ellipse fitting; comparing with the expert annotating (red colour). (Color figure online)

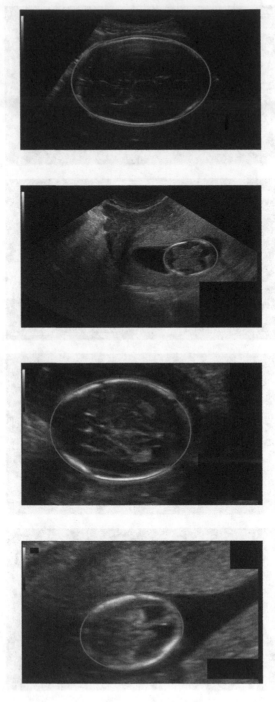

Fig. 2. Image examples show the ellipse fitted on unseen test data demonstrating the effectivity of our model.

References

1. Rueda, S., et al.: Evaluation and comparison of current fetal ultrasound image segmentation methods for biometric measurements: a grand challenge. IEEE Trans. Med. Imaging **33**(4), 797–813 (2014)
2. Loughna, P., Chitty, L., Evans, T., Chudleigh, T.: Fetal size and dating: charts recommended for clinical obstetric practice. Ultrasound **17**(3), 160–166 (2009)
3. Pemberton, L.K., Burd, I., Wang, E.: An appraisal of ultrasound fetal biometry in the first trimester. Rep. Med. Imaging **3**, 11–15 (2010)
4. Chervenak, F.A., et al.: How accurate is fetal biometry in the assessment of fetal age? Am. J. Obstet. Gynecol. **178**(4), 678–687 (1998)
5. Schmidt, U., et al.: Finding the most accurate method to measure head circumference for fetal weight estimation. Eur. J. Obstet. Gynecol. Reprod. Biol. **178**, 153–156 (2014)
6. Dudley, N.: A systematic review of the ultrasound estimation of fetal weight. Ultrasound Obstet. Gynecol.: Official J. Int. Soc. Ultrasound Obstet. Gynecol. **25**(1), 80–89 (2005)
7. Sarris, I., et al.: Intra-and interobserver variability in fetal ultrasound measurements. Ultrasound Obstet. Gynecol. **39**(3), 266–273 (2012)
8. Jardim, S.M., Figueiredo, M.A.: Segmentation of fetal ultrasound images. Ultrasound Med. Biol. **31**(2), 243–250 (2005)
9. Lu, W., Tan, J., Floyd, R.: Automated fetal head detection and measurement in ultrasound images by iterative randomized hough transform. Ultrasound Med. Biol. **31**(7), 929–936 (2005)
10. Yu, J., Wang, Y., Chen, P.: Fetal ultrasound image segmentation system and its use in fetal weight estimation. Med. Biol. Eng. Comput. **46**(12), 1227 (2008)
11. Carneiro, G., Georgescu, B., Good, S., Comaniciu, D.: Detection of fetal anatomies from ultrasound images using a constrained probabilistic boosting tree. IEEE Trans. Med. Imaging **27**(9), 1342–1355 (2008)
12. Yaqub, M., Kelly, B., Papageorghiou, A.T., Noble, J.A.: Guided random forests for identification of key fetal anatomy and image categorization in ultrasound scans. In: Navab, N., Hornegger, J., Wells, W.M., Frangi, A.F. (eds.) MICCAI 2015. LNCS, vol. 9351, pp. 687–694. Springer, Cham (2015). https://doi.org/10.1007/978-3-319-24574-4_82
13. Li, J., et al.: Automatic fetal head circumference measurement in ultrasound using random forest and fast ellipse fitting. IEEE J. Biomed. Health Inform. **22**(1), 215–223 (2018)
14. Long, J., Shelhamer, E., Darrell, T.: Fully convolutional networks for semantic segmentation. In: Proceedings of the IEEE Conference on Computer Vision and Pattern Recognition, pp. 3431–3440 (2015)
15. Wu, L., Xin, Y., Li, S., Wang, T., Heng, P.-A., Ni, D.: Cascaded fully convolutional networks for automatic prenatal ultrasound image segmentation. In: 2017 IEEE 14th International Symposium on Biomedical Imaging (ISBI 2017). IEEE, pp. 663–666 (2017)
16. Sinclair, M., et al.: Human-level performance on automatic head biometrics in fetal ultrasound using fully convolutional neural networks, arXiv preprint arXiv:1804.09102 (2018)
17. Looney, P., et al.: Fully automated, real-time 3D ultrasound segmentation to estimate first trimester placental volume using deep learning. JCI Insight **3**(11), e120178 (2018)

18. Ren, S., He, K., Girshick, R., Sun, J.: Faster R-CNN: towards real-time object detection with region proposal networks. IEEE Trans. Pattern Anal. Mach. Intell. **6**, 1137–1149 (2017)
19. van den Heuvel, T.L., de Bruijn, D., de Korte, C.L., van Ginneken, B.: Automated measurement of fetal head circumference using 2D ultrasound images. PLoS ONE **13**(8), e0200412 (2018)
20. He, K., Gkioxari, G., Dollár, P., Girshick, R.: Mask R-CNN. In: 2017 IEEE International Conference on Computer Vision (ICCV), pp. 2980–2988. IEEE (2017)
21. He, K., Zhang, X., Ren, S., Sun, J.: Deep residual learning for image recognition. In: Proceedings of the IEEE Conference on Computer Vision and Pattern Recognition, pp. 770–778 (2016)

Automated Fetal Brain Extraction from Clinical Ultrasound Volumes Using 3D Convolutional Neural Networks

Felipe Moser[1]([✉]), Ruobing Huang[1], Aris T. Papageorghiou[2],
Bartłomiej W. Papież[1,3], and Ana I. L. Namburete[1]

[1] Institute of Biomedical Engineering, Department of Engineering Science,
University of Oxford, Oxford, UK
`felipe.moser@univ.ox.ac.uk`
[2] Nuffield Department of Women's and Reproductive Health, John Radcliffe
Hospital, University of Oxford, Oxford, UK
[3] Big Data Institute, Li Ka Shing Centre for Health Information and Discovery,
University of Oxford, Oxford, UK

Abstract. To improve the performance of most neuroimage analysis pipelines, brain extraction is used as a fundamental first step in the image processing. However, in the case of fetal brain development for routing clinical assessment, there is a need for a reliable Ultrasound (US)-specific tool. In this work we propose a fully automated CNN approach to fetal brain extraction from 3D US clinical volumes with minimal preprocessing. Our method accurately and reliably extracts the brain regardless of the large data variations in acquisition (eg. shadows, occlusions) inherent in this imaging modality. It also performs consistently throughout a gestational age range between 14 and 31 weeks, regardless of the pose variation of the subject, the scale, and even partial feature-obstruction in the image, outperforming all current alternatives.

Keywords: 3D ultrasound · Fetal · Brain · Extraction · Automated · 3D CNN · Skull stripping

1 Introduction

Ultrasound (US) imaging is routinely used to study fetal brain development and to screen for central nervous system malformations, and has become standard clinical practice around the world thanks to its ability to capture the brain structures in the womb [1–3]. Three-dimensional (3D) US expands on this technique by allowing for the imaging of the whole brain at once, instead of one slice at a time. However, the positional variation of the brain inside the scan volume, as well as the large amount of extra-cranial tissue observed in the volume, constitute a serious challenge when analysing the data. This has led brain-extraction tools to become a fundamental first step in most neuroimage analysis pipelines

© Springer Nature Switzerland AG 2020
Y. Zheng et al. (Eds.): MIUA 2019, CCIS 1065, pp. 151–163, 2020.
https://doi.org/10.1007/978-3-030-39343-4_13

with several methods being developed for fetal Magnetic Resonance Imaging (MRI) data but a reliable US-specific tool has not yet been developed.

Another challenge is that during gestation, the fetus is constantly moving within the womb. This causes the position and orientation of the brain to vary drastically from measurement to measurement. In order to compensate for this high degree of variability, the standard clinical protocol is to position the soni-fication plane such that it is perpendicular to the midsagittal plane. While this reduces the pose variability to a degree, it is not consistent as it depends entirely on the clinician to be accurate. There is therefore a need for a method that can accurately and reliably determine the position, orientation, and volume of the brain from a 3D US scan.

Besides the variation in position of both the fetus and the probe, the development of the brain throughout gestation increases the variability of the data, since the scale, ossification, and structural characteristics inside the skull change for each gestational week. The physical interaction of the US beam with the increasingly ossifying skull also causes reverberation artefacts and occlusions. Most importantly, when imaging from one side of the skull, the brain hemisphere farthest from the probe (distal) is visible while the closest (proximal) hemisphere is mostly occluded [4].

Several methods have been developed for the purpose of brain extraction (a comprehensive comparison of them for neonatal brain extraction can be found in Serag et al. [5]), but very limited amount of work has been done in relation to the extraction of the brain volume from fetal imaging and the vast majority of it has been focused on MRI imaging. Publications such as [6] and [7] show the difficulty in developing a reliable method that can accurately locate the brain from the acquired images. To our knowledge, the only method developed for automated fetal brain extraction from 3D US is the one proposed by Namburete et al. [8]. This method uses a fully-convolutional neural network (FCN) to predict the position of the brain from 2D slices of the 3D US volume. This prediction is then used to generate a 3D probability mask of fetal brain localization. An ellipsoid is then registered to the mask, resulting in an approximation of the location, orientation, and spatial extent of the brain. While this method offers a solution to the problem of brain extraction, it still relies on a 2D approach and its ellipsoid approximation does not accurately represent the shape of the brain.

Here, we propose an end-to-end 3D Convolutional Neural Network (CNN) approach for automated brain localization and extraction from standard clinical 3D fetal neurosonograms. As opposed to Namburete et al. [8], this method is a fully 3D approach to brain extraction and requires minimal pre- or post-processing. We show that our network manages to accurately detect the complete brain with a high degree of accuracy and reliability, regardless of the gestational age (ranging from 14 to 30 weeks), brain and probe positions, and visible brain structures.

2   Brain Extraction Network

The general schematics of the CNN used for this work can be seen in Fig. 1. It is a variation of the network used in Huang et al. [9] and is similar in structure to the 3D U-Net network from [10]. This network design showed accurate and stable results for extracting brain-structures from 3D US scans and was therefore chosen as the starting point for this work. Our network comprises kernels of size k^3, l convolutional and down-sampling layers in hte encoder path, and l convolutional and up-sampling layers in the decoder path. The first two convolutional layers have f and $2f$ number of filters, with the remaining ones having $4f$. After each convolution, batch normalisation and ReLu activations were used. The network was trained end-to-end with the Adam optimizer, with a learning rate of 10^{-3} and a Dice-Loss function. Both input and output are of size $n \times n \times n$.

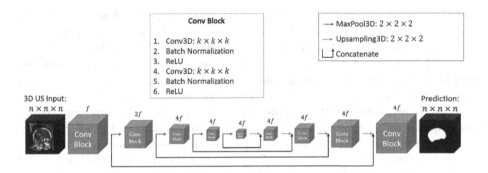

Fig. 1. Schematics of the 3D CNN used for brain extraction. Each convolution block performs the steps described, with the numbers of filters used being a multiple of the first bock's filters f and displayed above each block. This particular example shows four pooling layers (MaxPooling and UpSampling).

3   Experiments

3.1   Data

A total of $N = 1185$ fetal brain volumes spanning between 14.4 and 30.9 weeks of gestation were used for the training and testing of our networks. The distribution of the gestational ages is shown in Fig. 2. These volumes were obtained from the multi-site INTERGROWTH-21st study dataset [11] and are from healthy subjects that experienced a normal pregnancy. The original volumes have a median size of $237 \times 214 \times 174$ voxels. They were centre-cropped to a size of $160 \times 160 \times 160$ voxels and resampled to an isotropic voxel size of 0.6 mm $\times$ 0.6 mm $\times$ 0.6 mm.

In order to train our networks, labelled data representing the location of the brain within the 3D US volume were required. Manual labelling by a clinician

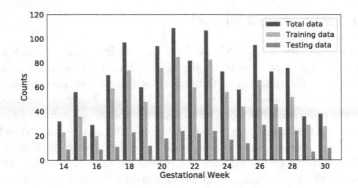

Fig. 2. Histogram of the number of volumes for each gestational week.

would be time-consuming and is likely to have a large degree of uncertainty due to the artefacts and occlusions that are intrinsic to this imaging modality.

To circumvent this problem, the spatiotemporal fetal brain atlases generated by Gholipour et al. [12] were used as templates. These atlases were generated from fetal MRI data. It represents the average brain shape for each gestational week and allows us to use it for semi-automated annotations.

To annotate our data, the complete dataset was aligned by hand and rigidly scaled to the same coordinate system. Each brain mask from the age-matched atlas was isotropically scaled to a selected reference using a similarity transformation to preserve the aspect ratio. The mask was then transformed back to the original position of the brain in the 3D US scan by performing the inverse transformation that was used to align the volumes in the first place. This resulted in a consistent brain annotation for the complete dataset. Since Gholipour et al. [12] atlases span a gestational age range of 21 to 38 weeks, a down-scaled version of the brain mask at 21 gestational weeks was used to represent gestational weeks 14 to 18 of our dataset.

3.2 Performance Evaluation

To determine the best network for fetal brain extraction, eight networks with different hyperparameters were created with the following characteristics: Input size n (same in all dimensions), number of pooling layers l, kernel size k (same in all dimensions), and number of filters of the first convolution f. For three-fold cross-validation experiments, 885 training volumes were partitioned into a subset of 685 for training and 200 for validation, three times. The best model was then retrained with the full 885 volumes and tested with the 300 hold-out testing set. A student's t-test result of t $= 1.26$, p $= 0.21$ confirmed no statistically significant difference in their age distributions.

To evaluate the performance of each network, we used four separate measures that could give an overall assessment of reliability. Firstly, the Euclidean Distance (ED) between the centre of mass of the binary masks was calculated.

This value gives a measure of the distance between the predicted centre and known centre of the brain. Since the position of the fetal brain relative to the 3D US volume varies throughout our dataset, the ED is a good indication of the accuracy of the predicted location of the brain. Secondly, the Hausdorff Distance (HD) between the masks was calculated. This metric determines the maximum distance between the prediction and the annotation and it allows for determination of local segmentation mismatch, regardless of the accuracy of the rest of the predictions. Since the masks used for annotation were derived from fetal MRI data which contains differences in observable structural information compared to 3D US, the use of this metric is critical to assess the reliability of our network. Thirdly, the Dice Similarity Coefficient (DSC) was calculated to determine the amount of overlap between the annotation and the predicted binary mask (different threshold were explored). This coefficient gives a good general assessment of the accuracy of the prediction. And fourthly, a Symmetry Coefficient (SC) was calculated to represent the symmetry of the prediction. Since the interaction between the 3D US beam and the skull causes the proximal hemisphere to be occluded, the structural information within the skull is asymmetric. However, the skull is generally symmetric about the sagittal plane, and therefore a prediction that reflects this is imperative. To calculate the SC, the prediction masks were aligned to a common set of coordinates with the sagittal, axial, and coronal planes being the mid-planes of the volume (same used for the data annotation described in Sect. 3.1). The right half of the brain mask prediction was then mirrored and the DSC between it and the left half is the value of the SC.

Using the fourth measures described (ED, HD, DSC, SC), we assessed the performance of our method under the intrinsic variability associated with fetal-brain 3D US. We first analysed the pose dependence of the accuracy of the prediction. As stated in Sect. 1, one of the main challenges of automatically extracting the brain from a 3D US volume is the pose variation of both the fetus relative to the probe. To assess the reliability of our network in regards to the orientation of the fetal brain in the volume, the Euler angles needed for the alignment of each volume to a set of common coordinates was compared to the DSC of that particular prediction. This would determine if there is a correlation between the orientation and the quality of the prediction. We then determined the dependency of the network to the gestational week of the subject. Since the goal is to develop a model that can reliably extract the brain regardless of the gestation week of the subject, their accuracy of the prediction should be consistent throughout the 14.4 to 30.9 weeks of gestation that comprises our dataset. To assess this, the results from the first four experiments were divided into gestational weeks and compared. Finally, we compared the predictions to the annotations by observing the regions of false-positives (predicted voxels not present in the annotation) and false-negatives (non-predicted voxels present in the annotations) results. This is a good qualitative method to analyse the regional accuracy of the prediction.

4 Results

4.1 Cross-validation and Network Selection

A description of each tested network, as well as the cross-validation results of each can be seen in Table 1. All networks managed to achieve a prediction with ED below 2 mm. The best score in terms of the ED between the centre of mass of the ground truth binary volume and the prediction of the network is from network D (1.16 mm), closely followed by network A. For reference, the median occipitofrontal diameter (OFD) for 22 weeks of gestation (the mean gestational age of our dataset) is 62 mm [13] so an ED of 2 mm corresponds to 3.2% of OFD at 22 weeks. The HD results show network D outperforming all other networks again with a result of 8.46 mm, followed by network F. Finally, the DSC results are very consistent among all networks, managing to stay above 0.90, and achieving a maximum of 0.94 for networks A, D, E, and F. Network D obtained the best cross-validation results across all tests and was therefore selected to be trained on the full dataset.

Table 1. Description of the eight tested networks as well as the cross-validation results. Network D, in bold, shows the best results. n: Input size (same in all dimensions). l: Number of pooling layers. k: Kernel size (same in all dimensions). f: Number of filters of the first convolution. ED: Euclidean Distance between centres of masses. HD: Hausdorff distance. DSC: Dice Similarity Coefficient. Param.: Number of trainable parameters in the network.

Network (n)	l	k	f	ED [mm]	HD [mm]	DSC	Param.
A (80)	4	3	16	1.17 ± 0.67	10.19 ± 6.40	0.94 ± 0.02	1.6 M
B	4	5	16	1.34 ± 0.95	9.53 ± 4.37	0.93 ± 0.03	7.4 M
C	4	7	16	1.33 ± 0.67	9.75 ± 4.30	0.93 ± 0.03	20.3 M
D	**4**	**3**	**8**	**1.16 ± 0.69**	**8.46 ± 3.66**	**0.94 ± 0.02**	**0.4 M**
E	4	3	4	1.25 ± 0.81	9.82 ± 6.01	0.94 ± 0.02	0.1 M
F (160)		3	4	1.24 ± 0.73	9.14 ± 4.53	0.94 ± 0.02	0.1 M
G	3	3	4	1.33 ± 0.72	12.12 ± 6.87	0.93 ± 0.02	0.07 M
H	2	3	4	1.81 ± 1.26	23.12 ± 6.99	0.90 ± 0.05	0.03 M

4.2 Testing

As shown in Sect. 4.1, network D outperformed other networks, and so it was used for further experiments. It was trained with the full 885 training volumes and tested with the independent test set of 300 volumes. The testing results are shown in Table 2. Five different prediction thresholds were tested to find the best results. HD and SC values of threshold 1 are the best ones. However, this threshold is too high and makes the DSC fall when compared to the overall best threshold of 0.5. The latter, while having a slightly worse HD and SC, has a slightly better ED, and it has a statistically significant improvement of the DSC. Considering the minimal variation of ED and HD throughout the thresholds, using threshold 0.5 is the most appropriate, since it achieves a high degree of overlap (0.94) with the annotations, while preserving symmetry (0.95).

Throughout the different thresholds, the results of our network are consistent. This shows the high degree of confidence of the predictions generated. The results are also very consistent with the ones observed during cross-validation (see Table 1), which confirms that the network works with new data.

Table 2. Testing results of the fully-trained network. A threshold of 0.5 (in bold) showed the most consistent results and was therefore the best. ED: Euclidean Distance between centres of masses. HD: Hausdorff distance. DSC: Dice Similarity Coefficient. SC: Symmetry Coefficient

Threshold	ED [mm]	HD [mm]	DSC	SC
0	8.34 ± 3.26	62.50 ± 7.91	0.27 ± 0.15	0.80 ± 0.02
0.25	1.43 ± 0.93	9.24 ± 4.77	0.94 ± 0.02	0.95 ± 0.02
0.5	**1.36 ± 0.72**	**9.05 ± 3.56**	**0.94 ± 0.02**	**0.95 ± 0.02**
0.75	1.36 ± 0.72	8.97 ± 3.54	0.93 ± 0.03	0.80 ± 0.02
1	1.42 ± 0.80	8.72 ± 4.23	0.90 ± 0.03	0.99 ± 0.01

4.3   Performance with Pose Variation

As mentioned in Sect. 3.2, it is crucial for our brain extraction network to work consistently regardless of the orientation of the brain within the US volume. This can be qualitatively observed in Fig. 3, which shows the outline of the brain-extraction prediction and the corresponding ground-truth, in red and green respectively, for six different 3D US volumes. These volumes have been selected to demonstrate the amount of variation between each scan, with the position of the fetus inside the womb as well as the position of the brain with respect to the probe varying drastically across scans.

As shown in Fig. 3, the network's prediction is remarkably close to the ground-truth, regardless of the position of the brain in the volume. It also manages to accurately predict the location of the brain when this is partially obscured either by the cropping or the shape of the ultrasound beam.

The distribution of the DSC in relation to the Euler angles is shown in Fig. 4. The Pearson's correlation coefficient was calculated for each Euler angle, with a $r_\alpha = 0.15$, $r_\beta = -0.20$, and $r_\gamma = -0.16$ respectively. The results show no significant correlation between the Euler angles and the DSC.

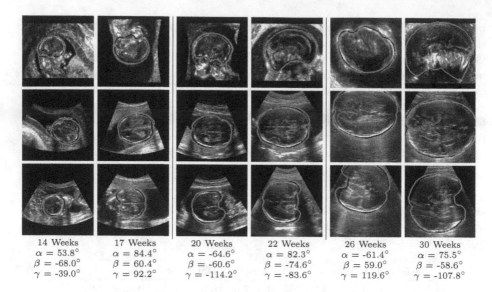

14 Weeks	17 Weeks	20 Weeks	22 Weeks	26 Weeks	30 Weeks
$\alpha = 53.8°$	$\alpha = 84.4°$	$\alpha = -64.6°$	$\alpha = 82.3°$	$\alpha = -61.4°$	$\alpha = 75.5°$
$\beta = -68.0°$	$\beta = 60.4°$	$\beta = -60.6°$	$\beta = -74.6°$	$\beta = 59.0°$	$\beta = -58.6°$
$\gamma = -39.0°$	$\gamma = 92.2°$	$\gamma = -114.2°$	$\gamma = -83.6°$	$\gamma = 119.6°$	$\gamma = -107.8°$

Fig. 3. Comparison of the brain-extraction prediction (red) and the ground-truth (green) superimposed onto the mid-planes of the 3D US volume. These volumes were selected to demonstrate the amount of variation between each scan. Top: XY-plane. Middle: XZ-plane. Bottom: YZ-plane. The gestation week of each volume and the Euler angles are displayed underneath. (Color figure online)

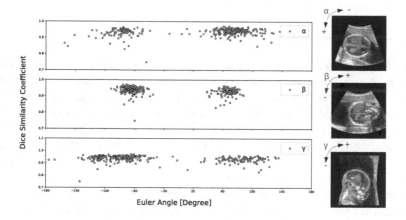

Fig. 4. Plot of Euler angles vs DSC. The angles cluster around 90° because of the manual alignment performed by the clinician being different to our coordinate system. No correlation between the Euler angles and the DSC was found.

4.4   Performance with Different Gestational Ages

The results describing the performance of the network for each gestational week are displayed in Fig. 5. The ED is consistently located between 1 mm and 3 mm throughout the data, with no observable correlation (r = 0.14) with the gestational age of the fetus. The HD is generally consistent around 10 mm for weeks 14 to 22, with its value increasing for each week with a maximum of 15 mm at week 30. This is likely a result of the different structural information observed in fetal brain MRI and US particularly around the brain stem (which is typically occluded in US images). The DSC show a slight increase between weeks 14 and 18, with a very consistent behaviour with a weak correlation to gestational week (r = 0.27). This is most likely caused by the atlas mask, since as mentioned in Sect. 3.1, the atlas of week 19 was used for that gestational age range. Regardless, with the exception of week 14, all DSCs are on average consistently above 0.90. Finally, the SC of the predicted volume is consistently high throughout the complete gestational period available in our data, with no significant correlation observed (r = 0.19).

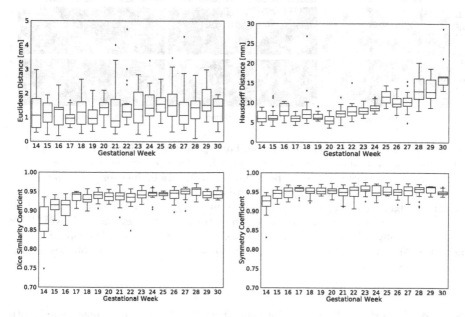

Fig. 5. Network performance for each gestational week. ED is consistent throughout all ages. HD increases after week 21 most likely due to the different structural information obtained from US in comparison to the MRI-based annotation. DSC drops for weeks earlier than 21, as expected due to the annotations for these weeks being based on the atlas for 21 gestational weeks. SC shows no correlation with gestational age.

4.5   Regional Performance

The regional performance of the network is shown as a map of false-positives and false-negatives for different gestational ages in Fig. 6. It can be observed that overall the network is consistent throughout all regions of the brain with the exception of the brain stem. This is most likely due to this structural infor-mation not being visible in the US scan. There is, however, one region of the brain that has worse regional performance for later gestational weeks: the space between the occipital cortex and the cerebellum. This is likely due to the fact that the annotations are based on the brain tissue and do not include the cerebrospinal fluid. As the brain develops, the separation between occipital cortex and cerebellum becomes more pronounced. However, since US does not offer a good contrast between the tissue and the cerebrospinal fluid, this separation is not visible and therefore appears as a false positive.

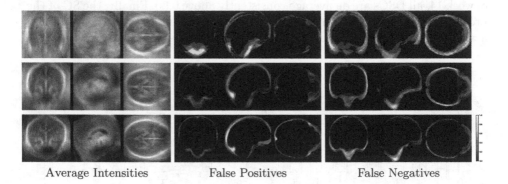

Average Intensities   False Positives   False Negatives

Fig. 6. Regional performance comparison. False-positives (predicted voxels not present in the annotation) and false-negatives (non-predicted voxels present in the annotations) give a qualitative assessment of the performance of the predictions for particular regions of the brain. Top: 14 Weeks of gestation. Middle: 22 Weeks. Bottom: 30 Weeks. These gestational ages have been selected to represent the age range in our dataset and the results shown are the averages of all volumes tested in that gestational week. They have all been aligned and rigidly scaled to the same coordinate set. The False Positives and False Negatives values have been normalized to the number of volumes used in that gestational week.

Table 3. Network results against method from [8]. Our network shows consistently better results accross all evaluations.

Method	ED [mm]	HD [mm]	DSC	SC
Our work	1.36 ± 0.72	9.05 ± 3.56	0.94 ± 0.02	0.95 ± 0.02
Namburete et al. [8]	3.68 ± 4.44	15.12 ± 5.24	0.85 ± 0.08	0.74 ± 0.06

4.6   Comparison with Previous Method

To quantitatively compare our network to the only other brain extraction method for fetal 3D US, the out test dataset was analysed with the method described in Namburete et al. [8]. The same experiments were performed and the results

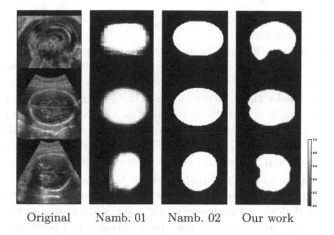

Original   Namb. 01   Namb. 02   Our work

Fig. 7. Comparison of our results with the method proposed by Namburete et al. [8]. Original: the three mid-planes of the original volume. Namb. 01 and Namb. 02: Probability mask generated from the 2D predictions and the ellipsoid fiting respectively [8]. Our work: The predictions obtained from our network. All columns but the first one share the same colormap.

are shown in Table 3. Our network manages considerably better results throughout all comparisons. This is expected, since the method proposed by Namburete et al. [8] relies on an approximation of the brain volume as ellipsoid, which does not accurately represent its shape. While an ellipsoid would be expected to have a high SC, its correspondence to the probability mask results in an inaccurate alignment, which is reflected in the low SC of 0.74, compared to our network's 0.95. This can be clearly seen in Fig. 7, where a comparison between the two methods is shown.

5   Discussion and Conclusion

In this work we present a method for automated brain extraction from fetal 3D US. A 3D CNN was developed and optimized to predict where the brain is located without any need for pre- or post-processing, and its performance was analysed over a variety of conditions usually presented in 3D US, such as the variation in brain position, rotation, scale, as well as a variation in developmental age and therefore internal brain structures.

Our network provided consistent results throughout all experiments, managing to accurately locate the brain to a mean ED of 1.36 mm. As a comparison, this distance represents only 2% of the mean occipitofrontal diameter of a brain of 22 weeks of gestation. The mean HD results of 9.05 mm is within the expected range due to the annotation masks being created from MRI data. The mean DSC of 0.94 shows that the predictions have a high level of overlap with the annotations and that our network manages consistently accurate predictions, which is also reflected on the mean SC of 0.95. The latter confirms that the network is not affected by the asymmetric structural information as a result of the beam-skull interactions.

In terms of pose variation, the network shows no correlation between the orientation of the brain and the accuracy of the prediction. The same consistency was shown throughout all gestational ages with the exception of HD for later gestational weeks (i.e. >25 weeks) and DSC for earlier gestational weeks (i.e. <17 weeks). This is caused by two limitations of the proposed brain extraction tool: the annotation of the data being adapted from fetal MRI data, and the earliest available gestational age for these annotations being 21 weeks (see Sect. 3.1). The fetal MRI data (and therefore our annotations) manage to separate the brain tissue and the cerebrospinal fluid, which is a distinction not visible in 3D US. This inherent difference between the two imaging modalities creates an annotation that does not accurately match our data around the edges of the mask. This is most pronounced in the space between the occipital cortex and the cerebellum, which becomes more pronounced as the gestational age increases, and is therefore reflected in the increased HD over time as well as in the regional performance analysis. The visibility of the brain stem in US is also not as good as in MRI, causing similar results. However, the fact that our network consistently extracts the brain including the cerebrospinal fluid is a good indication that our network has not been overfitted. As for the earliest gestational week available for annotations being week 21, it meant that our annotations for weeks 14–18 were less accurate. While the network results during this period is still very good (ED = 1.7 mm, HD = 7.14 mm, DSC = 0.92, SC = 0.95), it could be improved with a more accurate annotation.

The results of our network are significantly better than the ones obtained using the method proposed by Namburete et al. [8]. This was expected due to that method using a 2D slice approach to the brain extraction predictions, and requiring an ellipsoid approximation of the brain volume, which does not represent the real structural characteristics of the organ.

In this work we have shown that our 3D CNN solution to fetal brain extraction from 3D US works accurately, reliably, and consistently, regardless of the large data variation inherent in this imaging modality.

Acknowledgement. This work is supported by funding from the Engineering and Physical Sciences Research Council (EPSRC) and medical Research Council (MRC) [EP/L016052/1].

A. Namburete is grateful for support from the UK Royal Academy of Engineering under the Engineering for Development Research Fellowships scheme.

B. W. Papież acknowledges Rutherford Fund at Health Data Research UK.

We thank INTERGROWTH-21st for access to the dataset.

References

1. Kim, M.S., Jeanty, P., Turner, C., Benoit, B.: Three-dimensional sonographic evaluations of embryonic brain development. J. Ultrasound Med. **27**(1), 119–24 (2008)

2. Haratz, K.K., Lerman-Sagie, T.: Prenatal diagnosis of brainstem anomalies. Eur. J. Paediatr. Neurol. **22**(6), 1016–1026 (2018)
3. Namburete, A.I.L., van Kampen, R., Papageorghiou, A.T., Papież, B.W.: Multichannel groupwise registration to construct an ultrasound-specific fetal brain atlas. In: Melbourne, A., et al. (eds.) PIPPI/DATRA -2018. LNCS, vol. 11076, pp. 76–86. Springer, Cham (2018). https://doi.org/10.1007/978-3-030-00807-9_8
4. International society of ultrasound in obstetrics & gynecology education committee: sonographic examination of the fetal central nervous system: guidelines for performing the 'basic examination' and the 'fetal neurosonogram'. Ultrasound Obstet. Gynecol. **29**(1), 109–116 (2007)
5. Serag, A., et al.: Accurate Learning with Few Atlases (ALFA): an algorithm for MRI neonatal brain extraction and comparison with 11 publicly available methods. Sci. Rep. **6**, 23470 (2016)
6. Ison, M., Dittrich, E., Donner, R., Kasprian, G., Prayer, D., Langs, G.: Fully automated brain extraction and orientation in raw fetal MRI. In: Perinatal and Paediatric Imaging (PaPI 2012), MICCAI Workshop (2012)
7. Keraudren, K., Kyriakopoulou, V., Rutherford, M., Hajnal, J.V., Rueckert, D.: Localisation of the brain in fetal MRI using bundled SIFT features. In: Mori, K., Sakuma, I., Sato, Y., Barillot, C., Navab, N. (eds.) MICCAI 2013. LNCS, vol. 8149, pp. 582–589. Springer, Heidelberg (2013). https://doi.org/10.1007/978-3-642-40811-3_73
8. Namburete, A.I.L., Xie, W., Yaquba, M., Zisserman, A., Noble, J.A.: Fully-automated alignment of 3D fetal brain ultrasound to a canonical reference space using multi-task learning. Med. Image Anal. **46**, 1–14 (2018)
9. Huang, R., Noble, J.A., Namburete, A.I.L.: Omni-supervised learning: scaling up to large unlabelled medical datasets. In: Frangi, A.F., Schnabel, J.A., Davatzikos, C., Alberola-López, C., Fichtinger, G. (eds.) MICCAI 2018. LNCS, vol. 11070, pp. 572–580. Springer, Cham (2018). https://doi.org/10.1007/978-3-030-00928-1_65
10. Çiçek, Ö., Abdulkadir, A., Lienkamp, S.S., Brox, T., Ronneberger, O.: 3D U-Net: learning dense volumetric segmentation from sparse annotation. In: Ourselin, S., Joskowicz, L., Sabuncu, M.R., Unal, G., Wells, W. (eds.) MICCAI 2016. LNCS, vol. 9901, pp. 424–432. Springer, Cham (2016). https://doi.org/10.1007/978-3-319-46723-8_49
11. Papageorghiou, A.T., et al.: International fetal and newborn growth consortium for the 21st century (INTERGROWTH-21st): International standards for fetal growth based on serial ultrasound measurements: the fetal growth longitudinal study of the INTERGROWTH-21st project. Lancet **384**(9946), 869–879 (2014)
12. Gholipour, A., Rollins, C.K., Velasco-Annis, C., et al.: A normative spatiotemporal MRI atlas of the fetal brain for automatic segmentation and analysis of early brain growth. Sci. Rep. **7**(1), 476 (2017)
13. Moreira, N.C., et al.: Measurements of the normal fetal brain at gestation weeks 17 to 23: a MRI study. Neuroradiology **53**(1), 43–48 (2011)

Multi-task CNN for Structural Semantic Segmentation in 3D Fetal Brain Ultrasound

Lorenzo Venturini[1](✉), Aris T. Papageorghiou[2], J. Alison Noble[1],
and Ana I. L. Namburete[1]

[1] Institute of Biomedical Engineering, Department of Engineering Science, University
of Oxford, Oxford, UK
lorenzo.venturini@new.ox.ac.uk
[2] Nuffield Department of Obstetrics and Gynaecology,
University of Oxford, Oxford, UK

Abstract. The fetal brain undergoes extensive morphological changes throughout pregnancy, which can be visually seen in ultrasound acquisitions. We explore the use of convolutional neural networks (CNNs) for the segmentation of multiple fetal brain structures in 3D ultrasound images. Accurate automatic segmentation of brain structures in fetal ultrasound images can track brain development through gestation, and can provide useful information that can help predict fetal health outcomes. We propose a multi-task CNN to produce automatic segmentations from atlas-generated labels of the white matter, thalamus, brainstem, and cerebellum. The network as trained on 480 volumes produced accurate 3D segmentations on 48 test volumes, with Dice coefficient of 0.93 on the white matter and over 0.77 on segmentations of thalamus, brainstem and cerebellum.

1 Introduction

Fetal ultrasound scanning is a routine procedure during prenatal care, and is in many countries part of standard obstetric care. The scans are visually inspected to verify normal fetal development and to screen for disorders visible at specific gestational timepoints. These scans can be used to discern anatomical structures and track brain development [1]. Cortical structures first become visible within the fetal brain around 14 weeks of gestation and progressively develop throughout pregnancy [1].

Most studies that have analysed brain development have relied on MR imaging to perform segmentation and make quantitative measurements, due to its higher image resolution and signal-to-noise ratio [2]. However, MRI scans are relatively expensive and inaccessible, while ultrasound scans are a routine and widely available modality. Ultrasound displays artifacts which are difficult to interpret. Furthermore, due to the effects of increasing cranial ossification, the

Y. Zheng et al. (Eds.): MIUA 2019, CCIS 1065, pp. 164–173, 2020.
https://doi.org/10.1007/978-3-030-39343-4_14

cerebral hemisphere proximal to the probe tends to be indistinct and it is difficult to discern structural boundaries [3], while the distal hemisphere is more detailed.

A number of atlases have been constructed for fetal and neonatal brains using MRI. Kuklisova-Murgasova et al. [4] generated a publicly available 4D probabilistic atlas over a wider range of gestational ages (29–44 gestational weeks (GW)) that could be used to segment specific structures within the brain; however, this atlas was constructed using images of neonatal brains born preterm, and is therefore anatomically distinct from fetal brains. Most recently, Gholipour et al. [5] proposed a 4D spatiotemporal atlas of the fetal brain spanning 19–39 GW, using 3D MRI scans of fetuses and producing atlas labels of tissue type and structure.

Previous work on segmentation of fetal brain structures has proposed methods based on regression forests [6], and segmentation based on image atlases [7]. Machine-learning techniques such as convolutional neural networks (CNNs) can learn to distinguish important boundaries and artifacts and are increasingly popular in the segmentation of fetal ultrasound images [8], as they can learn to disregard some of the artifacts presented by ultrasound imaging and independently learn important segmentation features. Ronneberger et al. have developed the U-net [9], a CNN architecture for the segmentation of biomedical images.

We propose a machine learning-based method for automated segmentation of multiple fetal brain structures. We implement a CNN structure based on the U-net structure to perform multiple segmentations on 3D ultrasound volumes. To the best of our knowledge, this is the first work that demonstrates a CNN-based segmentation of individual fetal brain structures in 3D ultrasound. This is also the first work to demonstrate that segmentation in ultrasound can be achieved using a network trained exclusively on auxiliary generated labels.

2   Methods

2.1   Network Design

A 3D encoder-decoder network architecture based on the U-net architecture was used to perform multi-task segmentation. The size of the network was limited by memory constraints, so the top-level layer learned 16 $3 \times 3 \times 3$ feature maps. To satisfy memory constraints, the V-net architecture [10] was used. A softmax activation function was used at the output of the final convolutional layer to classify each voxel. The output was a five-class segmentation $\mathbb{Y} \in \mathbb{R}^{n \times N_x \times N_y \times N_z \times 5}$ where n is the number of volumes, and all volumes have dimensions $N_x \times N_y \times N_z$. Segmentation maps for the thalamus, white matter, brainstem, cerebellum and background were generated.[1] Multi-label Dice coefficient, the sum of the Dice coefficients of all classes, was used as the loss function, as this led to what visually

---

[1] Due to the size of this network, we used a batch size of 1 volume for training.

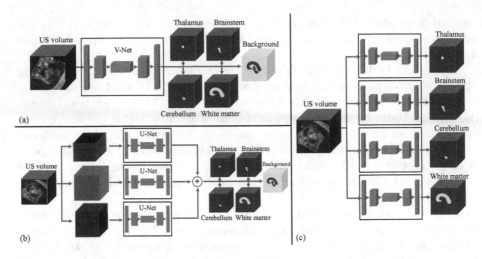

Fig. 1. The different network pipelines proposed. (a) The proposed 3D multi-task architecture, based on V-net. (b) A 2D multi-task framework, based on U-net, with QuickNAT-style merging of the different views. (c) A 3D single-task architecture, where a different network is trained per structure.

appeared to be the best results. Multi-label Dice is given by

$$DSC_{ml} = \sum_i \frac{2\left(GT_i \cap Seg_i\right)}{GT_i + Seg_i}$$

where GT and Seg are mappings of voxels corresponding to the ground truth and generated segmentation, respectively. The other parameters for the network's training were replicated from Milletari et al.'s V-net study [10], but ReLU activation functions were used instead of PReLU for simplicity.

The validation set was comprised of eight volumes per gestational week (for a total of 48 volumes). The remaining 480 volumes were used for training. A similar, single-task version of this network was also implemented for comparison. The architecture was identical, but the final layer was given a sigmoid activation function, similar to the original U-net architecture. Another technique which was found to slightly improve performance further was the application of a simple morphological operation (a $3 \times 3 \times 3$ morphological closing followed by an opening) to the resulting segmentation. This operation removes any small gaps from the segmentation, and weakly enforces smoothness near the edges of the segmentation. The edges are where the trained network shows the most uncertainty in its segmentations: Fig. 3 shows that the most misclassifications occur near the edge of the tissues of interest.

A classical 2D U-net architecture was also implemented for comparison. This network took 2D slices as input, and output a segmentation map for each slice. The segmentations of each slice were then stacked to obtain a full 3D semantic segmentation. To incorporate contextual information from other views, the data was sliced in 3 different ways, corresponding to the 3 canonical views, in a strategy similar to QuickNAT [11]. Each 2D network outputs "soft" segmentation masks for each structure, with each voxel given a value between 0 and 1 for each structure corresponding to the network's confidence. Combining the output of each network could exploit 3D information for segmentation, and therefore lead to a better accuracy than networks trained on individual views. Each network's output for every voxel was averaged and a threshold was applied to obtain a joint segmentation.

A comparison of all proposed network architectures can be seen in Fig. 1. All training was done on an Nvidia GeForce GTX 1080 GPU. The CNN converged to its highest Dice coefficient after 20 epochs, and after training each new volume could be segmented in 250 ms.

The data used for this study, to provide a dataset to train and test a CNN-based solution, were from the Fetal Growth Longitudinal Study (FGLS) of the INTERGROWTH-21st project. This was an international, multicentre, population-based project, conducted between 2009 and 2016. In the FGLS, serial 2-dimensional (2D) and 3-dimensional (3D) fetal scans were performed every 5 ± 1 weeks from $14 + 0$ weeks' gestation to delivery. Women participating in this study, who initiated antenatal care before 14 weeks' gestation, were selected based upon the WHO recommended criteria for optimal health, nutrition, education and socioeconomic status needed to construct international standards [12].

The volumes used in this investigation are all of healthy fetuses between 20 and 25 weeks' gestation. A total of 528 3D ultrasound volumes were selected within this age range, based on a visual inspection of the anatomy and the subjective visibility of brain structures of interest within each scan. This narrow range of gestational ages is of particular interest, as women have a routine ultrasound scan at 20 weeks of gestation, and sulci and gyri in the cortex become visible around this time in pregnancy [1].

All volumes were manually cropped to include just the cranium, and rotated to a canonical reference space [13]. Each brain was centered and resampled to a $160 \times 160 \times 160$ volume, with the mean voxel sampled at $0.6 \times 0.6 \times 0.6$ mm. The hemisphere distal to the ultrasound probe is always more detailed than the proximal hemisphere due to interactions between the concave skull and the ultrasound signal, but the acquisition protocol for this data was agnostic to which hemisphere would be more visible.

2.2 Label Generation

Given the size of this dataset and the visual artifacts and subject-specific characteristics inherent in ultrasound imaging, it is challenging and time-consuming for human experts to manually segment this dataset. We used an atlas-based method to generate a large amount of weak labels to compensate for this.

Gholipour et al. [5] recently proposed a 4D spatiotemporal atlas of the fetal brain spanning 21–37 GW, using 3D MRI scans of fetuses and producing atlas labels of tissue type and structure. This atlas can achieve segmentation quality comparable to human experts based on Dice coefficient [5]. This atlas was used to generate auxiliary labels for this dataset, similar to what was done by Guha Roy et al. [11] for the segmentation of brain structures in MRI with limited annotations.

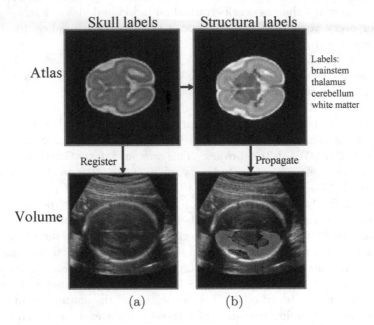

Fig. 2. The pipeline used to generate segmentation labels from the MRI atlas. The skull was segmented in each volume, a similarity transform - based registration was performed to find the correspondence, and then the structural labels were propagated.

To propagate the atlas labels to individual ultrasound volumes, a mask of the skull was manually fitted to each ultrasound volume: since the skull is a strong ultrasound reflector and has a predictable ellipsoidal shape, this could be done quickly. Registration based on a similarity transform (comprising translation, rotation and scaling) was then performed to find the transformation between the skull mask of an age-matched atlas and the manually labeled skull in each volume. The atlas-based segmentations of four structures, namely the thalamus, brainstem, cerebellum, and white matter were generated in this way. These structures were chosen because they are large and can be seen in ultrasound acquisitions: some, such as the cerebellum, are also inspected as part of routine clinical scans [14]. The transformation was applied to each of those structures, using nearest-neighbor interpolation to adjust to the new coordinate system. A schematic of the atlas-based segmentation framework can be seen in Fig. 2.

Since only the hemisphere distal to the ultrasound probe can be seen in any detail, for structures that extend far from the midsagittal plane (the white matter and the thalamus) only the label distal to the probe was segmented. The cerebellum and brainstem do cross the midsagittal plane, so the entire label was segmented.

3 Results

Each single view was trained for 20 epochs. After this the network showed a tendency to overfit and reduce validation accuracy.

Table 1. Segmentation performance of single-task and multi-task segmentation architectures, as measured by Dice coefficient (DSC), Euclidean distance of the centres of mass (ED) and Hausdorff distance (HD). Across measures and brain structures, the multi-task architecture outperforms the single-task network.

Network	DSC	ED (mm)	HD (mm)
Thalamus			
3D multi-task	**0.811 ± 0.061**	2.17 ± 1.35	**3.80 ± 1.95**
3D single-task	0.708 ± 0.070	2.82± 1.65	4.16 ± 2.39
2D	0.664 ± 0.081	**2.09 ± 1.74**	4.18 ± 2.87
Brainstem			
3D multi-task	**0.820 ± 0.081**	2.09 ± 1.26	**4.14 ± 1.29**
3D single-task	0.723 ± 0.098	1.96 ± 1.76	5.47 ± 2.37
2D	0.716 ± 0.066	**2.07 ± 1.95**	4.95 ± 3.20
Cerebellum			
3D multi-task	**0.773 ± 0.149**	2.42 ± 1.32	4.20 ± 2.39
3D single-task	0.689 ± 0.165	2.20 ± 1.72	4.42 ± 1.66
2D	0.681 ± 0.089	**2.15 ± 1.96**	**3.78 ± 2.77**
White matter			
3D multi-task	**0.921 ± 0.033**	2.27 ± 1.46	**5.93 ± 2.28**
3D single-task	0.865 ± 0.036	2.32 ± 1.72	5.90 ± 2.04
2D	0.819 ± 0.040	**2.23 ± 1.89**	14.40 ± 8.21

Table 1 shows the improvement in performance when doing multi-task segmentation compared to a single-task framework for each brain structure studied. For every brain structure, the multi-task segmentation framework performed better, with a mean improvement in Dice coefficient improvement of more than 33% over identical network trained with the same data on single-task segmentation. This is a substantial performance improvement, likely due to the fact that the brain structures analysed are spatially near to each other and often share

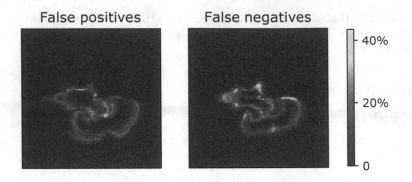

Fig. 3. A schematic showing the position of false negatives and false positives at a given axial slice for this data.

anatomical boundaries, meaning that the same features are useful to extract them. A richer training label effectively increases the amount of training data available, by providing important contextual information [15]. We expect that with larger training datasets, this difference should therefore decrease.

It is notable that segmentation of smaller structures, such as the thalamus, results in a significantly lower Dice coefficient than segmentation of the white matter on the same network. This can be explained by their differing physical characteristics: the thalamus is physically much smaller than the white matter label. In the dataset used, the white matter typically has a volume 15 times greater than the thalamus at 20 weeks, and 20 times greater at 24 weeks. The Dice coefficient is therefore biased by the larger number of interior voxels that can be predicted with high confidence, compared to voxels near the surface for which classification is more uncertain.. On the other hand, measures such as the Hausdorff distance are lower on smaller structures, showing that the overall subjective segmentation quality is similar across all structures.

Some examples of the resulting segmentations can be seen in Fig. 4. Where anatomical features are clearly visible in the ultrasound image, such as the boundaries of the white matter near the skull, the CNN appears to improve on the atlas-based labels: this is expected, as (beyond gestational age and skull shape) the atlas-based labels do not take individual variation into account. On the other hand, in regions where the ultrasound image is poor or subject to shadowing artifacts, such as the base of the medulla, the CNN appears to perform worse than the atlas.

Visually, the prediction seems to be significantly smoother than the atlas-based ground truth labels used for training, as seen in Fig. 6. This is likely due to the roughness of the original atlas-based segmentation: since nearest-neighbour interpolation is necessary, aliasing artifacts are likely to be introduced into the image. The resulting learned images, while smoother, do also appear to lose some of the detail available.

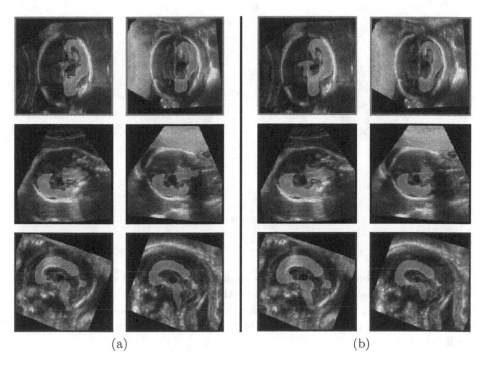

(a)   (b)

Fig. 4. (a) the atlas-generated labels used to train the CNN. (b) the resulting predictions on the same volumes (from the test set).

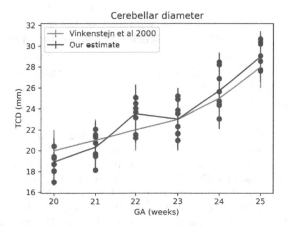

Fig. 5. Estimates of lengths, such as the transcerebellar diameter (TCD) derived from our data are in general agreement with the literature [14].

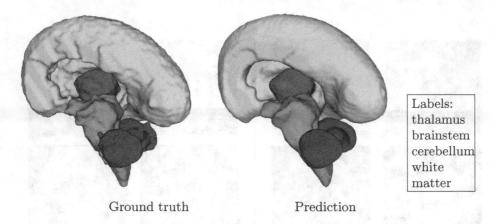

Ground truth   Prediction

Fig. 6. Comparison of the visual appearance in 3D of the atlas-based ground truth labels and the prediction for a volume.

It is also possible to compare the measurements we obtained to previous results in the literature. Figure 5 shows the transcerebellar diameter (TCD), a clinical biomarker often measured in scans [14]. Our proposed method finds segmentations with lengths that seem to be in agreement with others that have previously been done.

4   Conclusion

In this paper, we obtained multi-task segmentation maps of several brain structures from 3D ultrasound acquisitions, using only coarse atlas-based segmentations for training. The results show that a CNN can learn to segment these structures even from weak labels, and visually improve on the quality of the segmentation. A multi-task segmentation framework was also proposed that improves on the performance of a similar single-task network, and we showed that a natively 3D architecture outperforms a 2D architecture. The methods developed here are an interesting proof of concept, showing that this problem can be tackled with the proposed approach.

Acknowledgment. This work is supported by funding from the Engineering and Physical Sciences Research Council (EPSRC) and Medical Research Council (MRC) [grant number EP/L016052/1]. A. T. Papageorghiou is supported by the National Institute for Health Research (NIHR) Oxford Biomedical Research Centre (BRC). A. Namburete is grateful for support from the UK Royal Academy of Engineering under the Engineering for Development Research Fellowships scheme. We thank the INTERGROWTH-21st Consortium for permission to use 3D ultrasound volumes of the fetal brain.

References

1. Pistorius, L.R., et al.: Grade and symmetry of normal fetal cortical development: a longitudinal two-and three-dimensional ultrasound study. Ultrasound Obstet. Gynecol. **36**(6), 700–708 (2010)
2. Studholme, C.: Mapping fetal brain development in utero using magnetic resonance imaging: the Big Bang of brain mapping. Annu. Rev. Biomed. Eng. **13**(1), 345–368 (2011)
3. Namburete, A.I.L., Stebbing, R.V., Kemp, B., Yaqub, M., Papageorghiou, A.T., Noble, J.A.: Learning-based prediction of gestational age from ultrasound images of the fetal brain. Med. Image Anal. **21**(1), 72–86 (2015)
4. Kuklisova-Murgasova, M., et al.: Others: a dynamic 4D probabilistic atlas of the developing brain. NeuroImage **54**(4), 2750–2763 (2011)
5. Gholipour, A., et al.: A normative spatiotemporal MRI atlas of the fetal brain for automatic segmentation and analysis of early brain growth. Sci. Rep. **7**, 476 (2017)
6. Yaqub, M., et al.: Volumetric segmentation of key fetal brain structures in 3D ultrasound. In: Wu, G., Zhang, D., Shen, D., Yan, P., Suzuki, K., Wang, F. (eds.) MLMI 2013. LNCS, vol. 8184, pp. 25–32. Springer, Cham (2013). https://doi.org/10.1007/978-3-319-02267-3_4
7. Habas, P.A., et al.: A spatiotemporal atlas of MR intensity, tissue probability and shape of the fetal brain with application to segmentation. Neuroimage **53**(2), 460–470 (2010)
8. Schmidt-Richberg, A., et al.: Abdomen segmentation in 3D fetal ultrasound using CNN-powered deformable models. In: Cardoso, M.J., et al. (eds.) FIFI/OMIA - 2017. LNCS, vol. 10554, pp. 52–61. Springer, Cham (2017). https://doi.org/10.1007/978-3-319-67561-9_6
9. Ronneberger, O., Fischer, P., Brox, T.: U-Net: convolutional networks for biomedical image segmentation. In: Navab, N., Hornegger, J., Wells, W.M., Frangi, A.F. (eds.) MICCAI 2015. LNCS, vol. 9351, pp. 234–241. Springer, Cham (2015). https://doi.org/10.1007/978-3-319-24574-4_28
10. Milletari, F., Navab, N., Ahmadi, S.A.: V-Net: Fully convolutional neural networks for volumetric medical image segmentation, June 2016
11. Roy, A.G., Conjeti, S., Navab, N., Wachinger, C.: QuickNAT: a fully convolutional network for quick and accurate segmentation of neuroanatomy. NeuroImage **186**, 713–727 (2019)
12. Papageorghiou, A.T., et al.: International standards for fetal growth based on serial ultrasound measurements: the Fetal Growth Longitudinal Study of the INTERGROWTH-21st Project. Lancet **384**(9946), 869–879 (2014)
13. Namburete, A.I., Xie, W., Yaqub, M., Zisserman, A., Noble, J.A.: Fully-automated alignment of 3D fetal brain ultrasound to a canonical reference space using multi-task learning. Med. Image Anal. **46**, 1–14 (2018)
14. Vinkesteijn, A., Mulder, P., Wladimiroff, J.: Fetal transverse cerebellar diameter measurements in normal and reduced fetal growth. Ultrasound Obstet. Gynecol. **15**(1), 47–51 (2000)
15. Moeskops, P., et al.: Deep learning for multi-task medical image segmentation in multiple modalities. In: Ourselin, S., Joskowicz, L., Sabuncu, M.R., Unal, G., Wells, W. (eds.) MICCAI 2016. LNCS, vol. 9901, pp. 478–486. Springer, Cham (2016). https://doi.org/10.1007/978-3-319-46723-8_55

Towards Capturing Sonographic Experience: Cognition-Inspired Ultrasound Video Saliency Prediction

Richard Droste[1]([✉]), Yifan Cai[1], Harshita Sharma[1], Pierre Chatelain[1], Aris T. Papageorghiou[2], and J. Alison Noble[1]

[1] Department of Engineering Science, University of Oxford, Oxford, UK
richard.droste@eng.ox.ac.uk
[2] Nuffield Department of Women's and Reproductive Health,
University of Oxford, Oxford, UK

Abstract. For visual tasks like ultrasound (US) scanning, experts direct their gaze towards regions of task-relevant information. Therefore, learning to predict the gaze of sonographers on US videos captures the spatio-temporal patterns that are important for US scanning. The spatial distribution of gaze points on video frames can be represented through heat maps termed saliency maps. Here, we propose a temporally bidirectional model for video saliency prediction (BDS-Net), drawing inspiration from modern theories of human cognition. The model consists of a convolutional neural network (CNN) encoder followed by a bidirectional gated-recurrent-unit recurrent convolutional network (GRU-RCN) decoder. The temporal bidirectionality mimics human cognition, which simultaneously reacts to past and predicts future sensory inputs. We train the BDS-Net alongside spatial and temporally one-directional comparative models on the task of predicting saliency in videos of US abdominal circumference plane detection. The BDS-Net outperforms the comparative models on four out of five saliency metrics. We present a qualitative analysis on representative examples to explain the model's superior performance.

Keywords: Fetal ultrasound · Video saliency prediction · Gaze tracking · Convolutional neural networks

1 Introduction

Recently, it has been demonstrated that sonographer gaze tracking can aid standard plane detection in fetal ultrasound (US) imaging. Cai et al. [8] proposed the SonoEyeNet model for abdominal circumference plain (ACP) detection. Recorded gaze tracking heat maps—hereafter referred to as *saliency maps*—are used as attention on feature maps which are extracted with a fine-tuned SonoNet model [4]. Next, Cai et al. [9] proposed the Multi-task SonoEyeNet. Instead of relying on gaze data as an input, an attention module learns to predict

© Springer Nature Switzerland AG 2020
Y. Zheng et al. (Eds.): MIUA 2019, CCIS 1065, pp. 174–186, 2020.
https://doi.org/10.1007/978-3-030-39343-4_15

saliency maps so that no gaze tracking data is required for inference. Recently, Droste et al. [14] demonstrated that a saliency predictor trained entirely without manual annotations can be transferred to perform standard plane detection in routine clinical videos. These models perform standard plane detection and saliency prediction on single-frames only. However, ultrasound data and human eye movements are inherently spatio-temporal signals.

In this work we aim at improving ultrasound saliency prediction through spatiotemporal modeling, i.e. video saliency prediction. Therefore, we aim to bridge the gap between existing spatio-temporal models which do not leverage gaze information, e.g. for fetal cardiology [16,18] or US video partitioning [22], and models like SonoEyeNet that do not utilize temporal information. Chaabouni et al. [10] present an early convolutional neural network (CNN) based approach for video saliency prediction, adding optical flow as an additional input channel to a single-frame CNN. Bak et al. [2] propose to include optical flow via a two-stream architecture [23]. Bazzani et al. [5] achieve much larger temporal depth with a

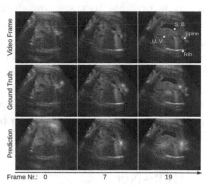

Fig. 1. BDS-Net predicted saliency maps. The key structures are marked. U. V. denotes umbilical vein and S. B. stomach bubble.

recurrent mixture density network by aggregating feature vectors with a long short-term memory (LSTM) model. Wang et al. [27] recently proposed a large video saliency benchmark (DHF1K) and show that existing video saliency predictors do not outperform the best single-frame saliency predictors. In contrast, Wang et al. achieve state-of-the-art on their benchmark with an architecture consisting of a CNN encoder and a convolutional LSTM decoder.

The above mentioned spatio-temporal models predict the saliency map of each video frame based on aggregated information of the previous frames. However, research in cognitive science suggests that human perception is not just reacting to past and present stimuli. Clark [13] argues that the brain is a 'prediction machine' that strives to minimize the prediction error between expectations versus sensory inputs. We transfer this insight to the problem of predicting sonographer visual saliency on US videos. Since the sonographer's expectations about future visual stimuli are unknown, we use future video frames as a proxy thereof, and ask the question: *To what extent are future video frames predictive for visual saliency?* Song et al. [25] recently proposed a bidirectional model for video salient object detection, which is a related application but aims at detecting and segmenting the most salient object in a scene rather than predicting the actual distribution of gaze points. Here, we propose an architecture combining a CNN and a temporally bidirectional recurrent neural network, BDS-Net, that predicts the visual saliency of each frame based on information of the entire video sequence, and compare the performance of the BDS-Net to an equivalent one-directional and a purely spatial model.

Contributions. The contributions of this study are three-fold: (1) To the best of our knowledge, this is the first study to propose a temporally bidirectional model for video saliency prediction, both in medical imaging and in computer vision more generally. Since the model considers the entire video sequence for saliency prediction of each frame, this approach is fundamentally different from previous models that only consider past and present frames. (2) We demonstrate that it is possible to train an effective video saliency predictor with few more than one hundred sequences, despite high inter-sequence variance. We achieve high data-efficiency by employing effective transfer learning and regularization techniques, and by reducing model complexity where possible, e.g. using a gated-recurrent-unit recurrent convolutional network (GRU-RCN) instead of a convolutional LSTM. (3) We demonstrate that the trained US video saliency predictor has learned meaningful aspects of sonographers' cognition in selecting the ACP. Therefore, we expect the model to be beneficial as part of architectures such as the Multi-task SonoEyeNet [9].

2   BDS-Net

The BDS-Net architecture consists of a *truncated SonoNet-64* model as frame-wise encoder, *adaptation I* that extracts the task-relevant features, a *bidirectional GRU-RCN* to aggregate the features temporally, and *adaptation II* to assemble the saliency map, followed by a *softmax* function (Fig. 2). In the following, we will use the vector notation $\mathbf{v}_t = [v_0^t, v_1^t, ..., v_n^t]^\top$.

Truncated SonoNet-64. The SonoNet model was recently proposed for US standard plane detection [4]. It is derived from the VGG-16 architecture [24], removing the final max-pooling and replacing the fully-connected layers with adaptation layers of 1×1-convolutions followed by global average pooling. Also, batch normalization [19] is added to each convolutional layer. The authors present three model variants with different numbers of convolutional kernels and train them on over 27 thousand US standard plane images. For this work, we use the largest variant, SonoNet-64, which was shown to achieve the highest overall precision. To use the model as a feature extractor, we remove the adaptation layers since they are classification-task specific. Further, we truncate the model by discarding the final three 3×3-convolutional layers to obtain lower-level features. We use the remaining 10 convolutional layers as frame-wise encoder of the BDS-Net. Since the SonoNet training data is substantially larger than the data available for this work, we use the model with fixed pre-trained weights.

Bidirectional GRU-RCN. We propose a bidirectional gated-recurrent-unit recurrent convolutional network (GRU-RCN) as the spatio-temporal decoder of the network. GRU networks [11] mitigate the exploding/vanishing gradient problem of a regular RNN similarly to LSTM networks [17] by updating the hidden state through element-wise additive and multiplicative gates instead of matrix multiplications. Compared to LSTM networks, however, they yield faster training convergence and higher accuracy on tasks like video captioning, despite reduced

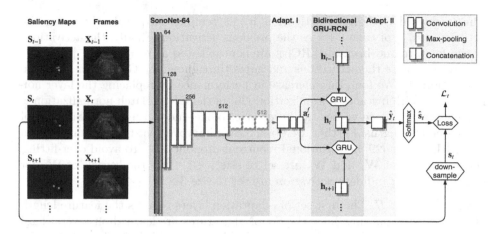

Fig. 2. Schema of the BDS-Net architecture and training procedure. For better readability, activation and normalization layers are not explicitly shown. The dashed part of the SonoNet-64 is only used for an ablation study.

complexity and fewer learned parameters [12]. While the standard GRU operates on 1D feature vectors, the GRU-RCN [3] is a straightforward extension for stacked 2D feature maps, replacing matrix products with convolutions. This modification vastly reduces the number of parameters compared to a fully-connected GRU and preserves the spatial feature topology. The bidirectional GRU-RCN is constructed from two separate GRU-RCN instances that propagate their hidden states forwards and backwards through time, respectively.

In the forward GRU-RCN, denoted by $\cdot^{\rightarrow}$, the candidate activation $\widetilde{\mathbf{h}}_t^{\rightarrow}$ at time t is computed from the feature activations $\mathbf{a}_t^f$ and the previous hidden state $\mathbf{h}_{t-1}^{\rightarrow}$ as:

$$\mathbf{r}_t^{\rightarrow} = \sigma(\mathbf{W}_r^{\rightarrow} * [\mathbf{h}_{t-1}^{\rightarrow} \,|\, \mathbf{a}_t^f] + \mathbf{b}_r^{\rightarrow}) \tag{1}$$

$$\widetilde{\mathbf{h}}_t^{\rightarrow} = \tanh\left(\mathbf{W}_h^{\rightarrow} * [\mathbf{r}_t^{\rightarrow} \circ \mathbf{h}_{t-1}^{\rightarrow} \,|\, \mathbf{a}_t^f] + \mathbf{b}_h^{\rightarrow}\right) , \tag{2}$$

where $\mathbf{r}_t^{\rightarrow}$ is the reset gate, $\mathbf{W}_{\cdot}^{\rightarrow}$ and $\mathbf{b}_{\cdot}^{\rightarrow}$ are the respective convolutional filters and biases, $\sigma(\cdot)$ is the logistic sigmoid function, $*$ is the convolution operator, $\circ$ denotes element-wise multiplication and $[\cdot|\cdot]$ denotes concatenation along the feature dimension. The reset gate controls the propagation of the previous hidden state into the current hidden state. Next, the activation $\mathbf{h}_t^{\rightarrow}$ is computed as a linear interpolation between the previous activation and the candidate activation, modulated by the update gate $\mathbf{z}_t^{\rightarrow}$:

$$\mathbf{z}_t^{\rightarrow} = \sigma(\mathbf{W}_z^{\rightarrow} * [\mathbf{h}_{t-1}^{\rightarrow} \,|\, \mathbf{a}_t^f] + \mathbf{b}_z^{\rightarrow}) \tag{3}$$

$$\mathbf{h}_t^{\rightarrow} = (1 - \mathbf{z}_t^{\rightarrow}) \circ \mathbf{h}_{t-1}^{\rightarrow} + \mathbf{z}_t^{\rightarrow} \circ \widetilde{\mathbf{h}}_t^{\rightarrow} . \tag{4}$$

The backward GRU-RCN activation $\mathbf{h}_t^{\leftarrow}$ is computed equivalently, using the activation $\mathbf{h}_{t+1}^{\leftarrow}$ of time $t+1$ as the previous activation. Finally, the activations of the forward and backward RCNs are concatenated as $\mathbf{h}_t = [\mathbf{h}_t^{\rightarrow} | \mathbf{h}_t^{\leftarrow}]$.

We normalize the activations and gates throughout the GRU with layer normalization [1]. We found no increase in performance by replacing the layer normalization with instance normalization [26] in the GRU. Batch normalization is not compatible since we set the batch size to one due to the high variability of the sequence lengths. Further, we observed no improvement through dropout on the GRU inputs [29] or variational recurrent dropout [15]. To avoid over-fitting, the kernel size of $\mathbf{W}_r$ and $\mathbf{W}_z$ are set to size 1×1. Only the kernels of $\mathbf{W}_h$ for computing the candidate activation are set to size 3×3.

Adaptations I & II. The first set of adaptation layers reduces the feature length of the SonoNet output. Since the SonoNet features are learned on several fetal anatomies, only a subset of the SonoNet features is likely to be relevant for the fetal abdomen. A convolutional layer of dimension $1 \times 1 \times 128 \times 512$ (2D kernel size × output dimension × input dimension) reduces the feature length, followed by a layer of dimension $3 \times 3 \times 128 \times 128$ to adapt the feature maps. Layer normalization [1] is performed after each layer. The final adaptation layer, a single convolutional layer of dimension $1 \times 1 \times 1 \times 128$, assembles the final saliency map.

Loss Function. At each time step $t \in \{0, 1, ..., T\}$ and output pixel $i \in \{0, 1, ..., P\}$, we obtain the predicted saliency $\hat{s}_i^t$ from the activations $\hat{y}_i^t$ of the final adaptation layer via the softmax function $\hat{s}_i^t = e^{\hat{y}_i^t} (\sum_{j=0}^{P} e^{\hat{y}_j^t})^{-1}$. Consequently, we implicitly treat the saliency maps as generalized Bernoulli distributions of fixations over the pixels of each frame [20]. We compute the loss as the sum of the Kullback-Leibler Divergences between the predicted distributions[1] and the downscaled ground truth distributions $\mathbf{s}_t$ as $\mathcal{L} = \sum_{t=0}^{T} \sum_{i=0}^{P} s_i^t \cdot (log(s_i^t) - log(\hat{s}_i^t))$.

3   Experiments

3.1   Data

The gaze data for this study had previously been recorded based on 33 fetal US videos, which were acquired according to a manual US sweep protocol, moving the probe from the bottom to the top of pregnant women's abdomen. Table 1 summarizes the data preparation procedure. *(1) Discarding of irrelevant frames:* From each of the 33 full sweeps an *ACP sweep* was extracted by discarding the frames which do not show the fetal abdomen. *(2) ACP selection by sonographers:* Each ACP sweep was presented to eight sonographers independently with the task of selecting the abdominal circumference plane (ACP). The sonographers were able to scroll through the frames using a keyboard until deciding on one

---

[1] In our implementation, for numerical stability, we compute $log(\hat{s}_i^t)$ with a log-softmax function instead of computing the softmax and logarithm sequentially.

Table 1. Data preparation procedure.

Data preparation step	Output
1) Discarding of irrelevant frames	ACP sweeps
For each sweep:	
2) ACP selection by sonographers	ACP search sequences
For each ACP search sequence:	
3) Gaze point aggregation	Gaze maps
4) Gaze map filtering	Saliency maps

ACP frame. The gaze of the sonographers was recorded at 30 Hz using an eye tracker (The EyeTribe) placed beneath the screen. In addition, the current sweep frame was registered for each gaze point. This yielded $33 \times 8 = 264$ sequences of gaze point-sweep frame pairs, which we will refer to as *ACP search sequences*. The ACP search sequences represent the way the sonographers moved through the frames to find the ACP. Therefore, they are a potentially useful approximation of freehand US video sequences where the sonographers find the ACP by moving the probe. The sequences were manually inspected and recordings with low-quality gaze data or miscalibration were discarded, leaving 116 sequences for further processing. *(3) Gaze point aggregation:* Since we want our model to learn the search strategy of sonographers from the first glance until the final ACP plane decision, we want to train the model on entire ACP search sequences. At 30 Hz, however, the sequences are too long (at least several hundred frames) and contain high redundancy among consecutive frames. Therefore, the gaze points were aggregated over intervals of 1000 ms at every 8th gaze sampling time, reducing the sampling rate from 30 Hz to 3.75 Hz. Gaze points outside the US image fan were discarded. A gaze map was computed for each aggregated set of gaze points by setting each pixel value to its corresponding number of gaze points. The frames were re-sampled at the same sampling times. The resulting sequences of gaze map-sweep frame pairs are of length 13 to 147 (avg. 33.6). *(4) Gaze maps filtering:* The saliency maps were computed by smoothing the gaze maps with a Gaussian kernel. The resulting final sequences of saliency map-sweep frame pairs are henceforth referred to as *saliency sequences*. The ACP sweeps were divided into 30 sweeps for training and 3 sweeps for validation with five-fold cross-validation.

3.2   Implementation Details

Preprocessing. The frames are preprocessed following Baumgartner et al. [4]. Data augmentation is performed by random rotation with an angle uniformly sampled from $[-25, 25]$ degrees and random horizontal flipping. Scale augmentation is omitted since it results in cropping of parts of the fetal abdomen. Next, the frames are normalized to zero-mean and unit-variance, multiplied by 255 and resized to 288×224 px. For the calculation of the loss, the ground truth saliency maps are transformed analogously and resized to 18×14 px, which is the output dimension of the network for inputs of 288×224 px.

Training. The model is trained over 140 epochs via stochastic gradient descent (SGD) with Nesterov momentum of 0.9 and initial learning rate 0.004. In accordance with Keskar et al. [21], we find that SGD yields better generalization to the validation set compared to ADAM. The learning rate is decayed by a factor of 0.5 each time the validation loss stagnates. The batch size is one to allow for varying sequence lengths. Sequences longer than 60 frames are truncated. The model is regularized via weight decay of $1 \cdot 10^{-6}$, dropout with rate 0.2 before the second and the last adaptation layers, as well as clipping the gradients outside the interval $[-5, 5]$. The z-gate bias is initialized to 1 to stabilize training by learning the spatial features first.

Comparative Models. Two comparative models are implemented: A one-directional GRU-RCN model and a purely spatial, single-frame model. The one-directional model is constructed by removing the backward GRU-RCN module from the BDS-Net. All other architectural and training parameters are identical. For the spatial model, the bidirectional GRU-RCN is simply replaced by an additional convolutional layer of dimension $3 \times 3 \times 128 \times 128$. Moreover, the layer normalization modules are removed and batch normalization is added to each layer. Training is performed on batches of 16 randomly selected frames and dropout with rate 0.5 is added to all layers. The initial learning rate is increased to 0.01 and no weight-decay is performed. Furthermore, we perform an ablation study to examine the effect of using the full SonoNet-64 (only adaptation layers removed) or the truncated SonoNet-32 models instead of the truncated SonoNet-64. The results are presented in Subsect. 4.3.

3.3 Evaluation Metrics

We evaluate the models on the metrics of the MIT Saliency Benchmark [7]. For this, we ported the MATLAB code published by the authors[2] to Python. We consider the fixation point (gaze map) based metrics Normalized Scanpath Saliency (NSS) and Area Under the ROC Curve by Judd (AUC-J), as well as the distribution (saliency map) based metrics Kullback-Leibler divergence (KLD), Linear Correlation Coefficient (CC) and Similarity (SIM). AUC-J, KLD and SIM are more sensitive to false negatives than to false positives, while NSS and CC treat them symmetrically [6]. To compute the average scores on each validation set, the scores are first averaged across time for each sequence and then across sequences. Thus, shorter and longer sequences weigh equally in the average. The differences between the respective cross-validated model scores are tested for statistical significance with the Wilcoxon test.

---

[2] https://github.com/cvzoya/saliency.

Table 2. Cross-validation scores (mean ± standard deviation) of the BDS-Net and the spatial and one-directional models. The best scores are marked in bold. The superscripts * and † denote an improvement with $p < 0.05$ over the spatial and one-directional models, respectively.

Model	NSS ↑	AUC-J ↑	KLD ↓	CC ↑	SIM ↑
Spatial	1.40 ± 0.36	0.83 ± 0.04	3.01 ± 0.37	0.26 ± 0.06	**0.23 ± 0.04**
One-directional	1.49 ± 0.34	0.85 ± 0.04 *	2.22 ± 0.25 *	0.27 ± 0.06	0.21 ± 0.03
BDS-Net	**1.61 ± 0.33** *†	**0.87 ± 0.03** *†	**2.16 ± 0.27** *†	**0.29 ± 0.06** *†	**0.23 ± 0.04** †

4 Results

4.1 Quantitative Results

Table 2 shows the validation scores of the BDS-Net and comparative models. The BDS-Net receives the best scores on all metrics except SIM. Moreover, both spatio-temporal models perform better on average than the spatial model on all metrics except SIM. For SIM, the BDS-Net scores better than the one-directional model but is on par with the spatial model.

4.2 Representative Examples

Figures 1 and 3 show examples of the predictions of the BDS-Net model and the comparative models on validation data. Since training and validation data were divided scan-wise, the frames are entirely unseen by the networks. Moreover, the networks are agnostic as to which sonographer is observing the scan.

Figure 1 shows input frames, ground truth saliency maps and BDS-Net predictions for three representative frames of one exemplary sequence. At frame zero, the prediction is highly uncertain. Areas throughout the middle and the upper boundary of the abdomen are predicted as fixation candidates. The highest probability is assigned to the area around the upper rib, which is not well visible. The ground truth fixation is between the spine and the upper rib. At frame seven, the BDS-Net assigns high saliency values to the spine and lower values to the umbilical vein. The ground truth fixations are at the spine. At frame nineteen, near the end of the sequence, the network assigns approximately equal probabilities to umbilical vein and spine. The ground truth fixations are indeed on umbilical vein and spine.

Figure 3 shows a more detailed example of BDS-Net predictions for five representative frames of another exemplary sequence. Additionally, the predictions of the spatial and one-directional models are shown. At frame zero, both spatio-temporal models predict uncertain saliency maps with a spread-out peak between spine and upper rib as denoted with *(a)* in the figure. The spatial model, which does not have information about the position of the frame in the sequence, predicts saliency at the spine with high certainty. The ground truth fixations are both at spine and the upper rib. Over the next frames, the recurrent models predict temporally smooth saliency maps with slightly varying maxima around the

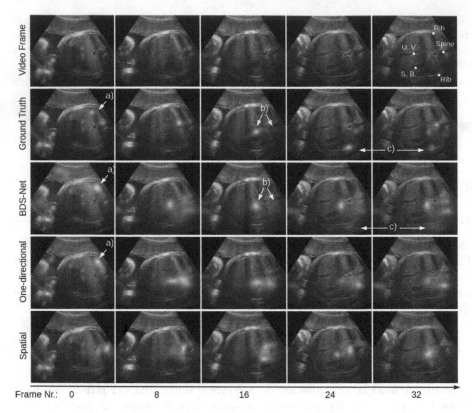

Fig. 3. Five frames from an exemplary ACP search sequence. The rows show the input frames, the ground truth saliency annotations, and the saliency predictions of the BDS-Net and the spatial and one-directional models, respectively. The relevant anatomical structures are denoted in the last input frame (top right).

center of the abdomen and the spine. The maxima of the spatial model fall into the same regions but the maps are less temporally smooth. Two key advantages of the BDS-Net predictions are denoted with *(b)* and *(c)*. At frame sixteen, in the middle of the sequence, the sonographer fixates the center of the abdomen. The spatial model predicts fixations around the spine and the one-directional model predicts fixations at either the spine or the center. Only the bidirectional model, which has information about both the previous and the subsequent frames, correctly predicts the fixation at the center and omits the spine, as denoted by *(b)*. On frames 24 and 32, towards the end of the search sequence, the sonographer fixates on the spine, the center of the abdomen and the lower rib, which is not well visible in these frames. Only the BDS-Net correctly assigns probability of fixation to the lower rib, indicated by *(c)*.

Table 3. Scores for the ablation study of the feature extractor on one randomly chosen validation set.

Feature extractor	NSS ↑	AUC-J ↑	KLD ↓	CC ↑	SIM ↑
Full SonoNet-64	1.50	0.86	2.21	0.29	0.22
Truncated SonoNet-32	2.01	0.90	2.01	0.37	0.25
Truncated SonoNet-64	**2.20**	**0.91**	**1.78**	**0.39**	**0.27**

4.3 Ablation Study

The quantitative results for the ablation study of the feature extractor are shown in Table 3. The ranking of the models is consistent across all five metrics: The full SonoNet-64 with higher-level features performs least favorable, the smaller truncated SonoNet-32 ranks second and the truncated SonoNet-64 performs best.

5 Discussion

The BDS-Net outperforms both comparative models on the AUC-J and NSS metrics, which are the default metrics of the MIT Saliency Benchmark [7]. Moreover, the one-directional model outperforms or matches the score of the spatial model on those metrics, despite the fact that training the spatial model is arguably easier for several reasons. First, gradient steps can be computed for each of the 3901 saliency maps per epoch separately. For the recurrent models, in contrast, gradient steps can only be computed for the 104 sequences per epoch. Second, the gradients are noisier for recurrent models in general. We mitigate this problem through gradient clipping, but it is an ad-hoc solution, and it does not resolve vanishing gradients. Finally, batch normalization is applied in the spatial model. For the recurrent models, since we set the batch size to one to account for the varying sequence lengths, we revert to layer normalization, which is known to stabilize training less for most tasks [28]. The fact that the recurrent models perform better despite the more difficult training conditions is a strong indication that spatial information (the current frame) alone is not sufficient to predict US video saliency accurately. This is in accordance with the results of Wang et al. [27] who have shown for natural videos that a recurrent architecture can outperform sophisticated single-frame saliency predictors.

Moreover, we have shown quantitatively and qualitatively that the bidirectional model performs better than the one-directional model. The added backwards GRU-RCN is the only difference between the two architectures, i.e. no other layers were added or removed and the training procedures are identical. This supports our hypothesis that sonographers are implicitly predicting future frames and focus their visual attention accordingly. Since predicting future frames requires domain expertise, we see this approach as a step towards modeling sonographer experience.

It is difficult to say how much room for improvement remains for the given dataset. Naturally, there is a certain inter-observer variability in the ACP search strategies. The model can only learn to predict the saliency corresponding to some average search strategy across all sonographers. To compute the actual maximum saliency scores, the inter-observer congruence (IOC) would need to be quantified. In our case, however, this is particularly difficult since each gaze sequence corresponds to a unique frame-sequence controlled by the sonographer. Therefore, there is no common reference frame for comparing the gaze sequences.

Nonetheless, despite the missing reference values for the quantitative evaluations, the qualitative analyses have shown that the BDS-Net has learned meaningful spatio-temporal patterns in the sonographers' search strategies. The analyses of Cai et al. [9] have shown that similar learned experience can significantly improve US standard plane detection on single frames. We expect that our video saliency predictor can further improve the performance of such models.

6   Conclusion and Outlook

We have presented a new model for predicting video saliency during ACP plane selection. We have shown that the temporally bidirectional BDS-Net model predicts saliency more accurately than single-frame and one-directional comparative models. The model has learned meaningful spatio-temporal patterns that attract sonographers' attention. Therefore, we expect the model to be beneficial for US standard plane detection tasks. In future work we will transfer the model to a larger dataset, which is currently being acquired. This will allow us to explore the limits of this approach for learning sonographic experience. Furthermore, we plan to integrate the model into architectures for US image analysis tasks such as standard plane detection and video partitioning.

Acknowledgements. This work is supported by the ERC (ERC-ADG-2015 694581, project PULSE) and the EPSRC (EP/R013853/1 and EP/M013774/1). AP is funded by the NIHR Oxford Biomedical Research Centre.

References

1. Ba, J.L., Kiros, J.R., Hinton, G.E.: Layer normalization. In: NIPS - Deep Learning Symposium (2016)
2. Bak, C., Kocak, A., Erdem, E., Erdem, A.: Spatio-temporal saliency networks for dynamic saliency prediction. IEEE Trans. Multimed. **20**(7), 1688–1698 (2018)
3. Ballas, N., Yao, L., Pal, C., Courville, A.: Delving deeper into convolutional networks for learning video representations. In: ICLR (2016)
4. Baumgartner, C.F., et al.: SonoNet: real-time detection and localisation of fetal standard scan planes in freehand ultrasound. IEEE Trans. Med. Imag. **36**(11), 2204–2215 (2017)
5. Bazzani, L., Larochelle, H., Torresani, L.: Recurrent mixture density network for spatiotemporal visual attention. In: ICLR (2017)
6. Bylinskii, Z., Judd, T., Oliva, A., Torralba, A., Durand, F.: What do different evaluation metrics tell us about saliency models? IEEE Trans. Pattern Anal. Mach. Intell. **41**(3), 740–757 (2019)

7. Bylinskii, Z., et al.: MIT Saliency Benchmark. http://saliency.mit.edu/
8. Cai, Y., Sharma, H., Chatelain, P., Noble, J.A.: SonoEyeNet: standardized fetal ultrasound plane detection informed by eye tracking. In: ISBI (2018)
9. Cai, Y., Sharma, H., Chatelain, P., Noble, J.A.: Multi-task sonoeyenet: detection of fetal standardized planes assisted by generated sonographer attention maps. In: Frangi, A.F., Schnabel, J.A., Davatzikos, C., Alberola-López, C., Fichtinger, G. (eds.) MICCAI 2018. LNCS, vol. 11070, pp. 871–879. Springer, Cham (2018). https://doi.org/10.1007/978-3-030-00928-1_98
10. Chaabouni, S., Benois-pineau, J., Hadar, O.: Deep Learning for Saliency Prediction in Natural Video. arXiv:1604.08010 (2016)
11. Cho, K., et al.: Learning phrase representations using RNN encoder-decoder for statistical machine translation. In: EMNLP (2014)
12. Chung, J., Gulcehre, C., Cho, K., Bengio, Y.: Empirical evaluation of gated recurrent neural networks on sequence modeling. In: NIPS (2014)
13. Clark, A.: Whatever next? predictive brains, situated agents, and the future of cognitive science. Behav. Brain Sci. **36**(03), 181–204 (2013)
14. Droste, R., et al.: Ultrasound Image Representation Learning by Modeling Sonographer Visual Attention. Accepted at IPMI (2019)
15. Gal, Y., Ghahramani, Z.: A theoretically grounded application of dropout in recurrent neural networks. In: NIPS (2016)
16. Gao, Y., Alison Noble, J.: Detection and characterization of the fetal heartbeat in free-hand ultrasound sweeps with weakly-supervised two-streams convolutional networks. In: Descoteaux, M., Maier-Hein, L., Franz, A., Jannin, P., Collins, D.L., Duchesne, S. (eds.) MICCAI 2017. LNCS, vol. 10434, pp. 305–313. Springer, Cham (2017). https://doi.org/10.1007/978-3-319-66185-8_35
17. Hochreiter, S., Schmidhuber, J.: Long short-term memory. Neural Comput. **9**(8), 1735–1780 (1997)
18. Huang, W., Bridge, C.P., Noble, J.A., Zisserman, A.: Temporal heartnet: towards human-level automatic analysis of fetal cardiac screening video. In: Descoteaux, M., Maier-Hein, L., Franz, A., Jannin, P., Collins, D.L., Duchesne, S. (eds.) MICCAI 2017. LNCS, vol. 10434, pp. 341–349. Springer, Cham (2017). https://doi.org/10.1007/978-3-319-66185-8_39
19. Ioffe, S., Szegedy, C.: Batch normalization: accelerating deep network training by reducing internal covariate shift. In: ICML (2015)
20. Jetley, S., Murray, N., Vig, E.: End-to-end saliency mapping via probability distribution prediction. In: CVPR (2016)
21. Keskar, N.S., Socher, R.: Improving Generalization Performance by Switching from Adam to SGD. arXiv:1712.07628 (2017)
22. Sharma, H., Droste, R., Chatelain, P., Drukker, L., Papageorghiou, A., Noble, J.A.: Spatio-temporal partitioning and description of full-length routine fetal anomaly ultrasound scans. Accepted at IEEE ISBI 2019 (2019)
23. Simonyan, K., Zisserman, A.: Two-stream convolutional networks for action recognition in videos. In: NIPS (2014)
24. Simonyan, K., Zisserman, A.: Very deep convolutional networks for large-scale image recognition. In: ICLR (2015)
25. Song, H., Wang, W., Zhao, S., Shen, J., Lam, K.-M.: Pyramid dilated deeper ConvLSTM for video salient object detection. In: Ferrari, V., Hebert, M., Sminchisescu, C., Weiss, Y. (eds.) ECCV 2018. LNCS, vol. 11215, pp. 744–760. Springer, Cham (2018). https://doi.org/10.1007/978-3-030-01252-6_44
26. Ulyanov, D., Vedaldi, A., Lempitsky, V.: Instance Normalization: The Missing Ingredient for Fast Stylization. arxiv:1607.08022 (2016)

27. Wang, W., Shen, J., Guo, F., Cheng, M.M., Borji, A.: Revisiting video saliency: a large-scale benchmark and a new model. In: CVPR (2018)
28. Wu, Y., He, K.: Group normalization. In: ECCV (2018)
29. Zaremba, W., Sutskever, I., Vinyals, O.: Recurrent Neural Network Regularization. arXiv:1409.2329 (2014)

Enhancement and Reconstruction

A Novel Deep Learning Based OCTA De-striping Method

Dongxu Gao[1](✉), Numan Celik[1], Xiyin Wu[1], Bryan M. Williams[1], Amira Stylianides[2], and Yalin Zheng[1](✉)

[1] Department of Eye and Vision Science, Institute of Ageing and Chronic Disease, University of Liverpool, Liverpool L7 8TX, UK
{dongxu.gao,yalin.zheng}@liverpool.ac.uk
[2] St. Pauls Eye Unit, Royal Liverpool University Hospital, Liverpool L7 8XP, UK
amira.stylianides@rlbuht.nhs.uk

Abstract. Noise in images presents a considerable problem, limiting their readability and hindering the performance of post-processing and analysis tools. In particular, optical coherence tomography angiography (OCTA) suffers from stripe noise. In medical imaging, clinicians rely on high quality images in order to make accurate diagnoses and plan management. Poor quality images can lead to pathology being overlooked or undiagnosed. Image denoising is a fundamental technique that can be developed to tackle this problem and improve performance in many applications, yet there exists no method focused on removing stripe noise in OCTA. Existing OCTA denoising methods do not consider the structure of stripe noise, which severely limits their potential for recovering the image. The development of artificial intelligence (AI) have enabled deep learning approaches to obtain impressive results and play a dominant role in many areas, but require a ground truth for training, which is difficult to obtain for this problem. In this paper, we propose a revised U-net framework for removing the stripe noise from OCTA images, leaving a clean image. With our proposed method, a ground truth is not required for training, allowing both the stripe noise and the clean image to be estimated, preserving more image detail without compromising image quality. The experimental results show the impressive de-striping performance of our method on OCTA images. We evaluate the effectiveness of our proposed method using the peak-signal-to-noise ratio (PSNR) and the structural similarity index measure (SSIM), achieving excellent results as well.

Keywords: OCTA · Stripe noise removal · Image decomposition · Deep learning

1 Introduction

In recent years, deep convolutional networks have achieved considerable success in image-level diagnostics in many areas of medical imaging [5,9], including ophthalmology [3,17]. Several deep learning (DL) algorithms have been very efficient

© Springer Nature Switzerland AG 2020
Y. Zheng et al. (Eds.): MIUA 2019, CCIS 1065, pp. 189–197, 2020.
https://doi.org/10.1007/978-3-030-39343-4_16

in detecting clinically significant features for ophthalmic diagnosis and prognosis of many diseases including diabetic retinopathy (DR) [6] and age related macular degeneration (AMD) [1].

Optical Coherence Tomography (OCT) is a non-invasive imaging technique, capable of providing tomographic images of the retina and contributing to the clinical diagnosis of several diseases including glaucoma, AMD and DR. Particularly, OCT angiography (OCTA), which is functional extension of OCT measuring motion of blood flow contrast, can provide a near-microscopic view of the retina in-vivo with high resolution. This results in a fast imaging modality that reveals structural detail of the retina vascular network [13]. Due to attractive qualities and capabilities of OCTA, it is widely used in ophthalmology studies to test for and predict DR and other diseases.

Although OCTA images provide high resolution retinal fundus information, the images are composed of strip data, resulting is visible striped artefacts, which hinder analysis and further processing. These strip OCTA artefacts can cause incorrect evaluation in segmentation for both traditional and more recent DL approaches, resulting in correct features not being detected with good accuracy and ultimately false predictions. In order to eliminate this strip noise data from OCTA images, with the help of Chang et al. [2], we propose a decomposition-based loss function to separate the desired, clean OCTA image from the stripe components.

The aim of this study is to build a model that is capable of reconstructing OCTA images by estimating and removing strip noise by incorporating a suitable loss function into a deep convolutional neural network. To the best of the authors' knowledge, this paper is first study to incorporate stripe-noise removal into deep learning framework.

With the recent improvements in deep neural networks and their excellent results in medical imaging, we focus on developing a DL technique in this study. The popular and widely used U-net, which is a fully convolutional network (FCN) variant, has demonstrated state-of-the-art performance in various medical image segmentation tasks [16], increasing sensitivity and prediction accuracy. In this paper, we propose a revised U-net for estimating clean OCTA images from noisy OCTA images. We estimate the strip noise within the model which enables us to use the low rank matrix of the stripe noise as a constraint in the loss function. In addition, the TV (total variational) norm of the estimated clean image is used as a constraint on the image.

The main contributions of this paper: 1. We develop a stripe noise removal framework based on U-net, introducing a new multi-outputs layer to estimate both the clean image and the noisy image. 2. We design a loss function which regularizes both the noise information and the predicted clean image. 3. We introduce a novel training process that removes the need for ground truth data which is important for applications where a ground truth is not available.

The rest of this paper is organized as follows. In the second section, we review related work from the different aspects including image de-noising, stripe noise removal, deep learning and loss functions. The third section introduces our method, and we present results with evaluation and discussion in the fourth section before concluding this work in Sect. 5.

2 Related Work

There have been several developments in de-noising with deep learning in recent years [10,11,21]. An adversarial and multi-scale feature extraction approach was used to remove image noise with a three-stage training procedure, and it is demonstrated that convolutional neural networks can be used for removing image noise [4]. Nam et al. [14] explored a noise modelling and analysing method and applied a cross-channel image noise method to show that the colour channels are independent. However, the existing methods only focus on removing noise while these noise types can be easily imitated.

Chang et al. introduced an image stripe noise removal method [2] on remote sensing image dataset and explored both the clean image and noise image quality from image decomposition perspective. Johnson et al. [8] considered that an input image can be transformed into an output image with training convolutional neural networks by introducing a perceptual loss function. Their experimental results proved that high-level features can be extracted from pre-trained networks by optimizing perceptual loss functions.

Zhao et al. researched the importance of perceptual loss for image restoration and explored the image quality correlation between humans and algorithms. Although the mean squared error plays a dominant role across diverse fields, it does not correlate with human's judgement of image quality. However, there is still no individual loss function that can achieve impressive results across different problems [22]. Yair et al. applied the weighted nuclear norm values of a whole image as a regularization term and considered the image restoration as an optimization problem, and it can be solved by introducing a unique variable splitting method and achieved leading results on deblurring and inpainting problems [20].

Plotz et al. contributed a benchmark for real photograph denoising algorithms when realistic ground truth data is lacking [15]. Generally speaking, realistic settings limit the relevance of de-noising techniques from a scientific evaluation perspective. Zhang et al. [21] introduced a residual deep learning method for removing Gaussian noise of an image. The residual learning strategy provides a certain to model different Gaussian noise level.

A fully convolutional net (FCN) has been shown to achieve impressive results on many different tasks such as classification and segmentation [12]. Built upon the FCN, Ronneberger et al. [16] proposed a fast neural network architecture (U-net) for medical image segmentation. Benefiting from symmetric and skip connections, one of the advantages of this architecture is that a large number of feature maps can be extracted. In addition, it can predict the image pixel's class. Therefore, it is a favourable network in medical image processing area because of the high-resolution nature.

3   Method

3.1   Image Model

In this paper, we will focus on stripe noise removal in OCTA images, below is the equation for describing the image model:

$$O = I + N \tag{1}$$

Where O represents an original image corrupted by stripe noise, I denotes the expected clean image without stripe noise, N is the stripe noise.

3.2   Model Architecture

The U-net, which is an extension of FCN [12], is used as the base network because it is an encoder-decoder neural network. We revise the model for OCTA de-striping and make it has two outputs. The advantages of our revised model are: 1. A decoder enables the parallel computing of different features representation at pixel level without changing the original image resolution. It is very important that all the fine-grained information of the OCTA image can be kept. 2. Multiresolution features and multilevel features (such as multiple scales and abstraction levels) representation can be computed effectively with an encoder. 3. We introduced a multi-outputs layer and pass both the original image and the noise image to the loss function, so there is no need to use the ground truth for the training process.

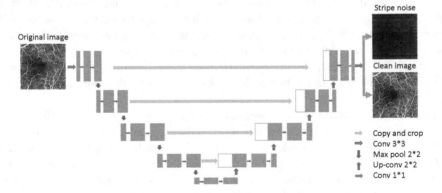

Fig. 1. The framework of Deep De-striping Net. The input is a measure OCTA image and the output is a de-noised image and a stripe noise image

Figure 1 demonstrates the framework we developed. At the final layer, we introduce a multi-outputs layer which enables the model to output both the de-striped image and the stripe noise image. We also use a fully convolutional network to freely use different image sizes as input and output two images with the same size. The estimated stripe noise can be used for exploring more accurate stripe noise removal methods in future works.

3.3   Loss Function

Different from previous loss functions for de-noising work, we build a loss function with the TV norm and the low rank matrix as regularization terms.

$$\min_{I,N} \left\{ \frac{1}{2} ||I + N - O||_F^2 + \tau ||I||_{TV} + \lambda \text{rank}(N) \right\} \qquad (2)$$

Equation 2 defines our loss function. I is the predicted image, O is the original image and N is the noise image.

The TV norm regularization is based on the principle of signal processing and has been applied in noise removal issues [19]. The introduction of TV norm has the advantage of being a close match to the desired image. There is a positive correlation between the total variation and the integral of the absolute gradient. The sharp boundaries of I can be preserved when minimizing the TV norm. The low rank matrix was introduced for the stripe component with the low-rank constraint. Therefore, the introduction of the low rank matrix and TV norm makes the estimated image preserve important detail information.

4   Experiments and Discussion

Our experiments are implemented using Keras 2.2.4 with Tensorflow 1.12 as backend and an Nvidia Titan XP GPU. The batch size is set as 64, the learning rate is 10^{-4}, τ is 0.005, λ is 0.005, epochs is 100, Adam is used as the optimizer, MSE (mean square error) loss function is used for comparison. The results using the method [2] are used for training our model with MSE loss function.

4.1   Dataset

We collected the OCTA images from 30 patients with 180 images (two eyes from 13 patients). Four images per eye from SVP (superficial vascular plexus), DVP (deep vascular plexus), AL (Avascular layer) and WR (whole retina) are collected from The Royal Liverpool University Hospital. In this paper, we treat each image separately for the purpose of training the deep learning models.

4.2   Results

Figure 2 shows the results of two selected examples including the original image, ground truth (predicted image with one stripe noise removal method [2] from the original image), predicted image with MSE loss function and our predicted image.

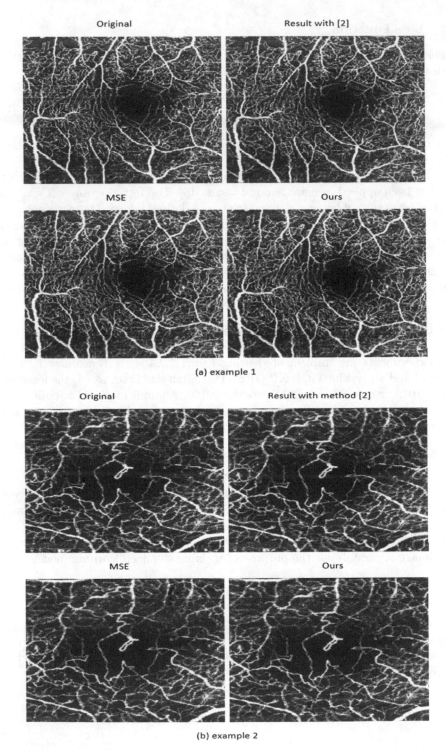

Fig. 2. Image denoising results on OCTA dataset

4.3 Evaluation

Two well-known image quality metrics are used for evaluating the effectiveness of our method, PSNR (peak-signal-to-noise) and SSIM (structural similarity index measure) [7].

With using the predicted image O and the ground truth image I, the PSNR is defined by:

$$PSNR(I, O) = 10log_{10}((255^2)/MSE(I, O)), \tag{3}$$

where MSE is

$$MSE(I, O) = \frac{1}{MN} \sum_{i=1}^{M} \sum_{j=1}^{N} (I_{ij} - O_{ij})^2. \tag{4}$$

The SSIM is defined by:

$$SSIM(I, O) = l(I, O)c(I, O)s(I, O), \tag{5}$$

where l(I,O), c(I,O), s(I,O) are luminance the comparison, contrast comparison and the structure comparison respectively [18].

Table 1. The performance in terms of image quality.

Methods (60 images)	PSNR ([2] results as reference images)	SSIM ([2] results as reference images)
MSE loss	33.09	0.949
Ours	33.29	0.951

4.4 Discussion

Our experimental results in Fig. 2 show our method can preserve detail information and result in improved image quality compared with the original image. Table 1 shows both PSNR and SSIM results calculated across 60 images. Our method has a slightly better performance in terms of both PSNR and SSIM compared with the MSE loss function. However, the ground truth is needed for training with the MSE loss function while it is not necessary to use the ground truth for our method. The introduction of the image decomposition and our proposed deep learning framework enables us to pass both the original image and the learned noise image to our loss function, thus the low rank matrix of the noise and the TV norm of the predicted clean image can be used as regularization terms of our loss function. The combination of our revised network and the image decomposition model can be applied in many applications where no ground truth is available.

5 Conclusion

In this paper, we developed a stripe noise removal method for OCTA images based on our revised U-net, using both the estimated clean image and noise image structure information in constraint terms for our proposed loss function. We compared our approach with a comparable approach using the MSE loss function to verify the effectiveness of our loss function. The experimental results showed that our estimated clean images preserved the image detail information. In addition, both PSNR and SSIM have been used as the evaluation metrics to prove our proposed method is effective in OCTA de-striping without the ground truth during the training process. It is believed our method can be applied in many deep learning applications where ground truth is not available.

References

1. Burlina, P.M., Joshi, N., Pekala, M., Pacheco, K.D., Freund, D.E., Bressler, N.M.: Automated grading of age-related macular degeneration from color fundus images using deep convolutional neural networks. JAMA Ophthalmol. (2017). https://doi.org/10.1001/jamaophthalmol.2017.3782
2. Chang, Y., Yan, L., Wu, T., Zhong, S.: Remote sensing image stripe noise removal: from image decomposition perspective. IEEE Trans. Geosci. Remote Sens. **54**(12), 7018–7031 (2016). https://doi.org/10.1109/TGRS.2016.2594080
3. De Fauw, J., et al.: Clinically applicable deep learning for diagnosis and referral in retinal disease. Nat. Med. **24**(9), 1342 (2018)
4. Divakar, N., Babu, R.V.: Image denoising via CNNs: an adversarial approach. In: IEEE Computer Society Conference on Computer Vision and Pattern Recognition Workshops, vol. 2017, pp. 1076–1083, July 2017. https://doi.org/10.1109/CVPRW.2017.145
5. Gibson, E., et al.: Niftynet: a deep-learning platform for medical imaging. Comput. Methods Programs Biomed. **158**, 113–122 (2018)
6. Gulshan, V., et al.: Development and validation of a deep learning algorithm for detection of diabetic retinopathy in retinal fundus photographs. JAMA - J. Am. Med. Assoc. (2016). https://doi.org/10.1001/jama.2016.17216
7. Hore, A., Ziou, D.: Image quality metrics: PSNR vs. SSIM. In: 2010 20th International Conference on Pattern Recognition, pp. 2366–2369. IEEE, August 2010. http://ieeexplore.ieee.org/document/5596999/
8. Johnson, J., Alahi, A., Fei-Fei, L.: Perceptual losses for real-time style transfer and super-resolution. In: Leibe, B., Matas, J., Sebe, N., Welling, M. (eds.) ECCV 2016. LNCS, vol. 9906, pp. 694–711. Springer, Cham (2016). https://doi.org/10.1007/978-3-319-46475-6_43
9. Litjens, G., et al.: A survey on deep learning in medical image analysis (2017). https://doi.org/10.1016/j.media.2017.07.005
10. Liu, P., Fang, R.: Learning Pixel-Distribution Prior with Wider Convolution for Image Denoising. CoRR arXiv abs/1707.0(c), pp. 1–11, July 2017. http://arxiv.org/abs/1707.05414arxiv.org/abs/1707.09135
11. Liu, P., Fang, R.: Wide Inference Network for Image Denoising via Learning Pixel-distribution Prior. ArXiv preprint, pp. 1–16, July 2017. http://arxiv.org/abs/1707.05414

12. Long, J., Shelhamer, E., Darrell, T.: Fully convolutional networks for semantic segmentation. In: 2015 IEEE Conference on Computer Vision and Pattern Recognition (CVPR), pp. 3431–3440. IEEE, 07–12 June 2015. http://ieeexplore.ieee.org/document/7298965/, https://doi.org/10.1109/CVPR.2015.7298965
13. Munk, M.R., et al.: OCT-angiography: a qualitative and quantitative comparison. PLoS ONE (2017). https://doi.org/10.1371/journal.pone.0177059
14. Nam, S., Hwang, Y., Matsushita, Y., Kim, S.J.: A holistic approach to cross-channel image noise modeling and its application to image denoising. In: 2016 IEEE Conference on Computer Vision and Pattern Recognition (CVPR), pp. 1683–1691. IEEE, June 2016. http://ieeexplore.ieee.org/document/7780555/, https://doi.org/10.1109/CVPR.2016.186
15. Plotz, T., Roth, S.: Benchmarking denoising algorithms with real photographs. In: Proceedings of the IEEE Conference on Computer Vision and Pattern Recognition, pp. 1586–1595 (2017)
16. Ronneberger, O., Fischer, P., Brox, T.: U-Net: convolutional networks for biomedical image segmentation. In: Navab, N., Hornegger, J., Wells, W.M., Frangi, A.F. (eds.) MICCAI 2015. LNCS, vol. 9351, pp. 234–241. Springer, Cham (2015). https://doi.org/10.1007/978-3-319-24574-4_28
17. Ting, D.S., Liu, Y., Burlina, P., Xu, X., Bressler, N.M., Wong, T.Y.: AI for medical imaging goes deep. Nat. Med. (2018). https://doi.org/10.1038/s41591-018-0029-3
18. Wang, Z., Bovik, A.C., Sheikh, H.R., Simoncelli, E.P., et al.: Image quality assessment: from error visibility to structural similarity. IEEE Trans. Image Process. 13(4), 600–612 (2004)
19. Wikipedia: Total variation denoising (2019)
20. Yair, N., Michaeli, T.: Multi-scale weighted nuclear norm image restoration. In: CVPR, pp. 3165–3174 (2018). https://doi.org/10.1109/CVPR.2018.00334
21. Zhang, K., Zuo, W., Member, S., Chen, Y., Meng, D.: Beyond a gaussian denoiser: residual learning of deep CNN for image denoising. IEEE Trans. Image Process. 26(7), 3142–3155 (2017). https://www4.comp.polyu.edu.hk/~cslzhang/paper/DnCNN.pdf
22. Zhao, H., Gallo, O., Frosio, I., Kautz, J.: Loss Functions for Image Restoration with Neural Networks. Arxiv XX(X), 1–11 (2015). http://arxiv.org/abs/1511.08861, https://doi.org/10.1109/TCI.2016.2644865

Edge Enhancement for Image Segmentation Using a RKHS Method

Liam Burrows[1], Weihong Guo[2], Ke Chen[1(✉)], and Francesco Torella[3(✉)]

[1] Department of Mathematical Sciences and Centre for Mathematical Imaging Techniques, University of Liverpool, Liverpool, UK
k.chen@liverpool.ac.uk
https://www.liv.ac.uk/~cmchenke
[2] Department of Mathematics, Applied Mathematics and Statistics, Case Western Reserve University, 2049 Martin Luther King Jr. Drive, Cleveland, OH 44106-7058, USA
[3] Department of Radiology, Royal Liverpool and Broadgreen University Hospitals, Liverpool L7 8XP, UK

Abstract. Image segmentation has many important applications, particularly in medical imaging. Often medical images such as CTs have little contrast in them, and segmentation in such cases poses a great challenge to existing models without further user interaction. In this paper we propose an edge enhancement method based on the theory of reproducing kernel Hilbert spaces (RKHS) to model smooth components of an image, while separating the edges using approximated Heaviside functions. By modelling using this decomposition method, the approximated Heaviside function is capable of picking up more details than the usual method of using the image gradient. Further using this as an edge detector in a segmentation model can allow us to pick up a region of interest when low contrast between two objects is present and other models fail.

Keywords: Image segmentation · RKHS · Heaviside function

1 Introduction

Image segmentation is a fundamental problem that has numerous applications in various fields. Variational segmentation models broadly fall into two categories: region based and edge based. Famous region based models include the Mumford-Shah model [8] and the Chan-Vese model [4], the latter being a simplified version of the former. Edge based methods aim to evolve a contour from some initial region towards edges in an image by making use of an edge detector. The model by Kass et al. [7] was among the first of this type, which was further developed by Caselles et al. in the Geodesic active contours model (GAC) [3].

Work supported by UK EPSRC grant EP/N014499/1.

Selective segmentation is a less studied subject and aims to identify a particular objects or objects of interest, and has particular important applications in medical imaging. In order to selectively segment an object, a set of marker points $\mathcal{M}$ are required from the user to indicate the object of interest. This was used by Gout et al. [6] who introduced a distance constraint to the GAC model, encouraging the contour to not evolve too far from the marker set. A more popular method to achieve selective segmentation is to combine elements from both edge based and region based models. Badshah and Chen [1] combined the model by Gout et al. with the Chan-Vese model, encouraging the contour to fit to the intensity of the object, as well as the edges. Rada and Chen [9] built on the Badshah-Chen model by introducing a new term which penalises the evolution of the contour from evolving too far away from the polygon formed by the marker points. Then, Spencer and Chen [11] further improved on the Rada-Chen model, incorporating the Euclidean distance as the constraint, as well as proposing a convex version of the model. The Euclidean distance, while achieving good results, is very parameter sensitive, and largely dependent on the placement of the marker points.

Most recently, Roberts and Chen [10] improved on the Spencer-Chen model, replacing the Euclidean distance with a more intuitive geodesic distance. The geodesic distance proposed by the authors increases when an edge is detected. This increases the robustness of marker placement and makes the parameter selection less sensitive. This model however heavily depends on detecting on edge detection. It is still a challenging problem when the contrast between the object of interest and the rest of the image is low, and when there is dominant noise. The commonly used image gradient information is not effective in detecting weak edges and thus causes leakage in segmentation.

In this work, we present a reproducible Kernel Hilbert space (RKHS) method to enhance certain features of an image, with the aim to make faint edges more prominent. We then aim to utilize their boundary features to segment regions of interest from images with low contrast.

2   RKHS

We employ methods from RKHS in order to detect weak edges where using the image gradient may fail to detect anything significant. This is typically a problem in images with noise and images in which seperate objects are present with little contrast between them. We suppose we can model an image to be composed of two parts: the smooth parts and the edges. Using our knowledge of RKHS, we can use a basis of kernel functions to model the smooth parts of an image, while edges can be represented by a set of Heaviside functions. In this section we will review some key details about Reproducing kernel Hilbert spaces (RKHS), show the kernel we use and lay out the details of the Heaviside function.

Consider an arbitrary set X, with a Hilbert space $\mathcal{H}$ of real functions on X. Let $\mathcal{F}$ be the family of functions $f : X \to \mathbb{R}$. Then $\mathcal{H}$ is an RKHS on X if there exists a symmetric function $K : X \times X \to \mathbb{R}$ such that: (i) for all

$x \in X$, $K \in \mathcal{H}$, (ii) there exists a reproducing relation $f(x) = \langle f, K(\cdot, x) \rangle$ for all $f \in \mathcal{H}$. The function K is called a Kernel function if, for every distinct n points $\{x_1, x_2, ..., x_n\} \subset X$, $K(x_i, x_j) \geq 0$, i.e. K is positive semi-definite function. For every RKHS there is a unique reproducing kernel, and conversely for every positive definite kernel $K : X \times X \to \mathbb{R}$, there is a unique RKHS on X such that K is its reproducing kernel.

Therefore, the choice of kernel can be picked accordingly to which space one wishes to work in. Choosing K to be polynomials of degree d: $K(\mathbf{x}, \mathbf{x}') = (1 + \langle \mathbf{x}, \mathbf{x}' \rangle)^d$ will correspond to a polynomial space. Whereas, if a Sobolev space is sought, we can use $K(\mathbf{x}, \mathbf{x}') = \frac{1}{2}e^{-\gamma|\mathbf{x}-\mathbf{x}'|}$. In addition, kernels such as $K(\mathbf{x}, \mathbf{x}') \propto |\mathbf{x} - \mathbf{x}'|, K(\mathbf{x}, \mathbf{x}') \propto |\mathbf{x} - \mathbf{x}'|^3$ correspond to $1D$ piecewise linear and cubic splines respectively.

For our case use the popular Gaussian kernel corresponding to C^∞ space to model the smooth parts, given by:

$$K(\mathbf{x}, \mathbf{x}') = \left(\frac{1}{\sqrt{2\pi}\sigma}\right)^2 e^{-\frac{|\mathbf{x}-\mathbf{x}'|^2}{2\sigma^2}}.$$

Let $I \in \mathbb{R}^{n \times m}$ be our discretised image intensity. We define the K on a discretised version of the image domain: $[0, 1] \times [0, 1]$ with step size $\frac{1}{n-1}, \frac{1}{m-1}$ respectively, so that $K \in \mathbb{R}^{n \times m}$.

Additionally, we extend the typical one dimensional approximated Heaviside function,

$$\psi(t) = \frac{1}{2} + \frac{1}{\pi} \arctan\left(\frac{t}{\xi}\right),$$

to two dimensions by considering $\mathbf{x} \in \mathbb{R}^2$ and the variation $\psi(\mathbf{v}_i \cdot \mathbf{x} + c_i)$. With $\mathbf{v}_i = (\cos\theta_i, \sin\theta_i)$, this two dimensional Heaviside function can describe an edge with orientation θ_i at position c_i. This will allow us to recover edges from an image at different orientations. We can then consider the edge part of an image to be modelled from a collection of these functions

$$g(\mathbf{x}) = \sum_{i=1}^{k} \beta_i \psi\left(\mathbf{v}_i \cdot \mathbf{x} + c_i\right) \tag{1}$$

where θ_j are equally partitioned into ℓ segments between $[0, 2\pi)$. We choose $\ell = 24$, given by

$$\theta_j \in \{0, \pi/12, 2\pi/12, ..., 23\pi/12\},$$

and c_j, describing the position of each edge, is taken from $c_j \in \{0, \frac{1}{N-1}, \frac{2}{N-1}, ..., 1\}$, where $N = nm$ is the number of pixels in our image and $k = \ell N$. It is useful to know that 1 can be written as $g = \Psi\beta$, where $\Psi \in \mathbb{R}^{\ell \times k}$ and $\beta \in \mathbb{R}^k$.

In 2016, Deng, Guo and Huang [5] used tools from RKHS and the above discussed Heaviside to obtain an image super-resolution model. Suppose L is a given low resolution image and we wish to recover a higher resolution image, H. Their model is given by the following

$$\min_{d,c,\beta} \frac{1}{2}||L - (T^l d + K^l c + \Psi^l \beta)||^2 + \lambda c' K^l c + \alpha||\beta||_1. \tag{2}$$

where the second terms ensures regularisation, the third term ensures sparsity in the edges, and T^l and K^l are matrices representing their choice of kernel and residual respectively. The l superscript denotes that they are basis functions in the low resolution space, and equivalent basis functions T^h, K^h, Ψ^h are used to find the high resolution image $H = T^h d + K^h c + \Psi^h \beta$ after recovering the three coefficients from the model.

We use a similar model to this to recover the smooth parts and edge parts, however since we are using the Gaussian kernel, we have two parts: $Kd + \Psi\beta$.

$$\min_{d,\beta} \frac{1}{2}||z - (Kd + \Psi\beta)||^2 + \lambda d' Kd + \alpha||\beta||_1 + \nu \int_\Omega g(\Psi\beta)|\nabla Kd + \Psi\beta|d\mathbf{x}, \tag{3}$$

where g is some edge detector close to 1 away from edges ($\Psi\beta \approx 0$) and 0 on edges ($\Psi\beta \neq 0$). This last term is the only major difference from (2), the purpose of it is too encourage the contrast to be large on edges.

We solve the model using an ADMM method and perform the computation of small patches of the image. By doing this, details which might otherwise go unnoticed to the human eye when zoomed out may become enhanced.

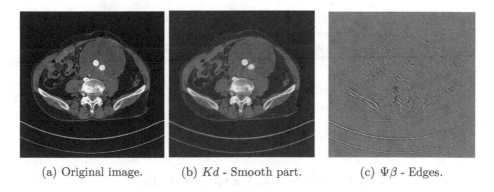

(a) Original image.   (b) Kd - Smooth part.   (c) $\Psi\beta$ - Edges.

Fig. 1. An example of the decomposition of an input image after using model (3).

Figure 1 shows the two parts of an image, Fig. 1b the smooth parts, whereas Fig. 1c is the $\Psi\beta$, representing the edges. This is of key interest as we aim to use this in a segmentation model to improve reliability.

3　Proposed Segmentation Model Using RKHS

In this section we propose that we can use our RKHS + Heaviside representation to achieve an improved segmentation result. In particular, since $\Psi\beta$ denotes the edges, we aim to use it in cases in where typically the image gradient is used in order to detect edges. The image gradient is sensitive to noise and low contrast. Particularly when both noise and regions of low contrast are present, the image gradient struggles to detect the region of low contrast without detecting noise. It is common to use a method of denoising when noise is present, however denoising runs the risk of smoothing out already sensitive edges. The $\Psi\beta$ term aims to improve on this flaw of the image gradient.

We demonstrate the idea using the framework of the Roberts-Chen model [10], but it could be applied to any segmentation model that relies on the gradient. We choose to use the Roberts-Chen framework as it employs a geodesic distance term which increases when an edge is detected. The model is given by minimising the following energy:

$$
\min_{u\in[0,1],c_1,c_2} \int_\Omega g(|\nabla z|)|\nabla u|d\mathbf{x} + \lambda_1 \int_\Omega (z-c_1)^2 u d\mathbf{x}
$$
$$
+ \lambda_2 \int_\Omega (z-c_2)^2(1-u)d\mathbf{x} + \theta \int_\Omega \mathcal{D}_G(\mathbf{x})u d\mathbf{x}. \tag{4}
$$

where $g(s) = \frac{1}{1+\iota s^2}$ is an edge detector, u represents our segmentation result, c_1 and c_2 are the average intensity inside and outside u respectively, and $\mathcal{D}_G$ is the geodesic distance. The authors in [10] define the geodesic distance to be,

$$
\mathcal{D}_G(\mathbf{x}) = \begin{cases} 0, & \mathbf{x}\in\mathcal{M} \\ \frac{\mathcal{D}_G^0(\mathbf{x})}{||\mathcal{D}_G^0(\mathbf{x})||_{L^\infty}}, & \text{otherwise} \end{cases} \tag{5}
$$

where $\mathcal{D}_G^0(\mathbf{x})$ is the solution of the Eikonal equation,

$$
|\nabla \mathcal{D}_G^0(\mathbf{x})| = \epsilon + \beta_G|\nabla z(\mathbf{x})|^2 + \theta_G \mathcal{D}_E(\mathbf{x}), \tag{6}
$$

where $\epsilon = 10^{-3}, \beta_G = 1000$ and $\theta_G = 0.1$ are fixed constants. We use a fast sweeping method proposed by [12] to solve Eikonal equations.

Overall, the model (4) makes use of the image gradient for edge detection in both the g term, and the geodesic distance term, $\mathcal{D}_G$. We propose to replace ∇z with $\Psi\beta$, so that our geodesic distance is now defined as follows:

$$|\nabla \mathcal{D}_G^0(\mathbf{x})| = \epsilon + \beta_G|\Psi\beta| + \theta_G \mathcal{D}_E(\mathbf{x}). \tag{7}$$

In Fig. 2, we see the comparison between the geodesic distance function. In particular, the version using the image gradient fails to penalise the region indicated by the red arrow where there is low contrast and not much of an edge to be detected. The geodesic distance using $\Psi\beta$ shows an improvement, showing a higher penalty to the bottom left region.

In addition to making use of the of the edge detection, we can use the reproduced image from the iterative model as our input in the segmentation. Let $I = Kd + \Psi\beta$, then the segmentation model we use will be given as follows

$$\min_{u\in[0,1],c_1,c_2} \int_\Omega g(|\Psi\beta|)|\nabla u|d\mathbf{x} + \lambda_1 \int_\Omega (I - c_1)^2 u d\mathbf{x}$$

$$+ \lambda_2 \int_\Omega (I - c_2)^2(1-u)d\mathbf{x} + \theta \int_\Omega \mathcal{D}_G(\mathbf{x})u d\mathbf{x} + \alpha \int_\Omega \nu_\epsilon(u)d\mathbf{x}. \tag{8}$$

where $\mathcal{D}_G(\mathbf{x})$ is computed using (7) and (5).

Roberts and Chen find the associated Euler-Lagrange equations to solve their model, using a modified AOS algorithm first introduced in [11] to solve the resulting time dependent PDE. Instead of taking that approach, we use a primal dual method similar to that detailed in [2].

4   Numerical Experiments

In this section we demonstrate how our $\Psi\beta$ used in our segmentation model can be an improvement over using the traditional $|\nabla z|$. We show the result of both to show the difference in performance.

We first show a synthetic image in Fig. 3 which shows two neighbouring ellipses with a slight edge between the two. The ellipse on the left has intensity value of 1, whereas the one on the right has value roughly 0.95. Our aim is to segment the ellipse on the left with the aid of a marker set placed inside, and we can demonstrate the methods robustness to noise by artifically incrementing the noise level. In the presence of no noise the image gradient is capable of selectively segmenting the left circle, however as we artificially increase the noise, the image gradient is unable to detect an edge even at 20% added noise. Figure 3 shows results after adding 20%, 40% and 60% noise. We use a slight preprocess on these images (such as TGV) before segmenting to just reduce the noise level slightly. After smoothing, we find that the image gradient is unable to detect the edge in all cases, whereas $\Psi\beta$ does a good job for 20% and 40%, but begins to struggle at 60%.

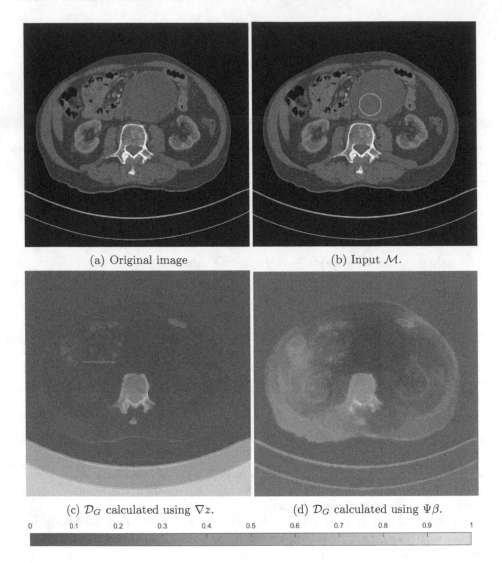

(a) Original image

(b) Input $\mathcal{M}$.

(c) $\mathcal{D}_G$ calculated using ∇z.

(d) $\mathcal{D}_G$ calculated using $\Psi\beta$.

Fig. 2. Geodesic distance comparison. Clearly the unwanted region indicated by the arrow has a higher penalty in (d) than (c). (Color figure online)

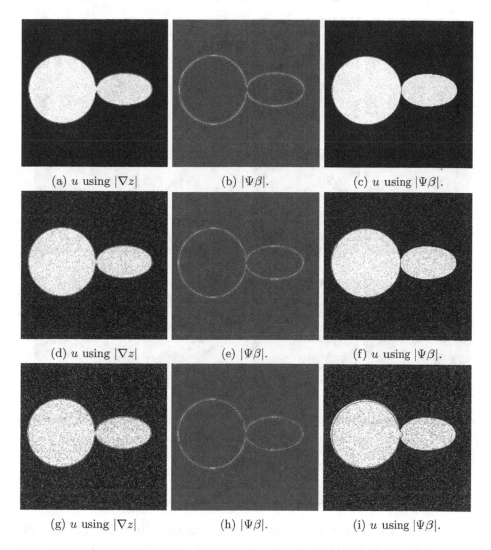

(a) u using $|\nabla z|$ (b) $|\Psi\beta|$.   (c) u using $|\Psi\beta|$.

(d) u using $|\nabla z|$ (e) $|\Psi\beta|$.   (f) u using $|\Psi\beta|$.

(g) u using $|\nabla z|$ (h) $|\Psi\beta|$.   (i) u using $|\Psi\beta|$.

Fig. 3. Illustration of the new RKHS method produces robust solutions in presence of noise. Row one: 20% noise. Row two: 40% noise. Row three: 60% noise.

Figure 4 demonstrates the advantages of using $\Psi\beta$ in a segmentation model on medical images - in particular focusing on abdominal aortic aneurysms. Obtaining an accurate segmentation result is often challenging for this particular problem, as there is often very little contrast between the boundary of the blood vessel and surrounding outside objects. Segmentation leaking is a common occurrence, and as demonstrated in Fig. 4, $\Psi\beta$ is capable of improved results in areas of low contrast.

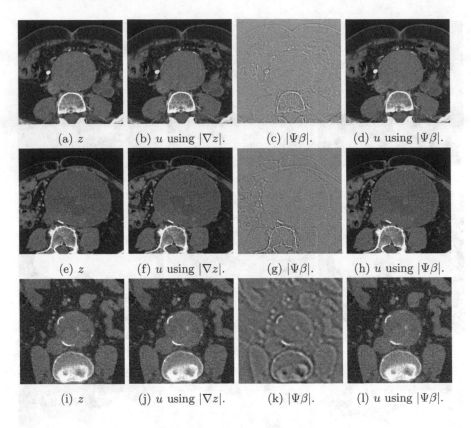

(a) z (b) u using $|\nabla z|$.   (c) $|\Psi\beta|$.   (d) u using $|\Psi\beta|$.

(e) z (f) u using $|\nabla z|$.   (g) $|\Psi\beta|$.   (h) u using $|\Psi\beta|$.

(i) z (j) u using $|\nabla z|$.   (k) $|\Psi\beta|$.   (l) u using $|\Psi\beta|$.

Fig. 4. Examples of low contrast segmentation of abdominal aortic aneurysms. Clearly segmentation using $\Psi\beta$ yields a better result.

5   Conclusion

Enhancing image edges is important in medical imaging. In this paper we have presented an alternative method to using the image gradient is a way of detecting edges in segmentation models. We have demonstrated this is particularly effective in images were low contrast is present between objects. The method involves running a pre-process in order to recover the edge part, $\Psi\beta$, from an image, and then replacing it with the image gradient in a segmentation model. Future work could be extended to more models than just the Roberts-Chen model discussed, as this work could easily be generalised into other models where the image gradient is commonly used by simply replacing it with $\Psi\beta$.

References

1. Badshah, N., Chen, K.: Image selective segmentation under geometrical constraints using an active contour approach. Commun. Comput. Phys. **7**(4), 759 (2010)
2. Bresson, X., Esedoḡlu, S., Vandergheynst, P., Thiran, J.-P., Osher, S.: Fast global minimization of the active contour/snake model. J. Math. Imaging Vis. **28**(2), 151–167 (2007)
3. Caselles, V., Kimmel, R., Sapiro, G.: Geodesic active contours. Int. J. Comput. Vision **22**(1), 61–79 (1997)
4. Chan, T.F., Vese, L.A.: Active contours without edges. IEEE Trans. Image Process. **10**(2), 266–277 (2001)
5. Deng, L.-J., Guo, W., Huang, T.-Z.: Single-image super-resolution via an iterative reproducing kernel hilbert space method. IEEE Trans. Circuits Syst. Video Technol. **26**(11), 2001–2014 (2016)
6. Gout, C., Le Guyader, C., Vese, L.: Segmentation under geometrical conditions using geodesic active contours and interpolation using level set methods. Numer. Algorithms **39**(1–3), 155–173 (2005)
7. Kass, M., Witkin, A., Terzopoulos, D.: Snakes: active contour models. Int. J. Comput. Vis. **1**(4), 321–331 (1988)
8. Mumford, D., Shah, J.: Optimal approximations by piecewise smooth functions and associated variational problems. Commun. Pure Appl. Math. **42**(5), 577–685 (1989)
9. Rada, L., Chen, K.: A new variational model with dual level set functions for selective segmentation. Commun. Comput. Phys. **12**(1), 261–283 (2012)
10. Roberts, M., Chen, K., Irion, K.L.: A convex geodesic selective model for image segmentation. J. Math. Imaging Vis. **61**, 1–22 (2018)
11. Spencer, J., Chen, K.: A convex and selective variational model for image segmentation. Commun. Math. Sci. **13**(6), 1453–1472 (2015)
12. Zhao, H.: A fast sweeping method for eikonal equations. Math. Comput. **74**(250), 603–627 (2005)

A Neural Network Approach for Image Reconstruction from a Single X-Ray Projection

Samiha Rouf[1], Chenyang Shen[2], Yan Cao[1(✉)], Conner Davis[1], Xun Jia[2], and Yifei Lou[1(✉)]

[1] University of Texas at Dallas, Richardson, TX 75080, USA
{yan.cao,Yifei.Lou}@utdallas.edu
[2] University of Texas Southwestern Medical Center, Dallas, TX 75235, USA

Abstract. Time-resolved imaging becomes popular in radiotherapy in that it significantly reduces blurring artifacts in volumetric images reconstructed from a set of 2D X-ray projection data. We aim at developing a neural network (NN) based machine learning algorithm that allows for reconstructing an instantaneous image from a single projection. In our approach, each volumetric image is represented as a deformation of a chosen reference image, in which the deformation is modeled as a linear combination of a few basis functions through principal component analysis (PCA). Based on this PCA deformation model, we train an ensemble of neural networks to find a mapping from a projection image to PCA coefficients. For image reconstruction, we apply the learned mapping on an instantaneous projection image to obtain the PCA coefficients, thus getting a deformation. Then, a volumetric image can be reconstructed by applying the deformation on the reference image. Experimentally, we show promising results on a set of simulated data.

Keywords: Time-resolved volumetric imaging · X-ray projection · Principal component analysis · Neural network

1 Introduction

Radiation therapy or radiotherapy is a type of cancer treatment that uses high-energy beams to destroy cancerous cells. Before the treatment, computerized tomography (CT) scans are taken to obtain anatomical images of a cancer patient so that physicians can determine how to deliver radiation beams to target at certain area. This preparatory process is called treatment planning and CT is often referred to as planning CT (pCT). Afterwards, the patient typically receives radiation sessions every day over several weeks. For each session, cone beam CT

Supported by the National Science Foundation's Enriched Doctoral Training Program, DMS grant #1514808.

Y. Zheng et al. (Eds.): MIUA 2019, CCIS 1065, pp. 208–219, 2020.
https://doi.org/10.1007/978-3-030-39343-4_18

(CBCT) scans are taken for the purpose of positioning the patient in the same way that the pCT is taken in order to match the treatment plan.

Both CT and CBCT aim to reconstruct a volumetric image based on a finite number of X-ray projection images. Although the image quality of CBCT is much worse than that of CT, it results in less radiation dose to patients so that it can be used on a daily basis. One challenging problem for CT/CBCT is respiratory movements that introduce blurring artifacts in the reconstructed images, greatly reducing the accuracy of radiation treatments. In fact, respiratory motion has been a critical issue in lung and abdomen radiotherapy [6].

A typical approach to mitigate motion blur is called *phase binning* [1]. In particular, a breathing signal is captured during the CT/CBCT scan in order to group the projection data into different breathing phases. Then these projection data are used to reconstruct one CT/CBCT image at each phase using some standard reconstruction algorithms such as Feldkamp, Davis, and Kress (FDK) [3]. Since the phase binning process involves temporal direction of a fourth dimension as opposed to 3D volume, the approach is often referred to as 4D-CT [11] and 4D-CBCT [14]. It is true that 4D-CT/4D-CBCT significantly reduces the motion blur, but the image quality is largely degraded by streaking artifacts due to an insufficient number of projections at each phase.

We aim at developing a machine learning algorithm for 4D-CBCT reconstruction from the corresponding projection image, which is so called as instantaneous image reconstruction [2,20]. We want to emphasize how difficult this problem is, as the current 4D-CBCT approach is using a number of projection images, while we only use one. We consider a dimension reduction technique via principal component analysis (PCA) [8,9]. Since breathing is roughly periodic, it is reasonable to assume that movements or deformations between phases have similar structures. Specifically, we represent each volumetric images as a deformation of a chosen reference image. It was reported in [9,21] that three principal components capture the majority of the respiratory motion. As a result, we can model the lung motion as a linear combination of three basis deformations, which successfully reduces a high dimension of volumetric images down to three. Then we consider to train an ensemble of neural networks (NNs) [5,7] to find a mapping from a projection image to PCA coefficients. When it comes to image reconstruction, we can obtain a volumetric image by applying the mapping on an instantaneous projection image.

Our contributions are three-fold. First, we successfully adapt the NN approach for CBCT image reconstruction from a single X-ray projection. Second, we propose a data augmentation technique that does not require a lot of real-patient data. Lastly, the proposed workflow is efficient enough to run on a personal laptop. This work also bears some clinical significance. On one hand, we offer a solution to reconstruct the volumetric image in almost real-time, which facilities other 4D treatment tasks such as tumor tracking. On the other hand, this proposed approach has the potential to adjust the treatment plan according to the in-treatment imaging, which may lead a better practice in radiotherapy.

The rest of the paper is organized as follows. As preliminaries, we provide the clinical feasibility for the proposed work, together with a brief review on PCA model and neural network in Sect. 2. We then detail the proposed approach in Sect. 3 including data augmentation, network structures, and training/testing workflow. The experiments on simulated data are conducted in Sect. 4. Finally, conclusions are given in Sect. 5.

2   Preliminaries

2.1   Clinical Feasibility

With the aforementioned clinical routine in radiotherapy, we have a set of 4D-CT images (pCT) available at the time of treatment planning, from which we can build an NN model for this specific patient. The training process includes data augmentation and training an ensemble of neural networks. During the treatment, the model is instantaneously applied to acquired CBCT projection image in order to obtain a volumetric image. Since pCT and CBCT are taken for the same patient, it is reasonable to assume that the lung motion model is consistent at training and testing stages. Once the NN model is learned, image reconstruction (the testing stage) can be realized in almost real-time.

2.2   PCA Deformation Model

Suppose there are N phases of volumetric images in one breath cycle, which can be obtained by applying 4D-CT reconstruction technique on a set of pCT images. One selects one image as a reference image and performs deformable image registration (DIR) between this reference image and the other images, thus leading to a set of deformation vector fields (DVFs). Specifically for a reference image $R(x)$ and a target image $T(x)$, the goal of DIR is to find a deformation vector field, denoted as $u(x)$, that satisfies $T(x) = R(x + u(x)), \forall x \in \Omega$ where Ω is the image domain. Among a variety of DIR algorithms (refer to a recent survey of [15]), we use Demon's algorithm [4,17] for its popularity and efficiency.

By putting all the DVFs together, we obtain a matrix, denoted as $X \in \mathbb{R}^{|\Omega| \times (N-1)}$, where each column represents the entire DVF for a corresponding phase. Following the convention, we center the DVFs by subtracting the mean and perform principal component analysis (PCA), thus leading to

$$X = \bar{\mathbf{x}} + \sum_{k=1}^{N-1} c_k \mathbf{u}_k, \tag{1}$$

where $\bar{\mathbf{x}}$ is the mean of X, $\{\mathbf{u}_k\}$ is a set of eigenvectors, and $\{c_k\}$ is a set of coefficients. Due to nearly periodic motion of the lungs, it is reasonable to assume that the DVF can be well approximated by only a few eigenvectors/coefficients from the PCA. In fact, it was reported in [9,21] that three largest eigenvectors are sufficient to describe the lung motion. Therefore, we assume that any DVF can be approximated as a weighted sum of three PCA components. We consider neural network to estimate the PCA coefficient, while similar works using the PCA model include 2D/3D registration approach [16] and sparse learning [20].

2.3   Neural Networks

Neural Networks (NN) [5] are information processing paradigms that are built to simulate the way brain nerves process information. Just like human brains, the more information one provides to the NN, the more the brain will "learn". As a result, once new information is passed to the NN, it will process the information based on what it already "learned". Mathematically speaking, the NN acts like a black box that takes the input data I and outputs the information O. In other words, NN aims to learn a mapping such that $O = f(I)$ using training pair of $\{I_j, O_j\}$. There are two challenges in this learning process. One is the requirement of large training data. On the other hand, the function is generally nonlinear and complex. Nowadays, deep learning allows a flexible representation of this nonlinear mapping f using neural networks with several layers, so called as *deep neural network* (DNN) [7]. In the context of CBCT reconstruction, Ma *et al.* [10] considered a multilayered neural network, while Wei *et al.* adopted a convolutional neural network (CNN) [19].

3   Our Method

The PCA deformation model successfully reduces a large dimension of the image to be reconstructed to only three coefficients. The goal of this work is to find a mapping from a projection image together with an angle where the projection is taken to produce three coefficients. Once we have the coefficients, we can obtain the corresponding DVF based on the PCA model (1) and hence a new volumetric image by applying the DVF on the reference image. For this purpose, our method consists of two stages: *training* and *testing*. We will detail the training stage in Sect. 3.1, followed by the testing stage and overall procedure summarized in Sect. 3.2.

3.1   Training

Since we only have limited volumetric images available ($N \sim 10$ from pCT), we need to generate training data on our own to learn the neural network. We elaborate on how to generate training data and how to train an ensemble of neural networks as follows.

Data Augmentation. From N phases of volumetric images in one breathing cycle, we select one image as reference and obtain three eigenvectors corresponding to the largest eigenvalues, as described in Sect. 2.2. We randomly generate a set of 3 coefficients that are deviated from corresponding principal coefficients with standard deviation proportional to their magnitudes. For each set of 3

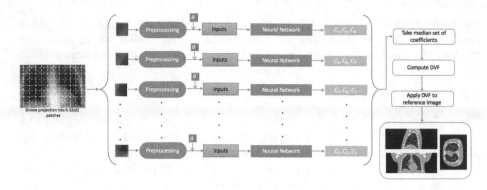

Fig. 1. An ensemble of neural networks, each is learned for a 32 × 32 patch of the X-ray projection. The structure of NN is given in Fig. 2.

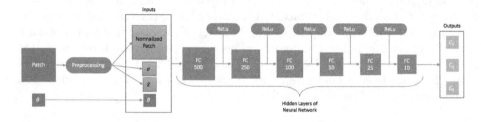

Fig. 2. The structure of a single neural network.

coefficients, we can obtain a DVF using the weighted sum of the principal components, which leads to a volumetric image by applying the DVF to the reference image. Finally, we simulate an X-ray projection corresponding to this volumetric image with a randomly chosen angle by using Siddon's ray tracing algorithm [13]. To summarize, the combination of X-ray projection, projection angle, and 3 coefficients form one training example. This process is repeated to generate as many training examples as we want. The goal for training is to find a mapping from X-ray projection and projection angle to the three coefficients.

Neural Networks. We adopt a projection partitioning technique used in [20] that considered a small patch of the projected data, which was found to be more efficiently (smaller dimension) and more effectively (avoid overfitting) than using the entire X-ray projection. In addition, we find empirically that patches of 32 by 32 pixels offer a good trade-off between information retrieval and computation time. Therefore, we divide the projection into a set of 32 × 32 patches and train a neural network for each patch. As a preprocessing step, a patch is normalized when the standard deviation σ and mean $\bar{x}$ are computed. A normalized patch along with $\sigma, \bar{x}$, and the angle θ of the projection is passed into the network, which outputs a set of three coefficients. Since we have a number of such patches, we obtain an ensemble of neural networks. We take the median of these coefficients to compute the DVF and hence obtain the volumetric image. The workflow of the ensemble is depicted in Figure 1.

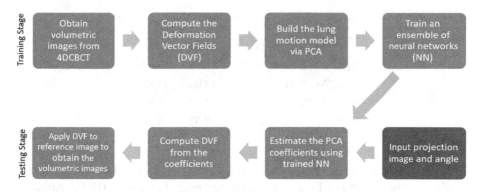

Fig. 3. The flowchart of the proposed approach.

For the network structure, we do not consider the classic convolutional neural network (CNN)[1] [7,19]. Instead, we consider full connected (FC) layer together with a rectified linear unit (ReLu). Specifically for FC, each unit of the previous hidden layer is connected to every unit in next hidden layer, while ReLu is a threshold operation to set any input value that is less than zero to be zero. Each neural network has six FC layers of sizes 500, 250, 100, 50, 25, and 10, each followed by ReLu as activation functions. The output layer has three outputs; one for each principal component. The networks are trained using batch gradient descent with momentum and L^2 regularization. Please refer to Fig. 2 for the network structure and MatConvNet [18] for implementation details.

3.2 Testing

For testing, we have an X-ray projection image together with an angle where it is taken. To be consistent with training, we decompose the projection image into a set of 32×32 patches. After the same preprocessing, each patch together with the angle is passed through the learned NN to yield three PCA coefficients. Using the predicted coefficients, we reconstruct the deformation vector field and apply it to the reference image to get the reconstructed volumetric image. Figure 3 summarizes the overall workflow of the proposed approach including both training and testing stages.

4 Experiment

4.1 Simulated Data Sets

We test our algorithm on a set of simulated data using a non-uniform rational B-spline (NURBS) based cardiactorso (NCAT) phanton [12]. Specifically in this work, we consider two breathing parameters: amplitude (diaphragm motion: 1 cm

---

[1] The projected image does not contain structural information, so CNN does not work very well.

Table 1. The RRE values with different training (column) and testing (row) pairs.

	D1RP3	D1RP5	D3RP3	D3RP5
D1RP3	5.79	4.98	13.90	13.44
D1RP5	6.04	5.18	14.09	13.63
D3RP3	12.31	12.37	7.77	7.02
D3RP5	12.11	12.15	7.61	6.82

and 3 cm) and respiratory period (3 s and 5 s). We denote the data set generated with diaphragm motion of 1 cm and respiratory period of 3 s as D1RP3; similarly we have D1RP5, D3RP3, and D3RP5. For each dataset, we generate 300 dynamic phantoms of size $256 \times 256 \times 100$ and corresponding cone beam projections using the Siddon's algorithm [13] from angles that are uniformly distributed over the angles of 0 to 180 degrees. Each projection image is of size 384×256 corresponding to a physical size of 40×30 cm^2. We use 10 volumetric images from the first period for training and test on the rest 290 images. We can also make two different datasets for training and testing. For example, D1RP3-D3RP5 refers to the case of using D1RP3 to train and testing on D3RP5.

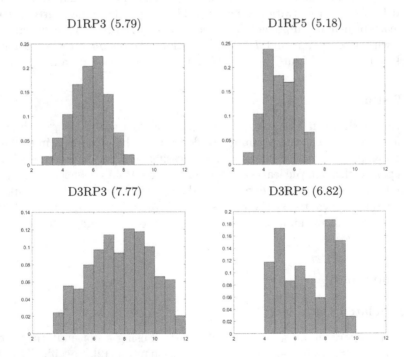

Fig. 4. Histograms of relative reconstruction errors. In each case, mean value is recorded in the parenthesis.

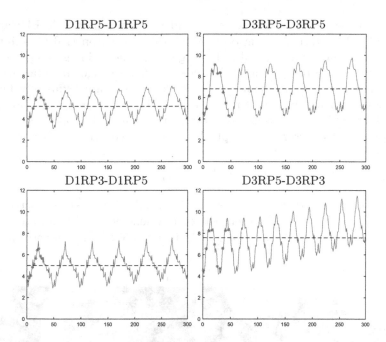

Fig. 5. Relative error of each image. The solid line marks the mean value and ten images used for training are denoted by *.

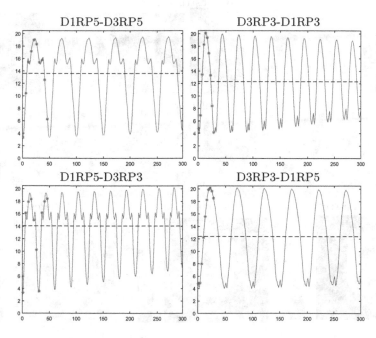

Fig. 6. Relative error of each image. The solid line marks the mean value and ten images used for training are denoted by *.

All the experiments are performed under Macintosh (OS X El Capitan) and MATLAB R2017b running on a Macbook Pro laptop (2.6 GHz Intel(R) Core(TM) i5). On average, it takes 1 s to generate one training image based on our data augmentation technique. We consider 5000 images in total, which is about 83 min. For training the neural networks, it takes approximately 1 h with 50 iterations (epochs). Finally it takes about 0.77 s to test on a single projection image, which achieves real-time processing on a personal laptop.

To quantify the error, we adopt the relative reconstruction error (RRE), defined as

$$\mathrm{RRE}(I, J) = \frac{\|I - J\|_2}{\|I\|_2},$$

where I is the ground-truth (volumetric) image and J is the reconstructed image from our approach. We first present the cases when the training and testing are the same dataset. In Fig. 4, we plot four histograms of RRE, each corresponding

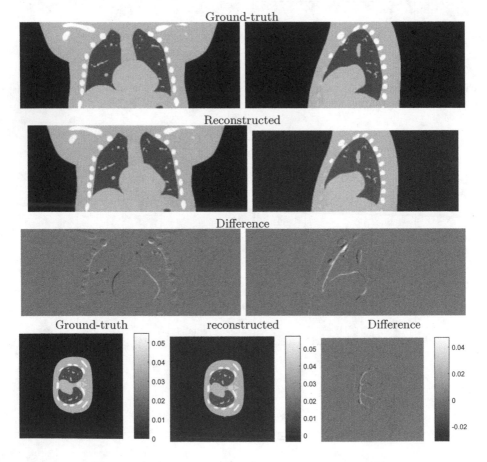

Fig. 7. Image reconstruction results of D1RP3-D1RP3.

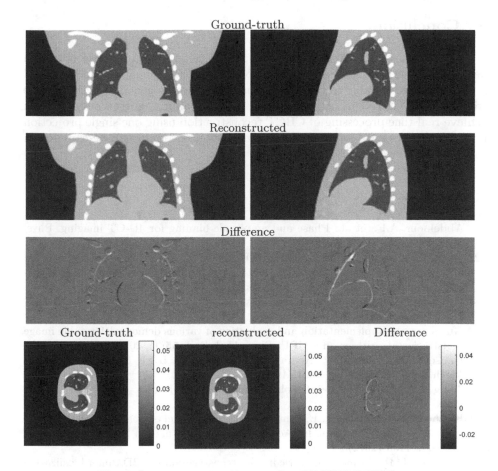

Fig. 8. Image reconstruction results of D3PR5-D1PR3.

to one dataset (training on ten images and testing on the rest 290). Figure 4 reveals that D1PR5 gives the smallest RRE and D3PR3 results in the highest. It is reasonable as quicker respiratory period and larger diaphragm motions tend to produce more motion artifacts, thus difficult to predict.

In Table 1, we provide RRE values with all the combinations of training and testing datasets. It shows that respiration period does not affect too much of the performance. In other word, if we can train on a relatively good level of diagram motion, our approach can be generalized to deal with different breathing periods.

In Figs. 5 and 6, we examine carefully the RRE values of each image in several datasets. Both Figs. 5 and 6 show a rough periodic pattern with error getting bigger when the images is away from the training period. We include the ten images used for training in Figs. 5 and 6, which follows the periodic pattern.

5   Conclusion

With a PCA-based lung motion model, we developed an ensemble of neural networks to reconstruct time-resolved volumetric CBCT images. The proposed method gave reasonable reconstruction error and can be generalized to the case when the testing setting is slightly different to training. Furthermore, we can achieve real-time processing of CBCT reconstruction using one single projection image on a personal laptop. The performance on real data will be explored in the future.

References

1. Abdelnour, A.F., et al.: Phase and amplitude binning for 4D-CT imaging. Phys. Med. Biol. **52**(12), 3515 (2007)
2. Cai, J.F., Jia, X., Gao, H., Jiang, S.B., Shen, Z., Zhao, H.: Cine cone beam CT reconstruction using low-rank matrix factorization: algorithm and a proof-of-principle study. IEEE Trans. Med. Imaging **33**(8), 1581–1591 (2014)
3. Feldkamp, L.A., Davis, L.C., Kress, J.W.: Practical cone-beam algorithm. J. Opt. Soc. Am. A: **1**(6), 612–619 (1984)
4. Gu, X., et al.: Implementation and evaluation of various demons deformable image registration algorithms on a GPU. Phys. Med. Biol. **55**(1), 207–219 (2010)
5. Hansen, L.K., Salamon, P.: Neural network ensembles. IEEE Trans. Pattern Anal. Mach. Intell. **12**(10), 993–1001 (1990)
6. Jiang, S.B.: Radiotherapy of mobile tumors. Semin. Radiat. Oncol. **16**, 239–248 (2006)
7. Krizhevsky, A., Sutskever, I., Hinton, G.E.: Imagenet classification with deep convolutional neural networks. In: Advances Neural Information Processing Systems, pp. 1097–1105 (2012)
8. Li, R., et al.: Real-time volumetric image reconstruction and 3D tumor localization based on a single x-ray projection image for lung cancer radiotherapy. Med. Phys. **37**(6), 2822–2826 (2010)
9. Li, R., et al.: On a PCA-based lung motion model. Phys. Med. Biol. **56**(18), 6009 (2011)
10. Ma, X.F., Fukuhara, M., Takeda, T.: Neural network CT image reconstruction method for small amount of projection data. Nucl. Instrum. Methods Phys. Res. A **449**(1), 366–377 (2000)
11. Pan, T., Lee, T.Y., Rietzel, E., Chen, G.: 4D-CT imaging of a volume influenced by respiratory motion on multi-slice CT. Med. Phys. **31**(2), 333–340 (2004)
12. Segars, W.P., Tsui, B.M., Lalush, D.S., Frey, E.C., King, M.A., Manocha, D.: Development and application of the new dynamic NURBS-based cardiac-torso NCAT phantom. J. Nucl. Med. **42**(5), 23 (2002)
13. Siddon, R.L.: Fast calculation of the exact radiological path for a three-dimensional CT array. Med. Phys. **12**(2), 252–255 (1985)
14. Sonke, J.J., Zijp, L., Remeijer, P., van Herk, M.: Respiratory correlated cone beam CT. Med. Phys. **32**(4), 1176–1186 (2005)
15. Sotiras, A., Davatzikos, C., Paragios, N.: Deformable medical image registration: a survey. IEEE Trans. Med. Imaging **32**(7), 1153–1190 (2013)

16. Staub, D., Docef, A., Brock, R.S., Vaman, C., Murphy, M.J.: 4D Cone-beam CT reconstruction using a motion model based on principal component analysis. Med. Phys. **38**(12), 6697–6709 (2011)

17. Thirion, J.P.: Image matching as a diffusion process: an analogy with Maxwell demons. Med. Image Anal. **2**, 243–260 (1998)

18. Vedaldi, A., Lenc, K.: Matconvnet: convolutional neural networks for matlab. In: ACM International Conference on Multimedia, pp. 689–692 (2015)

19. Wei, R., Zhou, F., Liu, B., Liang, B., Guo, B., Xu, X.: A CNN based volumetric imaging method with single x-ray projection. In: IEEE International Conference on Imaging Systems and Techniques (IST), pp. 1–6 (2017)

20. Xu, Y., et al.: A method for volumetric imaging in radiotherapy using single x-ray projection. Med. Phys. **42**(5), 2498–2509 (2015)

21. Zhang, Q., et al.: A patient-specific respiratory model of anatomical motion for radiation treatment planning. Med. Phys. **34**(12), 4772–4781 (2007)

Deep Vectorization Convolutional Neural Networks for Denoising in Mammogram Using Enhanced Image

Varakorn Kidsumran[1(✉)] and Yalin Zheng[2]

[1] Rajamangala University of Technology Isan, Surin Campus, Surin 32000, Thailand
varakorn.ki@rmuti.ac.th
[2] University of Liverpool, Liverpool L69 3GA, UK
yalin.zheng@liverpool.ac.uk

Abstract. Mammography is an X-ray image of the breast which has been widely used for the management of breast cancer. However, in many cases, it is not easy to identify a sign of cancer as tumour or malignancy due to clouding various noise patterns caused by the low dose radiation from the X-ray machine. Mammogram denoising is an important process to improve the visual quality of mammogram to help the radiologist's diagnosis when they screening mammogram. This paper introduces denoising deep vectorization convolutional neural networks using an enhanced image from direct contrast in a wavelet domain for training. Then, Denoised mammogram is obtained from mapping between the original and enhanced image. Mammogram image from the mini-MIAS database of mammograms was used in this experiment. The experimental results demonstrate that the proposed method can effectively suppress various noises in mammogram both qualitative and subjective test by comparison to traditional denoising methods.

Keywords: Mammography · Deep vectorization convolutional neural networks · Enhanced image · Wavelet domain

1 Introduction

Breast cancer is the most common cancer in the world. It was estimated that there were more than 522,000 deaths from breast cancer in 2012 [1] because the patient was diagnosed in the very late stages. At present, mammography screening is the most effective tool in the early detection stage. However, the misdiagnosis rate is approximately 20% [2] because mammograms from a low dose X-ray machine are corrupted by noise that makes their interpretation very difficult. Thus, ways to achieve robust reductions of noise in mammography has become a very important issue to improve the rate of correct diagnosis and decrease the breast cancer mortality rate.

Digital image processing has been widely used to improve the visual quality of images, and over the past decades, several denoising methods have been developed to reduce noise in mammograms. To adapt discrete scales to fit the size of the abnormal area, such as the size of micro-calcifications, Heinlein *et al.* [3] introduced an integrated

© Springer Nature Switzerland AG 2020
Y. Zheng et al. (Eds.): MIUA 2019, CCIS 1065, pp. 220–227, 2020.
https://doi.org/10.1007/978-3-030-39343-4_19

wavelet based on filter banks derived from continuous wavelet transformation, which has more flexibility to detect breast cancer. Mencattini *et al.* [4] developed a discrete dyadic wavelet transform to reduce variable noise estimated by a local iterative fuzzy method using adaptive thresholding. Elsherif *et al.* [5] introduced a wavelet packet to remove noise and enhance contrast in mammograms, but the effectiveness of this method depends on enhancement parameters. Matsuyama *et al.* [6] modified wavelet coefficients to remove noise in mammograms using hierarchical correlation based on an undecimated wavelet transform. This method is very simple, fast, and provides better visual quality compared to conventional undecimated wavelet transforms [7].

Some work has investigated noise suppression in natural images. Buades *et al.* [8] proposed a nonlocal means algorithm to preserve the structure in a digital image based on analysis of the noise model that is defined by the difference between a digital image and its denoised version. In this method, the visual quality of the image depends on filtering parameters. Dabov *et al.* [9] proposed principal component analysis (PCA) as part of a 3D transform that applied a shape adaptive transform to the input image. Their experimental results showed that this denoising method can preserve the detail of the image, but introduces some artifacts. Ender [10] developed block-matching and 3D filtering (BM3D) for Magnetic Resonance Imaging (MRI). The performance of this method is superior to traditional BM3D model. Gu *et al.* [11] proposed weighted nuclear norm minimization (WNNM) for image denoising by exploiting the image nonlocal self-similarity. The experimental results superior than state of the art denoising method both quantitative measure and visual perception quality. Luisier *et al.* [12] introduced a new Stein's unbiased risk estimator (SURE) by minimizing an estimate of the mean square error between a noisy and clean image. This approach illustrated better denoising performance in peak signal-to-noise ratio (PSNR) compared to the BayerShrink [13] and Bayesian least squares-Gaussian scale mixture (BLS-GSM) [14] methods. Blu *et al.* [15] modified the SURE method by adding a linear combination of the primary denoising process referred to as a linear expansion of thresholds (LET). The results suggested that the SURE-LET scheme led to improved images. Matsuyama *et al.* [16] proposed a SURE-LET image denoising method with directional lapped orthogonal transforms (DirLOTS), which differs from the SURE-LET method, and was used in [9] by adapting hierarchical tree construction of directional lapped orthogonal transforms as a shrinkage function in a wavelet transform to overcome the geometric problem in the SURE-LET method.

In addition, many researchers have investigated improvements to median filtering in denoising files. Wang *et al.* [17] presented the local statistical characteristics based on median filtering to remove noise. This method can be used to preserve edges in an image. Bhateja *et al.* [18] proposed a non- iterative adaptive median filter in which the experimental results were successful in suppressing impulses of high intensity noise. Wu *et al.* [19] improved the median filter by adding a filter function, which demonstrated good detail after filtering. Zhang *et al.* [20] modified the median filter further by designing comfortable direction templates to remove noise in ultrasound images. The advantage of this method is that it preserves edges and provides significant detail.

Recently, deep learning networks have become a role model to reduce noise in images. Burger *et al.* [21, 22] proposed multi layer perception (MLPs) for image denoising. The efficiency of MLPs depends on its architecture and the number of training examples. Jain and Seung [23] denoised a natural image successfully using convolutional neural networks (CNNs). This method demonstrated higher performance compared to the wavelet and Markov random field (MRF) methods. Xie *et al.* [24] introduced stacked sparse autoencoders that combine sparse coding and deep networks pre-trained with a denoising auto-encoder (DA). This method delivered performance comparable to the K-SVD algorithm. Agostinelli *et al.* [25] demonstrated a state of the art denoising method that uses adaptive multi column deep learning networks. This can reduce a variety of different noise types. Gondara [26] developed convolutional denoising autoencoders (CDA) for medical images. The denoising performance of this method produced high quality in objective tests such as structural similarity index (SSIM) but in subjective tests, the denoising quality decreased when the noise level increased. Ren *et al.* [27] proposed vectorization convolutional neural network (VCNN) to improve visuality of the image. The experimental results save time computing and can be applied to a different platform.

However, in the real world, noise in mammogram come from various sources such as quality of X-ray machine, the experience of user, even physical of breast. Then, exist denoising deep neural networks using noise model not suitable for ground truth mammogram. To overcome the limitations of prior work, denoising deep vectorization convolutional neural networks using enhanced image is proposed to robust noise in mammogram. Denoising neural networks can be estimate various noise from enhanced image. Noise free image is obtained by mapping enhanced image and original image. This scheme can decrease specific noise types in mammogram effectively.

The rest of this paper is organized as follows: Section II discusses several existing denoising methods that are to be compared with the proposed method. Section III describes the proposed method. Section IV presents the experimental results of the denoising method compared to state of the art methods, such as, BM3D-MRI, WNNM and VCNN, followed by conclusion in Section V.

2   Background

This section briefly describes state of the art denoising method such as block- matching and 3D filtering (BM3D) [10], weighted nuclear norm minimization (WNNM) [11], and vectorization convolutional neural network (VCNN) [27].

- Block-matching and 3D filtering (BM3D) [10]

The strategy of BM3D based on an image has a locally sparse representation in transform domain. The enhancement of the image is achieved by is grouping similar 2D image patches into 3D groups. Where, collaborative filtering is developed to decrease noise in 3D image groups.

- Weighted nuclear norm minimization (WNNM) [11]

The weighted nuclear norm is proposed to improve the flexibility of nuclear norm and study its minimization. This method is used to image denoising.

- Vectorization convolutional neural network (VCNN) [27]

The VCNN uses vectorized forms to replace convolution operators in deep CNNs which decreases time consuming on training and testing network. The VCNN can be apply in many image processing fields such as recognition, detection, denoise and image deconvolution.

3   The Proposed Method

Figure 1 illustrates the flowchart of proposed method for suppress noise in mammogram which consist of 5 steps are described as follows.

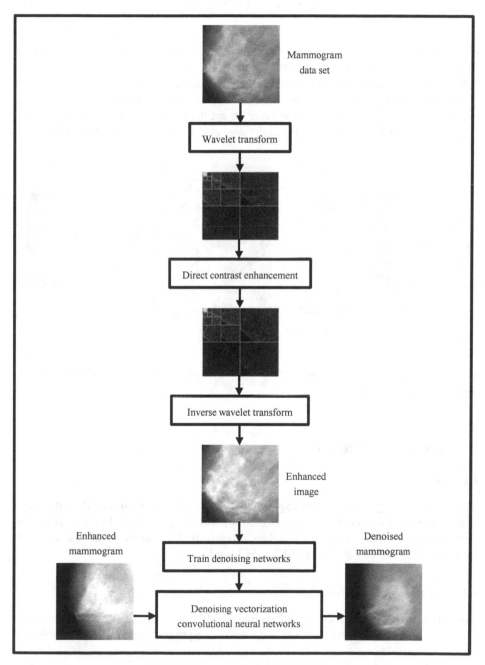

Fig. 1. The proposed method workflow

- The original mammogram is first cropped to 512×512 pixels and decomposed with 4-level Daubechies-4 discrete wavelet transform.
- Direct contrast technique is used to enhance contrast in original mammogram by multiplying the constant value (k) to all detail subbands in wavelet domain.

$$D_{l,global}^{n}(x, y) = k \cdot D_{l,original}^{n}(x, y) \qquad (1)$$

- The enhanced image is obtained from inverse wavelet transform. Then, all detail features in mammogram are boosted including noise.
- The enhanced mammogram is used to train in a training data set using VCNN model [27]. The training network contains 3 vectorization convolutional layers, followed by a ReLU layer for each convolution layer except the last one as shown in Fig. 2. The training network parameters as patch dimension is 64 and learning rate is 0.01.

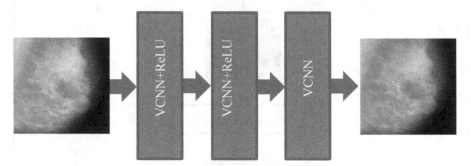

Fig. 2. The architecture of VCNN networks

- After training, denoised mammogram is obtained from mapping enhanced image to original mammogram image.

4   Experimental Results

4.1   Data

Mammogram images used in this experiment are provided by the mini-MIAS database of mammograms that contains 322 images [28]. 161 images are randomly selected for training data and 161 images for test data, repectively. Both training and test data sets are cropped from 1024×1024 to 512×512 pixels size.

4.2   Experimental Results

State of the art denoising method such as BM3D-MRI, WNNM and VCNN are selected to comparison with the proposed method which code programs are online available. To evaluate the performance of the proposed method, peak signal-to-noise ratio (PSNR) was used to measure the quality of the resulting image compared to well-known denoising algorithms.

Figure 3 shows an example denosing result obtained from various denosing methods compared to the original. It can be seen that BM3D-MRI, WNNM and VCNN can not preserve edges and significant detail as shown in the red square on the bottom left. In comparison, the denoising result obtained from the proposed method showed more higher visual quality among other methods.

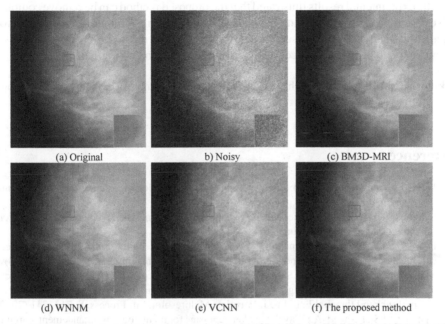

(a) Original	b) Noisy	(c) BM3D-MRI
(d) WNNM	(e) VCNN	(f) The proposed method

Fig. 3. Denoising results (a) Original, (b) Noisy, (c) BM3D-MRI, (d) WNNM, (e) VCNN (f) The proposed method

Table 1 demonstrates that the performance of denoising by comparing with PSNR of BM3D-MRI, WNNM, VCNN and the proposed method. The average PSNR of the proposed method is higher than the other methods. This clearifies that various noises in mammogram are effectively suppressed.

Table 1. Comparison of average PSNR results in different denoising methods.

	NOISY	BM3D-MRI	WNNM	VCNN	The proposed method
PSNR (dB)	20.17	23.01	33.17	34.11	**42.39**

5   Conclusion

Image denoising has been important to improve visual quality of an image. Especially, medical image such as mammogram that widely used to diagnose breast cancer in the early detection stage.

In this study, a denoising deep vectorization convolutional neural networks using enhanced image replace synthetic noisy image. This strategy breakdown the limitation existing denoising deep neural networks that suitable for noise model has been train in training data set.

The experimental results illustrated that the proposed method can be remove complex noise patterns and improved significant detailed features in mammograms, such as microcalcification and malignant tissue. The advantages of this method can help radiologists diagnose breast cancer more accurately during screening mammograms.

Acknowledgments. This research was funded by the Rajamangala University of Technology Isan, Surin Campus.

References

1. WHO   Homepage.   http://www.who.int/cancer/detection/breastcancer/en/index1.html. Accessed 27 June 2018
2. Diagnosisdelayed   Homepage.   http://www.diagnosisdelayed.com/breast-cancer-misdiagnosis,. Accessed 27 June 2018
3. Heinlein, P., Drexl, J., Schneider, W.: Integrated wavelets for enhancement of micro calcifications in digital mammography. IEEE Trans. Med. Imaging **22**(3), 402–413 (2003)
4. Mencattini, A., Rabottino, G., Salmeri, M., Sciunzi, B., Lojacono, R.: Denoising and enhancement of mammmographic images under the assumption of heteroscedastic additive noise by an optimal subband thresholding. Int. J. Wavelets Multiresolut. Inf. Process. **8**, 713–741 (2010)
5. Elsherif, M.S., Elsayad, A.: Wavelet packet denoising for mammogram enhancement. Circuits Syst. **1**, 180–183 (2001)
6. Matsuyama, E., Tsai, D.Y., Lee, Y., Tsurumaki, M.: A modified undecimated discrete wavelet transform based approach to mammographic image denoising. J. Digit. Imaging **26**, 748–758 (2013)
7. Starck, J.L., Fadili, J., Murtagh, F.: The undecimated wavelet decomposition and its reconstruction. IEEE Trans. Image Process. **16**(2), 297–309 (2007)
8. Buades, A., Coll, B., Morel, J.M.: A review of image denoising algorithms, with a new one. Multiscale Model. Simul. **4**, 490–530 (2005)
9. Dabov, K., Foi, A., Katkovnik, V., Egiazarian, K.: Image denoising by sparse 3-D transform-domain collaborative filtering. IEEE Trans. Image Process. **16**, 2080–2095 (2007)
10. Eksioglu, E.M.: Decoupled algorithm for MRI reconstruction using nonlocal block matching model: BM3D-MRI. J. Math. Imaging Vis. **56**, 430–440 (2016)
11. Gu, S., Zhang, L., Zuo, W., Feng, X.: Weighted nuclear norm minimization with application to image denoising. In: 2014 IEEE Conference on Computer Vision and Pattern Recognition, pp. 2862–2869 (2014)
12. Luisier, F., Blu, T., Unser, M.: A new SURE approach to image denoising: interscale orthonormal wavelet thresholding. IEEE Trans. Image Process. **16**, 593–606 (2007)

13. Chang, S.G., Yu, B., Vetterli, M.: Adaptive wavelet thresholding for image denoising and compression. IEEE Trans. Image Process. **9**, 1532–1546 (2000)
14. Portilla, J., Strela, V., Wainwright, M.J., Simoncelli, E.P.: Image denoising using scale mixtures of Gaussians in the wavelet domain. IEEE Trans. Image Process. **12**, 1338–1351 (2003)
15. Blu, T., Luisier, F.: The SURE-LET approach to image denoising. IEEE Trans. Image Process. **16**, 2778–2786 (2007)
16. Matsuyama, E.: SURE-LET image denoising with directional LOTS. In: Picture Coding Symposium, pp. 232–239 (2012)
17. Wang, J., Wang, Y., Li, Y., Liu, J.: Improved median filtering denoising algorithm and analysis. In: International Conference on Information Science and Control Engineering (IET) (2012)
18. Bhateja, V., Rastogi, K., Verma, A., Malhotra, C.: A non-iterative adaptive median filter for image denoising. In: International Conference on Signal Processing and Integrated Networks (SPIN), pp. 113–118 (2014)
19. Wu, S., Chen, H., Xu, X., Long, H., Jiang, W., Xu, D.: An improved median filter algorithm based on VC in image denoising. In: 10th International Conference on Computational Intelligence and Security (CIS), pp. 193–196 (2014)
20. Zhang, X., Cheng, S., Ding, H., Wu, H., Gong, N., Cheng, R.: Ultrasound medical image denoising based on multi-direction median filter. In: 8th International Conference on Information Technology in Medicine and Education (ITME), pp. 835–839 (2016)
21. Burger, H.C., Schuler, C.J., Harmeling, S.: Image denoising with multi-layer perceptrons, part 1: comparison with existing algorithms and with bounds. arXiv:1211.1544 (2012)
22. Burger, H.C., Schuler, C.J., Harmeling, S.: Image denoising with multi-layer perceptrons, part 2: training trade-offs and analysis of their mechanisms. arXiv:1211.1552. (2012)
23. Jain, V., Seung, S.: Natural image denoising with convolutional networks. In: Advances Neural Information Processing Systems, pp. 769–776 (2009)
24. Xie, J., Xu, L., Chen, E.: Image denoising and inpainting with deep neural networks. In: Advances Neural Information Processing Systems, pp. 341–349 (2012)
25. Agostinelli, F., Anderson, M.R., Lee, H.: Adaptive multi-column deep neural networks with application to robust image denoising. In: Advances in Neural Information Processing Systems, vol. 26, pp. 1493–1501 (2013)
26. Gondara, L.: Medical image denoising using convolutional denoising autoencoders. In: IEEE 16th International Conference on Data Mining Workshops (ICDMW), pp. 241–246 (2016)
27. Ren, J., Xu, L.: On vectorization of deep convolutional neural networks for vision tasks. In: AAAI, pp. 1840–1846 (2015)
28. Suckling, J., et al.: The mammographic image analysis society digital mammogram exerpta media. In: International Congress Sersis, vol. 1069, pp. 375–378 (1994)

Diagnosis, Classification and Treatment

A Hybrid Machine Learning Approach Using *LBP* Descriptor and *PCA* for Age-Related Macular Degeneration Classification in OCTA Images

Abdullah Alfahaid[1,2,3]($\boxtimes$), Tim Morris[1], Tim Cootes[4], Pearse A. Keane[2], Hagar Khalid[2], Nikolas Pontikos[2,5], Panagiotis Sergouniotis[6], and Konstantinos Balaskas[2]

[1] School of Computer Science, The University of Manchester, Oxford Road, Manchester M13 9PL, UK
abdullah.alfahaid@manchester.ac.uk
[2] Moorfields Eye Hospital, 162 City Road, London EC1V 2PD, UK
[3] College of Computer Science and Engineering at Yanbu, Taibah University, Medina, Kingdom of Saudi Arabia
[4] Centre for Imaging Sciences, The University of Manchester, Oxford Road, Manchester M13 9PL, UK
[5] UCL Genetics Institute, University College London, Gower Street, London WC1E 6BT, UK
[6] School of Biological Sciences, The University of Manchester, Oxford Road, Manchester M13 9PL, UK

Abstract. We propose a novel hybrid machine learning approach for age-related macular degeneration (AMD) classification to support the automated analysis of images captured by optical coherence tomography angiography (OCTA). The algorithm uses a Rotation Invariant Uniform Local Binary Patterns (*LBP*) descriptor to capture local texture patterns associated with AMD and Principal Component Analysis (*PCA*) to decorrelate texture features. The analysis is performed on the entire image without targeting any particular area. The study focuses on four distinct groups, namely, healthy; neovascular AMD (an advanced stage of AMD associated with choroidal neovascularisation (CNV)); non-neovascular AMD (AMD without the presence of CNV) and secondary CNV (CNV due to retinal pathology other than AMD). Validation sets were created using a Stratified K-Folds Cross-Validation strategy for limiting the overfitting problem. The overall performance was estimated based on the area under the Receiver Operating Characteristic (ROC) curve (AUC). The classification was conducted as a binary classification problem. The best performance achieved with the SVM classifier based on the AUC score for: (i) healthy vs neovascular AMD was 100%, (ii) neovascular AMD vs non-neovascular AMD was 85%; (iii) CNV (neovascular AMD plus secondary CNV) vs non-neovascular AMD was 83%.

Keywords: Optical coherence tomography angiography (OCTA) · Age-related macular degeneration (AMD) · Texture features

© Springer Nature Switzerland AG 2020
Y. Zheng et al. (Eds.): MIUA 2019, CCIS 1065, pp. 231–241, 2020.
https://doi.org/10.1007/978-3-030-39343-4_20

1   Motivation

Age-related macular degeneration (AMD) is a heterogeneous, multifactorial retinal condition and a leading cause of visual impairment in the elderly population [1,2]. AMD predominantly affects the macula, the central part of the retina and it is clinically categorised into non-neovascular (or dry) AMD and neovascular (or wet) AMD [3]. The hallmark of dry AMD is drusen, focal deposits of extracellular debris located under the retina and the retinal pigment epithelium (RPE); retinal pigment epithelial abnormalities including atrophy are common. Wet AMD is characterised by the presence of a common, vision-threatening complication of AMD called choroidal neovascularisation (CNV); this involves the growth of abnormal blood vessels typically originating from the choroid (a layer of tissue located underneath the retina and RPE) and involving the macular area [3]. Dry AMD is the more common subtype and it is associated with gradual visual loss whereas wet AMD is linked to a more acute presentation [3]. Notably, wet AMD can be successfully treated with intravitreal injection. Early detection and management are key and timely diagnosis is linked to improved outcomes [4]. Significant effort and healthcare resources are therefore put to the early identification of CNV and to the differentiation between individuals with wet and dry forms of AMD.

Different medical imaging modalities have been developed to help with this task. A promising recently introduced technique is Optical Coherence Tomography Angiography (OCTA) which combines dye-free angiography and non-invasive volumetric three-dimensional imaging. This modality has advantages over the widely used Optical Coherence Tomography (OCT) as it enables detailed visualisation of the retinal and choroidal circulation. Furthermore, OCTA is fast and non-invasive unlike other established modalities, such as Fundus Fluorescein Angiography (FFA) and Indocyanine Green Angiography (ICG) [5–7]. Importantly, OCTA enables characterisation of moving and static elements of retinal and choroidal blood flow and it allows visualisation of CNV and other abnormalities that can help distinguish between dry and wet AMD.

OCTA produces clear images of the retinal vasculature in different retinal layers including the superficial inner retina, the deep inner retina, the outer retina and the choriocapillaris layers. The current clinical standard for detecting CNV and evaluating the efficacy of the treatments for wet AMD involves visually examining the textural appearance of images from each of these layers. However, this is not a trivial task given the significant amount of data in each OCTA scan, the pattern variations between individuals, and the fact that neovascular and non-neovascular areas may appear similar [8]. It is therefore not uncommon for clinicians to request a second opinion due to the difficulties involved in the interpretation process. Figure 1 demonstrates the texture appearance of the retinal vasculature in the various retinal layers in the OCTA images for different eye conditions. Images from eyes with no pathology, dry AMD and wet AMD are shown. The complexity of the blood vessels pattern variations between the different retinal layers can be appreciated.

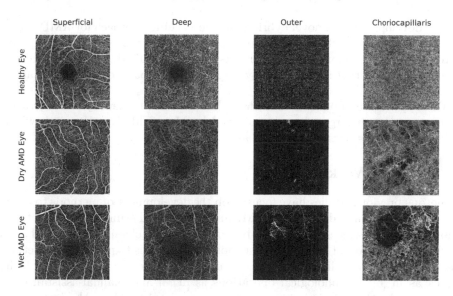

Fig. 1. The textural appearance of blood vessels network in the superficial inner retina, deep inner retina, outer retina and choriocapillaris layers in OCTA images. Each row illustrates a different eye condition from the various layers. The first shows a healthy eye, the second shows a dry AMD eye and the final row shows a wet AMD eye. It can be observed how the similarities appear in the patterns of the abnormalities in all layers for the dry and wet AMD eyes, while in some layers the patterns appear very similar, even in the healthy eye, namely the superficial inner and deep inner layers.

As seen in the previous figure, the texture of OCTA images is affected by AMD. Image texture is rich with very important information describing complex visual patterns that can be distinguished by colour, brightness, size or shape [9]. However, there is evidence to demonstrate that it is problematic for the human eye to recognise textural information which is related to higher-order statistics or to the spectral properties of an image [10]. Therefore, both quantifying the texture characteristics of OCTA images and building a predictive image classification algorithm that is capable of detecting the early stages of AMD is desirable. This could reduce the burden on ophthalmologists, remove the subjectivity due to personal interpretation and ensure a greater efficiency and reliability in the diagnosis process in daily clinical practice.

Image classification is an important component of computer-assisted medical diagnostic tools. Apart from an algorithm (based on Rotation Invariant Uniform Local Binary Patterns (*LBP*) as a texture descriptor) that was previously explored by our group [11], to the best of our knowledge, there has been no prior image classification work on AMD using OCTA images. The main contributions of the work undertaken are:

- The construction of new measurements that contribute the most to quantifying AMD presentation in OCTA images.

- The development of a novel hybrid machine learning approach for AMD classi-fication with less redundant and misrepresentative texture features compared to our previous algorithm [11] as it has the additional advantageous capacity to decorrelate texture features by applying Principal Component Analysis (PCA).
- The application of our previous algorithm [11] and the new hybrid algorithm to a new much larger dataset provided by Moorfields Eye Hospital, which includes early stages of AMD disease.

2   Related Work

Numerous previous studies have focused on the development of methods to auto-mate the analysis and detection of AMD in OCT or OCTA images. Most of these follow either an image segmentation-based or an image classification-based app-roach. The objective of image segmentation approaches is to partition the retinal vascular texture into disjunct regions. This includes the use of smoothing tech-niques as in [6], morphological operations as in [8] and manual-assistance by tracing the borders of the regions of interest as in [12]. Then, the images are labelled as healthy or AMD depending on either some measurements performed over the segmented regions or the visibility of the object of interest. This is in contrast with image classification, where the goal is to classify an unknown image into one of the pre-defined classes based on features derived from the image tex-ture using machine learning and pattern recognition techniques. Many ways of deriving the features have been used, including handcrafted texture descriptors as in [13–15] or potentially the features learned using deep learning technologies as in [16–18].

When assessing image segmentation-based methods as clinical diagnostic tools for AMD detection in OCTA images, there are several limitations. Impor-tantly, such approaches require an adequate image quality so that the abnormal blood vessel patterns are clearly visible; alternatively, the abnormalities can be difficult to segment. This problem is amplified by the fact that the measure-ments are likely to be derived from a deformed image texture structure due to the inclusion of pre-processing steps. These steps make use of morphological or smoothing operations as has occurred in [6] and [8] when part of the CNV, a key indicator of the presence of AMD, was excluded. Furthermore, the measurements may be influenced by human error/bias and often take considerable time when manual assistance is involved, as in [12]. To overcome the challenges associated with segmentation, an alternative path is to extract features from the whole image and use these features to build an image classification-based method.

While there are several image classification-based methods proposed, the vast majority were designed to be used on images produced from OCT rather than OCTA scans. However, OCT is not designed to produce images of the retinal vasculature and may fail to visualise/detect the abnormalities. What is more, handcrafted texture descriptor-based methods proposed in [13–15] are sensitive to noise such as image illumination variations and also include complex opera-tions and pre-processing steps to tackle image noise. However, that may change image details. On the other hand, it may be argued that deep learning-based methods using OCT images (including [16–18]) have greatly enhanced AMD detection performance despite the fact that a significant amount of training data was necessary to ensure that robust feature representations could be learned.

OCTA is an emerging imaging technique that enables visualisation of the retinal vasculature texture with an unprecedented level of detail. Given the recent introduction of robust OCTA imaging technologies only a limited amount of labelled training data is presently available. This and the complexity of OCTA images are important current limitations to the development of deep learning-based methods. Other issues with these methods include high computational complexity and memory requirements [19]. Moreover, they also present problems with the interpretation of outcomes due to the fact that the theoretical foundation is not well understood and the results are empirical [19].

3   Proposed Approach

The hybrid algorithm discussed in this paper follows the same pipeline of a previous algorithm reported by our group [11] but with an additional step of feature dimensionality reduction. Briefly, the new hybrid algorithm consists of three main steps. The first step is the texture feature extraction using the *LBP* descriptor to characterise all relevant variations in image texture patterns induced by AMD from the whole image. Subsequently, the feature dimensionality reduction step applies the *PCA*, which decorrelates the extracted features. Finally, there is a classification step, where the images are classified based on the new features represented by the *PCA*. Figure 2 shows a brief overview of the new hybrid algorithm pipeline for AMD classification.

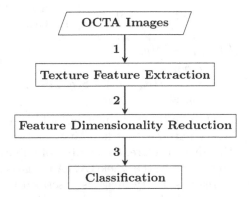

Fig. 2. Overview of the new analysis procedure for the hybrid classification algorithm. It begins with taking the OCTA images as an input, followed by feature extraction and dimensionality reduction respectively, and finally the classification.

3.1   Texture Feature Extraction

The study makes use of Rotation Invariant Uniform LBP, a handcrafted texture descriptor introduced by Ojala et al. [20]. Although there are several texture descriptors proposed in the literature, the choice of which one to use depends on the issues associated with the image texture to be measured and on how well the descriptor can cope with these issues [21]. Examples of common issues are the variations caused by rotation and illumination. Notably, OCTA image texture is affected by these changes [22]. Although the subjects' eyes are not purposefully rotated, there may be some orientation changes to the texture as the central, avascular region is orbited. Consequently, the main motivation of using the LBP in this study is that due to its various advantages (including ease of implementation) it can work effectively under limited resources. Also, it is capable of quantifying AMD in OCTA images while preserving the trade-off between two fundamental goals: (i) to provide a high-quality description with a balance between distinctiveness and robustness against the illumination and rotation changes; (ii) to have the lowest level of computational complexity.

To accurately measure the image texture using the LBP, the values of two important parameters need to be set up properly. The first parameter is the number of neighbouring points p spread on a circle and the second parameter is the radius r of the circle, which defines the length from the central point g_c to the neighbouring points g_n. The measurements are then derived by comparing a neighbouring point's g_n value, where $n = (0, 1, 2, 3, \ldots, p - 1)$, against the central point g_c value generating a binary pattern. The LBP values for each pixel within each image are constructed according to the following equations:

$$LBP = \begin{cases} \sum_{n=0}^{n=p-1} S(g_n - g_c) & \text{if } u(LBP_{p,r}) \leq 2 \\ p+1 & \text{Otherwise} \end{cases} \qquad (1)$$

$$\text{Where} \qquad S(x) = \begin{cases} 1 & \text{if } x \geq 0 \\ 0 & \text{Otherwise} \end{cases}$$

$$u(LBP_{p,r}) = |S(g_{p-1} - g_c) - S(g_0 - g_c)| \\ + \sum_{n=1}^{n=p-1} S(g_n - g_c) - S(g_{n-1} - g_c)| \qquad (2)$$

The $u(LBP_{p,r})$ is a procedure to count the number of bitwise transitions and to consider the uniform binary patterns that have at most two transitions, 1/0 or 0/1. When measuring the image texture, the number of uniform binary patterns that can occur is $p+1$. In this study, the whole OCTA image is processed without targeting any particular regions. Every image is described by a histogram with $p + 2$ bins that calculates the number of occurrences of the uniform LBP values within each image, while the supplementary bin in the histogram is to calculate the non-uniform binary patterns that occurred. Then, the generated histogram will form the feature vector that represents each image.

3.2   Feature Dimensionality Reduction

In image classification problems, the high dimensional and highly correlated aspects of the feature vectors have a critical impact on the performance of the machine learning algorithm used for conducting the classification. Techniques that can significantly overcome these issues in an interpretable fashion while maintaining most of the important information in the feature space are desirable. One of the most popular and commonly utilised techniques for this task is Principal Component Analysis (*PCA*) [23]. *PCA* is a statistical technique that uses a linear transformation to convert a high number of correlated features into a lower number of linearly uncorrelated features, named the principal components that successively maximise variance [23]. The use of *PCA* in our new algorithm was mainly motivated by the fact that when we increase the number of points around a particular circle (when calculating the *LBP* values), the dimensionality of the feature vectors that represent each OCTA image also increases. Consequently, this is likely to increase the chance of having correlated and therefore redundant features; hence, *PCA* was applied, as it provides the following advantages:

- It makes the method less biased since it eliminates redundant texture features;
- It improves the accuracy of the method as it reduces the occurrence of mis-representative texture features;
- It reduces the time taken for training the machine learning algorithm since it makes use of feature vectors of lower dimensionality.

In this step, the original dimensional feature vectors obtained from the OCTA images were reduced into lower dimensional feature vectors. These retained 95% of the variance which is a common percentage widely used when applying *PCA*.

3.3   Classification

Following the feature dimensionality reduction step, the newly constructed features are passed to a classifier for classification. Two different machine learning algorithms were tested in this work, namely K-Nearest Neighbour (KNN) and Support Vector Machine (SVM). The kernel type used for the SVM classifier is a linear kernel. The value of K neighbours for the KNN classifier was empirically set to one similar to our previous algorithm [11].

4   Evaluation

The hybrid algorithm described here and our previous algorithm [11] were evaluated based on their ability to distinguish between the various classes of images provided by Manchester Royal Eye Hospital and Moorfields Eye Hospital. The Manchester dataset included 23 healthy and 23 wet AMD samples. The Moorfields dataset included 166 wet AMD and 79 dry AMD cases; 25 secondary CNV cases were also included. In these secondary CNV samples the neovascularisation was due to causes other than AMD. Both datasets include four different images of each eye captured from four retinal layers, namely the superficial inner retinal layer, the deep inner retinal layer, the outer retinal layer and choriocapillaris layer. Table 1 summarises the number of images used. The study makes use of two-dimensional angiogram greyscale images captured from a 3×3 mm field of view and utilises all the pre-segmented images through the retinal layers by the default setting of the OCTA scan. This is because we wanted to avoid the additional complexity of manually segmenting the images. The addition of such a step would make the algorithm less user-friendly and would probably introduce bias.

Table 1. Summary for the number of images used in this study.

Hospital	Classes	Retinal layers				All layers
		Choriocapillaris	Outer	Deep	Superficial	
Manchester	healthy	23	23	23	23	92
	wet AMD	23	23	23	23	92
Moorfields	wet AMD	166	166	166	166	664
	non-CNV (dry AMD)	79	79	79	79	316
	secondary CNV	25	25	25	25	100

4.1   Evaluation Setup and Criteria

The classification was performed on each separate layer and in all layers combined as a binary classification problem. The motivations for performing the classification this way (on each separate layer) are to identify the predictive layer that has most information describing the abnormalities, and (in all layers combined) to investigate how well the algorithms operate on classifying the various layers at once by throwing all layers together. The classification was conducted as follows:

 I healthy vs wet AMD for the Manchester dataset;
 II wet AMD vs dry AMD for the Moorfields dataset;
III CNV (wet AMD plus secondary CNV) vs non-CNV (dry AMD) for the Moorfields dataset.

As the Moorfields dataset is imbalanced and all classes are important for us to detect, the following evaluation strategies for both algorithms were conducted on both datasets:

- Use the Stratified K-Folds Cross-Validation strategy to split the data into training and testing sets creating stratified folds; this means each fold is created by preserving the number of samples of each class. This is to ensure a consistent predictive performance for both algorithms and limitation of the overfitting problem.

- Use the Receiver Operating Characteristic (ROC) curve and compute the area under the curve (AUC). This is to provide equal weight for both classes in our binary classification problem.

These would give an accurate measure of insight into overall performance as well as ensuring enhanced validation for both algorithms.

4.2   Example Results

The performance of both algorithms was compared based on their AUC scores. The results are obtained by empirically choosing the same values of *LBP* parameters p and r that were used in our previous algorithm [11]. Two classifiers were tested on both approaches and the following two tables provide a comparison between their results. Table 2 provides the results of the SVM classifier and Table 3 shows the results of the KNN classifier. In both tables, the new hybrid algorithm is denoted as "Hybrid algorithm" while our previous algorithm [11] is denoted as "Previous algorithm".

Table 2. Classification results using both approaches with SVM classifier.

Algorithm	Binary classification	Retinal layers				All layers
		Choriocapillaris	Outer	Deep	Superficial	
Hybrid algorithm	healthy vs wet AMD	100% ± 0	99% ± 1	98% ± 3	95% ± 5	96% ± 2
	wet AMD vs dry AMD	83% ± 2	**85% ± 3**	80% ± 4	75% ± 3	78% ± 4
	CNV vs non-CNV	81% ± 3	**83% ± 1**	76% ± 2	69% ± 4	76% ± 2
Previous algorithm	healthy vs wet AMD	100% ± 0	96% ± 1	96% ± 3	91% ± 3	92% ± 3
	wet AMD vs dry AMD	81% ± 2	**83% ± 3**	79% ± 4	71% ± 2	75% ± 3
	CNV vs non-CNV	80% ± 3	**82% ± 3**	72% ± 4	67% ± 5	74% ± 3

Table 3. Classification results using both approaches with KNN classifier.

Algorithm	Binary classification	Retinal layers				All layers
		Choriocapillaris	Outer	Deep	Superficial	
Hybrid algorithm	healthy vs wet AMD	100% ± 0	99% ± 1	96% ± 2	93% ± 5	96% ± 3
	wet AMD vs dry AMD	81% ± 3	**84% ± 4**	75% ± 5	73% ± 2	75% ± 4
	CNV vs non-CNV	80% ± 4	**81% ± 5**	73% ± 3	68% ± 4	73% ± 2
Previous algorithm	healthy vs wet AMD	100% ± 0	98% ± 2	95% ± 4	90% ± 4	90% ± 2
	wet AMD vs dry AMD	80% ± 1	78% ± 4	71% ± 4	70% ± 3	71% ± 2
	CNV vs non-CNV	79% ± 4	**79% ± 3**	70% ± 5	66% ± 3	70% ± 1

The construction of discriminative features has a critical role in the performance of machine learning algorithms. This is due to the fact that when using misleading or highly correlated texture features, even with the use of the most sophisticated classifiers, attaining the desired performance level will not be possible.

The preliminary results in Tables 2 and 3 show that the new hybrid algorithm is robust against image noise and quality due to patient motion or illumination changes as compared to our previously reported algorithm [11]. This was confirmed by challenging two different classifiers, namely the SVM and KNN. Furthermore, the SVM classifier generally performs better than the KNN classifier on both algorithms. Moreover, the initial results of our classification algorithms show that the most predictable layers are the choriocapillaris and outer retinal layers.

5   Conclusion and Future Work

This paper reports a hybrid algorithm for AMD classification in OCTA images by combining the *LBP* descriptor with *PCA*. The *LBP* is responsible for capturing the local textural features from the OCTA images while the *PCA* is applied for eliminating the misrepresentative texture features. The algorithm is capable of capturing all related image variations induced by AMD as analysis is performed on the entire image. The results achieved so far have suggested that the proposed hybrid algorithm may be clinically useful for AMD classification in OCTA images. Deep learning methods might have superior performance, but the size of the dataset currently available is too small to investigate them.

Future work exploring the performance of deep learning methods in similar tasks would be of interest. To enable this, the collection of carefully curated data will be required. Data augmentation techniques to increase the number of images are generally inappropriate, since in this case they will distort the data in undesirable ways. Testing the presented and other algorithms on more complex tasks would also be of interest. Specifically, it would be clinically valuable to be able to distinguish variations in AMD in the same patient, namely active CNV (which requires treatment) from inactive CNV (which can be observed).

References

1. Colijn, J.M., et al.: Prevalence of age-related macular degeneration in Europe: the past and the future. Ophthalmology **124**, 1753–1763 (2017)
2. Bourne, R.R., et al.: Prevalence and causes of vision loss in high-income countries and in Eastern and Central Europe: 1990–2010. Br. J. Ophthalmol. **98**, 629–638 (2014)
3. Mitchell, P., Liew, G., Gopinath, B., Wong, T.Y.: Age-related macular degeneration. Lancet **392**, 1147–1159 (2018)
4. Mehta, H., et al.: Real-world outcomes in patients with neovascular age-related macular degeneration treated with intravitreal vascular endothelial growth factor inhibitors. Prog. Retin. Eye Res. **65**, 127–146 (2018)
5. Jia, Y., et al.: Quantitative optical coherence tomography angiography of vascular abnormalities in the living human eye. In: Proceedings of the National Academy of Sciences (2015). https://doi.org/10.1073/pnas.1500185112
6. Jia, Y., et al.: Quantitative optical coherence tomography angiography of choroidal neovascularization in age-related macular degeneration. Ophthalmology **121**(7), 1435–1444 (2014)

7. De Carlo, T.E., Romano, A., Waheed, N.K., Duker, J.S.: A review of optical coherence tomography angiography (OCTA). Int. J. Retin. Vitr. **1**(1), 5 (2015)
8. Liu, L., Gao, S.S., Bailey, S.T., Huang, D., Li, D., Jia, Y.: Automated choroidal neovascularization detection algorithm for optical coherence tomography angiography. Biomed. Opt. Express **6**(9), 3564–3576 (2015)
9. Simon, P., Uma, V.: Review of texture descriptors for texture classification. In: Satapathy, S.C., Bhateja, V., Raju, K.S., Janakiramaiah, B. (eds.) Data Engineering and Intelligent Computing. AISC, vol. 542, pp. 159–176. Springer, Singapore (2018). https://doi.org/10.1007/978-981-10-3223-3_15
10. Tourassi, G.D.: Journey toward computer-aided diagnosis: role of image texture analysis. Radiology **213**(2), 317–320 (1999)
11. Alfahaid, A., Morris, T.: An automated age-related macular degeneration classification based on local texture features in optical coherence tomography angiography. In: Nixon, M., Mahmoodi, S., Zwiggelaar, R. (eds.) MIUA 2018. CCIS, vol. 894, pp. 189–200. Springer, Cham (2018). https://doi.org/10.1007/978-3-319-95921-4_19
12. Talisa, E., et al.: Spectral-domain optical coherence tomography angiography of choroidal neovascularization. Ophthalmology **122**(6), 1228–1238 (2015)
13. Srinivasan, P.P., et al.: Fully automated detection of diabetic macular edema and dry age-related macular degeneration from optical coherence tomography images. Biomed. Opt. Express **5**(10), 3568–3577 (2014)
14. Wang, Y., Zhang, Y., Yao, Z., Zhao, R., Zhou, F.: Machine learning based detection of age-related macular degeneration (AMD) and diabetic macular edema (DME) from optical coherence tomography (OCT) images. Biomed. Opt. Express **7**(12), 4928–4940 (2016)
15. Sun, Y., Li, S., Sun, Z.: Fully automated macular pathology detection in retina optical coherence tomography images using sparse coding and dictionary learning. J. Biomed. Opt. **22**(1), 016012 (2017)
16. Treder, M., Lauermann, J.L., Eter, N.: Automated detection of exudative age-related macular degeneration in spectral domain optical coherence tomography using deep learning. Graef. Arch. Clin. Exp. Ophthalmol. **256**(2), 259–265 (2018)
17. Schlegl, T., et al.: Fully automated detection and quantification of macular fluid in OCT using deep learning. Ophthalmology **125**(4), 549–558 (2018)
18. Lee, C.S., Baughman, D.M., Lee, A.Y.: Deep learning is effective for classifying normal versus age-related macular degeneration OCT images. Ophthalmol. Retin. **1**(4), 322–327 (2017)
19. Guo, Y., Liu, Y., Oerlemans, A., Lao, S., Wu, S., Lew, M.S.: Deep learning for visual understanding: a review. Neurocomputing **187**, 27–48 (2016)
20. Ojala, T., Pietikainen, M., Maenpaa, T.: Multiresolution gray-scale and rotation invariant texture classification with local binary patterns. IEEE Trans. Pattern Anal. Mach. Intell. **24**(7), 971–987 (2002)
21. Ojala, T., Pietikainen, M.: Texture classification (2001). http://homepages.inf.ed.ac.uk/rbf/CVonline/LOCAL_COPIES/OJALA1/texclas.htm. Accessed 01 Jan 2019
22. Spaide, R.F., Fujimoto, J.G., Waheed, N.K.: Image artifacts in optical coherence angiography. Retina (Philadelphia, Pa.) **35**(11), 2163 (2015)
23. Jolliffe, I.: Principal component analysis. In: Lovric, M. (ed.) International Encyclopedia of Statistical Science, pp. 1094–1096. Springer, Heidelberg (2011). https://doi.org/10.1007/978-3-642-04898-2

Combining Fine- and Coarse-Grained Classifiers for Diabetic Retinopathy Detection

Muhammad Naseer Bajwa[1,2](✉) , Yoshinobu Taniguchi[3] ,
Muhammad Imran Malik[4], Wolfgang Neumeier[5], Andreas Dengel[1,2],
and Sheraz Ahmed[2]

[1] Technische Universität Kaiserslautern, 67663 Kaiserslautern, Germany
[2] German Research Center for Artificial Intelligence GmbH (DFKI),
67663 Kaiserslautern, Germany
bajwa@dfki.uni-kl.de, {andreas.dengel,sheraz.ahmed}@dfki.de
[3] Osaka Prefecture University, Naka, Sakai, Osaka 599-8531, Japan
taniguchi@m.cs.osakafu-u.ac.jp
[4] School of Electrical Engineering and Computer Science,
National University of Science and Technology (NUST), Islamabad 46000, Pakistan
malik.imran@seecs.edu.pk
[5] Opthalmology Clinic, Rittersberg 9, 67657 Kaiserslautern, Germany
dr.neumeier-kl@web.de

Abstract. Visual artefacts of early diabetic retinopathy in retinal fundus images are usually small in size, inconspicuous, and scattered all over retina. Detecting diabetic retinopathy requires physicians to look at the whole image and fixate on some specific regions to locate potential biomarkers of the disease. Therefore, getting inspiration from ophthalmologist, we propose to combine coarse-grained classifiers that detect discriminating features from the whole images, with a recent breed of fine-grained classifiers that discover and pay particular attention to pathologically significant regions. To evaluate the performance of this proposed ensemble, we used publicly available EyePACS and Messidor datasets. Extensive experimentation for binary, ternary and quaternary classification shows that this ensemble largely outperforms individual image classifiers as well as most of the published works in most training setups for diabetic retinopathy detection. Furthermore, the performance of fine-grained classifiers is found notably superior than coarse-grained image classifiers encouraging the development of task-oriented fine-grained classifiers modelled after specialist ophthalmologists.

Keywords: Computer-aided diagnosis · Medical image analysis · Automated diabetic retinopathy detection · Convolutional neural network · Deep learning in ophthalmology

---

Partially funded by Higher Education Commission (HEC) of Pakistan through Faculty Development Program of National University of Science and Technology (NUST), Pakistan.

Y. Zheng et al. (Eds.): MIUA 2019, CCIS 1065, pp. 242–253, 2020.
https://doi.org/10.1007/978-3-030-39343-4_21

1   Introduction

Diabetic patients are at constant risk of developing Diabetic Retinopathy (DR) that may eventually lead to permanent vision loss if left unnoticed or untreated. In such patients, increased blood sugar, blood pressure, and cholesterol can cause small blood vessels in retina to protrude and, in due course, haemorrhage blood into retinal layers and/or vitreous humour [4]. In severe conditions, scar tissues and newly proliferated fragile blood vessels blanket the retina and obstruct incoming light from falling on it. As a result, retina is unable to translate light into neural signals which leads to blindness. Diabetic retinopathy advances slowly and gradually and may take years to reach proliferative stage, however, almost every diabetic patient is potentially susceptible to this complication.

Timely diagnosis is the key to appropriate prognosis. Ophthalmologists usually detect DR by examining retinal fundus and looking for any signs of microaneurysms (bulging of blood vessels), blood leakage, and/or neovascularization [2]. While the indications of advanced stages of DR are rather prominent, these symptoms remain largely discrete in early stages. Figure 1 shows progress of DR from healthy to proliferative stage in Retinal Fundus Images (RFIs) taken from EyePACS dataset[1]. It can be observed from the figure that the difference between healthy and early stages of DR are very subtle and not readily discernible. Manual analysis of these images requires highly qualified and specialized ophthalmologists who may not be easily accessible in developing countries or remote areas of developed countries. Even when medical experts are available, large scale analysis of RFIs is highly time-consuming, labour-intensive and prone to human error and bias. Furthermore, manual diagnosis by clinicians is largely subjective and rarely reproducible and, therefore, inter-expert agreement for a certain diagnosis is generally very poor.

Computer-Aided Diagnosis (CAD) based on deep learning can provide easily accessible, efficient and economical solution for large-scale initial screening of many diseases including diabetic retinopathy. CAD can perform objective analysis of the given image and predict standardized and reproducible diagnosis, which is free from any bias or tiredness. It can not only help physicians by

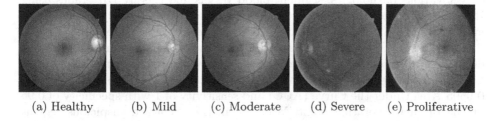

(a) Healthy   (b) Mild   (c) Moderate   (d) Severe   (e) Proliferative

Fig. 1. Progression of diabetic retinopathy from healthy to proliferative stage is subtle and gradual. Images are taken from EyePACS train set.

---

[1] https://www.kaggle.com/c/diabetic-retinopathy-detection/data.

reducing their workload but can also outreach to underprivileged population and afford them the opportunity of swift and cost-effective initial screening, which may effectively prevent advancement of disease into severer stage. Convolutional Neural Networks (CNNs) are computer algorithms inspired by biological visual cortex. They work especially well in visual recognition tasks. CNNs have been used to perform at par with or even outperform humans in various challenging image recognition problems [13,14]. Today automated image recognition can be divided into coarse-grained classification and fine-grained classification. In former case, images are classified into high-level categories like humans, animals, vehicles and other objects in a natural scene, for example. In later case, classification is focused on low-level categories like species of dogs or models of cars etc. Fine-grained classification is particularly challenging owing to high intra-class variations and low inter-class variations. Although DR is also a fine-grained classification task, it has normally been addressed using simple coarse-grained classification algorithms.

In this work we used a combination of general and fine-grained deep CNNs to analysed RFIs and predict automated diagnosis for DR. We used two of the most popular conventional image classification architectures i.e. Residual Networks [11] and Densely Connected Networks [12], a network search framework called NASNet [25] and two recently proposed methods for fine-grained classification namely NTS-Net [24] and SBS Layer [18]. We tried to harvest the combined potential of these two approaches by training them separately and taking their ensemble during inference. We used EyePACS and Messidor datasets for evaluation. Since previous researches have used vastly disparate experimental setups, we cannot directly compare our results with most of them. However, we performed a broad range of experiments, following the most common problem settings in the literature like normal vs abnormal, referable vs non-referable, ternary and quaternary classification in order to define benchmarks which will afford future works with an opportunity of fair comparison.

1.1 Related Work

Over the past decade, machine learning and deep learning have been used to detect various pathologies, segment blood vessels and classify DR grades using RFIs. Welikala et al. [23] detected proliferative DR by identifying neovascularization. They used an ensemble of two networks trained separately on 100 different patches for each network. The patches are taken from a selected set of 60 images collected from Messidor [8] and a private dataset. Since the dataset had only 60 images they performed leave-one-out cross validation and achieved 0.9505 Area Under the Curve (AUC) and sensitivity of 1 with specificity of 0.95 at the optimal operating point. Wang et al. [22] identified suspicious regions in RFIs and classified DR into normal (nDR) vs abnormal (aDR) and referable (rDR) vs non-referable (nrDR). They developed a CNN based model called Zoom-in-Network to identify important regions. To classify an image the network uses the overview of the whole images and pays particular attention to important regions. They took 182 images from EyePACS dataset and had a trained ophthalmologist draw bounding boxes around 306 lesions. On Messidor dataset they achieved 0.921 AUC, 0.905 accuracy and 0.960 sensitivity at 0.50 specificity for nDR vs aDR.

Gulshan et al. [10] conducted a comprehensive study to distinguish rDR from nrDR grades. They trained a deep CNN on 128175 fundus images from a private dataset and tested on 9963 images from EyePACS-I and 1748 images of Messidor-2. They achieved AUC of 0.991 on EyePACS-I and 0.990 on Messidor-2. Guan et al. [9] proposed that modelling each classifier after individual human grader instead of training a single classifier using average grading of all human experts improves classification performance. They trained 31 classifiers using a dataset of total 126522 images collected from EyePACS and three other clinics. The method is tested on 3547 images from EyePACS-I and Messidor-2, and achieved 0.9728 AUC, 0.9025 accuracy, and 0.8181 specificity at 0.97 sensitivity. However, it would have been more interesting if they had provided comparison of their suggested approach with ensemble of 31 networks modelled after average grading. Costa et al. [7] used adversarial learning to synthesize colour retinal images. However, the performance of their classifier trained on synthetic images was less than the classifier trained on real images. Aujih et al. [5] found that blood vessels play important role in disease classification and fundus images without blood vessels resulted in poor performance by the classifier.

The role of multiple filter sizes in learning fine-grained features was studied by Vo et al. [21]. To this end they used VGG network with extra kernels and combined kernels with multiple loss networks. They achieved 0.891 AUC for rDR vs nrDR and 0.870 AUC for normal vs abnormal on Messidor dataset using 10-fold cross validation. Somkuwar et al. [20] performed classification of hard exudates by exploiting intensity features using 90 images from Messidor dataset and achieved 100% accuracy on normal and 90% accuracy on abnormal images. Seoud et al. [19] focused on red lesions in RFIs, like haemorrhages and microaneurysms, and detected these biomarkers using dynamic shape features in order to classify DR. They achieved 0.899 and 0.916 AUC for nDR vs aDR and rDR vs nrDR, respectively on Messidor. Rakhlin et al. [16] used around 82000 images taken from EyePACS for training and around 7000 EyePACS images and 1748 images from Messidor-2 for testing their deep learning based classifier. They achieved 0.967 AUC on Messidor and 0.923 AUC on EyePACS for binary classification. Ramachandran et al. [17] used 485 private images and 1200 Messidor images to test a third party deep learning based classification platform, which was trained on more than 100000 images. Their validation resulted in 0.980 AUC on Messidor dataset for rDR vs nrDR classification. Quellec et al. [15] capitalized a huge private dataset of around 110000 images and around 89000 EyePACS images to train and test a classifier for rDR vs nrDR grades and achieved 0.995 AUC on EyePACS.

2   Materials and Methods

This section provides details on the datasets used in this work and the ensemble methodology employed to perform classification.

2.1   Datasets

We used EyePACS dataset published publicly by Kaggle for a competition on Diabetic Retinopathy Detection. Table 1 gives overview of EyePACS dataset. Although this dataset is very large in size, only about 75% of its images are of reasonable quality that they can be graded by human experts [16]. EyePACS is graded on a scale of 0 to 4 in accordance with International Clinical Diabetic Retinopathy (ICDR) guidelines [3]. However, low gradability of this dataset raises suspicions on the fidelity of labels provided with each image. We pruned the train set to get rid of 657 completely uninterpretable images. For testing on EyePACS we used 33423 images randomly taken from test set.

Table 1. Overview of EyePACS Dataset. IRMA stands for IntraRetinal Microvascular Abnormalities

Severity grade	Criterion	Train set		Test set	
		Images	Percentage	Images	Percentage
0	No abnormalities	25810	73.48	39533	73.79
1	Microaneurysms only	2443	6.95	3762	7.02
2	More than just microaneurysms but less than Grade 3	5292	15.07	7861	14.67
3	More than 20 intraretinal hemorrhages in each of 4 quadrants OR definite venous beading in 2+ quadrants OR prominent IRMA in 1+ quadrant AND no signs of proliferative retinopathy	873	2.48	1214	2.27
4	Neovascularization OR Vitreous/preretinal hemorrhage	708	2.02	1206	2.25
	Total	35126	100	53576	100

Table 2. Overview of messidor dataset

Severity grade	Criterion	Images	Percentage
0	No microaneurysms AND No haemorrhages	546	45.50
1	Microaneurysms $<=$ 5 AND No haemorrhages	153	12.75
2	5 $<$Microaneurysms $<$15 AND 0 $<$Haemorrhages $<$5 AND No Neovascularization	247	20.58
3	Microaneurysms $>=$ 15 OR Haemorrhages $>=$ 5 OR Neovascularization	254	21.17
	Total	1200	100

Messidor dataset consists of 1200 images collected at three different clinics in France. Each clinic contributed 400 images. This dataset is graded for DR on a scale of 0 to 3 following the criteria given in Table 2. Messidor dataset is validated by experts and is, therefore, of higher quality than EyePACS in terms of both image quality and labels.

2.2 Methodology

Figure 2 illustrates complete pipeline of the system combining coarse-grained and fine-grained classifiers. Before feeding an image to the network, we first applied Otsu Thresholding to extract and crop retinal rim from RFI and get rid of irrelevant black background. Since the images in both datasets are taken with different cameras and under different clinical settings, they suffer from large brightness and colour variations. We used adaptive histogram equalization to normalize brightness and enhance the contrast of visual artefacts which are critical for DR detection. Since the images are in RGB colour space, we first translate them into YCbCr colour space to distribute all luminosity information in Y channel and colour information in Cb and Cr channels. Adaptive histogram equalization is then applied on Y channel only and the resultant image is converted back to RGB colour space. We further normalized the images by subtracting local average colour from each pixel to highlight the foreground and help our network detect small features. Figure 3 shows the effects of preprocessing steps on RFIs. These pre-processed images are then used to train all five networks individually. During inference, each network gives diagnosis which are ensembled to calculate the final prediction.

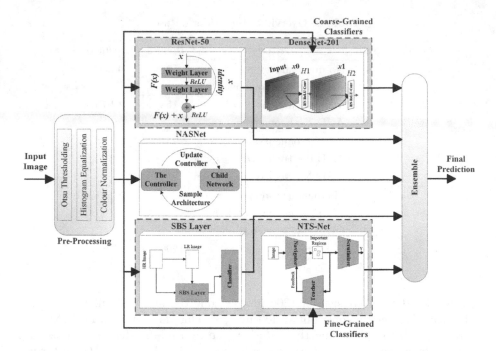

Fig. 2. System overview: combining coarse-grained and fine-grained classifiers

Experimental Setup. From EyePACS train set, we randomly selected 30000 images for training and rest of the 4469 images were used for validation. Test set of EyePACS was used for reporting results on this dataset. From Messidor, we used 800 images for training and 400 images from Lariboisière Hospital for testing (as done by Lam et al. [6]). We employed a broad range of hyper parameters during training. All networks are initialized with pre-trained weights and

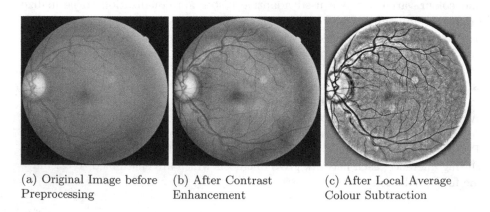

(a) Original Image before Preprocessing

(b) After Contrast Enhancement

(c) After Local Average Colour Subtraction

Fig. 3. Effects of preprocessing steps on retinal fundus images

fine-tuned on ophthalmology datasets. To evaluate these models on EyePACS and Messidor datasets under similar problem settings, we first parallelized DR grades of both datasets using criteria given in Fig. 4.

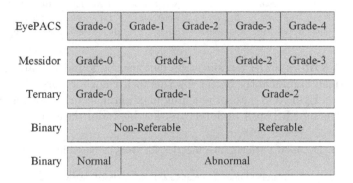

Fig. 4. Conversion of EyePACS grades to quaternary, ternary and binary classification

3   Results and Analysis

From Sect. 1.1 we observe that previous works on EyePACS and Messidor have used disparate train and test splits and different classification tasks for example Quaternary, Ternary and Binary (rDR vs nrDR and nDR vs aDR). Furthermore, different researchers use different performance metrics to evaluate their method. Therefore, in such scenario comparison of any two works in not directly possible [1]. However, we conducted extensive experiments to perform all four classification tasks mentioned above and report comprehensive results to allow a rough comparison with some of the published state-of-the-art results on these datasets.

3.1   Results of Binary Classification

As discussed above, many previous works focus primarily on binary classification as nDR vs aDR or rDR vs nrDR grading. The criteria to convert 4 or 5 grades into binary grades is given in Fig. 4. For our binary classification, the number of images used for training, validation and testing from EyePACS and Messidor are given in Tables 3 and 4. It can be seen from the tables that there is extensive class imbalance between both classes. Table 5 provides detailed performance metrics for all classification tasks including nDR vs aDR classification. Our results are competitive to that of Wang et al. in terms of accuracy and for all other metrics we outperform them. It should be noted here that Wang et al. performed 10-fold cross validation and although their sensitivity of 96 is higher than our 89.75, it is calculated at 50% specificity while ours is at 90% specificity.

Table 3. Class distribution for normal vs abnormal classification

Grade	EyePACS			Messidor		
	Train	Validate	Test	Train	Validate	Test
Normal	22668	2744	24741	346	49	151
Abnormal	7332	1725	8682	354	51	249
Total	30000	4469	33423	700	100	400

Table 4. Class distribution for referable vs non-referable classification

Grade	EyePACS			Messidor		
	Train	Validate	Test	Train	Validate	Test
Non-Referable	28825	4177	31937	453	65	181
Referable	1175	292	1486	247	35	219
Total	30000	4469	33423	700	100	400

Results of rDR vs nrDR classification can also be found in Table 5. All networks performed significantly better for this task than for normal vs abnormal classification on EyePACS dataset reaching maximum accuracy around 96% with 99.44% AUC using SBS layer architecture. For Messidor dataset, both NTS-Net

Table 5. Detailed performance metrics for various classification settings

Model	Accuracy (%)		AUC (%)		Sensitivity (%)		Specificity (%)	
	EyePACS	Messidor	EyePACS	Messidor	EyePACS	Messidor	EyePACS	Messidor
Results of binary (normal vs abnormal) classification								
NTS-Net	**88.19**	88.00	92.72	95.20	88	88.00	72	87.51
SBS Layer	80.11	89.50	86.20	95.17	80	89.50	54	**92.07**
ResNet-50	82.86	87.75	89.46	95.06	83	87.75	75	90.49
DenseNet-201	82.66	87.75	89.69	95.89	83	87.75	**77**	88.14
NASNet	82.19	87.25	88.49	95.04	82	87.25	73	89.14
Ensemble	87.74	**89.75**	**93.44**	**96.50**	**88**	**89.75**	75	91.44
Vo et al.	N/A	87.10	N/A	87.00	N/A	88.2	N/A	85.7
Wang et al.	N/A	90.50	N/A	92.10	N/A	96	N/A	50
Soud et al.	N/A	N/A	N/A	89.90	N/A	N/A	N/A	N/A
Results of binary (referable vs non-referable) classification								
NTS-Net	94.93	**93.25**	99.10	**96.56**	95	**93**	75	**94**
SBS Layer	**95.89**	88.75	**99.44**	94.90	**96**	89	67	90
ResNet-50	95.08	86.75	98.97	94.95	95	87	81	89
DenseNet-201	94.70	89.25	99.05	95.33	95	89	82	91
NASNet	91.98	87.50	97.45	95.16	92	88	**85**	89
Ensemble	95.34	89.25	99.23	96.45	95	89	81	91
Lam et al.	N/A	74.5	N/A	N/A	N/A	N/A	N/A	N/A
Vo et al.	N/A	89.70	N/A	89.10	N/A	89.3	N/A	90
Wang et al.	N/A	91.10	N/A	95.70	N/A	97.8	N/A	50
Seoud et al.	N/A	74.5	N/A	91.60	N/A	N/A	N/A	N/A
Results of ternary classification								
NTS-Net	84.43	84.50	94.89	94.61	84	85	72	**94**
SBS Layer	76.93	84.50	90.95	94.12	77	85	50	91
ResNet-50	81.23	80.50	93.51	93.79	81	81	74	92
DenseNet-201	79.20	80.25	92.87	94.25	79	80	**77**	93
NASNet	78.95	81.75	91.93	94.00	79	82	71	89
Ensemble	**84.94**	**85.25**	**95.28**	**95.40**	**85**	**85**	73	92
Lam et al.	N/A	68.8	N/A	N/A	N/A	N/A	N/A	N/A
Results of quaternary classification								
NTS-Net	82.53	74.50	95.72	91.84	83	75	**76**	**92**
SBS Layer	82.00	65.00	95.69	88.43	82	65	67	88
ResNet-50	81.82	70.25	95.53	91.31	82	70	71	89
DenseNet-201	79.38	74.00	95.04	92.26	79	74	75	91
NASNet	73.73	71.75	92.06	90.84	74	72	74	86
Emsemble	**83.42**	**76.25**	**96.31**	**92.99**	**83**	**76**	73	91
Lam et al.	N/A	57.2	N/A	N/A	N/A	N/A	N/A	N/A

and SBS Layer stand out from traditional classifiers. NTS-Net outperforms all other methods in all metrics, whereas ensemble of all methods gives sub-optimal performance than individual fine-grained methods. This can happen when majority of classifiers used for ensemble have a skewed performance towards downside and only a few give standout results.

3.2 Results of Multi-class Classification

The complexity of classification task was gradually increased from binary to ternary and quaternary classification. Tables 6 and 7 show the class distribution in train, validation and test splits for this multi-class setting. For ternary classification we used the criterion used by [6], as shown in Fig. 4.

Table 6. Class distribution for 4-class classification

Grade	EyePACS			Messidor		
	Train	Validate	Test	Train	Validate	Test
0	22668	2744	24741	346	49	151
1	6157	1433	7196	107	16	30
2	685	166	753	155	22	70
3	490	126	733	92	13	149
Total	30000	4469	33423	700	100	400

Table 7. Class distribution for 3-class classification

Grade	EyePACS			Messidor		
	Train	Validate	Test	Train	Validate	Test
0	22668	2744	24741	346	49	151
1	6157	1433	7196	107	16	30
2	1175	292	1486	247	35	219
Total	30000	4469	33423	700	100	400

Performance of individual networks and their ensemble for ternary and quaternary classification is given in Table 5. Ensemble of all models gave better performance in this case. We also observe that the performance of NTS-Net is higher than all other individual networks. Our accuracies for both ternary and quaternary classification are superior than accuracies reported by Lam et al. [6]. Figure 5 provides a detailed overview of classification performance of ensemble.

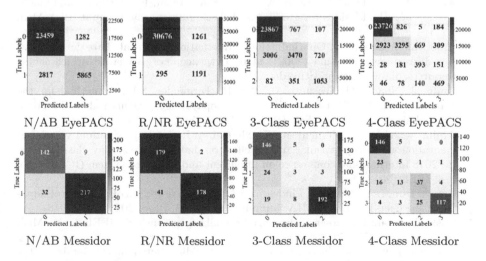

Fig. 5. Confusion matrices for EyePACS and messidor for all classification tasks. N/AB refers to normal vs abnormal; R/NR refers to referable vs non-referable

4   Conclusion

Diabetic Retinopathy detection using retinal fundus images is a fine-grained classification task. The biomarkers of this disease on retinal images are usually very small in size, especially for early stages, and are scattered all across the image. The ratio of pathologically important region to the whole input volume is therefore minuscule. Due to this reason traditional deep CNNs usually struggle to identify regions of interest and do not learn discriminatory features well. This problem of small and distributed visual artefacts coupled with unavailability of large publicly available high-quality dataset with reasonable class imbalance makes DR detection particularly challenging for deep CNN models. However, fine-grained classification networks have high potential to provide standardized and large-scale initial screening of diabetic retinopathy and help in prevention and better management of this disease. These networks are equipped with specialized algorithms to discover the important region from the image and pay particular heed to learn characterizing features from those regions.

We achieved superior performance for diabetic retinopathy detection on binary, ternary and quaternary classification tasks than many previously reported results. However, due to hugely different experimental setups and choice of performance metrics, it is unfair to draw a direct comparison with any of the cited research. Nevertheless, we have provided a wide spectrum of performance metrics and detailed experimental setup for comparison by any future work.

References

1. Abràmoff, M.D., et al.: Improved automated detection of diabetic retinopathy on a publicly available dataset through integration of deep learning. Investig. Ophthalmol. Vis. Sci. **57**(13), 5200–5206 (2016)
2. Akram, M.U., Khalid, S., Tariq, A., Khan, S.A., Azam, F.: Detection and classification of retinal lesions for grading of diabetic retinopathy. Comput. Biol. Med. **45**, 161–171 (2014)
3. American Academy of Ophthalmology: International clinical diabetic retinopathy disease severity scale (2002). http://www.icoph.org/dynamic/attachments/resources/diabetic-retinopathy-detail.pdf
4. Amin, J., Sharif, M., Yasmin, M., Ali, H., Fernandes, S.L.: A method for the detection and classification of diabetic retinopathy using structural predictors of bright lesions. J. Comput. Sci. **19**, 153–164 (2017)
5. Aujih, A., Izhar, L., Mériaudeau, F., Shapiai, M.I.: Analysis of retinal vessel segmentation with deep learning and its effect on diabetic retinopathy classification. In: 2018 International conference on intelligent and advanced system (ICIAS), pp. 1–6. IEEE (2018)
6. Carson Lam, D.Y., Guo, M., Lindsey, T.: Automated detection of diabetic retinopathy using deep learning. AMIA Summits Transl. Sci. Proc. **2017**, 147 (2018)
7. Costa, P., et al.: End-to-end adversarial retinal image synthesis. IEEE Trans. Med. Imaging **37**(3), 781–791 (2018)
8. Decencière, E., et al.: Feedback on a publicly distributed image database: the messidor database. Image Anal. Ster. **33**(3), 231–234 (2014)

9. Guan, M.Y., Gulshan, V., Dai, A.M., Hinton, G.E.: Who said what: modeling individual labelers improves classification. In: Thirty-Second AAAI Conference on Artificial Intelligence (2018)

10. Gulshan, V., et al.: Development and validation of a deep learning algorithm for detection of diabetic retinopathy in retinal fundus photographs. JAMA **316**(22), 2402–2410 (2016)

11. He, K., Zhang, X., Ren, S., Sun, J.: Deep residual learning for image recognition. In: Proceedings of the IEEE Conference on Computer Vision and Pattern Recognition, pp. 770–778 (2016)

12. Huang, G., Liu, Z., Van Der Maaten, L., Weinberger, K.Q.: Densely connected convolutional networks. In: Proceedings of the IEEE Conference on Computer Vision and Pattern Recognition, pp. 4700–4708 (2017)

13. Lee, C.Y., Gallagher, P.W., Tu, Z.: Generalizing pooling functions in convolutional neural networks: mixed, gated, and tree. In: Artificial intelligence and statistics, pp. 464–472 (2016)

14. Liu, C., et al.: Progressive neural architecture search. CoRR abs/1712.00559 (2017). http://arxiv.org/abs/1712.00559

15. Quellec, G., Charrière, K., Boudi, Y., Cochener, B., Lamard, M.: Deep image mining for diabetic retinopathy screening. Med. Image Anal. **39**, 178–193 (2017)

16. Rakhlin, A.: Diabetic retinopathy detection through integration of deep learning classification framework. bioRxiv, p. 225508 (2018)

17. Ramachandran, N., Hong, S.C., Sime, M.J., Wilson, G.A.: Diabetic retinopathy screening using deep neural network. Clin. Exp. Ophthalmol. **46**(4), 412–416 (2018)

18. Recasens, A., Kellnhofer, P., Stent, S., Matusik, W., Torralba, A.: Learning to zoom: a saliency-based sampling layer for neural networks. In: Proceedings of the European Conference on Computer Vision (ECCV), pp. 51–66 (2018)

19. Seoud, L., Hurtut, T., Chelbi, J., Cheriet, F., Langlois, J.P.: Red lesion detection using dynamic shape features for diabetic retinopathy screening. IEEE Trans. Med. Imaging **35**(4), 1116–1126 (2016)

20. Somkuwar, A.C., Patil, T.G., Patankar, S.S., Kulkarni, J.V.: Intensity features based classification of hard exudates in retinal images. In: 2015 Annual IEEE India Conference (INDICON), pp. 1–5. IEEE (2015)

21. Vo, H.H., Verma, A.: New deep neural nets for fine-grained diabetic retinopathy recognition on hybrid color space. In: 2016 IEEE International Symposium on Multimedia (ISM), pp. 209–215. IEEE (2016)

22. Wang, Z., Yin, Y., Shi, J., Fang, W., Li, H., Wang, X.: Zoom-in-Net: deep mining lesions for diabetic retinopathy detection. In: Descoteaux, M., Maier-Hein, L., Franz, A., Jannin, P., Collins, D.L., Duchesne, S. (eds.) MICCAI 2017. LNCS, vol. 10435, pp. 267–275. Springer, Cham (2017). https://doi.org/10.1007/978-3-319-66179-7_31

23. Welikala, R., et al.: Automated detection of proliferative diabetic retinopathy using a modified line operator and dual classification. Comput. Methods Programs Biomed. **114**(3), 247–261 (2014)

24. Yang, Z., Luo, T., Wang, D., Hu, Z., Gao, J., Wang, L.: Learning to navigate for fine-grained classification. In: Proceedings of the European Conference on Computer Vision (ECCV), pp. 420–435 (2018)

25. Zoph, B., Vasudevan, V., Shlens, J., Le, Q.V.: Learning transferable architectures for scalable image recognition. In: Proceedings of the IEEE conference on computer vision and pattern recognition, pp. 8697–8710 (2018)

Vessel and Nerve Analysis

Automated Quantification of Retinal Microvasculature from OCT Angiography Using Dictionary-Based Vessel Segmentation

Astrid M. E. Engberg[1]([✉]), Jesper H. Erichsen[2], Birgit Sander[2], Line Kessel[2,3], Anders B. Dahl[1], and Vedrana A. Dahl[1]

[1] Technical University of Denmark, Kgs. Lyngby, Denmark
{asteng,abda,vand}@dtu.dk
[2] The Eye Clinic, Rigshospitalet-Glostrup, Glostrup, Denmark
[3] Faculty of Health and Medical Sciences, University of Copenhagen, Copenhagen, Denmark

Abstract. Investigations in how the retinal microvasculature correlates with ophthalmological conditions necessitate a method for measuring the microvasculature. Optical coherence tomography angiography (OCTA) depicts the superficial and the deep layer of the retina, but quantification of the microvascular network is still needed. Here, we propose an automatic quantitative analysis of the retinal microvasculature. We use a dictionary-based segmentation to detect larger vessels and capillaries in the retina and we extract features such as densities and vessel radius. The method is validated on repeated OCTA scans from healthy subjects, and we observe high intraclass correlation coefficients and high agreement in a Bland-Altman analysis. The quantification method is also applied to pre- and postoperative scans of cataract patients. Here, we observe a higher variation between the measurements, which can be explained by the greater variation in scan quality. Statistical tests of both the healthy subjects and cataract patients show that our method is able to differentiate subjects based on the extracted microvascular features.

Keywords: OCTA · Dictionary-based segmentation · Quantification

1 Introduction

An important function of the retinal microvasculature is to supply the inner retinal layers with nutrients and oxygen. Disturbances in the vasculature occur in many eye diseases. Hence, the state of the retinal microvasculature could be an important health indicator for certain eye-related conditions. One of the specific clinical problems motivating our work is assessing the risk for developing pseudophakic macular edema after cataract surgery. Our future objective is to investigate the influence of parameters related to the risk of edema development. We are therefore in need of a method to characterize the retinal microvasculature.

© Springer Nature Switzerland AG 2020
Y. Zheng et al. (Eds.): MIUA 2019, CCIS 1065, pp. 257–269, 2020.
https://doi.org/10.1007/978-3-030-39343-4_22

The retinal microvasculature is visible in images captured using the non-invasive modality, Optical Coherence Tomography Angiography (OCTA). The visualisation of the microvasculature is often susceptible to noise, potentially leading to misinterpretation by the operator. For the same reason, automated, clinically useful methods for extracting the vasculature from OCTA images are scarce. Methods based on user-provided annotations of various objects in large image data sets are often capable of handling a significant noise level. Unfortunately, obtaining sufficient pre-annotated data is time consuming and therefore costly.

In this paper, we present an approach for quantifying retinal microvasculature from OCTA images which utilizes a very limited amount of pre-annotated data. We present an investigation of the method with a focus on reproducibility and feasibility of extracting microvascular features.

1.1   Related Work

Since the first clinical studies on OCTA were published in 2014, several quantification schemes for OCTA images have been invented with varying complexity and quality. Most studies use their own quantification algorithm, where the vessel detections vary from using global thresholding [11], binarization and skeletonization [12], to more complex filtering and thresholding approaches [3,14]. More advanced methods include local fractal dimension [9], and a probabilistic model utilizing the 3D spatial information from both the superficial and the deep layer [8]. Some OCTA devices provide proprietary software for extracting vessel densities, e.g. the AngioVue OCTA system (Optovue, Inc., Fremont, CA, USA) [16], Cirrus (Carl Zeiss Meditec, Inc., Dublin, CA, USA) [7], and RS-3000 Advance (NIDEK, Gamagori, Japan) [1].

Two groups compare available macular vessel density algorithms [18,20]. Shoji et al. [20] assess binarization algorithms in ImageJ (National Institutes of Health, Bethesda, MD, USA) on data obtained from two OCTA systems, while Rabiolo et al. [18] compare seven previously published quantification algorithms, concluding that the estimated densities are significantly different for each algorithm. Both studies therefore recommend using the same quantification algorithm in longitudinal follow-up, and note that vessel density is dependent on both the OCTA device, the acquisition size, and the post-processing algorithm, and hence comparisons to other studies or databases should be done with care.

Due to the lack of a standardized quantification method, we choose to develop an algorithm that in later work can be used for assessing the risk of edema development. In our detection we differ between capillaries and larger vessels. This is because larger vessels influence the estimate of the capillary density. Other groups try to overcome this by filtering [2] or computing the skeleton density [14,18]. Our approach allows for more precise description of the vasculature.

2   Proposed Approach

We propose using a dictionary-based method for segmenting and detecting the vascular network in the parafoveal region, i.e. the region around fovea. Our approach is motivated by the assumption that the appearance of the vascular network may be described as a combination of a limited number of characteristic image features.

2.1   Dictionary-Based Segmentation

The success of the quantification highly relies on its ability to detect capillaries and larger vessels. We are using a probabilistic pixel classification method for segmenting the OCTA images [5,6], where pre-segmented data is used to learn a dictionary of image patches with corresponding label information. The method was originally developed for texture segmentation [4].

In short, the dictionary-based method involves two steps: building the dictionary by using pre-annotated data, and processing previously unseen images. Building the dictionary relies on two processes: defining dictionary clusters which encode local appearance of OCTA images, and learning the desired segmentation for the dictionary clusters by incorporating manual annotations. In our case, the pre-annotated data consists of one training image. For the clustering process, we extract local features from the training image. We choose to use the following features: the intensities from a patch around a pixel, and the first and the second degree image derivatives computed in a patch around a pixel. To reduce the dimensionality of the feature vectors, we perform principal component analysis (PCA) on the set of features from the training image. Once we have extracted the local features, we perform k-means clustering to obtain cluster centers, which in our context constitute the dictionary. The assumption behind clustering is that the variability in the local appearance of retinal images may be explained using a limited set of clusters. The learning process is performed by extending the dictionary with user-provided labeling. For each cluster in the dictionary, we access the pixels belonging to this cluster, and average the available pre-annotated information. The assumption behind learning is that pixels belonging to the same cluster should have a similar segmentation.

The segmentation of the unseen image is now performed by extracting local features and assigning them to the dictionary clusters. Then we obtain the desired segmentation information from the dictionary. We refer the reader to [6] for details.

Choosing the Appropriate Parameters. In our dictionary for retinal microvasculature, we work with three classes: capillaries, larger vessels (arterioles and venules), and background. Larger vessels are defined as vessels of twice (or more) the radius of the capillaries. Similarly to the study by Eladawi et al. [8], manual annotations are produced and used as ground truth (GT). Two images acquired as described in Sect. 3.1 were annotated manually by AMEE.

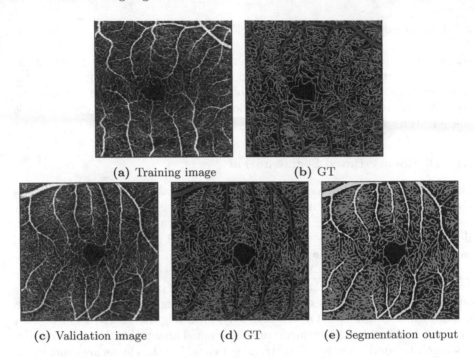

(a) Training image   (b) GT

(c) Validation image   (d) GT   (e) Segmentation output

Fig. 1. Subfigures (a) and (b) show the image used for building the dictionary and corresponding ground truth annotation. Subfigures (c) and (d) show the validation image and corresponding manual annotation used for choosing model parameters. Subfigure (e) shows the segmentation output of the validation image.

One image is used for creating the dictionary (Fig. 1a and b), and the other is used for choosing the parameters and for validation (Fig. 1c–e).

The dictionary is built using the features from both the image seen in Fig. 1a and from a 90 degree rotated version. We use 50.000 patches of size 7×7 pixels to build the PCA co-variance matrix. The 10 biggest PCA features are kept. Afterwards, we build a k-means tree with 5 layers and a branching factor of 5 from 100.000 training patches of size 13×13 pixels. Further, we compute a 3×3 weight matrix such that the resulting class probabilities of the training image equal the annotated class labels.

We optimize the parameters for the PCA patch size and the cluster patch size with respect to the validation image and GT image shown on Fig. 1c and d. Dice scores between the manual segmentation (Fig. 1d) and the resulting segmentation (Fig. 1e) are 0.82 for larger vessels, 0.71 for capillaries, and 0.76 for background. The corresponding Jaccard scores are 0.70, 0.56, and 0.62.

2.2 Quantitative Analysis (Values VD, CD, BD, and VR)

We extract different measures in order to quantify the vascular network in the superficial retinal layer (SRL) and the deep retinal layer (DRL) of OCTA data.

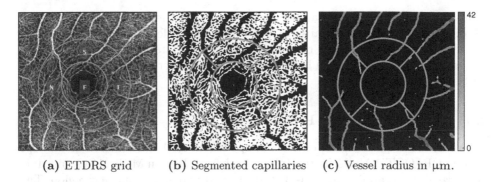

(a) ETDRS grid   (b) Segmented capillaries   (c) Vessel radius in μm.

Fig. 2. Illustration of the quantification. Subfigure (a) shows the modified ETDRS grid where the foveal center is defined as the center of the avascular zone, the radius of the central area is 0.5 mm, and the parafoveal zone is defined as the annulus with radius 0.5 mm and 1 mm. Letters F, S, I, N and T stand for fovea, superior, inferior, nasal and temporal, respectively. Subfigure (b) shows detected capillaries used to compute the capillary density. Subfigure (c) shows the radius of the larger vessels in μm.

We divide the parafoveal area into sections using the Early Treatment Diabetic Retinopathy Study (ETDRS) grid [10], see Fig. 2a. Due to the size of the OCTA images, the area of interest was limited to the circular radius of 1 mm from the foveal center. In each section, we analyze the densities of capillaries, larger vessels and background, and compute the mean radius of the larger vessels.

The area densities of each class are computed for each section as a unit-less ratio between the amount of pixels belonging to that specific class and the total amount of pixels in the section. Figure 2b shows the binary map of the capillaries in the five sections. We compute the vessel density (VD), the capillary density (CD) and the background density (BD) in the SRL. These abbreviations will be used from now on in this paper. In the DRL, we assume that only capillaries are present, so here we use the same segmentation model as in the SRL but combine the classes for capillaries and larger vessels. In the SRL, we compute the vessel radius of the larger vessels by first median filtering the binary class mask with a 3 × 3 neighborhood. We then compute the skeletonization of the output and compute the signed Euclidean distance field to the pixel center in order to obtain the distance to the closest boundary. In order to visualize the vessel thickness (Fig. 2c), the radius is plotted by dilating it with a disk shaped structure.

For every scan, the mentioned metrics in the five areas and in the two layers are concatenated into a feature vector, which is used in the analysis. The complete approach was implemented in MATLAB v9.5 (Mathworks, Inc.).

3 Validation of Approach

The robustness of the proposed method is validated by applying it to data from healthy subjects acquired under three different scenarios but with an expectancy of no significant changes in the capillary network. We therefore hypothesize that there will be limited variation in the extracted features for each subject throughout the different scans. In order to test the feasibility of the method for abnormal eyes, we apply the method on OCTA data from cataract patients before and after surgery. We here hypothesize that there will be more variation in the extracted features.

3.1 Data and Scanning Protocol

All data are acquired using either a swept source DRI OCT Triton or a swept source DRI OCT-1 Model Triton (plus), both from Topcon Medical Systems, Inc. The size of the scanning frame is $3 \times 3\,mm^2$ centered on the macular region. The participants are instructed to fixate on a central object in order for the fovea to be located in the center. The retinal tissue layers are automatically segmented by the Topcon Advanced Boundary Algorithm (Topcon Medical Systems, Inc.). However, if the operator deems it necessary, the retinal layers are adjusted manually in IMAGEnet 6, a browser-based software provided by Topcon Medical Systems, Inc. The layers are accessed in order to find the foveal center for the post-processing. Due to displacements from motion artefacts, different scans of the same patient are inspected in order to determine the center across scans. En face angiograms of the superficial retinal layer (SRL) and the deep retinal layer (DRL) are exported in JPG format. The OCTA images have dimension 320×320 pixels with a pixel resolution of $9.38\,\mu m$. The OCTA images are included if the image quality provided by the Triton OCT is 50 or more, if there are no blink- or movement artefacts, and if an area with a radius of 1 mm from the foveal center is visible.

Healthy Data Set (Scenarios H1, H2, and H3). To develop and validate the analysis method, we have chosen to acquire a high quality data set containing OCTA images from 10 healthy subjects at three different scenarios denoted H1, H2 and H3. In scenario **H1**, the pupil is undilated, in scenario **H2**, the pupil is medically dilated (using eye drops; Tropicamid (Mydriacyl) 1%), and in scenario **H3**, the pupil is dilated and the subject has just walked briskly 5–7 flight of stairs. Between the first and second scenario, the optical quality of the eye changes because dilating the pupil allows more light to enter and exit the eye. We can hence test the effect of the amount of entered light. Between the second and the third scenario, we can test the robustness of the method with respect to potential perfusion changes from the exercise. The OCTA recordings are repeated until the quality-requirements described above are fulfilled. All subjects have normal visual acuity and no systemic or ocular disease. The visual acuity is measured in ETDRS letters. When converted into Snellen equivalent, the mean best corrected visual acuity is 1.25 (range 1.0 to 2.0). Maximal refraction is within $+/- 3$ D. We examine one eye from each of the 10 subjects, where 5 are male and 5 are female, and there are 5 left eyes and 5 right eyes. The mean age is 32.4 years with a standard deviation of 8.5.

Cataract Data Set (Scenarios C1 and C2). For investigating how the method performs with lower quality data, OCTA images from cataract patients who underwent cataract surgery are obtained preoperatively (denoted **C1**) and three weeks postoperatively (denoted **C2**). The pupil is medically dilated using eye drops (Tropicamid (Mydriacyl) 1% and Phenylephrine (Metaoxedrin) 10%) on the preoperative images but not on the postoperative ones. OCTA images from 44 consecutive patient records are reviewed by AMEE and JHE, and 10 of these are found to comply with the above mentioned quality criteria on both pre- and postoperative images. Of the 10 included patients, 8 are female and 2 are male, and there are 5 left eyes and 5 right eyes. The mean age is 70.3 years with a standard deviation of 5.5.

3.2 Statistical Analysis

In order to estimate the reproducibility of the method, we compare the different scanning scenarios of the healthy subjects. The intraclass correlation coefficient (ICC) is computed along with its 95% confidence intervals. The ICC describes both the correlation and the agreement of the measurements [17] and is used in most repeatability studies [14, 18, 20]. We use a single-rating, absolute-agreement and 2-way mixed-effect model using a MATLAB function provided by Salarian [19]. ICC values under 0.5 indicate poor reliability, between 0.5 to 0.75 moderate reliability, between 0.75 and 0.9 good reliability, and over 0.9 excellent reliability [15]. We use Bolt-Altman analysis to evaluate the agreement between the measurements, where we report the mean difference and the limits of agreement, which are set to 1.96 standard deviation, as done in [18]. We also perform a t-test for paired measurements as in [1, 20], where p-values less than 0.05 are considered significant.

We use factor analysis for analyzing the healthy data set and the cataract data set. Since we only have few observations and variables of high dimensions (the feature vector for every scan), we apply a principal axis factoring in order to decrease the number of dimensions. We use the varimax rotation to ease the interpretation of the factors. The sample means of each factor score are then compared using a multivariate two-sided analysis of variance (MANOVA) for the different groups as well as between all subjects. These statistical analyses are performed in SAS Studio, 3.8, SAS Institute Inc., Cary, NC, USA, and the reported p-values are the Wilks' lambda statistic.

4 Results

Figure 3a–d show a good quality OCTA scan from a healthy subject and the resulting segmentation and radius information. Similarly, Fig. 3e–h show a reduced quality OCTA scan from a healthy subject included in the study. The average of the extracted metrics (used in the feature vector) of the healthy patients can be seen in Table 1. The ICC and Bolt-Altman analysis for the healthy subjects and cataract patients are reported in Table 2.

Table 1. Mean and standard deviation of the features for all healthy subjects in all three scenarios (F: Fovea, I: Inferior, S: Superior, N: Nasal, T: Temporal, VD: Vessel density, CD: Capillary density, VR: Vessel radius in μm).

	SRL-F	SRL-I	SRL-S	SRL-N	SRL-T
VD	0.0025 (0.0031)	0.047 (0.018)	0.053 (0.020)	0.026 (0.015)	0.034 (0.029)
CD	0.26 (0.067)	0.44 (0.069)	0.47 (0.048)	0.45 (0.054)	0.45 (0.069)
VR	5.89 (5.52)	13.66 (1.62)	14.73 (2.01)	13.24 (2.91)	12.70 (2.28)
	DRL-F	**DRL-I**	**DRL-S**	**DRL-N**	**DRL-T**
CD	0.21 (0.075)	0.61 (0.062)	0.61 (0.055)	0.57 (0.061)	0.55 (0.067)

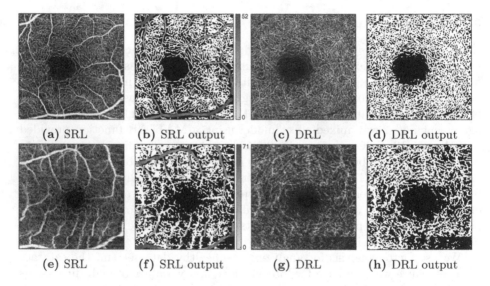

(a) SRL	(b) SRL output	(c) DRL	(d) DRL output
(e) SRL	(f) SRL output	(g) DRL	(h) DRL output

Fig. 3. Examples of segmentation output. Subfigures (a)–(d) illustrate the quantification of a good and clear OCTA, and (e)–(h) illustrate the quantification of a blurry OCTA. Subfigures (b) and (f) show the larger vessel radius in μm.

We apply a factor analysis on the healthy subjects containing 30 observations where each has a 20-dimensional variable. Six factors are selected, which account for 84.65% of the variation in the data. An arbitrary threshold of ±0.5 is applied to ease the interpretation of the rotated factor pattern. The linear combinations of the resulting variables explaining each factor are shown in Table 3. When applying the MANOVA to the factor scores, we do not observe an effect in-between the three pupil scenarios (p = 0.9513), but there is a highly significant difference between the different subjects (p < 0.0001). We also perform a factor analysis of the cataract patients in a similar way. Here, we obtain different

Table 2. Comparisons between the different groups, H1: Healthy with undilated pupil, H2: Healthy with dilated pupil, H3: Healthy with dilated pupil after exercise, C1: Cataract patients preoperative, C2: Cataract patients postoperative. Reporting the MD: mean difference, LoA: limits of agreement, ICC: Intraclass correlation coefficient, CI: confidence interval. For the variables VD: Vessel density, CD: Capillary density, BD: Background density, VR: Vessel radius.

Comparison			Agreement		Reliability	Pairwise comparison
Group 1	Group 2	Metric	MD	LoA	ICC (95% CI)	P-value
H1	H2	SRL CD	−0.0057	−0.11/0.10	0.86 (0.76, 0.92)	0.47
H1	H2	SRL VD	−0.0014	−0.020/0.018	0.93 (0.88, 0.96)	0.32
H1	H2	SRL BD	0.0071	−0.11/0.12	0.85 (0.78, 0.92)	0.41
H1	H2	SRL VR	−0.68	−6.32/4.97	0.79 (0.65, 0.87)	0.10
H1	H2	DRL CD	−0.011	−0.14/0.12	0.92 (0.86, 0.95)	0.23
H2	H3	SRL CD	0.0039	−0.10/0.11	0.86 (0.77, 0.92)	0.67
H2	H3	SRL VD	−0.00050	−0.018/0.017	0.94 (0.89, 0.96)	0.69
H2	H3	SRL BD	−0.0027	−0.11/0.11	0.88 (0.79, 0.93)	0.74
H2	H3	SRL VR	−0.50	−4.36/3.36	0.89 (0.81, 0.94)	0.079
H2	H3	DRL CD	0.0040	−0.12/0.11	0.93 (0.89, 0.96)	0.63
C1	C2	SRL CD	0.049	−0.12/0.22	0.65 (0.36, 0.81)	**<0.001**
C1	C2	SRL VD	0.0026	−0.029/0.035	0.80 (0.67, 0.88)	0.27
C1	C2	SRL BD	−0.052	−0.23/0.13	0.68 (0.41, 0.83)	**<0.001**
C1	C2	SRL VR	0.65	−6.90/8.20	0.73 (0.57, 0.84)	0.13
C1	C2	DRL CD	−0.018	−0.18/0.14	0.90 (0.82, 0.94)	0.24

linear combinations describing 7 factors (accounting for 89.15% of the variation in the data). We do not observe a significant difference between the pre- and postoperative scans ($p = 0.2471$), but the difference between the subjects is still very significant ($p < 0.0001$).

5   Discussion

The relatively high Dice and Jaccard scores of the validation image in Fig. 1e indicate an adequate segmentation performance of the proposed dictionary model. A typical misclassification seen in our results is when bigger capillaries and smaller vessels are not distinguished. However, this is to be expected, as it is also difficult to manually set the border between the two classes. The quality of the OCTA image in Fig. 3e–h makes it difficult to distinguish the capillaries manually, and the algorithm is challenged as well.

Table 3. Factors for healthy subjects, where SRL: Superficial retinal layer, DRL: Deep retinal layer, F: Fovea, I: Inferior, S: Superior, N: Nasal, T: Temporal, VD: Vessel density, CD: Capillary density, BD: Background density, VR: Vessel radius.

Factor 1	Factor 2	Factor 3	Factor 4	Factor 5	Factor 6
SRL-F CD: 0.64	SRL-S VR: −0.72	SRL-F CD: 0.54	SRL-S CD: 0.73	SRL-I CD: 0.85	SRL-N CD: 0.84
SRL-T VD: 0.83	SRL-N VR: 0.66	SRL-N VD: 0.83	SRL-T CD: 0.82	DRL-S CD: −0.53	SRL-I VD: −0.52
SRL-F VD: 0.95	SRL-I VR: −0.68	SRL-N VR: 0.55	SRL-S VD: −0.68	DRL-I CD: 0.83	
SRL-T VR: 0.60	DRL-S CD: 0.57	SRL-T VR: 0.61			
SRL-F VR: 0.85	DRL-N CD: 0.67	DRL-F CD: 0.66			
	DRL-T CD: 0.91				

5.1   Discussion of Healthy Subjects

The retinal capillary density is generally reported to be between 30–60%, and the great variation in different studies is due to different OCTA devices and methods [13]. Our capillary density estimates lie within this range (see Table 1) and are hence realistic, although cross-study comparisons should be done with care. The ICC values in Table 2 between H1 and H2 all vary between moderate to excellent, being highest for SRL VD and lowest for SRL VR. Similar trends can be seen when comparing H2 and H3, although here all ICC values are higher. We also see that the mean differences all lie close to zero, although the radius seems to have a small bias as well as a higher dispersion in the measurements. When looking at the pairwise comparison, no significant difference is observed between the different scanning scenarios. One reason why the vessel radius stands out in high dispersion could be due to a greater variation in the segmentation at the border between larger vessels and capillaries, and because of the hard margins between the ETDRS-sections.

When investigating the estimated factors from the factor analysis seen in Table 3, one can observe anatomical patterns. E.g. Factor 1 explains the capillary and vessel densities as well as the radius in the foveal and temporal region in the superficial layer. The MANOVA-test on the factor scores shows no difference between the three scenarios (H1, H2 and H3), and hence we can conclude that neither the dilation nor the exercise has an effect on the extracted metrics. We did however observe a difference between subjects, which means that each feature vector is representative of its subject, and hence that the extracted metrics are subject-specific.

5.2   Discussion for Cataract Patients

In Table 2, we see more variation in the quantitative metrics for the cataract patients, and decreased agreement and reliability compared to the healthy subjects. The quality of the cataract OCTA data is generally lower compared to the healthy subjects. Although it should be noted that all images used in this study were subjected to high quality requirements (compared to the common state of cataract patient data) in our selection process. The quality did however vary in both scenarios (pre- and postoperative). One reason for the great variation could be the optical deterioration of the lens preoperatively which results in a more blurry image, whereas the postoperative images were performed on an undilated pupil. Even though we did not observe a difference between undilated and dilated pupils for the healthy subjects, this could still have an effect for the older population. Obtaining an OCTA image with good image quality is challenging in cataract patients due to several reasons. The cataract itself worsens the optic quality of the image and patients may have difficulty focusing, which in turn may lead to small movements of the eye when the image is taken, causing artefacts. Furthermore, many cataract patients suffer from dry eyes and/or blepharitis, which causes irritation and blinking during the acquisition. Finally, the quality of the procedure is also dependent on the operator in terms of giving good instructions to the patient and securing a well focused image.

Finer details, like capillaries, typically suffer most by the reduced quality in the cataract patients. This is supported by the significantly different SRL CD and SRL BD in the pairwise comparison. We did not observe any significant difference between the pre- and postoperative scans in the factor analysis, which can also be explained by the varying data quality in both scenarios – although a smaller p-value ($p = 0.2471$) suggests using a larger sample size.

Only a small circular area is used in the quantitative analysis. The method and the analysis could be improved by extending the area of interest through technical adjustments such that more information would be included. Ongoing improvements in OCTA systems to obtain a larger acquisition size would also facilitate this. Furthermore, additional metrics describing the microvasculature could be interesting to include. This can easily be added in the analysis pipeline, since the factor analysis is able to handle more variables than observations, while still explaining most of the variability in the data. This is only a preliminary investigation of the approach, and therefore there is still a need to evaluate it in other pathological conditions and on images captured in different devices.

6   Conclusion

This paper presents a new method that automatically quantifies the retinal vasculature in the superficial and deep layer. This is done through feature-based segmentation of the larger vessels and capillaries and through extracting metrics like densities and vessel radius from the segmentation. The method is validated on three repeated scans on healthy subjects in normal state (undilated pupil), with dilated pupil, and dilated pupil combined with exercise. The repeated scans are highly correlated showing that the reproducibility of our method is good. We do not see a difference between the scanning scenarios, although there is a difference between each subject demonstrating that the method is sensitive to variation in the structure of the vasculature. We also show that the method can be applied to cataract operated patients. This shows great promise for extending the method in order to investigate if there are features in the microvasculature that are important for the risk of developing macular edema after cataract surgery.

Acknowledgements. We would like to thank Professor Emeritus Knut Conradsen, DTU Compute, and Assistant Professor Anders Nymark, DTU Compute, for valuable assistance and guidance in the statistical analysis.

References

1. Al-Sheikh, M., et al.: Repeatability of automated vessel density measurements using optical coherence tomography angiography. Br. J. Ophthalmol. **101**(4), 449–452 (2017)
2. Campbell, J., et al.: Detailed vascular anatomy of the human retina by projection-resolved optical coherence tomography angiography. Sci. Rep. **7**, 42201 (2017)
3. Chu, Z., et al.: Quantitative assessment of the retinal microvasculature using optical coherence tomography angiography. J. Biomed. Opt. **21**(6), 066008 (2016)
4. Dahl, A., Larsen, R.: Learning dictionaries of discriminative image patches. In: Proceedings of BMVC, pp. 77.1–77.11. BMVA Press (2011)
5. Dahl, A.B., Dahl, V.A.: Dictionary snakes. In: 2014 22nd International Conference on Pattern Recognition (ICPR), pp. 142–147. IEEE (2014)
6. Dahl, A.B., Dahl, V.A.: Dictionary based image segmentation. In: Paulsen, R.R., Pedersen, K.S. (eds.) SCIA 2015. LNCS, vol. 9127, pp. 26–37. Springer, Cham (2015). https://doi.org/10.1007/978-3-319-19665-7_3
7. Lei, J., et al.: Repeatability and Reproducibility of Superficial Macu-lar Retinal Vessel Density Measurements Using Optical Coherence Tomography Angiography En FaceImages. JAMA Ophthalmol. **135**(10), 1092–1098 (2017). https://doi.org/10.1001/jamaophthalmol.2017.3431
8. Eladawi, N., et al.: Early diabetic retinopathy diagnosis based on local retinal blood vessel analysis in optical coherence tomography angiography (OCTA) images. Med. Phys. **45**(10), 4582–4599 (2018)
9. Gadde, S.G., et al.: Quantification of vessel density in retinal optical coherence tomography angiography images using local fractal dimension. Invest. Ophthalmol. Vis. Sci. **57**(1), 246–252 (2016)
10. Early Treatment Diabetic Retinopathy Study Research Group: Classification of diabetic retinopathy from fluorescein angiograms: ETDRS report number 11. Ophthalmology **98**(5), 807–822 (1991)
11. Hwang, T.S., et al.: Automated quantification of capillary nonperfusion using optical coherence tomography angiography in diabetic retinopathy. JAMA Ophthalmol. **134**(4), 367–373 (2016)
12. Iafe, N.A., et al.: Retinal capillary density and foveal avascular zone area are age-dependent: quantitative analysis using optical coherence tomography angiography. Invest. Ophthalmol. Vis. Sci. **57**(13), 5780–5787 (2016)
13. Kashani, A.H., et al.: Optical coherence tomography angiography: a comprehensive review of current methods and clinical applications. Prog. Retin. Eye Res. **60**, 66–100 (2017)
14. Kim, A.Y., et al.: Quantifying microvascular density and morphology in diabetic retinopathy using spectral-domain optical coherence tomography angiography. Invest. Ophthalmol. Vis. Sci. **57**(9), OCT62–OCT370 (2016)
15. Koo, T.K., Li, M.Y.: A guideline of selecting and reporting intraclass correlation coefficients for reliability research. J Chiropr. Med. **15**(2), 155–163 (2016)
16. Mastropasqua, R., et al.: Foveal avascular zone area and parafoveal vessel density measurements in different stages of diabetic retinopathy by optical coherence tomography angiography. Int. J. Ophthalmol. **10**(10), 1545 (2017)

17. McGraw, K.O., Wong, S.P.: Forming inferences about some intraclass correlation coefficients. Psychol. Methods **1**(1), 30 (1996)
18. Rabiolo, A., et al.: Comparison of methods to quantify macular and peripapillary vessel density in optical coherence tomography angiography. PLoS ONE **13**(10), e0205773 (2018)
19. Salarian, A.: Intraclass Correlation Coefficient (ICC) Matlab Central File Exchange (2016). https://se.mathworks.com/matlabcentral/fileexchange/22099-intraclass-correlation-coefficient-icc. Accessed 5 Feb 2019
20. Shoji, T., et al.: Reproducibility of macular vessel density calculations via imaging with two different swept-source optical coherence tomography angiography systems. Transl. Vis. Sci. Technol. **7**(6), 31 (2018)

Validating Segmentation of the Zebrafish Vasculature

Elisabeth Kugler[1,2]($\boxtimes$), Timothy Chico[1,2], and Paul A. Armitage[1]

[1] Faculty of Medicine, Department of Infection,
Immunity and Cardiovascular Disease, University of Sheffield, Sheffield S10 2JF, UK
ekugler1@sheffield.ac.uk
[2] The Bateson Centre, Firth Court, Western Bank, University of Sheffield,
Sheffield S10 2TN, UK

Abstract. The zebrafish is an established vertebrate model to study development and disease of the cardiovascular system. Using transgenic lines and state-of-the-art microscopy it is possible to visualize the vascular architecture non-invasively, *in vivo* over several days. Quantification of the 3D vascular architecture would be beneficial to objectively and reliably characterise the vascular anatomy. So far, no method is available to automatically quantify the 3D cardiovascular system of transgenic zebrafish, which would enhance their utility as a pre-clinical model. Vascular segmentation is essential for any subsequent quantification, but due to the lack of a segmentation "gold standard" for the zebrafish vasculature, no in-depth assessment of vascular segmentation methods in zebrafish has been performed. In this study, we examine vascular enhancement using the Sato et al. enhancement filter in the Fiji image analysis framework and optimise the filter scale parameter for typical vessels of interest in the zebrafish cranial vasculature; and present methodological approaches to address the lack of a segmentation gold-standard of the zebrafish vasculature.

Keywords: 3D · Analysis · *in vivo* · LSFM · Sato filter · Segmentation · Vasculature · Zebrafish

1 Introduction

1.1 Zebrafish as a Model in Cardiovascular Research

Zebrafish are an increasingly used vertebrate model to study mechanisms of cardiovascular development and disease. Characteristics, such as genomic similarity to human, high fecundity, and rapid *ex utero* development allow use of zebrafish in various fields of translational biology [1].

In zebrafish, establishment of a basic cardiovascular system occurs within 24 hours post fertilization (hpf) in parallel to basic body-plan establishment [2], allowing the study of vascular development in a short time window.

© Springer Nature Switzerland AG 2020
Y. Zheng et al. (Eds.): MIUA 2019, CCIS 1065, pp. 270–281, 2020.
https://doi.org/10.1007/978-3-030-39343-4_23

The availability of cell-specific fluorescent transgenic reporter lines allows visualization of structures of interest, such as endothelial cells which outline the vascular lumen, and imaging of these in living animals. Moreover, with the emergence of new microscopy techniques, such as light sheet fluorescence microscopy (LSFM), it is now possible to acquire vascular information in greater anatomical detail and over prolonged periods of time [3]. Together, transgenic lines and LSFM have replaced the need for laborious microangiography to visualize the vasculature and helped overcome limitations of tissue penetration during image acquisition. Hence, datasets acquired are highly rich in information as well as anatomical depth and detail. This means that the limitation of experimental throughput and assessment has become data analysis, rather than data acquisition.

1.2 Previous Work Aimed at Quantifying the Vascular System in Zebrafish

While some aspects of the zebrafish vascular architecture are obvious without quantification (such as increasing network complexity during development), others may be too subtle for human perception (such as diameter changes). Computational quantification of the vascular architecture in 3D is not just more comprehensive (eg. vessel diameter, length, branching, etc.), but also reproducible (eg. overcoming subjective bias or inter-observer variability) than human assessment. However, while quantification of vascular geometry is widely applied in the medical field, it has received less attention in pre-clinical models such as zebrafish.

Analysis of the zebrafish trunk vasculature was previously investigated by Feng et al. [4–7] using microangiography to visualize perfused vessels with segmentation based on active contours with the assumption of locally tubularity of vessels. Segmentation of the trunk vasculature in a transgenic reporter line visualized with LSFM was presented in a technical report using a vascular enhancement method based on the assumption that vessels were locally tubular [8,9]. This report did not give further specification about parameter optimization or applicability of the presented method to other vascular beds or transgenic reporter lines. Quantification of left hind-brain vessels in zebrafish was previously performed by Tam et al. [10]. Measurements of vessel density and diameter were performed using commercial software. Unfortunately, no further specification was given about pre-processing steps or parameter settings, so replication of this method is not possible. Quantification of the mid-brain vasculature in zebrafish was performed by Chen et al. [11]. Measurements obtained included branching points, diameter, and vascular hierarchy. However, the use of commercial software and lack of documentation again prevented any further assessment of the proposed approach. In addition, the implemented method focused only on a sub-region of the cranial vasculature (mid-brain vessels called *middle mesencephalic cerebral arteries* and *posterior mesencephalic cerebral arteries*), rather than the whole cranial vasculature.

Previously we presented methods to enhance the zebrafish cranial vasculature using general filtering (GF; Median Filter and Rolling ball) [12] and enhancement utilizing the Hessian matrix with the assumption of local vessel tubularity, based on the filter proposed by Sato et al. [9,13]. This was further complemented by investigation of different segmentation approaches which were readily implemented in the Fiji image analysis framework [14]. We showed that Otsu thresholding delivered more robust segmentation results than the tested k-means clustering, statistical region merging, or fast-marching level set method implementations.

We showed that lumenized vessels displayed a cross-sectional double-peak intensity distribution, while small/unlumenized vessels had a single-peak distribution. The appearance of both within the same sample cautioned us to produce an image analysis pipeline, which would enhance and segment lumenized and unlumenized vessels to a similar degree.

1.3   Contributions of This Work

In this work, we extend our previous study by providing an analysis of the impact of input shapes (modelled tubes) as well as of the scale (sigma) of the Sato filter on the segmentation outcome; and also quantitative measures using the Full-Width Half Maximum (FWHM).

While reasonably extensive anatomical knowledge and some phantom models are available in the medical field for validation of vessel segmentation and quantification approaches, these points of reference are lacking for zebrafish. Hence, we present several methodological approaches to allow the assessment of segmentation robustness and sensitivity without the availability of a gold-standard or phantom model for the zebrafish cranial vasculature. This was achieved by analysis of datasets which challenged segmentation robustness and sensitivity to understand whether true biological quantifications could be extracted. These examined datasets were composed as follows: **(i)** Investigation of data with a controlled decrease of image quality by reduction of laser power during repeated image acquisition to assess noise sensitivity. **(ii)** Vascular volume quantification in a double-transgenic line with two reporter lines displaying different image properties (signal levels and fluorophore expression pattern), to assess whether similar segmentation robustness could be achieved. **(iii)** Segmentation and volume quantification in samples prior to and after exsanguination to assess segmentation sensitivity to subtle biological changes.

2   Materials and Methods

2.1   Zebrafish Husbandry

Conduction of experiments was done conforming to the rules and guidelines of institutional and UK Home Office regulations under the Home Office Project Licence 70/8588 held by TC. Maintenance of adult transgenic zebrafish

Tg(fli1a:eGFP)^{y1} [15] and *Tg(kdrl:HRAS-mCherry)^{s916}* [16] was conducted according to previously described standard husbandry protocols [17]. Embryos, obtained from controlled mating, were kept in E3 medium (5mM NaCl, 0.17mM KCl, 0.33mM CaCl and 0.33mM MgSO$_4$ diluted to 1X E3 with distilled H$_2$O) buffer with methylene blue and staged according to Kimmel et al. [2].

2.2   Image Acquisition Settings, Properties and Data Analysis

Anaesthetized embryos were embedded in 2% LM-agarose with 0.01% Tricaine in E3 (MS-222, Sigma). Data acquisition of the cranial vasculature was performed using a Zeiss Z.1 light sheet microscope, Plan-Apochromat 20x/1.0 Corr nd = 1.38 objective, dual-side illumination with online fusion, activated Pivot Scan, image acquisition chamber incubation at 28 °C, and a scientific complementary metaloxide semiconductor (sCMOS) detection unit. The properties of acquired data were as following: 16bit image depth, 1920 × 1920 × 400-600 voxel (x, y, z) image with an approximate voxel size of 0.33 × 0.33 × 0.5 μm, respectively. Multi-colour images in double-transgenic embryos were acquired in sequential mode. All image analysis, pre-processing and segmentation were performed using the open-source software Fiji [14]. Motion correction was performed as in [12].

2.3   Datasets

Modelled Tubes. Modelled tubes (hollow, filled and Gaussian blurred) were produced manually with uniform signal intensity of 255 against zero background intensity circular ROI selection using Fiji [14]. Tubes were produced to resemble the following biological settings: **(i) hollow tubes** - 20 μm outer diameter (1.13 μm wall thickness), 8.3 μm outer diameter (0.8 μm wall thickness), 5 μm outer diameter (0.6 μm wall thickness); resembling lumenized vessels; **(ii) filled tubes** - resembling unlumenized vessels (with the same outer diameter as above); **(iii) Gaussian blurred tubes** - Gaussian filter with sigma 5vx; resembling a more realistic intensity distribution of fluorescence; **(iv) increasing noise** - Gaussian white noise with standard deviation 25, 50 or 100 (zero background intensity); resembling autofluorescence and background noise.

Transgenic Zebrafish. Data were acquired as described in Sect. 2.2. Data to test vascular enhancement approaches were acquired at 4 days post fertilization (dpf). Data analysed for assessment of segmentation robustness included the following: **(i)** Dataset with controlled decrease of vascular contrast-to-noise ratio (CNR) by decrease of laser power during repeated acquisition [12] (laser power 1.2%, 0.8% and 0.4%; exposure 30 ms for all; n = 10 embryos from 2 experimental repeats); **(ii)** 4dpf double-transgenic *Tg(fli1a:eGFP)^{y1}*, *Tg(kdrl:HRAS-mCherry)^{s916}* (n = 21 embryos from 2 experimental repeats); **(iii)** Exsanguination by mechanical opening of heart cavity with forceps was done in 4dpf *Tg(kdrl:HRAS-mCherry)^{s916}* (n = 16 embryos from 2 experimental repeats).

2.4 Image Enhancement, Segmentation and Total Vascular Volume Measurement

The following vascular enhancement methods were studied: (**i**) General filtering (GF): 2D median filter with a radius of 6 voxels (13-by-13 neighbourhood) [18] and rolling ball algorithm of size 200 [19], as presented in [12]. (**ii**) Tubular Filtering (TF): Enhancement based on the line enhancement filter presented by Sato et al. [9] with the assumption of local vessel tubularity, as implemented into the Fiji Tubeness Plugin by Mark Longair, Stephan Preibisch and Johannes Schindelin [14], as presented in [13]. Segmentation of enhanced images was performed using global Otsu thresholding [20] to distinguish vascular from non-vascular information. Following segmentation, the total dorsal cranial vascular volume was quantified in the cranial region of interest (ROI) as described in [12].

2.5 Tubular Filtering Enhancement Evaluation

The modelled tubes were used to establish the Sato filter response for different types of tubes when the filter is applied with varying scale parameters. We were particularly interested in establishing how the filter responds to a double-peak distribution from a hollow tube, as typically found in lumenized vessels. At present, these responses were judged by visual assessment. Manual diameter measurements were taken from a variety of cranial vessels using original images in the dataset consisting of 4dpf embryos (single measurement in z-plane at the middle of vessel; $n = 12$ 4dpf embryos). Vessels studied in this work were the central artery (CtA), middle mesencephalic central artery (MMCtA), primordial midbrain channel (PMBC) and basal aorta (BA). Intensity profiles were plotted for each vessel using a line ROI in original, enhanced, and segmented images (obtained from same measurement position as manual measurement). FWHM was quantified from extracted cross-sectional line ROIs using Matlab. Images were segmented using the global Otsu thresholding approach [13] to extract the vasculature and produce binary masks from which corresponding intensity profiles could be plotted and vessel diameters identified. For data representation intensity profiles and vessel diameter measurements were averaged.

2.6 Statistics and Data Representation

Gaussian distribution conformation was evaluated using the D'Agostino-Pearson omnibus test [21]. Statistical analysis was performed using One-way ANOVA or paired Students t-test in GraphPad Prism Version 7 (GraphPad Software, La Jolla California USA). Statistical significance was represented as: $p < 0.05$ *, $p < 0.01$ **, $p < 0.001$ ***, $p < 0.0001$ ****. Graphs show mean values ± standard deviation. Image representation and visualization was done with Inkscape Version 0.48 (https://www.inkscape.org). Images were visualized as maximum intensity projections (MIPs) using false-colour representation and intensity inversion where appropriate.

3   Results and Discussion

3.1   Sato Enhancement in Fiji

The Sato filter was originally developed for vessels visualized with MRI, which display a single-peak cross-sectional intensity profile that can be modelled with a Gaussian distribution. In our data, we found that enhancement of vessels with a filter scale (sigma) equivalent to the vessel diameter changed the original double-peak distributions to a single-peak (Fig. 1A), while this was not the case when vessels exceeded sigma size (Fig. 1b). To understand how the filter would enhance possible scenarios of the vasculature in transgenic zebrafish we examined the following modelled tubes:

(i) hollow tubes: edges were successfully enhanced at a smaller scale (5 μm). Hollow tubes were converted to filled tubes (double-to-single peak conversion) when enhancement scale was at the size of tubes (10 μm; Fig. 1C). *(ii) filled tubes:* tubes were successfully enhanced (Fig. 1D). *(iii) Gaussian blurred tubes:* enhancement was similar to unblurred tubes (Fig. 1E). *(iv) increasing noise:* addition of artificial noise at levels of 25 and 50 did significantly not alter enhancement (Fig. 1F), while addition at a level of 100 noticeably decreased enhancement quality.

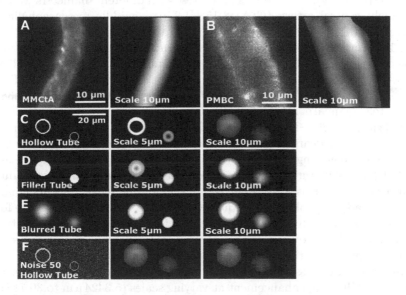

Fig. 1. (A, B) Vessel enhancement based on the filter proposed by Sato et al. [9] successfully enhanced vessels. Filter responses were tested on modelled tubes with (C) 3D hollow tubes, (D) 3D filled tubes, (E) 3D filled tubes with Gaussian blur, and (F) increasing noise levels.

Together, these data suggested that enhancement at the scale of tubes of interest would return single-peak intensity distributions regardless of the starting model. Thus, lumenized and unlumenized vessels should be equally well enhanced when enhancement filter size was at the size of vessels of interest. Our data also suggest that the Sato filter resulted in equivalent enhancement over a reasonable range of noise levels. However, visual assessment of the enhanced modelled tubes does suggest that there may be some enlargement of the apparent vessel size, when the filter is applied at a scale equivalent to the vessel diameter, although this is not obvious in the zebrafish data. This finding requires further investigation.

3.2 Influence of Sato Vessel Enhancement Scale on Measured Vessel Diameter

We next wanted to assess quantitatively the impact of the Sato enhancement scale parameters on measured vascular diameter. This is particularly important given the possible enlargement of apparent vessel size identified in the previous section. We also examined whether Full-Width at Half Maximum (FWHM) was an appropriate measure to estimate vessel diameter in our datasets and could be used in the future to automatically quantify vessel diameters. Hence, we compared manually measured vessel diameters to FWHM from original cross-sectional intensity distributions (henceforth referred to as manual and original FWHM, respectively). Four vessels of interest with different diameters were chosen. Namely, central artery (CtA) $8.15 \pm 1.27\,\mu m$, middle mesencephalic central artery (MMCtA) $9.78 \pm 2.09\,\mu m$, primordial midbrain channel (PMBC) $11.14 \pm 1.68\,\mu m$, basal aorta (BA) $22.28 \pm 3.89\,\mu m$ (manual measurements). No statistically significant differences between manual and original FWHM measurements were found (Fig. 2A; CtA $p > 0.9999$, MMCtA $p > 0.9999$, PMBC $p > 0.9999$, BA p 0.9879) and absolute error was as follows: CtA $1.47 \pm 0.38\,\mu m$ (17.98% from manual), MMCtA $0.92 \pm 0.23\,\mu m$ (9.39% from manual), PMBC $1.38 \pm 0.44\,\mu m$ (12.29% from manual), BA $3.89 \pm 1.23\,\mu m$ (17.41% from manual). Assessment of Pearson Correlation and Bland-Altman test showed no systematic error, suggesting that the observed differences were random. We anticipate that the encountered differences were mainly due to human error during manual measurements as well as outliers in the original measurements (Fig. 2A, black arrowhead). The observed outliers (3 out of 48 measurements) were found to be caused by strongly asymmetric cross-sectional double-peak intensity distributions. Together, we suggest that the FWHM can be used as a measure of vascular diameter with the caveat that outliers need to be considered.

Quantification of vessel FWHM after Sato enhancement and diameter after segmentation following enhancement at varying scales ($5.3424\,\mu m$ to $30.718\,\mu m$) was performed. For the CtA, MMCtA and PMBC, the highest correlation between manual measurements and vessel diameter after enhancement and segmentation were found when the enhancement scale was at the size of the vessel of interest (Fig. 2C, D, E). When the enhancement scale exceeded the size of the vessel of interest, segmented vessel diameter was increased and led to overestimation (Fig. 2C, E; green and red). The only exception for this was found in the BA, for which segmented vessel diameter was not drastically changed by changes of scale size (Fig. 2E). Thus, we examined the cross-sectional intensity distributions in more detail and found that double-to-single peak conversions were observed at the following scales in the CtA 8.0232, MMCtA 9.3604 and PMBC 10.6848, while in the BA this conversion was only observed at a scale of 23.718.

To further examine the overall impact of enhancement scales on the segmentation outcome, we quantified the dorsal cranial vascular volume after enhancement at varying scales. We found that the vascular volume increased with enhancement scale size (Fig. 2G); highlighting the necessity for in-depth parameter assessments tp avoid introducing systematic errors.

As the scale 10.6848 showed double-to-single peak conversion of most examined vessels, the following experiments were performed at this scale. In future, it needs to be examined whether optimum scales for individual vessels can be combined into a multi-scale approach to further improve enhancement and segmentation outcomes.

3.3   Assessment of Segmentation Robustness and Sensitivity

Images with Decreasing Image Quality. To evaluate quantitatively whether TF was more accurate and robust than the previously suggested GF approach [12], both pre-processing methods were applied to a dataset with decreasing image quality (decreasing CNR by decrease of laser power (LP) during image acquisition). After segmentation, the vascular volume was quantified. For both pre-processing methods, the returned volume was not statistically significantly changed by the CNR decrease (Fig. 3). However, the coefficient of variation (CoV), within groups was higher after GF (**LP 0.4:** GF 17.56%, TF 9.14%; **LP 0.8:** GF 14.63%, TF 6.97%; **LP 1.2:** GF 18.76%, TF 7.81%). Visual assessment showed that GF returned more noise-speckles as vascular-positive-voxels in images with lower quality. This was probably the reason for an increase of volume with decrease of image quality after GF, which was not observed after TF.

Double-Transgenic Samples. To further quantify segmentation robustness, images were acquired in the double-transgenic $Tg(kdrl{:}HRAS\text{-}mCherry)^{s916}$, $Tg(fli1a{:}eGFP)^{y1}$. For these, it has been previously shown that CNR levels are statistically significantly lower [12], but, also, that the *fli1a* construct has a broader expression pattern (pan-endothelial; not exclusively vascular).

Hence, segmentation was expected to be more challenging for this transgenic due to lower image quality and that it would return a slightly larger vascular volume, as additional non-vascular anatomy would be segmented. Again, GF and TF pre-processing were applied and segmentation performed. Visual assessment showed that $Tg(fli1a{:}eGFP)^{y1}$ was markedly less well extracted after GF than TF. This was confirmed by quantitative comparison of the extracted vascular volume (Fig. 3B). The volume extracted from the $Tg(kdrl{:}HRAS\text{-}mCherry)^{s916}$ was not statistically significantly different between GF and TF approaches (p > 0.9999), but the extracted volume after GF showed a higher CoV (CoV 24.37%) than after TF (CoV 10.84%). Extracted volume from

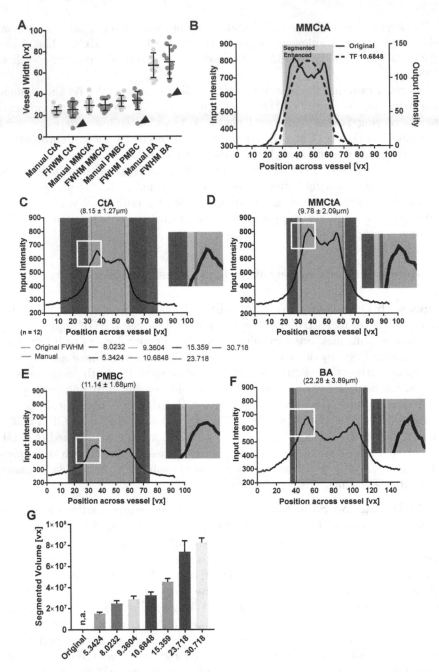

Fig. 2. (A) FWHM was found to be a suitable measurement of vessel width (n = 12). (B) FWHM of enhanced and segmented profiles highly correlated. (C–F) Vessel width after enhancement and segmentation increased with enhancement scale size (black line is original intensity profile). Best agreement of manual measurements and vessel width was found when enhancement scale was at the size of the vessel of interest. (G) Vascular volume quantification after enhancement and segmentation was found to increase with scale size during the enhancement step. (Color figure online)

the $Tg(fli1a{:}eGFP)^{y1}$ was not statistically significantly different (p = 0.1016), but was more variable after GF (CoV 35.97%) than TF (17.28%; p 0.1016). These data suggested that TF prior to segmentation was more accurate in comparison to GF, especially in data with lower CNR.

Exsanguinated Samples. To test whether subtle biological changes could successfully be detected, the vascular volume was quantified in embryos with decreased vascular volume (exsanguinated) but preserved anatomy. We therefore compared data acquired in the same animals pre- and post-exsanguination. Quantification after GF showed no statistically significant difference between control and exsanguinated samples (p = 0.2596; Fig. 4A), while volumes were statistically significant different after TF (p<0.0001; Fig. 4B; n = 16 4dpf larvae; 2 experimental repeats). Again, high CoVs were found for volumes after GF (control 38.26%; exsanguinated 26.28%), while lower CoVs were found after TF (control 10.22%; exsanguinated 9.65%). These data suggested that TF is more sensitive, allowing the detection of true biological differences, while the high variance after GF would not allow for this.

3.4   Conclusion

In this work, we presented an in-depth investigation of the Sato filter implementation, in the Fiji image analysis framework, when applied to the enhancement of the zebrafish cranial vasculature. We presented an analysis of the impact of scale size on modelled tubes as well as real data, finding that Sato enhancement worked equally well on hollow tubes, filled tubes and was robust to the effect of noise, suggesting that lumenized and unlumenized vessels would be enhanced

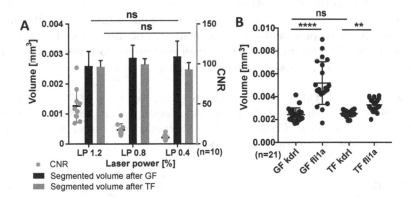

Fig. 3. (A) Vascular volume in images with decreasing CNR was statistically significantly different for GF (black bars p 0.3248) or TF (gray bars p 0.9981; n = 10 4dpf larvae). (B) The vascular volume of $Tg(kdrl{:}HRAS\text{-}mCherry)^{s916}$ was not statistically significant different (p > 0.9999; n = 21 4dpf larvae from 2 experimental repeats, Kruskal-Wallis test). Segmentation of the $Tg(fli1a{:}eGFP)^{y1}$ resulted in highly variable segmentation after GF (35.97%), while TF had a lower coefficient of variance with 17.28%.

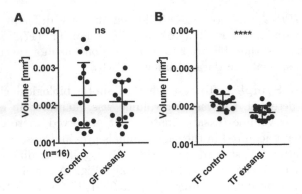

Fig. 4. Vascular volume compared in the same sample before and after exsanguination. It was found that GF prior to segmentation delivered more variable results (A), when compared to TF (B); allowing to extract true biological differences between samples. *Abbr.: exsang. - exsanguinated*

similarly. We optimised the enhancement scale parameter for typical vessels of interest in the zebrafish cranial vasculature and demonstrated that our automated segmentation approach produced vessel diameter measurements that were in excellent agreement with manual measurements. As neither phantom data nor a gold-standard for the zebrafish vasculature exist, we suggested methodological approaches to further evaluate our segmentation performance, studying its robustness to the artificial decrease of image quality, assessment of transgenic reporter lines with different CNR levels, as well as vascular volume quantification pre- and post-exsanguination. Concluding, the proposed segmentation approach allowed for the objective quantification of cranial vascular volume as a read-out of the vascular phenotype in embryonic transgenic zebrafish; facilitating the study of mechanisms of vascular development and disease, as well as the effect of drugs or chemical components. This shall contribute to the use of zebrafish as a pre-clinical model in cardiovascular research.

Acknowledgments. This work was supported by a University of Sheffield, Department of Infection, Immunity and Cardiovascular Disease, Imaging and Modelling Node Studentship awarded to EK. The Zeiss Z1 light-sheet microscope was funded via British Heart Foundation Infrastructure Award awarded to TC.

References

1. Gut, P., Reischauer, S., Stainier, D.Y.R., Arnaout, R.: Little fish, big data: zebrafish as a model for cardiovascular and metabolic disease. Physiol. Rev. **97**(3), 889–938 (2017)
2. Kimmel, C.B., Ballard, W.W., Kimmel, S.R., Ullmann, B., Schilling, T.F.: Stages of embryonic development of the zebrafish. Dev. Dyn. **203**(3), 253–310 (1995)

3. Huisken, J., Swoger, J., Del Bene, F., Wittbrodt, J., Stelzer, E.H.K.: Optical sectioning deep inside live embryos by selective plane illumination microscopy. Science **305**(5686), 1007–1009 (2004). (New York)
4. Feng, J., Ip, H.H.S., Cheng, S.H., Chan, P.K.: A relational-tubular (ReTu) deformable model for vasculature quantification of zebrafish embryo from microangiography image series. Comput. Med. Imaging Graph. Off. J. Comput. Med. Imaging Soc. **28**(6), 333–344 (2004)
5. Feng, J., Cheng, S.H., Chan, P.K., Ip, H.H.S.: Reconstruction and representation of caudal vasculature of zebrafish embryo from confocal scanning laser fluorescence microscopic images. Comput. Biol. Med. **35**(10), 915–931 (2005)
6. Feng, J., Ip, H.H.S.: A statistical assembled deformable model (SAMTUS) for vasculature reconstruction. Comput. Biol. Med. **39**(6), 489–500 (2009)
7. Ip, H., Feng, J., Cheng, H.: Automatic segmentation and tracking of vasculature from confocal scanning laser fluorescence microscope images sequences using multi-orientation dissection sections. In: Proceedings of IEEE, pp. 249–252 (2002)
8. Schneider, S.: Segmentation of zebrafish vasculature. Technical report, MOSAIC group (2015)
9. Sato, Y., et al.: 3D multi-scale line filter for segmentation and visualization of curvilinear structures in medical images. In: Troccaz, J., Grimson, E., Mösges, R. (eds.) CVRMed/MRCAS -1997. LNCS, vol. 1205, pp. 213–222. Springer, Heidelberg (1997). https://doi.org/10.1007/BFb0029240
10. Tam, S., et al.: Death receptors DR6 and TROY regulate brain vascular development. Dev. Cell **22**(2), 403–417 (2012)
11. Chen, Q., et al.: Haemodynamics-driven developmental pruning of brain vasculature in zebrafish. PLoS Biol. **10**(8), 1–18 (2012)
12. Kugler, E., Chico, T., Armitage, P.: Image analysis in light sheet fluorescence microscopy images of transgenic zebrafish vascular development. In: Nixon, M., Mahmoodi, S., Zwiggelaar, R. (eds.) MIUA 2018. CCIS, vol. 894, pp. 343–353. Springer, Cham (2018). https://doi.org/10.1007/978-3-319-95921-4_32
13. Kugler, E., Plant, K., Chico, T., Armitage, P.: Enhancement and segmentation workflow for the developing zebrafish vasculature. J. Imaging **5**(1), 14 (2019)
14. Schindelin, J., et al.: Fiji - an open source platform for biological image analysis. Nat. Methods **9**(7), 676–682 (2012)
15. Lawson, N.D., Weinstein, B.M.: In vivo imaging of embryonic vascular development using transgenic zebrafish. Dev. Biol. **248**(2), 307–318 (2002)
16. Chi, N.C., et al.: Foxn4 directly regulates tbx2b expression and atrioventricular canal formation. Genes Dev. **22**(6), 734–739 (2008)
17. Westerfield, M.: The Zebrafish Book: A Guide for Laboratory Use of Zebrafish (Brachydanio Rerio), 2nd edn. University of Oregon Press, Corvallis (1993)
18. Lim, J.: Two-Dimensional Signal and Image Processing, pp. 469–476. Prentice Hall, Englewood Cliffs (1990)
19. Sternberg, S.: Biomedical image processing. Computer **16**, 22–34 (1983)
20. Otsu, N.: A threshold selection method from gray-level histograms. Trans. Sys. Man Cybern. **9**(1), 62–66 (1979)
21. D'Agostino, R.B., Belanger, A.: A suggestion for using powerful and informative tests of normality. Am. Stat. **44**(4), 316–321 (1990)

Analysis of Spatial Spectral Features of Dynamic Contrast-Enhanced Brain Magnetic Resonance Images for Studying Small Vessel Disease

Jose Bernal[1]([✉])(iD), Maria del C. Valdés-Hernández[1](iD), Javier Escudero[2](iD),
Paul A. Armitage[3](iD), Stephen Makin[4](iD), Rhian M. Touyz[4](iD),
and Joanna M. Wardlaw[1](iD)

[1] Centre for Clinical Brain Sciences, University of Edinburgh, Edinburgh, UK
{jose.bernal,m.valdes-hernan}@ed.ac.uk
[2] School of Engineering, Institute for Digital Communications,
University of Edinburgh, Edinburgh, UK
[3] Academic Unit of Radiology, University of Sheffield, Sheffield, UK
[4] Institute of Cardiovascular and Medical Sciences,
University of Glasgow, Glasgow, UK

Abstract. Cerebral small vessel disease (SVD) comprises all the pathological processes affecting small brain vessels and, consequently, damaging white and grey matter. Although the cause of SVD is unknown, there seems to be a dysfunction of the small vessels. In this paper, we propose a framework comprising tissue segmentation, spatial spectral feature extraction, and statistical analysis to study intravenous contrast agent distribution over time in cerebrospinal fluid, normal-appearing and abnormal brain regions in patients with recent mild stroke and SVD features. Our results show the potential of the power spectrum for the analysis of dynamic contrast-enhanced brain MRI acquisitions in SVD since significant variation in the data was related to vascular risk factors and visual clinical variables that characterise the burden of SVD features. Thus, our proposal may increase sensitivity to detect subtle features of small vessel dysfunction. A public version of our framework can be found at https://github.com/joseabernal/DynamicBrainMRIAnalysis.git.

Keywords: Spatial spectral analysis · Functional principal component analysis · Dynamic brain magnetic resonance image · Cerebral small vessel disease

1 Introduction

Small vessel disease of the brain (SVD) comprises multiple pathological processes affecting small cerebral arteries leading to damage of white and grey matter (WM and GM, respectively) [17]. SVD is a serious problem that has been associated with cognitive decline, physical fragility, depression, dementia, and stroke [17].

© Springer Nature Switzerland AG 2020
Y. Zheng et al. (Eds.): MIUA 2019, CCIS 1065, pp. 282–293, 2020.
https://doi.org/10.1007/978-3-030-39343-4_24

Although the cause(s) of SVD remain unclear, there seems to be dysfunction of the small vessels. This issue may be studied using dynamic contrast-enhanced MRI (DCE-MRI). However, factors such as scanner signal drift, tissue variations, and imaging artefacts introduce systematic errors hampering assessing small vessel dysfunction accurately.

Dynamic brain MRI acquisitions are commonly temporally sparse and spatially dense. Thus, proposals usually find alternative data representations to reduce the dimensionality of the data while retaining critical information [7,8]. The power spectrum [1] is a common approach in digital signal processing to describe the strength of frequency components into the overall signal. It has been successfully applied in dynamic susceptibility contrast brain MRI to characterise neurophysiological and hemodynamic patterns of Alzheimer's disease [8]. We hypothesise that spatial spectral feature analysis of dynamic post-contrast signal changes can identify tissue differences in tissues that relate to the burden of SVD features and clinical factors.

In this paper, we propose a framework to study contrast signal-time trajectory in healthy and pathological brain regions in dynamic contrast-enhanced (DCE) brain MRI acquisitions. The framework comprises segmentation, spatial spectral and functional data analyses, and statistical group comparison. The contributions of this work are: (i) we showcase a fully functional framework to analyse DCE-MRI acquisitions using spatial spectral and functional data analyses jointly, and (ii) we describe an application of our framework to the study of DCE-MRI signals of a relatively large cohort ($n = 201$) with various extents of SVD features.

2   Methods

2.1   Analysis Framework

The processing pipeline consists of three steps. First, all regions of interest are segmented for each patient in the cohort. Second, signals measured in each brain region are described using the power spectrum. Third, spectral features are examined using statistical tests to establish whether they vary with any of the clinical variables. A scheme of the pipeline applied per patient is displayed in Fig. 1. Further details of each step are provided in the following sections.

2.2   Subjects and Clinical Scores

We used data from a prospective study of patients with recent mild stroke and SVD features ($n = 201$ subjects, 79 women). The sample clinical characteristics have been published previously [13,15]; those relevant to this study are summarised in Table 1. The DCE-MRI acquisition parameters have been detailed in [6]. Patients were scanned at approximately one month after the first stroke presentation. Following a pre-contrast acquisition, an intravenous bolus injection of gadolinium was administered with the start of 20 further acquisitions

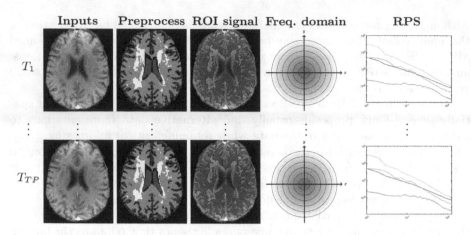

Fig. 1. High level schematic of our processing pipeline per patient. RPS refers to the radial power spectrum. On the left, T_1, T_2, ..., T_{TP} refer to each one of the time points. The inputs are the post intravenous gadolinium contrast brain MRI sequences, T1-w, T2-w, FLAIR and the susceptibility-weighted images. Initially, we segment the regions of interest using the different imaging sequences. Then, we study the dynamic signals per region of interest using the obtained segmentation masks. Next, we convert the signals to the spatial frequency domain using the Fourier transform. Finally, we average power values over all frequencies in concentric rings of a specific width. In the RPS column, the lines represent the RPS compute for cerebrospinal fluid (blue), deep grey matter (orange), normal-appearing white matter (yellow), and white matter hyperintensities (purple). (Color figure online)

with a temporal resolution of 73s, leading to a DCE-MRI duration of about 24 minutes ($\approx$ 21 time points). We considered only those after the 4th time point to minimise initial perfusion effects that occur before the contrast agent is well mixed within the blood/intravascular compartment.

The following baseline clinical and demographic variables were obtained from the study database: biological sex (m/f), smoker (y/n), diabetes (y/n), hyperlipidemia (y/n) defined as a previous diagnosis, or diagnoses at time of stroke of a total cholesterol over 5mmol/L, hypertension (y/n) defined as a previous history of hypertension, or hypertension diagnosed at presented of stroke. Additionally, we considered visual clinical ratings recorded at inclusion. In particular, we utilised Fazekas [2], basal ganglia perivascular spaces (BGPVS) [10], and total SVD [12] scores to account for the location, presence and size of WMH, the existence of enlarged PVS on the basal ganglia, and the burden of four MRI features of the SVD (lacunes, microbleeds, PVS, and WMH). We summed up periventricular and deep WM scores to obtain a total Fazekas score that ranged from zero to six [4,14]. A senior and experienced neuroradiologist generated all visual scores.

Table 1. Clinical variables from the sample relevant for this study. The first column lists the variables of interest, the second one shows the frequency and approximated relative frequency of that variable.

Clinical variable	No. of patients (% of the total)
Hypertension	150 (74.6%)
Diabetes	25 (12.4%)
Hyperlipidemia	120 (59.7%)
Smoker	130 (64.7%)
Fazekas score	
0	6 (3.0%)
1	16 (8.0%)
2	74 (36.8%)
3	23 (11.4%)
4	32 (14.9%)
5	20 (10.0%)
6	32 (15.9%)
BGPVS score	
0	3 (1.5%)
1	102 (50.7%)
2	54 (26.9%)
3	26 (12.9%)
4	16 (8.0%)
Total SVD score	
0	67 (33.3%)
1	48 (23.9%)
2	46 (22.9%)
3	27 (13.4%)
4	13 (6.5%)

2.3   Segmentation of Regions of Interest

We examined four regions of interest comprising both normal-appearing and abnormal brain regions: cerebrospinal fluid (CSF), deep GM (DGM), normal-appearing WM (NAWM), and WMH. Each segmentation was performed following the protocol described in [13]. Trained analysts double-checked and manually edited all segmentation masks under the supervision of an experienced neuroradiologist. To avoid partial volume effects, we eroded the resulting binary masks. The segmentation methods were evaluated previously against manually obtained reference segmentations, in images acquired with similar scanning parameters and on the same scanner like the ones this study uses [13]. On 150 individuals, the mean difference for ICV segmentations was 2.7% (95% CI ±7%). On 20 individuals, the Jaccard Index was 0.98 (95% CI = ±0.03) for WM, 0.46 (95% CI = ±0.12) for CSF, and 0.61 (95% CI = ±0.37) for WMH. In a test-retest analysis on 14 cases comprising volunteers and patients with mild non-disabling stroke, the coefficient of variation for repeated measurements of the segmentation technique was 0.21 [5]. Further information concerning inter-analyst agreements can be found in [13].

2.4   Power Spectral Features of the Regions of Interest

We characterised the signals gauged within normal-appearing and abnormal tissues in dynamic brain MRI acquisitions using the radial power spectrum (RPS) [1]. The steps to compute it for each time point and each region of interest are three-fold. First, we selected the signal in the region of interest using the segmentation masks. Second, we used the 2D discrete Fourier transform to obtain a representation of each axial slice forming the region of interest in the frequency domain. Let $I \in \mathbb{R}^{N \times N \times K}$ be a brain MR volume with K axial slices and $f_k(x, y)$ be its k-th axial slice, the corresponding discrete Fourier transform for each slice, $F_k(u, v)$, is expressed as follows

$$F_k(u, v) = \sum_{i=0}^{N-1} \sum_{j=0}^{N-1} f_k(i, j) \exp\left(-2\iota\pi\left(\frac{ui}{N} + \frac{vj}{N}\right)\right). \tag{1}$$

Third, we computed the magnitude spectra and averaged all the frequencies over concentric rings of width 1 using the following formula

$$R^{(s)} = \frac{1}{K} \sum_{k=1}^{K} \frac{1}{2\pi} \int_0^{2\pi} |F_k(s\cos(\theta), s\sin(\theta))|\, d\theta, \tag{2}$$

where $s = \sqrt{u^2 + v^2}$ and $\theta = \tan^{-1}\left(\frac{v}{u}\right)$ are polar coordinates. For each time point and each region of interest, the signal was described using 129 frequencies.

2.5   Functional Data Analysis

Each one of the elements of the RPS can be seen as a function in time. In such a way, we could find the eigenvalues and eigenfunctions that better describe the different observations. We followed the method proposed by Happ & Greven [3]. Let $D = 129$ be the number of elements under study at each time point, $P = 201$ the number of patients, and $R = \{R^{(1)}, ..., R^{(D)}\}$ the set of elements, each of them described by the corresponding measurements, $r_1^{(j)}$, $r_2^{(j)}$, ..., $r_P^{(j)}$, the overall process is four-fold. First, each variable was centred by subtracting its mean value. Second, eigenfunctions and scores were calculated for each variable using the functional PCA. The principal component functions were obtained constructively by finding orthogonal functions $\Phi_k^{(j)}$, $k = 1, ..., M^{(j)}$, for which principal component scores $\xi_{ik}^{(j)}$, $i = 1, ..., P$, mathematically expressed as

$$\xi_{ik}^{(j)}(t) = \int \Phi_k^{(j)}(t)\, r_i^{(j)}(t) dt, \tag{3}$$

maximised $\sum_i \xi_{ik}^{(j)^2}$, subject to $||\Phi_k^{(j)}||^2 = 1$, were $t \in [4, 5, 6, ..., 21]$ represents each time point. We set $M^{(j)}$ to five as resulting eigenvectors accounting for the 99% of the univariate variation. Third, all of these scores $\xi_{ik}^{(j)}$ were arranged in a matrix form, $\Xi \in \mathbb{R}^{P \times \sum M^{(j)}}$, such that the ith row contained $(\xi_{i1}^{(1)}, ..., \xi_{iM^{(1)}}^{(1)}, ..., \xi_{i1}^{(p)}, ..., \xi_{iM^{(p)}}^{(p)})$. Fourth, scores were calculated using eigenanalysis on the covariance matrix of Ξ. We resorted to using the first three modes of variation which explained around the 99% of the data variation. The output at this stage was a score for per mode of variation for each subject.

2.6 Validation Against Clinical Parameters

We used the Kruskal-Wallis test to determine whether the features vary with any of the clinical parameters. We verified that the analysed variables were not normally distributed using the Shapiro-Wilk test. The null hypothesis was that subjects with different clinical scores exhibit similar feature values. We computed the test in R version 3.5.1 and corrected the p-values for multiple comparisons using the Benjamini-Hochberg false discovery rate control method.

3 Experiments and Results

The application of our framework to the case study was performed as follows. We segmented each of the 201 DCE-MRI scans, measured the signal gauged in each region of interest using the resulting tissue masks, calculated the RPS of each time point and each region of interest, and extracted spectral measurements. Once we obtained the spectral features for each patient, we compared them using statistical analysis to explore whether they vary with clinical variables. An example of the radial power spectra calculated from signals gauged in the WM hyperintense (WMH) regions with the lowest and highest WMH burden is shown in Fig. 2. As it can be observed, the two groups of patients exhibit different spectra.

We compared two functional data analysis approaches: a time-averaged and a dynamic RPS, the latter taking signal over time into account. The corresponding significance values obtained from the Kruskal-Wallis test are presented in Tables 2 and 3, respectively. Apart from considering the RPS of each region of interest independently, we analysed them jointly as well. The corresponding results are displayed in the column "All" of both tables. In most of the cases, both approaches coincided. However, there were slight differences. The spectral features computed in the CSF region varied with Fazekas scores, but the method considering the temporal dimension exhibited stronger evidence (p-value < 0.0001) than the time-averaged scheme (p-value < 0.001). Similarly, the scores yielded by the dynamic RPS showed more evidence of variation with the

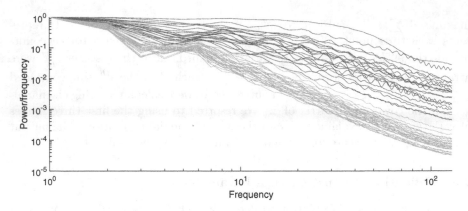

Fig. 2. Radial power spectra calculated from real signals in the white matter hyperintense regions for patients with Fazekas score 6 (red) and Fazekas score 0 and 1 (blue). Each line corresponds to the spectrum of a patient in the time point 4. The values were normalised by the DC value for visualisation purposes. (Color figure online)

Table 2. Kruskal-Wallis test results obtained when comparing the functional principal component (PC) scores extracted from the time-averaged RPS of subjects grouped by clinical variables. The percentages under each one of the PC correspond to the portion of data variability each of them describe. The column "All" refers to the result obtained when analysing principal components of all regions of interest jointly. Significant values are shown in bold. For each univariate test, the number of samples was 201.

Clinical variable	CSF			DGM			NAWM			WMH			All		
	PC1	PC2	PC3	PC1	PC2	PC3	PC1	PC2	PC3	PC1	PC2	PC3	PC1	PC2	PC3
	98%	1%	1%	98%	1%	1%	98%	1%	1%	99%	1%	1%	84%	15%	1%
Biological sex	**0.003**	**0.001**	**0.001**	**0.002**	**0.031**	**0.008**	**0.021**	**0.032**	**0.018**	0.562	0.379	0.297	**0.009**	0.137	**0.001**
Smoker	0.386	0.949	0.184	0.055	0.238	0.063	0.281	0.310	0.330	0.274	0.392	0.297	0.195	0.211	0.733
Diabetes	0.060	0.697	**0.042**	0.825	0.874	0.892	0.281	0.168	0.287	**0.032**	0.392	0.596	0.259	0.938	**0.004**
Hyperlipidemia	0.785	0.921	0.618	**0.035**	0.460	0.136	0.374	0.495	0.414	0.508	0.392	0.297	0.259	0.124	0.503
Hypertension	0.060	0.380	**0.047**	0.471	0.874	0.752	**0.021**	**0.012**	**0.022**	**0.021**	0.379	0.297	**0.019**	0.938	0.091
Fazekas	0.060	0.921	**0.002**	**0.007**	0.238	0.091	**0.001**	**0.001**	**0.001**	**0.001**	**0.001**	**0.001**	**0.001**	0.137	**0.001**
BGPVS	0.060	0.697	**0.002**	0.106	0.419	0.313	**0.021**	**0.012**	**0.034**	**0.001**	**0.013**	**0.001**	**0.006**	0.162	**0.001**
Total SVD	0.090	0.921	**0.002**	0.054	0.669	0.313	**0.001**	**0.001**	**0.001**	**0.001**	**0.001**	**0.001**	**0.001**	0.137	**0.001**

burden of SVD features in the DGM region (p-value < 0.05) compared to the time-averaged RPS (p-value > 0.05). These outcomes suggest that the temporal component is of relevance, consistent with our hypothesis [16].

The power spectrum (expressed through PC1, PC2, and PC3) varied (p-value < 0.05) when grouped by visual clinical ratings regardless of the region of interest, indicating that burden of SVD influenced the spectral findings.

Table 3. Kruskal-Wallis test results obtained when comparing the functional principal component (PC) scores extracted from the RPS of subjects grouped by clinical variables. The percentages under each one of the PC correspond to the portion of data variability each of them describe. The column "All" refers to the result obtained when analysing principal components of all regions of interest jointly. Significant values are shown in bold. For each univariate test, the number of samples was 201.

Clinical variable	CSF			DGM			NAWM			WMH			All		
	PC1	PC2	PC3	PC1	PC2	PC3	PC1	PC2	PC3	PC1	PC2	PC3	PC1	PC2	PC3
	96%	3%	1%	90%	8%	2%	94%	5%	1%	93%	5%	1%	84%	15%	1%
Biological sex	**0.001**	**0.001**	0.563	**0.001**	**0.001**	0.849	**0.019**	**0.001**	0.573	0.554	0.273	0.102	0.060	0.121	**0.001**
Smoker	0.313	0.209	0.939	0.066	0.110	0.769	0.292	0.346	0.263	0.257	0.985	0.740	0.216	0.239	0.456
Diabetes	0.051	0.991	0.376	0.828	0.317	0.607	0.292	0.319	0.941	**0.035**	0.985	0.979	0.251	0.994	**0.048**
Hyperlipidemia	0.784	0.209	0.563	**0.044**	0.980	0.607	0.403	0.425	0.573	0.481	0.985	0.583	0.251	0.098	0.862
Hypertension	0.051	0.991	0.563	0.486	0.801	0.849	**0.023**	0.708	0.538	**0.019**	0.261	0.372	**0.017**	0.994	0.686
Fazekas	**0.041**	0.257	**0.001**	**0.009**	**0.026**	0.941	**0.001**	**0.001**	0.573	**0.001**	**0.001**	**0.001**	**0.001**	0.098	**0.001**
BGPVS	**0.041**	0.828	**0.001**	0.124	**0.026**	0.159	**0.023**	**0.001**	0.751	**0.001**	**0.001**	0.102	0.066	0.142	**0.001**
Total SVD	0.051	0.164	**0.001**	0.067	**0.013**	0.496	**0.001**	**0.001**	0.836	**0.001**	**0.001**	**0.001**	**0.001**	0.121	**0.001**

The measurements gathered from all the regions of interest varied significantly with biological sex (p-value < 0.05), overall WMH burden (p-value < 0.001), BGPVS scores (p-value < 0.05), and total SVD (p-value < 0.001). An example of the distribution of principal component scores in CSF, DGM, and NAWM, according to deep and periventricular Fazekas scores, is shown in Fig. 3. The tendency overall was that scores decreased significantly (p-value < 0.05) with increased WMH burden.

The power spectrum appeared to change concerning covariates such as diabetes, hypertension, and hyperlipidemia (p-value < 0.05), but the variations were observed in specific regions: hypertension in NAWM, WMH and All, diabetes in WMH and All, and hyperlipidemia in DGM. When stratified by smoker vs non-smoker, the signals did not differ (p-value > 0.05). In general, the power spectrum calculated from normal-appearing and abnormal regions of interest varied significantly mostly with clinical SVD ratings.

In terms of computational time, the manual rectification of the segmentation boundaries consumed most of the time, being it reported to take between 20 and 60 min per patient for WM and WMH depending on the expertise of the analyst. [13]. The calculation of the RPS per patient can take up to one minute, and the computation of the PCA scores for a region of interest for all 201 patients takes approximately two minutes on a Microsoft Windows 10 machine with 8GB RAM (i5-4590 CPU Intel(R) processor @ 3.30 GHz).

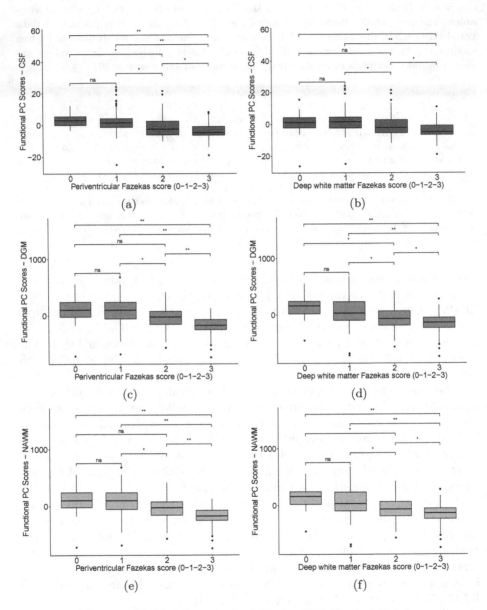

Fig. 3. Distribution of PCA scores obtained for the CSF (blue), DGM (orange), and NAWM (yellow) regions grouped by the Fazekas periventricular WM (left) and deep WM (right) scores of the patients. The scores correspond to PC3 for CSF, PC2 for DGM, and PC2 for NAWM. Significant differences between pairs of groups have been highlighted with $^*p < 0.05$ $^{**}p < 0.001$. (Color figure online)

4   Discussion

In this paper, we propose, for the first time, a framework incorporating power spectrum analysis to study dynamic brain MRI signals of brain-related pathological processes. Our team implemented the segmentation protocols and methods, the computation of the power spectrum features, and the integration of each processing module into the proposed framework. In particular, we applied our processing pipeline to the study of SVD tissue changes using DCE-MRI acquisitions. The power spectrum gauged from normal-appearing and abnormal brain regions of a population with features of SVD of differing extents was analysed to explore whether patients with biological sex, hypertension and visual ratings (namely Fazekas, BGPVS, total SVD scores) exhibited distinctive spectra. To the best of our knowledge, this is the first time that the power spectrum has been examined for this purpose in a relatively large cohort ($n = 201$) with a wide range of SVD features. Of note, we applied our framework to the study of DCE-MRI signals. Nonetheless, this does not prevent it from being used for analysing other dynamic and non-dynamic brain MR acquisitions.

Due to the relative temporal sparsity and spatial density of the dynamic brain MRI acquisitions, the number of techniques that can be used to process them is reduced. Thus, finding adequate alternative representations is crucial. In this paper, we explored the spectral density for studying dynamic brain MRI signals in SVD. According to our evaluation results, we found that spectral features of the DCE-MRI acquisitions vary significantly with biological sex, hypertension, and clinical visual ratings (namely, Fazekas, BGPVS, and total SVD scores). This outcome suggests that the use of the power spectrum is suitable for examining these types of acquisitions.

We evaluated two approaches for extracting features out of the RPS. The first one contemplated computing the mean RPS over time and studying it through univariate functional PCA. The second one consisted of describing each element of the dynamic RPS through multivariate functional PCA. In both cases, we retrieved the scores resulting in the direction of the first three modes of variation. We observed that the two approaches coincide most of the time. However, as the former method summarises the temporal changes in the data, the latter should be preferred. Our experimental results support this claim and are consistent with our hypothesis.

The spatial spectral features extracted from the signal in the NAWM region differed significantly with SVD features and their extents. This observation agrees with previous finding in which the differentiation between "normal" and "abnormal" tissues becomes less evident with increased age and SVD feature severity [14]. Interestingly, a small percentage of the variations of the spectral features in the CSF region differed significantly with SVD features and their extents. This situation might be related to leakage of gadolinium in CSF with increased SVD burden [14]. The Fazekas score has been found to be associated with increasing BBB leakage [9,16] and as the spectral features of patients grouped by this visual rating were statistically different, this outcome suggests that our framework can be used in the study of small vessel dysfunction.

Differences regarding biological sex (and presumably brain size as well) are expected as a direct consequence of the scaling theorem of the Fourier transform in which the size of the region of interest is linked to its Fourier representation [11]. Thus, the size of brain structures within a region of interest is expected to be, in principle, encoded in the spectral features used. This problem may be alleviated by processing the regions of interest in a short-time Fourier transform or sliding-window processing fashion or by manually selecting matching and relevant regions of interest with exactly the same area/volume on all scans. In the future, we plan to explore alternatives in these regards.

In the present work, we showed that a strategy based on the study of the power spectrum could be used to investigate dynamic post-contrast signal changes in tissues that relate to the extents of SVD features, vascular risk factors and clinical visual ratings. The outcomes of our experiments add confidence to previous findings in which DCE-MRI signals from patients with different age, health status, and premorbid brain condition exhibited different tendencies [16]. For instance, a pre- and post-contrast texture analysis concluded that local signal variations, measured in terms of homogeneity and contrast, differed significantly depending on the extents of SVD features [14]. Future work should consider understanding what physiopathological processes cause these power spectrum variations and which aspects of the spectrum are different among patient groups.

Acknowledgements. JB holds an MRC Precision Medicine Doctoral Training Programme studentship from the University of Edinburgh. This work was supported by the Row Fogo Charitable Trust (MVH) grant no. BRO-D.FID3668413, Wellcome Trust (patient recruitment, scanning, primary study Ref No. WT088134/Z/09/A), Fondation Leducq (Perivascular Spaces Transatlantic Network of Excellence), and EU Horizon 2020 (SVDs@Target) and the MRC UK Dementia Research Institute (Wardlaw programme). The authors thank participants in the study, the radiographers and staff at the Edinburgh Imaging Facilities (www.ed.ac.uk/clinical-sciences/edinburgh-imaging/research/facilities-and-equipment/edinburgh-imaging-facilities), the UK Dementia Research Institute at the University of Edinburgh, and the Row Fogo Centre for Ageing and the Brain.

References

1. Naidu, P.S., Mathew, M.: Chapter 3 Power spectrum and its applications. In: Naidu, P.S., Mathew, M. (eds.) Analysis of Geophysical Potential Fields, Advances in Exploration Geophysics, vol. 5, pp. 75–143. Elsevier, Amsterdam (1998)
2. Fazekas, F., et al.: White matter signal abnormalities in normal individuals: correlation with carotid ultrasonography, cerebral blood flow measurements, and cerebrovascular risk factors. Stroke **19**(10), 1285–1288 (1988)
3. Happ, C., Greven, S.: Multivariate functional principal component analysis for data observed on different (dimensional) domains. J. Am. Statistical Assoc. **113**, 649–652 (2018)
4. Hernández, M.D.C.V., et al.: Metric to quantify white matter damage on brain magnetic resonance images. Neuroradiology **59**(10), 951–962 (2017)

5. Hernández, M.D.C.V., Ferguson, K.J., Chappell, F.M., Wardlaw, J.M.: New multi-spectral MRI data fusion technique for white matter lesion segmentation: method and comparison with thresholding in FLAIR images. Eur. Radiol. **20**(7), 1684–1691 (2010)
6. Heye, A.K., et al.: Tracer kinetic modelling for DCE-MRI quantification of subtle blood-brain barrier permeability. Neuroimage **125**, 446–455 (2016)
7. Khalifa, F., et al.: Models and methods for analyzing DCE-MRI: a review. Med. Phys. **41**(12), 124301 (2014)
8. Mattia, D., et al.: Quantitative EEG and dynamic susceptibility contrast MRI in Alzheimer's disease: a correlative study. Clin. Neurophysiol. **114**(7), 1210–1216 (2003)
9. Muñoz Maniega, S., et al.: Integrity of normal-appearing white matter: influence of age, visible lesion burden and hypertension in patients with small-vessel disease. J. Cereb. Blood Flow Metab. **37**(2), 644–656 (2017)
10. Potter, G., Doubal, F., Jackson, C., Sudlow, C., Dennis, M., Wardlaw, J.: Associations of clinical stroke misclassification ('clinical-imaging dissociation') in acute ischemic stroke. Cerebrovasc. Dis. **29**(4), 395–402 (2010)
11. Smith, J.O.: Mathematics of the Discrete Fourier Transform (DFT). W3K Publishing, Palo Alto (2007)
12. Staals, J., Makin, S.D., Doubal, F.N., Dennis, M.S., Wardlaw, J.M.: Stroke subtype, vascular risk factors, and total MRI brain small-vessel disease burden. Neurology **83**(14), 1228–1234 (2014)
13. Valdés Hernández, M.D.C., et al.: Rationale, design and methodology of the image analysis protocol for studies of patients with cerebral small vessel disease and mild stroke. Brain Behav. **5**(12), e00415 (2015)
14. Valdés-Hernández, M.D.C., et al.: Application of texture analysis to study small vessel disease and blood-brain barrier integrity. Front. Neurol. **8**, 327 (2017)
15. Wardlaw, J.M., et al.: White matter hyperintensity reduction and outcomes after minor stroke. Neurology **89**(10), 1003–1010 (2017)
16. Wardlaw, J.M., et al.: Blood-brain barrier failure as a core mechanism in cerebral small vessel disease and dementia: evidence from a cohort study. Alzheimer's Dementia **13**(6), 634–643 (2017)
17. Wardlaw, J.M., Smith, C., Dichgans, M.: Mechanisms of sporadic cerebral small vessel disease: insights from neuroimaging. Lancet Neurol. **12**(5), 483–497 (2013)

Segmentation of Arteriovenous Malformation Based on Weighted Breadth-First Search of Vascular Skeleton

Zonghan Wu[1,2], Baochang Zhang[1,2], Jun Yang[3], Na Li[1,2], and Shoujun Zhou[1(✉)]

[1] Shenzhen Institutes of Advanced Technology, Chinese Academy of Sciences, Shenzhen, China
sj.zhou@siat.ac.cn
[2] Shenzhen Colleges of Advanced Technology, University of Chinese Academy of Sciences, Shenzhen, China
[3] The Rockets Military Nursing Home in Guangzhou, Guangzhou 510515, China

Abstract. Cerebral arteriovenous malformations (AVM) are prone to rupture, which will lead to life-threatening conditions. Because of the complexity and high mortality and disability rate of AVM, it has been a severe problem in surgery for many years. In this paper, we propose a new method of AVM location and segmentation based on graph theory. A weighted breadth-first search tree is created from the result of vascular skeletonization, and the AVM is automatically detected and extracted. The feeding arteries, draining veins and the AVM nidus are segmented according to the topological structure of the vessel. We evaluate the proposed method on clinical data sets and achieve an average accuracy of 95.14%, sensitivity of 82.28% and specificity of 94.88%. The results show that our method is effective and is helpful for the treatment of vascular interventional surgery.

Keywords: Arteriovenous malformation · Image segmentation · Graph theory

1 Introduction

The pathological characteristics of cerebral arteriovenous malformations are that there are no capillaries between arteries and veins, but a group of abnormal vascular structure with uneven diameter and thickness of vessel wall, which lacks elastic layer and muscular layer and is prone to rupture and bleed [1]. At present, the main treatment methods for cerebral arteriovenous malformations include surgical resection of malformations, endovascular embolization and stereotactic radiotherapy [2]. Therefore, a detailed knowledge of the structure of the feeding arteries, draining veins and the AVM nidus is helpful to optimize the surgical planning.

Bullitt et al. [3] have proposed a kind of software for three-dimensional (3D) visualization of vessels to display previously manually segmented AVM nidus by volume rendering technology. Saring et al. [4] measured and counted the temporal intensity curve of blood flow and obtained the 3D blood flow parameters of spatial dynamic characteristies by using the 4D magnetic resonance angiography (MRA) image sequences

© Springer Nature Switzerland AG 2020
Y. Zheng et al. (Eds.): MIUA 2019, CCIS 1065, pp. 294–301, 2020.
https://doi.org/10.1007/978-3-030-39343-4_25

and used the registration of the parameter space and 3D MRA data to support the visual evaluation of the AVM.

Nyui et al. used factor analysis method [5] and principal component analysis (PCA) method [6] to analyze dynamic X-ray CT images with contrast and opaque media for AVM extraction. Babin et al. used pixel profiling approach [7] and centerline analysis method [8] to detect and segment the AVM region. So far, there have been several methods to analyze the internal structure of AVM in detail, which including feeding arteries, draining veins and the AVM nidus. Forkert et al. [9] used blood flow velocity features, vessel radii, vascular tubular features and spherical features to train support vector machine (SVM) classifier to segment AVM nidus, and combined with the time intensity parameter diagram of blood flow in the 4D MRA images to divide vessels within a certain radius around the nidus into feeding arteries or draining veins. Using the information of vascular density and variance combined with manual tagging, Clarencon et al. [10] described a method for label propagation on 3D digital rotational angiography (3DRA) images to segment AVM nidus, feeding arteries and draining veins, respectively. This process requires a lot of manual interaction and a high degree of experience and is not suitable for clinical practice. Babin et al. [11], based on centerline analysis [8], constructed a vascular tree to locate and segment the AVM nidus and then segmented the feeding arteries and draining veins according to the radii and gray values of the vessels connected to the nidus. However, the skeleton method they used will produce cavities or holes, destroy the original topological structure, and their AVM extraction method is only valid for the compact spherical nidus, but in fact, most AVM nidi are not like this.

In this paper, we propose a novel method of AVM location and segmentation based on graph theory. We segment cerebral vessels from time-of-flight magnetic resonance angiography (TOF-MRA) images based on our previous method [12]. Then, a weighted breadth-first search (BFS) method is used to create a spanning tree from the results of skeletonization, and a topological path is constructed from the root point of the carotid artery through the feeding artery to the AVM nidus and then to the draining vein. We implement the automatic detection and extraction of AVM and segment the feeding arteries and the draining veins according to the sequence of the paths connecting the AVM nidus in the spanning tree.

The paper is organized as follows. The second part introduces our method. In the third part, we use 10 sets of clinical data to verify our method. Finally, we summarize and discuss in the fourth part.

2 Methodology

The process of localization and segmentation of AVM Nidus is shown in Fig. 1. Firstly, we segment the cerebral vessels from the 3DMRA image based on our previous method [12] and generate the vascular skeletonization image using the method of Lee et al. [13]. After that, we use the weight-based breadth-first search method to create a spanning tree. The vertex with the highest degree is obtained from the adjacency matrix. And we use our extraction method to locate the AVM nidus in the skeleton image. A topological path is established to segment the feeding artery and draining vein connected to the AVM nidus.

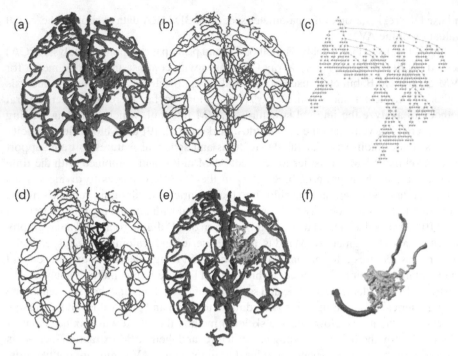

Fig. 1. The process of localization and segmentation of AVM Nidus, (a) the segmentation of cerebrovascular, (b) the skeletonization image, (c) the spanning tree, (d) the localization of AVM nidus in skeleton image, (e) the result of vessel reconstruction and (f) the segmentation result of feeding artery and draining vein.

2.1 Weighted Breadth-First Search Tree

To analyze the vascular topology, we need to transform the vascular skeletonization image represented by voxels into an undirected graph represented by vertices and edges. We connect each voxel with all its adjacent voxels (26-connected voxels), and take all voxels with degree 1 or greater than 2 as vertices of the graph, discarding isolated voxels with degree 0. And the node degree is the number of edges associated with the node. A series of voxels with degree 2 connected between two adjacent vertices are used as the edge of the graph, and the number of voxels is used as the length of the edge.

Then we use a weighted breadth-first search method to create a spanning tree of the undirected graph, and construct a topological path from the root point of the carotid artery through the feeding artery to the AVM nidus and then to the draining vein. The process of search method is as follows:

Algorithm 1 Weighted Breadth-first Search

Input: The undirected graph $G = (V, E)$, where V is a set of vertices and E is a set of edges.

1. For each vertex $v_i (i = 2, 3, \ldots, n)$, compute the minimum path length to the root vertex v_1 (carotid artery root point) as the weight W_i.
2. Put v_1 into the queue *Vertex*.
3. Take the first vertex v_f of *Vertex* and add it to the array *Visited*.
4. Add the new vertices adjacent to v_f that has not been visited into *Vertex*, and add the edges between v_f and the new vertices into the adjacency matrix *Graph*.
5. Delete v_f in *Vertex* and rank the vertices in *Vertex* from small to large according to W_i.
6. Loop 3-5 steps until there are no vertices in *Vertex*.

Output: The adjacency matrix *Graph*, which is the spanning tree.

Due to the existing imaging techniques and segmentation algorithms, the branches of cerebral arteries and veins are directly connected. In addition to the feeding artery, other arterial branches can also be connected to the AVM nidus through the venous sinus and draining veins. Therefore, it is necessary to disconnect the loop in the cerebrovascular network to ensure that only the arterial branch where the blood supply artery is located can be connected to the AVM nidus, and it is also necessary to preserve the integrity of the AVM nidus and the draining vein. The weight-based breadth-first search method is used to disconnect the loops in the undirected graph at the location far from the root vertex, ensuring that a topological path is from the root of the carotid artery to the AVM nidus through the feeding artery and then to the draining vein. As illustrated in Fig. 2, the circle is equivalent to the loop in the cerebrovascular system, and the green part with dense nodes is equivalent to the AVM area. The breadth-first search method will disconnect the loop at where the number of nodes on both sides is nearly average. However, due to a large number of nodes in the AVM, the use of BFS may destroy the topological structure from feeding artery to AVM nidus and then to the draining vein, and the use of weight-based BFS can avoid such a situation, instead.

2.2 Localization and Extraction of AVM Nidus

As shown in Fig. 3, the higher the degree of vertices in the skeleton image, the redder the color. It can be observed that the internal vertices of the AVM nidus are dense and the degrees of them are relatively higher. Therefore, the position and density of vertices with higher degrees can be used to locate and extract AVM nidus. Since the node with the highest degree is observed inside the AVM, we define the area of the AVM nidus as all vertices within the range of 8 neighborhood of the vertex with the highest degree in the spanning tree:

$$V_{AVM} = \{v | (Graph + Graph')^{1 \sim 8} (v, v_H) \neq 0\} \tag{1}$$

where V_{AVM} is the set of AVM vertices, $Graph^n$ is the nth power of the adjacency matrix Graph, and v_H is the vertex with the highest degree. The degree represents the

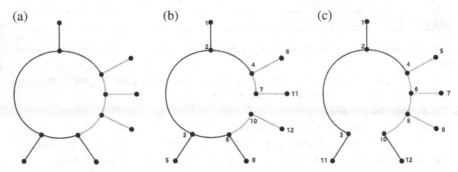

Fig. 2. The difference between the BFS method and weight-based BFS method, (a) the original image, (b) the result of the BFS method, (c) the result of the weight-based BFS method. The circle is equivalent to the loop in the cerebrovascular system, the green part with dense nodes is equivalent to the AVM area, the red and blue arcs represent the feeding artery and the draining vein, respectively. And the numbers represent the order in which the nodes are visited. (Color figure online)

Fig. 3. The axial, sagittal, and coronal views of skeleton image. The color of vertices represents its degree, and the higher the degree, the redder the color. (Color figure online)

number of incident links of the node. $Graph + Graph'$ transforms the directed spanning tree into an undirected graph to facilitate the acquisition of neighborhood vertices. If the position of coordinates (x, y) in $Graph^n$ is not zero, it represents that there is a path from x to y through n edges. The paths in the skeleton image between v_H and all the vertices v satisfying the conditions constitute the AVM nidus. 8 is the best neighborhood range we have obtained through experiments.

2.3 Segmentation of Feeding Arteries and Draining Veins

Through Sects. 2.1 and 2.2, we have found a topological path from the root point of the carotid artery through the feeding artery to the AVM nidus and then to the draining vein. So the path connecting the ancestor vertex of the AVM vertex in the spanning tree is the feeding artery, and the path connecting its descendant vertex is the draining vein.

$$\text{artery} = \{P_{v,v_H} | Graph^{10}(v, v_H) \neq 0\} \tag{2}$$

$$\text{vein} = \{P_{v,v_H} | Graph^{10}(v_H, v) \neq 0\} \tag{3}$$

where P_{v,v_H} is the path in the skeleton image between v_H and all the vertices v satisfying the conditions, and we use the Dijkstra shortest path method [14] here. Our method can find a feeding artery, one or more draining veins. The range of ancestor vertex and descendant vertex can be adjusted to control the length of the feeding artery and draining vein.

3   Experiments

Our experimental data consists of 10 time-of-flight magnetic resonance angiography (TOF-MRA) clinical data, which are collected using a GE Signa HDx 3.0T MRI scanner (TR = 25.0, TE = 3.5, flip angle = 20). Each volume has the size of $512 \times 512 \times 381$ (voxels), and the resolution is uniformly 0.34 mm $\times$ 0.34 mm $\times$ 0.25 mm.

3.1   Localization and Segmentation of AVM Nidus

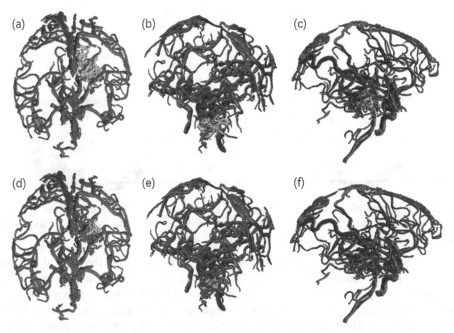

Fig. 4. The result of localization and segmentation of AVM Nidus, (a)–(c) the result of our method, (d)–(f) the result of Babin's method.

The proposed method achieved good segmentation results on these 10 sets of data, accurately located and segmented AVM nidus. The method of Babin et al. only works well for compact spherical structures, but in fact, there are gaps between the segmented vessels affected by MRA imaging and vascular segmentation algorithm. As shown in Fig. 4, our method can segment AVM nidus with gaps or even sparse vessels very well.

To derive a quantitative comparison of the segmentation results, the manual segmentation results of AVM nidus by three experts were treated as the ground truth. We used three segmentation evaluation metrics to evaluate the proposed method, i.e. accuracy (Acc), sensitivity (Sen) and specificity (Spe). They were defined as:

$$\begin{cases} Acc = \frac{TP+TN}{TP+TN+FP+FN} \\ Sen = \frac{TP}{TP+FN} \\ Spe = \frac{TN}{TN+FP} \end{cases} \tag{4}$$

where TP, FP, TN, and FN denote the number of voxel segmentation results of true-positive, false-positive, true-negative and false-negative, respectively. The proposed method showed an average Acc of 95.14%, Sen of 82.28% and Spe of 94.88% comparing to ground truth.

3.2 Segmentation of Feeding Artery and Draining Vein

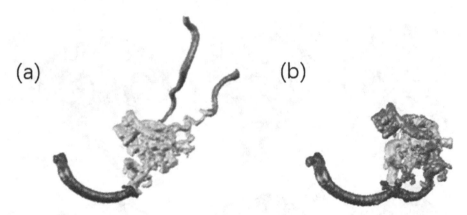

Fig. 5. The segmentation results of feeding artery and draining vein, (a) the result of our method, (b) the result of Babin's method. The red part is the feeding artery, the green part is the AVM nidus, and the blue part is the draining vein. (Color figure online)

Segmentation of feeding artery and draining vein based on the topological path of the spanning tree is shown in Fig. 5. Our method can segment the feeding artery and the draining vein very well, while Babin's method is similar to the region growing method based on gray value and vascular radius. However, due to the uneven thickness of AVM vessels, their method sometimes fails to find the correct feeding artery and draining vein, which is prone to over-segmentation and mis-segmentation.

4 Conclusion

In this work, we proposed a novel weight-based breadth-first search method for AVM location and segmentation. The AVM nidus was located in the spanning tree of the skeleton image using our method. A topological path was established to segment the feeding artery and draining vein connected to the AVM nidus. The AVM localization and extraction method was validated on 3DMRA clinical data sets of cerebral vessels and showed an accuracy of 95.14%, sensitivity of 82.28% and specificity of 94.88%. The results show that our method is effective and helpful to optimize the operation plan.

Acknowledgment. This work is supported by Key Laboratory of Health Informatics, Chinese Academy of Sciences and by the following funds: National Science Foundation of China (No. 81827805), Shenzhen Engineering Laboratory for Key Technologies on Intervention Diagnosis and Treatment Integration.

References

1. Cho, S.K., et al.: Arteriovenous malformations of the body and extremities: analysis of therapeutic outcomes and approaches according to a modified angiographic classification. J. Endovasc. Ther. **13**(4), 527–538 (2006)
2. Ogilvy, C.S., et al.: Recommendations for the management of intracranial arteriovenous malformations - a statement for healthcare professionals from a special writing group of the stroke council, American stroke association. Stroke **32**(6), 1458–1471 (2001)
3. Bullitt, E., et al.: Computer-assisted visualization of arteriovenous malformations on the home personal computer. Neurosurgery **48**(3), 576–583 (2001)
4. Saring, D., et al.: Visualization and analysis of cerebral arteriovenous malformation combining 3D and 4D MR image sequences. Int. J. Comput. Assist. Radiol. Surg. **2**(1 Suppl.), 75–79 (2007)
5. Nyui, Y., Ogawa, K., Kunieda, E.: Extraction of arteriovenous malformation with factor analysis. In: Proceedings 2000 International Conference on Image Processing, pp. 621–624 (2000)
6. Nyui, Y., Ogawa, K., Kunieda, E.: A new approach to the visualization of intracranial arteriovenous malformation with principal component analysis. In: IEEE Nuclear Science Symposium, p. 23-11 (2000)
7. Babin, D., et al.: Pixel profiling for extraction of arteriovenous malformation in 3-D CTA images. In: Proceedings ELMAR-2014, pp. 1–4 (2014)
8. Babin, D., et al.: Skeleton calculation for automatic extraction of arteriovenous malformation in 3-D CTA images. In: IEEE 11th International Symposium on Biomedical Imaging, pp. 425–428 (2014)
9. Forkert, N.D., et al.: Computer-aided nidus segmentation and angiographic characterization of arteriovenous malformations. Int. J. Comput. Assist. Radiol. Surg. **8**(5), 775–786 (2013)
10. Clarençon, F., et al.: Elaboration of a semi-automated algorithm for brain arteriovenous malformation segmentation: initial results. Eur. Radiol. **25**(2), 436–443 (2015)
11. Babin, D., et al.: Skeletonization method for vessel delineation of arteriovenous malformation. Comput. Biol. Med. **93**, 93–105 (2018)
12. Zhou, S., et al.: Segmentation of brain magnetic resonance angiography images based on MAP–MRF with multi-pattern neighborhood system and approximation of regularization coefficient. Med. Image Anal. **17**(8), 1220–1235 (2013)
13. Lee, T.C., et al.: Building skeleton models via 3-D medial surface axis thinning algorithms. CVGIP Graph. Model. Image Process. **56**(6), 462–478 (1994)
14. Dijkstra, E.W.: A note on two problems in connexion with graphs. Numerische mathematic **1**(1), 269–271 (1959)

Image Registration

A Variational Joint Segmentation and Registration Framework for Multimodal Images

Adela Ademaj[1], Lavdie Rada[2($\boxtimes$)], Mazlinda Ibrahim[3], and Ke Chen[4]

[1] Biomedical Engineering Department, Heidelberg University, Heidelberg, Germany
[2] Biomedical Engineering Department, Bahcesehir University,
Besiktas, Istanbul, Turkey
lavdie.rada@eng.bay.edu.tr
[3] National Defence University of Malaysia, 57000 Kuala Lumpur, Malaysia
[4] Centre for Mathematical Imaging Techniques, University of Liverpool,
Liverpool, UK

Abstract. Image segmentation and registration are closely related image processing techniques and often required as simultaneous tasks. In this work, we introduce an optimization-based approach to a joint registration and segmentation model for multimodal images deformation. The model combines an active contour variational term with a mutual information smoothing fitting term and solves in this way the difficulties of simultaneously performed segmentation and registration models for multimodal images. This combination takes into account the image structure boundaries and the movement of the objects, leading in this way to a robust dynamic scheme that links the object boundaries information that changes over time. Comparison of our model with state of art shows that our method leads to more consistent registrations and accurate results.

1 Introduction

In the last decades, the development of imaging devices significantly increased image data collections and consecutively the need of human interpretation. To reduce labor hours of specialists new image processing techniques are developed. Two main image processing techniques are image registration and segmentation. Image registration provides understanding of the data behavior in two or more different scenarios taken in different times. Meanwhile, image segmentation is the process of partitioning objects or features considering characteristics of each pixel. In a wide range of problems, such as comparison of images taken at different time or modalities, image registration depend on image segmentation and vice-versa. Erdt et al. [10] indicates that more than 20% of scientific research in medical imaging area requires a combined registration and segmentation scheme. The combined registration and segmentation models can

© Springer Nature Switzerland AG 2020
Y. Zheng et al. (Eds.): MIUA 2019, CCIS 1065, pp. 305–316, 2020.
https://doi.org/10.1007/978-3-030-39343-4_26

be divided into two different categories: simultaneous registration and segmentation or joint segmentation and registration which lastly has been referred as *regmentation*. In the last years, *regmentation* functional models have been introduced on monomodal imaging frameworks. These models are variational rigid registration based models [9,27], variational nonrigid registration based models [2,24], or atlas based models [12]. Recently, Ibrahim-Rada-Chen [14] present a linear curvature regmentation variational based model which gets profit on the global and local deformation and easily can cope with images which contain more than one object.

In difference with monomodal images, multimodal image processing requires spatial correspondence estimation or extract a certain information from images with different modalities (or protocols). Such example, can be from imaging using CT and MRI protocols, which brings different image perspective into the medical field of clinical and pre-clinical diagnostics. Even though, assessing image similarity becomes challenging for multimodal images, multimodal images are desired as they bring different image perspectives and properties. In difference with monomodal images registration, where simple image similarity measures between image pairs can provide a good spatial transformation, dealing with multimodalities is substantially more difficult. Registering multimodal images, which are acquired through different mechanisms and have distinct modalities, involves the alignment of both images in terms of shapes and salient components while preserving the modality of one given image. For a comprehensive overview of registration techniques in a systematic manner, the work of Sotiras et al. [23] is referred. Results performed simultaneous registration and segmentation are shown in Fig. 1. The same images are processed with the new proposed joint regmentation model and the results shown in the last section indicate that the

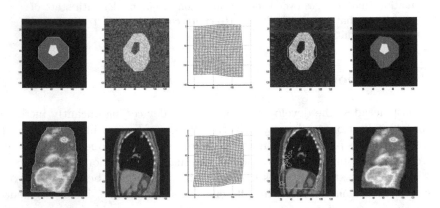

Fig. 1. Simultaneous segmentation and linear curvature registration based on MI model noisy synthetic image and CT image. First column segmented template image, T & $\phi_0(\mathbf{x})$; Second column reference image R; Third column the deformation field; Forth column the transformed template image R & $\phi_0(\mathbf{x} + u(\mathbf{x}))$; Last column the template image after transformation, $T(\mathbf{x} + u(\mathbf{u}))$.

combination of segmentation and registration tasks into a single framework is relevant as avoids the errors produced from simultaneously tasks. The existing registration models for mono modal-images can be classified as unsupervised and supervised methods. Among different registration similarity measure, the most popular one is mutual information (MI), introduced by Viola et al. [25]. Recently, learning-based approaches to measure image similarity have been proposed as well. Techniques involving KL-divergence [4], max-margin structured learning [6], boosting [11], or deep learning [22] have been investigated. However, their large training sets make them limited by the lack of generalization to previously unseen object classes and the registration performance depends on numerical optimization of the optimal registration parameters. There are a few unsupervised works to solve the joint segmentation and registration problem. The work proposed by Wang and Vemuri [26] introduces a registration driven by segmentation of a reference image without varying degrees of non-rigidity. The model applies cross cumulative residual entropy as a distance measure [26] and piecewise constant Chan-Vese model for segmentation [5]. Although, the algorithm can accommodate image pairs having very distinct intensity distributions but fails in deforming large objects. Droske et al. [8] presented variational method based on a local energy density which uses a generalized motion of image morphology for multimodal image registration. Similar idea was lastly proposed by Aganj and Fischl [1] where segmentation score was applied to register two multimodal brain images. This model is not designed for general image deformation as the model applies rigid image transformation. Recently, a joint segmentation and registration model has been proposed by Debroux and Le Guyader [7], which is formulated based on the combination of nonlocal total variation and nonlocal shape descriptors. The fitting term of this model uses a weighted sum of squared difference distance measure, similar to [14] and the model depends on the many parameters such as window size, patch size, number of neighbors pixels, etc. To overcome the difficulties mentioned above we propose a new variational model which combines linear curvature, known for its ability to generate smooth transformations [3,18], and mutual information (MI) as a distance measure [16,25] for the spatial transform.

The outline of this paper is as follows. In Sect. 2, we review the existing models for monomodal image *regmentation* and multimodal image registration. In Sect. 3, we introduce our proposed new model for multimodal image *regmentation*. In the last section, we show some numerical tests including comparison.

2   Related Work

Deformable multimodal registration attempts to find spatial correspondence between image pairs that are acquired with distinct scanning protocols. In registration process, these correspondences are estimated by determining the spatial transformation which is applied to transformed image to make similar to the reference image. Generally, image registration is driven by the chosen similarity measure and the chosen deformation model. For images taken from the same modality, the image similarity usually is consider to be high, thus simple and direct image similarity measures are used, for instance the sum of squared intensity differences (SSD). We start this section with an overview of the work introduced by Ibrahim-Rada-Chen [14] for monomodal *regmentation* of two given images. The model jointly merges the segmentation and registration into a variational function represented in a level set. This method improves the Guyader-Vese model [13] by replacing the nonlinear elastic term with linear curvature model and adding a weighted Heaviside sum of SSD term. Let R be a reference image and T template image in a given domain $\Omega \subset R^d$, where d is the spatial dimensional of the given data. The registration problem is formulated as a transformation vector $\phi(x)$ such that $T(\phi(x)) = R(x)$. In the variational approach, image registration, the transformation is $\phi(x) = x + u(x)$, with $u(x)$ the displacement vector field $u(x) = (u_1(x), u_2(x))$. The aim of the Ibrahim-Rada-Chen [14] model is to match the contour of the template image and at the same time segment the reference image to show the deformation of displacement field lead by segmentation process. The segmentation of the template image T is represented by the zero level line $\phi_x : \Omega \to \Re$ which is the template contour Γ, denoted as in [14]. The joint functional proposed by Ibrahim-Rada-Chen [14] monomodal image *regmentation* is given as follows:

$$
\min_{c_1, c_2, u(x)} \mathcal{J} = \lambda_3 \mathcal{D}^{\mathrm{SSDH}}(T, R, \phi_0(x), u(x)) + \lambda_1 \int_\Omega |R(x) - c_1|^2 H_\epsilon(\phi_0(x + u(x))) \, dx
$$
$$
+ \lambda_2 \int_\Omega |R(x) - c_2|^2 (1 - H_\epsilon(\phi_0(x + u(x)))) \, dx + \alpha \mathcal{S}^{\mathrm{LC}}(u) \tag{1}
$$

where λ_1, λ_2, λ_3 are numerical constants. $u(x) = (u_1(x), u_2(x))$ is the displacement vector field, ϕ_0 is a zero-level-set, $\mathcal{D}^{\mathrm{SSDH}}$ is the sum of squared difference distance measure

$$
\mathcal{D}^{\mathrm{SSDH}} = \int_\Omega (T(x + u(x)) - R(x))^2 H_\epsilon(\phi_0(x + u(x))) \, dx \tag{2}
$$

which is weighted by the regularised Heaviside function

$$
H_\epsilon(\phi_0(x + u(x))) = \frac{1}{2}(1 + \frac{2}{\pi} \arctan \frac{\phi_0(x + u(x))}{\epsilon}). \tag{3}
$$

The term $\mathcal{S}^{\mathrm{LC}}(u)$ in (1) is the curvature regulariser term

$$
\mathcal{S}^{\mathrm{LC}}(u) = \int_\Omega (\Delta u_1)^2 + (\Delta u_2)^2 \, dx. \tag{4}
$$

The values of c_1 and c_2 in Eq. (1) present the average intensity values inside and outside the boundary $\phi_0(x)$ in the reference image. By minimizing the functional (1) containing the SSD term and linear curvature term, we get:

$$
c_1 = \frac{\int_\Omega R(x) H_\epsilon(\phi_0(x + u)) \, dx}{\int_\Omega H_\epsilon(\phi_0(x + u)) \, dx}, c_2 = \frac{\int_\Omega R(x)(1 - H_\epsilon(\phi_0(x + u))) \, dx}{\int_\Omega 1 - H_\epsilon(\phi_0(x + u)) \, dx}. \tag{5}
$$

In multilevel representation, the registration problem is solved using quasi-Newton method by updating the deformation field $u(x)$ on each level. Even though the model shows a good performance, it cannot cope with multimodal images.

Referring to registration for multimodality, different techniques have been introduced [15,20,21]. A significant work, proposed by Modersitzki [18], based on mutual information and curvature regularizer, competes the other registration models. The joint functional for multimodal registration is given as follows:

$$J(x + u(x)) = \mathcal{D}[T, R] + \alpha \mathcal{S}^{\mathrm{LC}} \tag{6}$$

where $\mathcal{S}^{\mathrm{LC}}$ is the curvature regularization term, α is a numerical constant and $\mathcal{D}[T, R]$ presents the distance measures considered as functionals in multimodal image pair T and R. Modersitzki [18] suggested to use mutual information (MI) (shortly reviewed below as CASE 1) or Normalized Gradient Field (NGF) (shortly reviewed below as CASE 2) to measure the similarity distance between two multimodal images.

CASE 1: The joint functional based on MI, $\mathcal{D}[T, R] = \mathcal{D}^{\mathrm{MI}}$

$$J(x + u(x)) = \mathcal{D}^{MI} + \alpha \mathcal{S}^{\mathrm{LC}} \tag{7}$$

where $\mathcal{D}^{MI}$ presents mutual information as a distance measure between the given images T and R and it is given with the formula:

$$\mathcal{D}^{\mathrm{MI}} = \int_\Omega \rho_{[T]} \log \rho_{[T]} dt + \int_\Omega \rho_{[R]} \log \rho_{[R]} dr - \int_{\Omega^2} \rho_{[T,R]} \log \rho_{[T,R]} d(t, r) \tag{8}$$

where $\rho_{[T]}(t)$ and $\rho_{[R]}(r)$ are marginal densities which are expressed as:

$$\rho_{[T]}(t) = \int_\Omega \rho_{[T,R]}(t, r) dr, \quad \rho_{[R]}(r) = \int_\Omega \rho_{[T,R]}(t, r) dt \tag{9}$$

and $\rho_{[T,R]}(t, r)$ presents joint gray value distributions. Its discretized form is expressed as:

$$\rho_{[T,R]}(t, r) = \frac{1}{m} \sum_{j=1}^{m} K(t - T(x + u(x)), \sigma) K(r - R(x), \sigma) \tag{10}$$

with σ that stands for the width of Parzen density estimator and K for kernel and m presents the size of the sample image. The kernel positions in template and reference image are represented with t and r respectively in the gray value space for $j = 1$.

CASE 2: The joint functional based on NGF, $\mathcal{D}[T, R] = \mathcal{D}^{\mathrm{NGF}}$

$$J(x + u(x)) = \mathcal{D}^{NGF} + \alpha \mathcal{S}^{\mathrm{LC}}, \tag{11}$$

$\mathcal{D}^{NGF}$ presents Normalized Gradient Force distance measure for two given multimodal images, T and R which is defined as follows

$$\mathcal{D}^{\mathrm{NGF}} = \int_\Omega 1 - (n[T(x + u(x))]^T n[R(x)])^2 dx \tag{12}$$

where

$$n[T(\boldsymbol{x} + \boldsymbol{u}(\boldsymbol{x}))]^T n[R(\boldsymbol{x})] \approx \frac{\nabla T_i^T \nabla R_i}{\sqrt{||\nabla T_i||^2 + \eta^2}\sqrt{||\nabla R_i||^2 + \eta^2}} \tag{13}$$

where the term η is an important constant parameter because it determines the edge and what has to be within the specified noise level. Moreover, the terms ∇T_i and ∇R_i present gradient intensity of template image and reference image, respectively which are computed as

$$\nabla T_i = [(\partial_1^h T(\boldsymbol{x} + \boldsymbol{u}(\boldsymbol{x})))_i, (\partial_2^h T(\boldsymbol{x} + \boldsymbol{u}(\boldsymbol{x})))_i]^T, \tag{14}$$

$$\nabla R_i = [\partial_1^h R(\boldsymbol{x})_i, \partial_2^h R(\boldsymbol{x})_i]^T, \tag{15}$$

The NGF can be considered as the L2 norm of r, the residual of the alignment of the normalised gradients of two images at a pixel position $\boldsymbol{x}$,

$$r_{\boldsymbol{x}}^h = 1 - (n[T(\boldsymbol{x} + \boldsymbol{u}(\boldsymbol{x}))]^T n[R(\boldsymbol{x})])^2 \tag{16}$$

for discrete images T and R of size N × N using finite difference method. The images are discretised on a uniform mesh using vertex centred discretisation where $\boldsymbol{x}_{i,j}$ denotes the pixel position or on a non-uniform mesh with finite difference method. The gradient is calculated using

$$\partial_{x_1} T^h(\boldsymbol{x}_{i,j}) = \frac{T^h(\boldsymbol{x}_{i+1,j}) - T^h(\boldsymbol{x}_{i-1,j})}{2h}, \ \partial_{x_2} T^h(\boldsymbol{x}_{i,j}) = \frac{T^h(\boldsymbol{x}_{i,j+1}) - T^h(\boldsymbol{x}_{i,j-1})}{2h},$$

where the first order central finite difference scheme is used to approximate the first order derivatives. The discretized form of NGF distance measure:

$$NGF = hd \sum_i^{N^2} (1 - (r_i)^2) \tag{17}$$

where $hd = h^1...h^d$. In short words, NGF is based on the alignment of the edges in the reference and template images. The gradient is normalised with its magnitude.

3   The Proposed New Joint Regmentation Model for Multimodality

Our joint functional model for multimodal image *regmentation* combines an active contour without edges, a curvature regulariser, and a mutual information distance measure. Our proposed *regmentation* functional is given with the formula:

$$\min_{c_1, c_2, \boldsymbol{u}(\boldsymbol{x})} \mathcal{J} = \lambda_1 \int_\Omega |R(\boldsymbol{x}) - c_1|^2 H_\epsilon(\phi_0(\boldsymbol{x} + \boldsymbol{u}(\boldsymbol{x}))) \, \mathrm{d}\boldsymbol{x}$$

$$+ \lambda_2 \int_\Omega |R(\boldsymbol{x}) - c_2|^2 (1 - H_\epsilon(\phi_0(\boldsymbol{x} + \boldsymbol{u}(\boldsymbol{x})))) \, \mathrm{d}\boldsymbol{x} + \gamma \mathcal{S}^{\mathrm{LC}}(\boldsymbol{u})$$

$$+ \mu \mathcal{D}^{\mathrm{MIH}}(T, R, \phi_0(\boldsymbol{x}), \boldsymbol{u}(\boldsymbol{x}))$$

$$\tag{18}$$

where $\mathcal{S}^{\mathrm{LC}}$ is the linear curvature, $\mathcal{D}^{\mathrm{MIH}}$ represents the mutual information distance term which is weighted by the parameter μ. The $\mathcal{D}^{\mathrm{MIH}}$ term is evaluated as the mutual information between foreground of the template and the reference, TH_ϵ and RH_ϵ, respectively:

$$\mathcal{D}^{\mathrm{MIH}} = \int_\Omega \rho_{[TH_\epsilon]} \log \rho_{[TH_\epsilon]} dt + \int_\Omega \rho_{[RH_\epsilon]} \log \rho_{[RH_\epsilon]} dr - \int_{\Omega^2} \rho_{[TH_\epsilon, RH_\epsilon]} \log \rho_{[TH_\epsilon, RH_\epsilon]} d(t, r)$$

The joint functional is solved with respect to the displacement field using discretise then optimise approach based on the quasi-Newton method in multilevel framework for faster implementation. The grid points are located at the center of the cell $\Omega^h = \{\boldsymbol{x}_{i,j} = (x_{1,i}, x_{2,j}) = ((i - 0.5)h, (j - 0.5)h)\}$, $1 \leq i, j \leq N$. We re-define the solution vector $\boldsymbol{U} = \begin{bmatrix} \boldsymbol{u}_1 \\ \boldsymbol{u}_2 \end{bmatrix}_{2N^2 \times 1}$, $\boldsymbol{x} = \begin{bmatrix} \boldsymbol{x}_1 \\ \boldsymbol{x}_2 \end{bmatrix}_{2N^2 \times 1}$, where $\boldsymbol{u}_1$, $\boldsymbol{u}_2$ present displacement vectors and $\boldsymbol{x}_1$, $\boldsymbol{x}_2$ position vectors. Thus, the discretised form of the joint functional in (18), by a finite difference method is:

$$\min_{c_1, c_2, U} \mathcal{J} = \lambda_1 \sum_{i,j=1}^{N} |R(\boldsymbol{x}_{i,j}) - c_1|^2 H_\epsilon(\phi_0(\boldsymbol{x}_{i,j} + \boldsymbol{u}(\boldsymbol{x}_{i,j})))$$

$$+ \lambda_2 \sum_{i,j=1}^{N} |R(\boldsymbol{x}_{i,j}) - c_2|^2 (1 - H_\epsilon(\phi_0(\boldsymbol{x}_{i,j} + \boldsymbol{u}(\boldsymbol{x}_{i,j}))))$$

$$+ \mu \eta_t \eta_r \sum_{i=1}^{n_t} \sum_{j=1}^{n_r} \rho_{i,j} \log(\rho_{i,j} + \epsilon) + \gamma \sum_{l=1}^{2} \sum_{i,j=1}^{N} \Big(-4u_l(\boldsymbol{x}_{i,j}) + u_l(\boldsymbol{x}_{i+1,j})$$

$$+ u_l(\boldsymbol{x}_{i-1,j}) + u_l(\boldsymbol{x}_{i,j+1}) + u_l(\boldsymbol{x}_{i,j-1}) \Big)^2.$$

where $\eta_t = (t_n - t_0)/n_t$ and $\eta_r = (r_n - r_0)/n_r$, n_t and n_r present the number of grid. The ϵ term is just a small numerical constant to prevent extra considerations such as "$0 log 0$". Furthermore, we are using homogeneous Neumann boundary conditions approximated by one side differences $u_l(\boldsymbol{x}_{i,1}) = u_l(\boldsymbol{x}_{i,2}), u_l(\boldsymbol{x}_{1,j}) = u_l(\boldsymbol{x}_{2,j}), u_l(\boldsymbol{x}_{i,N-1}) = u_l(\boldsymbol{x}_{i,N}), u_l(\boldsymbol{x}_{N-1,j}) = u_l(\boldsymbol{x}_{N,j}), l = 1, 2$.

Minimizing Eq. (3) brings a system of nonlinear equation with unknown $\boldsymbol{U}$:

$$\Delta \mathcal{J}(c_1, c_2, \boldsymbol{U}) = 0 \tag{19}$$

where

$$\Delta \mathcal{J} = \lambda_1 \sum_{i,j=1}^{N} |R(\boldsymbol{x}_{i,j}) - c_1|^2 H_\epsilon(\phi_0(\boldsymbol{x}_{i,j} + \boldsymbol{u}(\boldsymbol{x}_{i,j})))$$

$$+ \lambda_2 \sum_{i,j=1}^{N} |R(\boldsymbol{x}_{i,j}) - c_2|^2 (1 - H_\epsilon(\phi_0(\boldsymbol{x}_{i,j} + \boldsymbol{u}(\boldsymbol{x}_{i,j})))) \tag{20}$$

$$+ \mu \mathcal{D}^{\mathrm{MIH}} + \gamma \mathbf{u}^T \mathbf{B} \mathbf{u}).$$

The gradient of the regularization term is computed as the multiplication of $\mathbf{B}$ matrix(constant matrix) of size $2N_1 N_2 \times 2N_1 N_2$ that contains the coefficients of $\mathbf{U}$. The matrix $\mathbf{B}$ is written as

$$\mathbf{B} = \begin{bmatrix} L^T L & 0 \\ 0 & L^T L \end{bmatrix}, \tag{21}$$

where L is a a block tridiagonal matrix from the regularization term. For finding the solution of our minimization problem, we start with zero initial guess, $U = 0$, solving

$$H\delta U = -G \tag{22}$$

where $G = \nabla_u TH(x + u) = \frac{\partial \mathcal{D}^{MIH}}{\partial u(x)} \frac{\partial TH(x+u)}{\partial u(x)}$, for δU and update $U \leftarrow U + \tau \delta U$ with τ as the Armijo line search parameter [19]. H and G are the Hessian and gradient matrix for the functional $\mathcal{J}$ in Eq. (3) with respect to the displacement vector U.

4   Numerical Results and Conclusions

In this section, we present several examples for synthetic as well as real data in comparison with the linear curvature model based on MI applied on a synthetic image and a set of real images. In addition, results of our proposed joint regmentation are compared to image registration applying mutual information as a distance measure and image segmentation simultaneously. The image pairs used in all our experiments have significantly different intensity profiles, deformation and presence of noise. In each iteration we compute the Jacobian matrix of the transformation,

$$J = \begin{bmatrix} 1 + \frac{\partial u_1}{\partial x_1} & \frac{\partial u_1}{\partial x_2} \\ \frac{\partial u_2}{\partial x_1} & 1 + \frac{\partial u_2}{\partial x_2} \end{bmatrix} \tag{23}$$

and the minimal value of Jacobian matrix $\mathcal{F} = \min(\det(J))$ is also calculated to make sure there is no folding or cracking in the deformation field if its value is greater than 0. Figures 2 and 3 present the results of the proposed model in comparison with Ibrahim et al. [14] regmentation model, and linear curvature image registration model based on MI and NGF proposed by Modersitzki et al. [17]. The results show that the Ibrahim et al. [14] model, which uses the SSD similarity measure, is not capable to register and segment the given synthetic multimodal images whereas the proposed model can successfully cope with it. The fail of the method proposed by Ibrahim et al. [14] is expected as the model is not designed for multimodal images. On the other hand, the linear curvature registration model which applies MI, proposed by Modersitzki et al. [17], provides larger deformation than the same model using NGF, referring to Fig. 3. Even though, this model still has a poor registration in comparison with the proposed method, shown in Fig. 2 last three columns.

Figures 4 and 5 show the results on real multimodal data images. In the first row of both figures we show comparison results between the proposed model and linear curvature model [17] of two chest images (T1 and T2 weighted images). The second and the third row have the thorax (PET and CT images [17]) and brain images with high deformation and presence of noise. We clearly see that the proposed model delivers good results, referring to the fifth, sixth, and seventh column of Fig. 4, whereas the linear curvature model [17] shown in the third and fourth column stuck in local minima due to highly non convexity of MI functional.

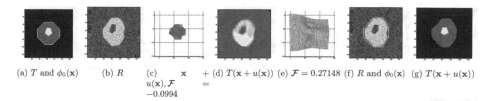

(a) T and $\phi_0(\mathbf{x})$ (b) R (c) $\mathbf{x}$ + (d) $T(\mathbf{x}+u(\mathbf{x}))$ (e) $\mathcal{F} = 0.27148$ (f) R and $\phi_0(\mathbf{x})$ (g) $T(\mathbf{x}+u(\mathbf{x}))$
$u(\mathbf{x}), \mathcal{F}$ =
-0.0994

Fig. 2. Comparison between the proposed model and monomodal image *regmentation* [14]. The proposed model successfully deforms the given template image, whereas the monomodal joint regmentation fails to segment and register with parameters $\lambda_1 = \lambda_2 = \lambda_3 = 1$ and $\alpha = 5$. The proposed joint functional is calculated with $\lambda_1 = \lambda_2 = \lambda_3 = 0.0005$ and $\mu = 2.0e - 05$.

Table 1. The first, second and third columns show the linear curvature model [17] parameters and distance measurements before and after, while the other columns are referring to our model, shown in Fig. 4.

Figure 4	α	Before $\mathcal{D}^{\mathrm{MI}}$	After $\mathcal{D}^{\mathrm{MI}}$	λ_1	λ_2	μ	γ	Before $\mathcal{D}^{\mathrm{MIH}}$	After $\mathcal{D}^{\mathrm{MIH}}$
Chest	0.05	-1.4938	-1.3204	0.01	0.01	0.01	0.00045	-1.1677	-1.6247
Thorax	0.5	-0.2478	-0.2608	2	2	0.1	2	-0.40906	-0.88796
Brain	0.003	-1.1994	-1.2234	0.002	0.002	0.002	0.000035	-1.4477	-1.5911

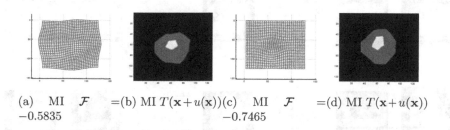

(a)   MI $\mathcal{F}$ =(b) MI $T(\mathbf{x}+u(\mathbf{x}))$(c)   MI $\mathcal{F}$ =(d) MI $T(\mathbf{x}+u(\mathbf{x}))$
-0.5835 -0.7465

Fig. 3. Comparison between the proposed model and Modersitzki et al. [17] models which use linear curvature in combination with MI measure and NGF for synthetic images. First and second column shows the deformation for MI model, The third and the forth column NGF model. The distance measure of mutual information before of linear curvature model is $D^{MI} = -1.0303$ and our proposed model is $D^{MIH} = -0.22806$, whereas the value of distance measure after the registration is $D^{MI} = -1.0426$ and after image registration and segmentation is $D^{MIH} = -0.53424$

Our multimodal joint regmentation provides accurate segmentation and registration in comparison also to linear curvature based on NGF model [17]. These results are shown in Fig. 5. The gradient field distance measure involves second order image derivatives, as a consequence there can be a problem dealing with noisy images. Table 1 shows the parameters and distances measurements before and after the performance of linear curvature model [17] and our model.

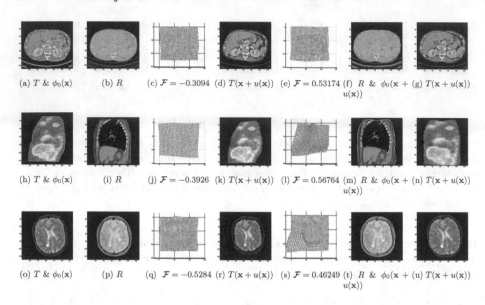

(a) T & $\phi_0(\mathbf{x})$ (b) R (c) $\mathcal{F} = -0.3094$ (d) $T(\mathbf{x}+u(\mathbf{x}))$ (e) $\mathcal{F} = 0.53174$ (f) R & $\phi_0(\mathbf{x}+$ (g) $T(\mathbf{x}+u(\mathbf{x}))$
                                                                                                      $u(\mathbf{x}))$

(h) T & $\phi_0(\mathbf{x})$ (i) R (j) $\mathcal{F} = -0.3926$ (k) $T(\mathbf{x}+u(\mathbf{x}))$ (l) $\mathcal{F} = 0.56764$ (m) R & $\phi_0(\mathbf{x}+$ (n) $T(\mathbf{x}+u(\mathbf{x}))$
                                                                                                       $u(\mathbf{x}))$

(o) T & $\phi_0(\mathbf{x})$ (p) R (q) $\mathcal{F} = -0.5284$ (r) $T(\mathbf{x}+u(\mathbf{x}))$ (s) $\mathcal{F} = 0.46249$ (t) R & $\phi_0(\mathbf{x}+$ (u) $T(\mathbf{x}+u(\mathbf{x}))$
                                                                                                       $u(\mathbf{x}))$

Fig. 4. Comparison between linear curvature model using MI [17] and the proposed model for real images data.

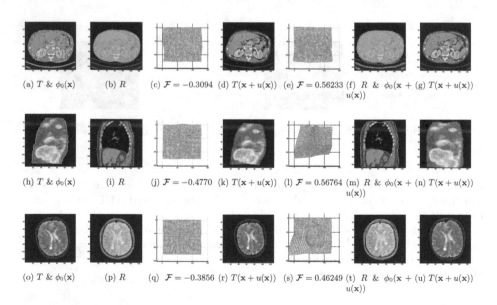

(a) T & $\phi_0(\mathbf{x})$ (b) R (c) $\mathcal{F} = -0.3094$ (d) $T(\mathbf{x}+u(\mathbf{x}))$ (e) $\mathcal{F} = 0.56233$ (f) R & $\phi_0(\mathbf{x}+$ (g) $T(\mathbf{x}+u(\mathbf{x}))$
                                                                                                      $u(\mathbf{x}))$

(h) T & $\phi_0(\mathbf{x})$ (i) R (j) $\mathcal{F} = -0.4770$ (k) $T(\mathbf{x}+u(\mathbf{x}))$ (l) $\mathcal{F} = 0.56764$ (m) R & $\phi_0(\mathbf{x}+$ (n) $T(\mathbf{x}+u(\mathbf{x}))$
                                                                                                      $u(\mathbf{x}))$

(o) T & $\phi_0(\mathbf{x})$ (p) R (q) $\mathcal{F} = -0.3856$ (r) $T(\mathbf{x}+u(\mathbf{x}))$ (s) $\mathcal{F} = 0.46249$ (t) R & $\phi_0(\mathbf{x}+$ (u) $T(\mathbf{x}+u(\mathbf{x}))$
                                                                                                      $u(\mathbf{x}))$

Fig. 5. Comparison between linear curvature model using NGF [17] and the proposed model for real images data.

In conclusion, the proposed model is suitable for jointly segmenting and registering images with intensity difference, severe deformation, and presence of noise. The model avoids in this way the need of pre-or post-processes such as pre-registration step which might be required for the segmentation task or vice-versa.

References

1. Aganj, I., Fischl, B.: Multimodal image registration through simultaneous segmentation. IEEE Signal Process. Lett. **24**(11), 1661–1665 (2017). https://doi.org/10.1109/LSP.2017.2754263
2. Alvarez, L., Weickert, J., Snchez, J.: Reliable estimation of dense optical flow fields with large displacements. Int. J. Comput. Vis. **39**(1), 4156 (2000)
3. Fischer, B., Modersitzki, J.: Curvature based image registration. J. Math. Imaging Vis. **18**(1), 81–85 (2003)
4. Guetter, C., Xu, C., Sauer, F., Hornegger, J.: Learning based non-rigid multi-modal image registration using kullback-leibler divergence. In: Duncan, J.S., Gerig, G. (eds.) MICCAI 2005. LNCS, vol. 3750, pp. 255–262. Springer, Heidelberg (2005). https://doi.org/10.1007/11566489_32
5. Chan, T., Vese, L.: An active contour model without edges. In: Nielsen, M., Johansen, P., Olsen, O.F., Weickert, J. (eds.) Scale-Space 1999. LNCS, vol. 1682, pp. 141–151. Springer, Heidelberg (1999). https://doi.org/10.1007/3-540-48236-9_13
6. Lee, D., Hofmann, M., Steinke, F., Altun, Y., Cahill, N.D., Scholkopf, B.: Learning similarity measure for multi-modal 3D image registration. CVPR (2009)
7. Debroux, N., Guyader, C.L.: A joint segmentation/registration model based on a nonlocal characterization of weighted total variation and nonlocal shape descriptors. SIAM Imaging Sci. **11**(2), 957–990 (2018)
8. Droske, M., Rumpf, M.: Multiscale joint segmentation and registration of image morphology. IEEE Trans. Pattern Anal. Mach. Intell. **29**(12), 2181–2194 (2007). https://doi.org/10.1109/TPAMI.2007.1120
9. Dydenko, I., Friboulet, D., Magnin, I.: A variational framework for affine registration and segmentation with shape prior: application in echocardiographic imaging. In: VLSM Workshop-ICCV, pp. 201–208 (2003)
10. Erdt, M., Steger, S., Sakas, G.: Regmentation: a new view of image segmentation and registration. J. Radiat. Oncol. Inform. **4**(1), 1–23 (2012)
11. Michel, F., Bronstein, A.M., Paragios, N.: Boosted metric learning for 3D multi-modal deformable registration. ISBI (2011)
12. Gooya, A., Pohl, K.M., Bilello, M., Biros, G., Davatzikos, C.: Joint segmentation and deformable registration of brain scans guided by a tumor growth model. In: Fichtinger, G., Martel, A., Peters, T. (eds.) MICCAI 2011. LNCS, vol. 6892, pp. 532–540. Springer, Heidelberg (2011). https://doi.org/10.1007/978-3-642-23629-7_65
13. Guyader, L., La, V.: A combined segmentation and registration framework with a nonlinear elasticity smoother. Comput. Vis. Image Understand. **11**, 1689–1709 (2011)
14. Ibrahim, M., Chen, K., Rada, L.: An improved model for joint segmentation and registration based on linear curvature smoother. J. Algorithms Comput. Technol. **10**(4), 314–324 (2016)
15. Jiang, D., Shi, Y., Chen, X., Wang, M., Song, Z.: Fast and robust multimodal image registration using a local derivative pattern. Int. J. Med. Phys. Res. Pract. **44**(2), 497–509 (2017)

16. Maes, F., Collignon, A., Vandermeulen, D., Marchal, G., Suetens, P.: Multi-modality image registration by maximization of mutual information. IEEE Trans. Med. Imaging **16**(2), 187–198 (1997)
17. Modersitzki, J.: Numerical Methods for Image Registration. Oxford University Press, Oxford (2004)
18. Modersitzki, J.: Flexible Algorithms for Image Registration. SIAM publications, Philadelphia (2009)
19. Nocedal, J., Wright, S.: Numerical Optimization, vol. 2. Springer, New York (1999). https://doi.org/10.1007/b98874
20. Ruhaak, J., et al.: Intensity gradient based registration and fusion of multi-modal images. In: IEEE 10th International Symposium on Biomedical Imaging, pp. 572–575 (2013)
21. Shakir, H., Ahsan, S.T., Faisal, N.: Multimodal medical image registration using discrete wavelet transform and Gaussian pyramids. In: IEEE International Conference on Imaging Systems and Techniques, pp. 1–6 (2015)
22. Simonovsky, M., Gutiérrez-Becker, B., Mateus, D., Navab, N., Komodakis, N.: A deep metric for multimodal registration. In: Ourselin, S., Joskowicz, L., Sabuncu, M.R., Unal, G., Wells, W. (eds.) MICCAI 2016. LNCS, vol. 9902, pp. 10–18. Springer, Cham (2016). https://doi.org/10.1007/978-3-319-46726-9_2
23. Sotiras, A., Christos, D., Paragios, N.: Deformable medical image registration: a survey. IEEE Trans. Med. Imaging **32**(7), 1153–1190 (2013)
24. Unal, G., Slabaugh, G.: Coupled PDEs for non-rigid registration and segmentation. In: IEEE Computer Society Conference on Computer Vision and Pattern Recognition, vol. 1, pp. 37–42 (2002)
25. Viola, P., Wells III, W.M.: Alignment by maximization of mutual information. IJCV **24**(2), 137–154 (1997)
26. Wang, F., Vemuri, B.C.: Simultaneous registration and segmentation of anatomical structures from brain MRI. In: Duncan, J.S., Gerig, G. (eds.) MICCAI 2005. LNCS, vol. 3749, pp. 17–25. Springer, Heidelberg (2005). https://doi.org/10.1007/11566465_3
27. Yezzi, A., Zollei, L., Kapur, T.: A variational framework for joint segmentation and registration. In: IJCV, pp. 44–51 (2001)

An Unsupervised Deep Learning Method for Diffeomorphic Mono-and Multi-modal Image Registration

Anis Theljani and Ke Chen[✉]

Department of Mathematical Sciences and EPSRC Liverpool Centre for Mathematics
in Healthcare, University of Liverpool, Liverpool, UK
{a.theljani,k.chen}@liverpool.ac.uk

Abstract. Different from image segmentation, developing a deep learning network for image registration is less straightforward because training data cannot be prepared or supervised by humans unless they are trivial. In this paper we present an unsupervised deep leaning model in which the deformation fields are self-trained by an image similarity metric and a regularization term. The latter consists of a smoothing constraint on the derivatives and a constraint on the determinant of the transformation in order to obtain spatially smooth and plausible solution.

The proposed algorithm is first trained and tested on synthetic and real mono-modal images. The results show how it deals with large deformation registration problems and leads to a real time solution with no folding. It is then generalised to multi-modal images. Although any variational model may be used to work with our the deep learning algorithm, we present a new model using the reproducing Kernel Hilbert space theory, where an initial given pair of images, which are assumed non-linearly correlated, are first processed and optimized to serve the purpose of "intensity or edge correction" and to yield intermediate new images which are more strongly correlated and will be used for training the model. Experiments and comparisons with learning and non-learning models demonstrate that this approach can deliver good performance and simultaneously generate an accurate diffeomorphic transformation.

Keywords: Deep learning · Image registration · Variational model · Multi-modality images · Similarity measures · Mappings

1 Introduction

Image registration consists of constructing a reasonable geometrical correspondence between given two or more images of the same object taken at different times or using the same or different devices in order to locate different or complementary information. Applications of image registration include diverse fields

Supported by the UK EPSRC grant EP/N014499/1.

such as astronomy, optics, biology, chemistry, medical imaging and remote sensing and particularly in medical imaging. For an overview of image registration methodology, approaches and applications, we refer to [7,8,14–16]. Though the topic is actively studied and useful models exist, there remain many challenges to be tackled mathematically, particularly in the registration of images from different modalities. There exist various deformable variational models for image registration where the unknown **u** is sought in a properly chosen functional space [2,5,10,13,17]. Generally speaking, the variational problem consists of solving the optimization problem

$$\min_{\mathbf{u}} \left\{ \mathcal{L}(\mathbf{u}) = S(\mathbf{u}) + \frac{\lambda}{2} D(T(\varphi), R) \right\} \tag{1}$$

where $\varphi(\boldsymbol{x}) = \boldsymbol{x} + \mathbf{u}(\boldsymbol{x})$ and **u** is a displacement field. In (1), $S(\mathbf{u})$ is a regularisation term which controls the smoothness of **u** and reflects our expectations by penalising unlikely transformations. The second part $D(T(\varphi), R)$ is a similarity term which measures the goodness of the registration. These models are called non-learning based models as the optimization problem (1) should be solved for each pair for images T and R. Although various non-learning-based models have been proposed in the recent years and many numerical and computational algorithms have been developed to accelerate the numerical resolution of these models, it remains a very challenging question of achieving both an accurate solution and fast speed for real time applications.

In deep learning approaches proposed in the recent years where the aim is to optimize and learn spatial transformations between pairs of images to be registered [3,4,9,11,12], often, they require ground-truth deformation fields for the training task. They are called supervised models and their main drawback is the inability to predict transformations that may not be in the same range or class of the training transformations. As example, a deep learning model, which is learned and trained on a dataset where the ground-truth contains only small displacement fields, fails to predict and to give accurate results for large displacement.

In order to remedy these drawbacks, another class of deep leaning models was proposed. These unsupervised models do not require ground-truth deformation fields for training. The deformation fields are self-trained and driven by image similarity metrics computed on the input data. In [11], a spatial transformer network is developed to learn transformations for 2D images; however only affine and thin plate spline transformations were used. More general non-parametric transformations were considered in [9,12] for mono-modal images where the loss function is regularized by penalizing the first derivatives of the displacement u, e.g, the total variation of u, which promotes smoothness of predicted displacement fields during training. However, two issues are outstanding: (i) extension of the work to multi-modal images; and (ii) control of folding in the deformations by suitable regularizations leading to physically accurate transformations. For this latter reason, such models are generally not suitable for real life problems where deformation is large and folding can occur.

2   A Learning Model

In this work, we present a learning based diffeomorphic model for both mono- and multi-modal image registration. Our focus is on the multi-modal case. The idea is that the deformation fields are self-trained by an image similarity metric and a regularization term. In Fig. 1, we present an overview of the method:

(i)   The pair of images T and R are fed as input to the network which computes the transformation $\varphi(\boldsymbol{x})$. Any network that can captures the image features may work well in our model. Here, we used a network with four blocks containing different kernels with different sizes and PReLU activation is used at each of the block. The last block is responsible for generating the deformation filed $\mathbf{u}$.

(ii) An estimated transformation $\varphi(\boldsymbol{x})$ is then computed based on these features.

(iii) The loss function $\mathcal{L}(\cdot)$ in our model takes the form of the energy (1) i.e.

$$\mathcal{L}(\mathbf{u}) = S(\mathbf{u}) + \frac{\lambda}{2} D(T(\varphi), R). \qquad (2)$$

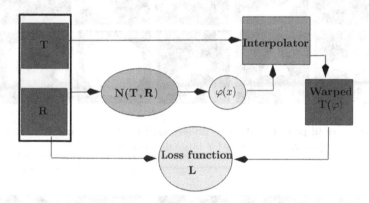

Fig. 1. The model architecture in mono-modality

Generally, only the similarity measures depend on the image modality. We discuss in the sequel the choice of the regularizer and the similarity measures for both mono- and multi-modality registration.

Regularization $S(\mathbf{u})$. The regularization that we use consists of a smoothing constraint on the derivatives and a constraint on the determinant of the transformation in order to obtain spatially smooth and plausible solutions [2,6]. The transformations are constrained by a regularization term which combines first- and second order derivatives and promotes smoothness. The second-order derivatives allow getting smooth transformations and penalize affine linear transformations which are not included in the kernel of the first-order derivatives based regularizers [18]. To ensure the map to be locally invertible, we also add an additional regularization term depending on $\det(\nabla\varphi(x))$ to help avoiding the mesh folding problem. More precisely, we consider

$$S(\mathbf{u}) = \|\nabla\mathbf{u}\|_2^2 + \|\nabla^2\mathbf{u}\|_2^2 + \|\phi(\det(\nabla\varphi))\|_2^2 \tag{3}$$

where $\phi(v) = \frac{v^2}{(v-1)^2}$. This term is originally used in the non-learning hyperelastic model and is known to be very efficient in getting diffeomorphic maps. To compare between different common regularizers, we consider 10 synthetic images drawn in order to compare the regularizers for possible large deformations. We trained our network with these images where each image can serve as a template and reference at the same time (so that we have $9! = 362880$ pairs of images). As shown in Fig. 2, different $S(\mathbf{u})$ leads to different results that we can distinguish between them visually and by also checking the value of $\mathcal{Q}_{min} = \min\det(\nabla\varphi)$. This part of the work is to tune our model.

Fig. 2. Dataset of 9! pairs based on 10 images used for the training, testing and comparing between different regularizers for large displacement.

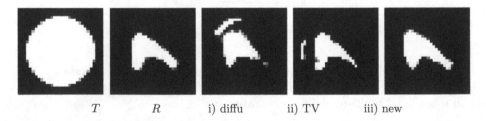

$\qquad T \qquad\qquad R \qquad\qquad$ i) diffu $\qquad\qquad$ ii) TV $\qquad\qquad$ iii) new

Fig. 3. Comparison between different regularizers—registration of an unseen pair of images. From left to right, Template T, Reference R, registered using different regularization terms: (i) the diffusion ($\mathbf{Q}_{min} = 0.38$), (ii) Total-Variation ($\mathbf{Q}_{min} = -0.24$) and (iii) $S(\mathbf{u})$ in (3) ($\mathbf{Q}_{min} = 0.71$), respectively. Clearly (iii) gives the best result.

Similarity $D(T(\varphi), R)$. Various similarity terms can be used to measure the goodness of the registration. In this work, we use an alternative measure to the correlation coefficient which is well suitable for measuring linear dependence between images, hence mono-modal images. More precisely, we use the following combined correlation-like measure (**CLM**):

$$\mathbf{CLM}(T, R) = \left\| \frac{(T - m_T)}{\sqrt{\mathbf{Var}(T)}} - \frac{(R - m_R)}{\sqrt{\mathbf{Var}(R)}} \right\|_2^2 \tag{4}$$
$$+ \left(\sqrt{\mathbf{E}(T - m_T)^2} + \sqrt{\mathbf{E}(R - m_R)^2} - \sqrt{\mathbf{E}(T + R - m_{T+R})^2} \right)^2$$

where $m_T = \frac{1}{|\Omega|} \int_\Omega T \, dx$, $\mathbf{Var}(T) = \mathbf{E}((T - m_T)^2)$ is the variance of T and $\mathbf{Var}(R)$ is defined similarly. For multi-modal images, i. e., non-linearly correlated images, we consider a preprocessing approach in which the given pair of images T, R are first processed and optimized to serve the purpose of 'intensity or edge correction', to yield intermediate images $T^* = g_1(T)$, $R^* = g_2(R)$ (which are more strongly correlated than T, R) respectively. The model architecture in this case is represented by the following diagram (Fig. 4):

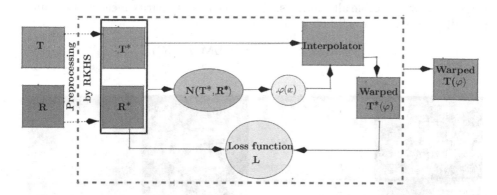

Fig. 4. Our model architecture for multi-modality registration

The optimization step to obtain $g_1(T), g_2(R)$ is a standalone step. We aim to find g_1 and g_2 as the minimum **CLM** (**MCLM**):

$$\mathbf{MCLM}(T, R) := \inf_{g_1, g_2 \in V} \mathbf{CLM}(g_1(T), g_2(T)), \tag{5}$$

where V is the space of Borel measurable functions. The **MCLM** measure promotes stronger dependence than linear ones. The following proposition shows the utility of the **MCLM** in describing general dependence between T and R.

Proposition 1. *The* **MCLM** *measure has the following properties:*

1. **MCLM**$(T, R) = 0$ *if* $T = \phi(R)$ *or* $T = \phi(R)$ *where* $\phi(\cdot)$ *is not necessary invertible;*
2. **MCLM**$(T, R) = 0$ *implies* $g_1(T) = g_2(R)$ *where* $g_1, g_2 \in V$.

Approximation of (5) *by RKHS.* In order to approximate the infimum in (5), we employ the theory of reproducing kernel Hilbert space [1]. For given T, R, we first employ a k-means algorithm to find suitable n respective clusters $\{t_i\}, \{r_j\}$ for images T, R. Then these intensity values $\{t_i\}, \{r_j\}$ will define the RKHS bases and we denote by such RKHS images $g_1(T)$ and $g_2(R)$ respectively

$$T^* = g_1(T(\boldsymbol{x})) := \sum_{i=1}^{n} \alpha_i K(T(\boldsymbol{x}), t_i),$$

$$R^* = g_2(R(\boldsymbol{x})) := \sum_{j=1}^{n} \beta_j K(R(\boldsymbol{x}), r_j). \quad (6)$$

In this work, the Gaussian kernel $K(x, y) = \frac{1}{\sqrt{2\pi}\sigma} \exp\{-\frac{(x-y)^2}{2\sigma^2}\}$ is used but there are many possible alternatives. Hence the new similarity measure for multi-modality registration will be denoted as

$$\mathbf{MCLM}(T, R) := \inf_{\alpha, \beta} \mathbf{CLM}(T^*, R^*). \quad (7)$$

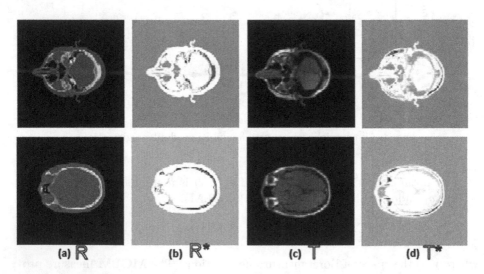

(a) R   (b) R*   (c) T   (d) T*

Fig. 5. The effect of RKHS approximation applied to two pairs of images T and R. (a) Initial CT image R. (b) Intermediate R^*. (c) Initial MRI image T. (d) Intermediate T^*. For both tests, the transformed intermediate images R^* and T^* and more strongly correlated than R and T. Here, we used $m = 74, \sigma = 0.05$ in the Gaussian Kernel and α and β are obtained by minimizing (7).

In Fig. 2, we illustrate the benefits of using the RKHS approximation in transforming the image T and R into new images T^* and R^* which are visually more suitable and may be seen as linearly correlated (Fig. 5).

3   Numerical Tests

In the numerical validation, we assess the performance of the proposed model that we call **DL Model** in registering mono-and multi-modal images. We compare with the classical variational model that optimizes the same loss function, i.e., the same regularizer and similarity measure, in a non-learning approach which we call **NL Model**. Registration quality is evaluated using the MI (the larger the better) between the two images $T(\varphi)$ and R. We also assess if the map φ is diffeomorphic by checking the minimum of the Jacobian determinant $\det(\nabla\varphi(x))$ i.e. yes if positive.

Mono-modal Images: For mono-modality registration, we train our model on 160 pairs of MRI heat images. We test our model on 20 unseen images and we display in Table 1 4 (typical) comparison results with the **NL model** in term of CPU time and accuracy. For the CPU time, **DL Model** is by far faster and predict the transformation in 1 s for a pair of images with a resolution of 256×256. The **NL model** achieves the same result in term of accuracy but it takes more than 1 min, to 100 times difference.

Table 1. Comparison between learning **DL Model** and non-learning **NL Model** for the images in Fig. 3 in Speed (time) and Quality (MI). One sees that **DL** is about 100 times faster for a similar result.

	Examples			
	Exp 1	Exp 2	Exp 3	Exp 4
Time (s) for **DDLmodel**	1.18 ± 0.6			
Time (s) for **NLmodel**	112.54	120.31	107.11	112.5
MI for **DLmodel**	1.49	1.66	1.53	1.57
MI for **NLmodel**	1.52	1.64	1.54	1.59
Q_{min} for **DLmodel**	0.45	0.63	0.65	0.67
Q_{min} for **NLmodel**	0.61	0.47	0.69	0.54

Multi-modal Images: For multi-modal images, we train the network on 120 pairs of CT and MRI images. In this example, we test the registration of a MRI image to a CT with much noise and texture, where prominent edges do not correspond to each other. We present the prediction result for registering 4 pairs

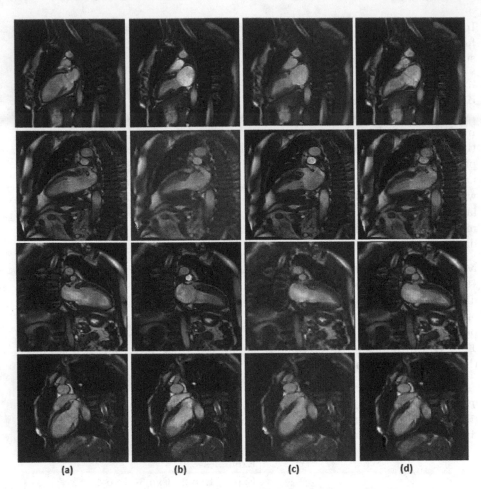

(a)   (b)   (c)   (d)

Fig. 6. Pairwise registration results for 4 pair of MRI images. Each row represents the result for a pair test. (a) Moving images T. (b) Fixed images R. (c) and (d) are the registered images using **DL Model** and **NL Model**, respectively. The mutual information errors and the values of $\mathbf{Q}_{min}$ related to these tests are given in Table 1

of MRI and CT images. To see the quality visually, we show the fused CT and MRI images for the four examples before and after registration. Clearly after the registration the images are well aligned while the learning approach is about 100 times faster for a comparable quality of **NL** (Figs. 6 and 7).

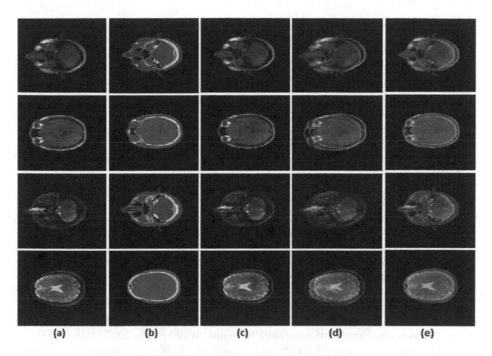

Fig. 7. Pairwise registration results for 4 pair of MRI-CT images by the learning model. Each row represents the result for a pair where: (a) Fixed images R. (b) Moving images T. (c) Registered images using **DL Model**. (d) and (e) are the fused images before and after the registration, receptively.

4   Conclusions

We have developed and presented an unsupervised deep learning approach for mono-and multi-modal images registration. We tested and compared different choices of regularization constraints on the deformation fields. The results have shown that control on the Jacobian determinant of the deformation is necessary in the loss function in order to get a diffeomorphic map, mainly for large displacements. The learning model was first designed and tested for mono-modal images. Adding a preprocessing step based on the reproducing kernels Hilbert space techniques, we found that the same learning approach works effectively for multi-modal images. Future work will consider generalizations to 3 dimensional images registration.

References

1. Aronszajn, N.: Theory of reproducing kernels. Trans. Am. Math. Soc. **68**, 337–404 (1950)
2. Burger, M., Modersitzki, J., Ruthotto, L.: A hyperelastic regularization energy for image registration. SIAM J. Sci. Comput. **35**, 132–148 (2013)

3. Cao, X., Yang, J., Wang, L., Xue, Z., Wang, Q., Shen, D.: Deep learning based inter-modality image registration supervised by intra-modality similarity. In: Shi, Y., Suk, H.-I., Liu, M. (eds.) MLMI 2018. LNCS, vol. 11046, pp. 55–63. Springer, Cham (2018). https://doi.org/10.1007/978-3-030-00919-9_7

4. de Vos, B.D., Berendsen, F.F., Viergever, M.A., Sokooti, H., Staring, M., Isgum, I.: A deep learning framework for unsupervised affine and deformable image registration. Med. Image Anal. **52**, 128–143 (2019)

5. Droske, M., Ring, W.: A mumford-shah level-set approach for geometric image registration. SIAM J. Appl. Math. **66**, 2127–2148 (2006)

6. Droske, M., Rumpf, M.: A variational approach to nonrigid morphological image registration. SIAM J. Appl. Math. **64**, 668–687 (2004)

7. Fischer, B., Modersitzki, J.: Ill-posed medicine - an introduction to image registration. Inverse Prob. **24**, 034008 (2008)

8. Gigengack, F., Ruthotto, L., Burger, M., Wolters, C.H., Jiang, X., Schafers, K.P.: Motion correction in dual gated cardiac pet using mass-preserving image registration. IEEE Trans. Med. Imaging **31**, 698–712 (2012)

9. Haskins, G., Kruger, U., Yan, P.: Deep learning in medical image registration: a survey. arXiv preprint arXiv:1903.02026 (2019)

10. Henn, S.: A multigrid method for a fourth-order diffusion equation with application to image processing. SIAM J. Sci. Comput. **27**, 831–849 (2005)

11. Jaderberg, M., Simonyan, K., Zisserman, A., et al.: Spatial transformer networks. In: Advances in Neural Information Processing Systems, pp. 2017–2025 (2015)

12. Li, H., Fan, Y.: Non-rigid image registration using fully convolutional networks with deep self-supervision. arXiv preprint arXiv:1709.00799 (2017)

13. Mang, A., Biros, G.: Constrained h^1-regularization schemes for diffeomorphic image registration. SIAM J. Imaging Sci. **9**, 1154–1194 (2016)

14. Modersitzki, J.: FAIR: Flexible Algorithms for Image Registration. SIAM, Philadelphia (2009)

15. Oliveira, F., Tavares, J.M.: Medical image registration: a review. Comput. Methods Biomech. Biomed. Eng. **17**, 73–93 (2014)

16. Sotiras, A., Davatzikos, C., Paragios, N.: Deformable medical image registration: a survey. IEEE Trans. Med. Imaging **32**, 1153–1190 (2013)

17. Theljani, A., Chen, K.: An augmented Lagrangian method for solving a new variational model based on gradients similarity measures and high order regulariation for multimodality registration. Inverse Prob. Imaging **13**, 309–335 (2019)

18. Zhang, J., Chen, K., Yu, B.: A novel high-order functional based image registration model with inequality constraint. Comput. Math. Appl. **72**, 2887–2899 (2016)

Automatic Detection and Visualisation of Metastatic Bone Disease

Nathan Sjoquist[1](✉), Tristan Barrett[2], David Thurtle[3],
Vincent J. Gnanapragasam[3], Ken Poole[4], and Graham Treece[1]

[1] Department of Engineering, University of Cambridge, Cambridge, UK
ns706@cam.ac.uk
[2] Department of Radiology, University of Cambridge, Cambridge, UK
[3] Department of Urology, University of Cambridge, Cambridge, UK
[4] Department of Medicine, University of Cambridge, Cambridge, UK

Abstract. This paper presents a novel method of finding and visualising metastatic bone disease in computed tomography (CT). The approach we suggest locates disease by comparing trabecular bone density between symmetric bony regions. Areas of strong difference could indicate metastatic bone disease as bone lesions either increase or decrease bone density. Our detection method is completely automatic and only requires raw CT data as input. Results are visualised in an interactive 3-dimensional viewer which displays a polygonal mesh of the bone structure overlaid with colour combined with resliced CT data. Diseased regions are clearly highlighted in both the mesh and in the resliced CT data. We test our method on both healthy and diseased CT data to demonstrate the validity of the technique. Experimental results show that our method can detect metastatic bone disease, although further work is needed to improve the robustness of the technique and to decrease false positives.

Keywords: Metastatic bone disease · Disease asymmetry · CT analysis · CT segmentation · Articulated registration · Computer aided detection

1 Introduction

Metastatic bone disease (MBD) is a common secondary feature of prostate cancer, breast cancer and many other malignancies. Bone scintigraphy is the usual imaging detection technique but has limitations in detecting changes over time and is only 2-dimensional. With improvements in bone-targeting therapies, better evaluation of therapeutic response is now a Europe-wide research priority. Positron-emission tomography is promising, but the expense and inaccessibility means CT is increasingly used for serial assessment. This requires time-consuming reviews of many CT cross-sectional images looking for small and diffuse bony changes – the summary of which is difficult to demonstrate in a

Supported by the W. D. Armstrong Trust.

multidisciplinary review setting as well as to demonstrate to the patient. To address these challenges, we have designed and implemented a new method for the detection and visualisation of MBD.

The human skeleton is nearly symmetric in both shape and bone density. When two healthy symmetric bony regions are compared with each other, the regions will hence contain little difference. However, metastatic bone disease causes sclerotic and lytic bone lesions to form, increasing the density (sclerotic lesions) or decreasing the density (lytic lesions) of the bone. When the bone density of symmetric regions containing bone lesions is compared, the difference is greatly increased as the lesions do not generally form symmetrically across both regions.

Our algorithm can automatically find these bone lesions by symmetrically comparing the bone density in the human skeleton and grouping areas of large difference. The method displays these differing densities through a novel visualisation. We believe this method has the potential to improve accuracy and to reduce the time-consuming review of the individual inspection of hundreds of cross-sectional images. We envisage our technique being used to identify potential areas of concern, although the final reporting of a bone lesion will require multi-planar assessment of the source data by the radiologist, oncologist and surgical team. In this paper, we describe our method in detail and show that it can automatically locate and visualise MBD in CT.

2   Background

Bone is a common site for metastasis by malignant tumours such as the lung, breast, and prostate cancer. This is significant as MBD can cause more devastating effects than the primary cancer does. MBD substantially increases morbidity due to its complications which include pain, impaired mobility, pathological fracture, spinal cord compression, cranial nerve palsies, nerve root lesions, hypercalcaemia and suppression of bone marrow [6] making it an important area of research.

In normal bone, development and maintenance of bone tissues is sustained through a balance of osteoclasts and osteoblasts resorbing and depositing bone tissue. With MBD, this process is disrupted as cancerous cells from the primary site contribute to the establishment of bone metastatic lesions. Most metastases are osteoblastic, causing an increase in bone density, although some metastases are osteolytic causing a reduction of bone density. In both cases, MBD causes an abnormal change in the density of bone.

Publications describe methods that can semi-automatically or automatically detect bone lesions by comparing learned features (generated by hand or through an automated machine learning technique) to raw and segmented CT data [3–5,10–12,16,18]. While these methods are promising, they have a limited spatial scope (only the vertebral column), offer poor visualisation of the results (only 2-dimensional) and often have a low accuracy. Machine learning in general also requires large quantities of accurately marked training data which can be difficult to obtain in a medical imaging setting due to patient agreements and lack of experts to mark the data.

Instead of using a machine learning approach, our paper explores a completely new approach to the problem. Rather than using training data to learn features distinct to bone lesions, we compare each CT data set to itself through symmetry in order to find irregularities. This removes the need for training data and bypasses the problem often encountered with machine learning algorithms – poor results when the training data does not contain a specific, never before encountered pattern. Also, instead of searching directly for 2-dimensional bone regions, we compare the modelled bone density between each symmetric pair of points in our 3-dimensional mesh, which reduces the dimensionality of the problem (when compared to searching for all combinations of 2-dimensional slices through an entire CT data set).

Using symmetry to locate disease has been shown to be a successful approach in a variety of situations [13,17]. Although results from these papers could be improved, they demonstrate the feasibility of the approach. We have extended these ideas and have applied them to the problem of locating MBD. When compared to the previous approaches [13,17], the rigidity of the 3-dimensional bone improves symmetric matching as the bone surface is not malleable like soft tissue. We do not compare shape differences directly (which is a harder problem) but only the bone density between each pair of points. This leads to improved accuracy. We chose to compare the less dense inner trabecular bone density instead of the more dense outer cortical bone, as the trabecular density measurements are more often affected by the disease.

3   Methodology

Our method first extracts regions containing bone in the CT data. It converts these 2-dimensional segmentations into a 3D polygonal (target) mesh that encapsulates the bone structure. Using the Cortical Bone Mapping (CBM) method [15], we measure the trabecular bone density. Our method then solves for symmetric point correspondence between left and right sides by registering the segmented target mesh with an atlas mesh. Finally, we display differences in trabecular bone across symmetric regions in an interactive 3-dimensional viewer.

3.1   Segmentation

In order to measure trabecular bone density and to find bone symmetry, the contours that surround regions of bone are first located within the CT images through the process of segmentation. In this process, pixels are classified as being either of a bone type or of a non-bone type.

Pixels labelled as bone generally contain a higher intensity than that of non-bone (unless the CT data contains a metal implant or a contrast agent). Bone has a radiodensity typically greater than 1200 Hounsfield units (HU) while soft tissue and fat have a radiodensity typically between -70 HU and 100 HU. However, in the region between 100 HU and 1200 HU, there is much overlap between the radiodensity of trabecular bone, fat and soft tissue. Because of this, a single intensity threshold cannot be used to separate bone from non-bone, as a threshold set low enough to include all bone will include also fat and soft tissue. A threshold set high enough to exclude fat and soft tissue will also exclude bone. To overcome this issue, context from surrounding pixels must be combined with pixel intensity to correctly classify a pixel as being bone or non-bone.

Many approaches have been developed that can segment bone in CT. However, the problem has yet to be fully and accurately solved. Approaches can be placed into general groups including adaptive thresholding, hysteresis thresholding, region growing, watershed, active contour, edge based, level set, graph cuts, statistical shape models, articulated registration based and machine learning based.

We chose to use a hysteresis thresholding technique as regions of bone and non-bone can be generally divided into high intensity (bone) and low intensity (non-bone) regions making it a natural thresholding type problem. Hysteresis thresholding addresses the overlapping regions of bone and non-bone by provided context which greatly improves the accuracy when compared to global thresholding. Furthermore, the Cortical Bone Mapping (CBM) method used for measuring bone density, produces the best results when the contour lines surrounding the bone lie exactly between the bone and non-bone pixels. Because we use a thresholding technique, we can produce contour lines with sub-pixel accuracy by interpolating between threshold pixels, improving both the density calculation and the segmentation accuracy.

First, we select all pixels that are very likely bone (dense outer cortical bone). Then, a second lower threshold is used to select all pixels that may be bone but also may contain some soft tissue and fat. The pixels segmented at this lower level are only kept if they are connected to a pixel segmented at the higher threshold.

Additionally, we developed a new method to improve accuracy based on the gap filling ideas from Gelaude [9]. We found that using a third threshold slightly lower than the second was useful in filling gaps. These gaps occur as the hard cortical bone is sometimes non-existent in the CT. They can severely decrease segmentation accuracy as the contour traces the inside of the bone structure instead of the outside. Figure 1a shows an example.

To address this, we first generate a contour line using a third slightly lower threshold. We then compare this contour line with the contour line generated by the hysteresis thresholding technique. As the third threshold is lower than both the hysteresis thresholds, its contour line lies on the outside of the hysteresis contour at all points. By following these contour lines in counter clockwise direction, each time these two contours diverge, the path lengths are compared. The contour lines lie relatively close to each other until gaps in the cortical bone are reached. In these locations, the contours often diverge as the hysteresis contour traces the gap and then re-joins with the lower contour. The hole is filled if the path length of the hysteresis threshold contour is more than three times longer than that of the lower contour. Figure 1b shows an example of a hole being filled. The green (hysteresis) contour follows the blue (lower threshold) contour until they diverge at the gap. The yellow contour shows the divergence. Its length is compared to the length of the blue line during the divergence, causing the final hysteresis contour to skip the gap.

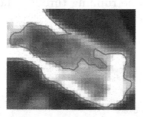

(a) Gaps in the contour can severely decrease segmentation accuracy.

(b) Joining a gap in the contour to improve segmentation accuracy.

Fig. 1. Generic segmentation of bone from CT data. (Color figure online)

3.2 Bone Mesh Creation and Cortical Bone Mapping (CBM)

Once contours are found in each original CT image, the marching tetrahedra algorithm is used to create a polygonal mesh of the bone structure. We used the Stradwin software implementation [14] for this. Figure 2a shows the results of our segmentation and mesh

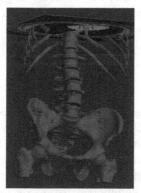
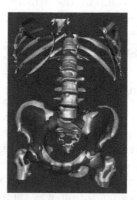

(a) This figure shows what a typical fully automatic segmentation looks like, as well as indicating the typical range over which we attempt to detect MBD.

(b) This figure shows the target mesh (in yellow) being registered to the atlas mesh (multi-coloured) using articulated registration.

Fig. 2. Segmentation and registration. (Color figure online)

Once the 3D polygonal meshes are formed, the Cortical Bone Mapping (CBM) technique is used to accurately estimate trabecular bone density [15]. In CBM, the trabecular bone density is estimated by following the vector normal at each triangle in the mesh into the bone. The distribution of bone density along this line is modelled, which crucially allows average trabecular density to be estimated without any bias from the nearby, and much denser, cortical bone. Figure 3a shows the outline of the segmented bone in red running parallel to the cortical bone and a line normal to the cortical bone running through the bone surface in cyan while Fig. 3b shows the density being measured.

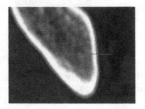
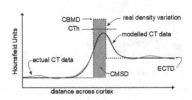

(a) Measuring the cortical thickness along the cyan line through the segmented bone.   (b) Measuring the bone density by modelling the cortical thickness.

Fig. 3. CBM can measure cortical bone mineral density (CBMD), trabecular bone mineral density (ECTD), cortical thickness (CTh), and cortical mass surface density (CMSD). (Color figure online)

3.3   Finding Symmetry

In order to compare the density of symmetric regions, we search for a symmetric point correspondence for each point in the segmented target mesh. To do this we register the target segmentation to an atlas containing a polygonal mesh of a healthy skeleton where each point has been symmetrically paired with another.

3.4   Atlas Creation

To create our atlas, a healthy CT data set was initially segmented using the threshold feature in the Stradwin software [14]. The segmentation was then thoroughly corrected by hand again using the Stradwin software. The full segmentation of the skeleton was also split into multiple regions by labelling general bone regions (pelvis, left femur, right femur, individual ribs and vertebrae).

Each segmentation was then paired with a symmetric region by mirroring one to match the other (e.g. left femur matched to right femur). In the case of a bone region without a corresponding symmetry match (vertebrae and pelvis), the region is matched with an x-axis inverted copy of itself. Similarity-based rigid registration was first used to register each pair of symmetric regions using the wxRegSurf software [8]. Where necessary, deformable registration was used to improve accuracy.

3.5   Articulated Registration

The atlas is registered to the target segmentation using an articulated registration technique. Our registration method was inspired by the techniques used to segment skeletons in mouse CT [1] and human CT [7]. Registration in general produces the best results when initial placement is as accurate as possible. Articulated registration takes advantage of this through a series of steps. The entire skeleton as a whole is first registered for initial placement. Then smaller and smaller sub-regions of the skeleton are registered following a hierarchical anatomical tree atlas. This guides the small subsections of the skeleton into a generally correct location before final registration. This greatly improves accuracy as the whole skeleton guides the registration but each bony region can move independently of each other. For each registration step, a rotation, translation and scale is found that minimizes the distance between each point on the atlas (or bony sub-region) and the target mesh. We implemented the classic registration method described by Besl [2]. Figure 2b shows the atlas mesh being registered to the target mesh using articulated registration.

Our method differs from the papers above [1,7] as their purpose is to segment the skeleton. Instead, we use this registration technique to find the symmetry between each point. At the end of the registration, for each point in the target mesh, we look up the closest point in the atlas mesh. Each point in the atlas has a symmetric point, so we use that symmetric point to lookup the closest point in the target mesh. In this way we can find the symmetric mappings between all points in the target mesh. We compare vector normals and point distances (between atlas and target) and only set symmetric mappings if these are within a reasonable range. Our current registration method is only rigid with scaling—in future we plan to implement a non-rigid registration to improve upon the performance shown in Fig. 2b.

4   Visualisation of Disease

Once every point on the left side has been mapped to a corresponding point on the right side, the trabecular bone density at every point on one side of the skeleton is compared with the other side. As long as one side is diseased, the difference in trabecular bone density between left and right is significant. We wrote a specialized viewer to visualise the results by displaying a polygonal mesh of the skeleton and uniquely colouring each vertex. Figure 4a shows the trabecular bone density (i.e. not using symmetrical difference) overlaid on the mesh. A darker blue indicates a more dense bone. Figure 4b shows the segmented mesh with a raw CT slice.

Areas of large difference are grouped together, and each group has the potential to be either healthy or diseased. Healthy regions are still included with diseased regions at this point, as all groups with large differences in trabecular bone density between left and right sides are selected without taking into account which side caused the differences. Intuitively, trabecular bone density in healthy tissue will be relatively constant, while diseased bone that has increased

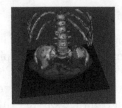
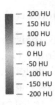

(a) Polygonal mesh overlaid with colour map.

(b) CT slice data included with mesh.

(c) Colour scale of trabecular bone density.

Fig. 4. The segmentation result is overlaid with a colour map representing the trabecular bone density measurements. (Color figure online)

or decreased bone density will have certain intense changes in trabecular bone density. When the gradient of trabecular bone density of healthy bone is compared to the gradient of its difference in trabecular bone density, the differences are generally high. However, when the gradient of trabecular bone density of diseased bone is compared to the gradient of its difference in trabecular bone density, the differences are generally low. In this way, regions are determined to be either diseased or healthy. Figure 5a shows the trabecular bone being subtracted from each side. In Fig. 5b, the density differences greater or less than 200 HU are grouped together. In Fig. 5c, only the diseased areas are shown.

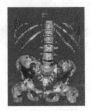
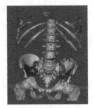
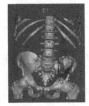

(a) The density of each side is subtracted from the other side.

(b) Regions of strong density differences are grouped together.

(c) Healthy regions are removed by comparing gradients.

Fig. 5. Disease is visualized by comparing trabecular bone density differences and gradients of symmetric regions.

A user can click on any point in the mesh and find it's symmetric point pair. A green line is drawn between the points, and their surface normals are drawn in red and blue. A new, re-slice image plane is defined using the vector pointing from first point to the second, and the vector created by adding the two normals together. In Fig. 6a, a user has clicked on the diseased right side. The CT is automatically re-sliced to show the diseased region. Figure 6b shows MBD on the right side of the pelvis. In Fig. 6b, it can be seen that the right side contains a much greater trabecular bone density than the left side does.

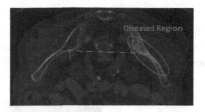

(a) Re-slicing the CT data at the user selected location.

(b) The diseased right side can be seen in the re-sliced CT image.

Fig. 6. A user can click on the mesh to calculate a re-sliced CT image at that location.

5   Results and Discussion

We tested our method using eleven diseased CT data sets and one healthy CT data set. All of our CT data sets are under an ethics agreement enabling us to use them for research and publication. We compared the diseased areas found in our viewer (areas of high colour contrast) to metastatic scoring sheets from conventional CT review marked by experts. We also compared our results to 3-dimensional bone lesions marked by an expert which can be seen in Fig. 7a. Areas were marked as being either malignant or suspicious. Malignant areas are diseased regions and suspicious areas are abnormal regions that could be disease. Table 1 shows our results.

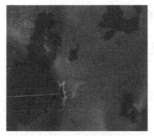

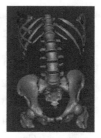

(a) 3-dimensional lesions marked by an expert are displayed in red inside the mesh and compared to our results displayed on the surface of the mesh.

(b) Results correctly show our healthy atlas to be mostly disease free.

Fig. 7. Results are visually compared to 3-dimensional lesions marked by an expert to determine accuracy.

5.1   Evaluation

Our method found 73.1% of suspicious areas and 79.2% of malignant areas correctly. It found 77.0% of all diseased areas. Our results also contained a number of false positives in each data set - most of which are small in size. This can be seen in Fig. 8. However, there are limitations in finding the true accuracy of our results as the clinician-defined areas of suspicion may not be perfect. This may be the cause of some of the false positives as they may actually be abnormal regions missed by the expert.

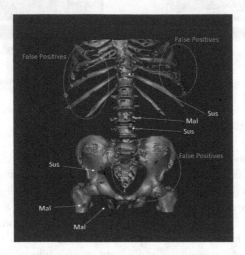

Fig. 8. A typical result showing suspicious areas (Sus), malignant areas (Mal) and false positives.

Table 1. Sensitivity of the viewer finding diseased regions.

Name	Suspicious areas	Malignant areas
Case 1	3/4	1/1
Case 2	4/4	1/1
Case 3	2/3	5/8
Case 4	1/2	None
Case 5	1/2	None
Case 6	1/2	1/1
Case 7	None	4/4
Case 8	2/3	3/4
Case 9	2/3	3/4
Case 10	3/3	9/13
Case 11	None	11/12

We also tested our method on a healthy CT data set (our atlas) by processing its raw CT data. The method found the healthy data mostly disease free although it incorrectly displays small spots of disease in a few locations. This can be seen in Fig. 7b. Some of these spots of disease are due to natural and healthy asymmetry. This can be seen in the sacroiliac joint which is shown as partially diseased. These areas of healthy asymmetry will need to be addressed in the future.

5.2 Discussion

These results demonstrate that our technique can successfully detect and visualise metastases in bone. Our method already has a relatively high sensitivity in finding areas of diseased bone, but can be improved. It can find bone metastases in the vertebra, pelvis and upper femurs, which is an improvement over existing methods which are mostly limited to the vertebra.

Our results currently contain many false positives. These are due to errors in the automatic segmentation, registration and symmetry point matching although in some cases, our method may be correctly identifying abnormal regions missed by the expert. One important next step to reduce false positives, is to use a deformable registration technique that can both preserve atlas symmetry but also provide more accurate matching with the target mesh. After this step, the deformed atlas mesh can be used to measure the trabecular bone density instead of using the target segmentation mesh. This should greatly improve the accuracy, as errors will be reduced both in the trabecular bone density measurement and in the symmetric point matching.

Our method produces poor results in the ribs. This is due to significant errors in the registration as there is much variability between rib shape and size. Poor registration produces poor symmetric point matching severely decreasing the accuracy of this method. Deformable registration as well as improvements to the rib components of the articulated registration should help with this problem.

6 Conclusions

We have introduced a new algorithm that can automatically detect and display metastasized cancer in bones. Preliminary results demonstrate that the algorithm can successfully find areas of bone disease and visualise them in a unique way. It already has a relatively high sensitivity in finding areas of diseased bone but it can still be improved. The technique currently suffers from a high false positive rate, which will need to be addressed. It is fully automatic, and successfully works as a diagnostic tool providing a unique way to visualise disease.

References

1. Baiker, M., et al.: Atlas-based whole-body segmentation of mice from low-contrast Micro-CT data. Med. Image Anal. **14**(6), 723–737 (2010)
2. Besl, P.J., McKay, N.D.: Method for registration of 3-D shapes. In: Sensor Fusion IV: Control Paradigms and Data Structures, vol. 1611, pp. 586–607. International Society for Optics and Photonics (1992)

3. Burns, J.E., Yao, J., Wiese, T.S., Muñoz, H.E., Jones, E.C., Summers, R.M.: Automated detection of sclerotic metastases in the thoracolumbar spine at CT. Radiology **268**(1), 69–78 (2013)
4. Chmelik, J., Jakubicek, R., Jan, J.: Tumorous spinal lesions: computer aided diagnosis and evaluation based on CT data-a review. Current Med. Imaging Rev. **14**(5), 686–694 (2018)
5. Chmelik, J., et al.: Deep convolutional neural network-based segmentation and classification of difficult to define metastatic spinal lesions in 3D CT data. Med. Image Anal. **49**, 76–88 (2018)
6. Diel, I.J., Kaufmann, M., Bastert, G.: Metastatic Bone Disease: Fundamental and Clinical Aspects. Springer, Heidelberg (2012)
7. Fu, Y., Liu, S., Li, H.H., Yang, D.: Automatic and hierarchical segmentation of the human skeleton in CT images. Phys. Med. Biol. **62**(7), 2812 (2017)
8. Gee, A.: wxregsurf software (2019). http://mi.eng.cam.ac.uk/~ahg/wxRegSurf/
9. Gelaude, F., Vander Sloten, J., Lauwers, B.: Semi-automated segmentation and visualisation of outer bone cortex from medical images. Comput. Methods Biomech. Biomed. Eng. **9**(1), 65–77 (2006)
10. Jan, J., Novosadova, M., Demel, J., Ouředníček, P., Chmelik, J., Jakubíček, R.: Combined bone lesion analysis in 3D CT data of vertebrae. In: 2015 37th Annual International Conference of the IEEE Engineering in Medicine and Biology Society (EMBC), pp. 6374–6377. IEEE (2015)
11. Peter, R., Malinsky, M., Ourednicek, P., Jan, J.: 3D CT spine data segmentation and analysis of vertebrae bone lesions. In: 2013 35th Annual International Conference of the IEEE Engineering in Medicine and Biology Society (EMBC), pp. 2376–2379. IEEE (2013)
12. Roth, H.R., Yao, J., Lu, L., Stieger, J., Burns, J.E., Summers, R.M.: Detection of sclerotic spine metastases via random aggregation of deep convolutional neural network classifications. In: Yao, J., Glocker, B., Klinder, T., Li, S. (eds.) Recent Advances in Computational Methods and Clinical Applications for Spine Imaging. LNCVB, pp. 3–12. Springer, Cham (2015). https://doi.org/10.1007/978-3-319-14148-0_1
13. Tahmoush, D., Samet, H.: Using image similarity and asymmetry to detect breast cancer. In: Medical Imaging 2006: Image Processing, vol. 6144, p. 61441S. International Society for Optics and Photonics (2006)
14. Treece, G.M.: Stradwin software (2019). http://mi.eng.cam.ac.uk/~rwp/stradwin/
15. Treece, G.M., Poole, K.E., Gee, A.H.: Imaging the femoral cortex: thickness, density and mass from clinical CT. Med. Image Anal. **16**(5), 952–965 (2012)
16. Wiese, T., Yao, J., Burns, J.E., Summers, R.M.: Detection of sclerotic bone metastases in the spine using watershed algorithm and graph cut. In: Medical Imaging 2012: Computer-Aided Diagnosis, vol. 8315, p. 831512. International Society for Optics and Photonics (2012)
17. Yang, L., Xie, Y., Li, B., Xie, M., Wang, X., Zhang, J.: Symmetry based prostate cancer detection. Br. J. Radiol. **88**(1050), 20150132 (2015)
18. Yao, J., Burns, J.E., Summers, R.M.: Computer aided detection of bone metastases in the thoracolumbar spine. In: Li, S., Yao, J. (eds.) Spinal Imaging and Image Analysis. LNCVB, vol. 18, pp. 97–130. Springer, Cham (2015). https://doi.org/10.1007/978-3-319-12508-4_4

High-Order Markov Random Field Based Image Registration for Pulmonary CT

Peng Xue, Enqing Dong$^{(\boxtimes)}$, and Huizhong Ji

School of Mechanical, Electrical and Information Engineering, Shandong University,
Weihai 264209, China
enqdong@sdu.edu.cn

Abstract. Deformable image registration is an important tool in medical image analysis. In the case of lung four dimensions computed tomography (4D CT) registration, there is a major problem that the traditional image registration methods based on continuous optimization are easy to fall into the local optimal solution and lead to serious misregistration. In this study, we proposed a novel image registration method based on high-order Markov Random Fields (MRF). By analyzing the effect of the deformation field constraint of the potential functions with different order cliques in MRF model, energy functions with high-order cliques form are designed separately for 2D and 3D images to preserve the deformation field topology. For the complexity of the designed energy function with high-order cliques form, the Markov Chain Monte Carlo (MCMC) algorithm is used to solve the optimization problem of the designed energy function. To address the high computational requirements in lung 4D CT image registration, a multi-level processing strategy is adopted to reduce the space complexity of the proposed registration method and promote the computational efficiency. Compared with some other registration methods, the proposed method achieved the optimal average target registration error (TRE) of 0.93 mm on public DIR-lab dataset with 4D CT images, which indicates its great potential in lung motion modeling and image guided radiotherapy.

Keywords: Image registration · 4D CT · Markov Random Field · Topology preservation

1 Introduction

In the lung 4D CT image registration, due to the influence of heart beats and respiratory movements, the local intensity inhomogeneity of the 4D CT images and the large deformation of the fine textures are caused. In this case, some traditional continuous optimization-based image registration methods are easy to fall into local optimal solutions [1], and obtain unacceptable results, which are not suitable for the registration of lung images.

The image registration method based on MRF model is a non-rigid registration method using discrete optimization. For the MRF-based image registration method, according to whether the high-order clique structure is included in the energy function,

© Springer Nature Switzerland AG 2020
Y. Zheng et al. (Eds.): MIUA 2019, CCIS 1065, pp. 339–350, 2020.
https://doi.org/10.1007/978-3-030-39343-4_29

it can be divided into a low-order MRF-based image registration method and a high-order MRF-based image registration method [2]. In general, the low-order MRF-based image registration method only considers the energy function based on pairwise interactions between the variables of the field. Therefore, most of the modern discrete optimization algorithms such as message passing [3, 4] and graph cuts [5] are used to solve the optimization problem.

In high-order MRF-based image registration, the energy functions are composed of complex high-order cliques above two-element cliques. Although the high-order cliques can impose constraints more effectively on the deformation field and can further improve the registration accuracy, but the discrete optimization algorithms introduced above can't solve the energy functions with the high-order cliques. In response to the above problem, the Quadratic Pseudo-Boolean Optimization (QPBO) algorithm [6] was proposed to improve the graph cuts algorithm, so that QPBO can solve the non-submodular optimization problems. After that, Cordero-Grande et al. [7] proposed a 2D image registration method that could maintain the topology by using parameter estimation and MCMC-based optimization algorithm.

In lung 4D CT image registration, in order to solve the problem of large motion of small features in the lung, Han et al. [8] proposed an image registration method based on robust 3D SURF (Speeded Up Robust Features) descriptors for feature detection and matching. For lung images with local intensity inhomogeneity, Normalized Gradient Fields (NGF) [9] or Modality Independent Neighborhood Descriptors (MIND) [10] are usually used. In recent related researches, Vishnevskiy et al. [11] proposed an isopTV method that used an isotropic total variation regularization term to constrain the deformation field, it's result is the best in the DIR-lab dataset with 4D CT images. In the same year, Rühaak et al. [12] proposed a method for estimate large motion in lung CT by integrating regularized keypoint correspondences into dense deformable registration which is best in the COPD dataset. In 2018, Eppenhof et al. [13] used 3D convolutional neural networks for 4D CT lung image registration, although the registration time can be greatly shortened while ensuring a certain registration accuracy, its registration accuracy has no advantage compared with other methods.

In order to further improve the accuracy of lung 4D CT image registration, a high-order MRF-based 4D CT image registration method (HO-MRF) is proposed in this paper. Based on high-order MRF, an energy function with high-order cliques is designed to maintain the topology of deformation field and MCMC algorithm is used to solve the optimization problem of the energy function. In view of the fact that the lung 4D CT image has many voxels and large local motion range, an effective multi-level processing strategy is adopted to reduce the complexity of the algorithm and improve the computational efficiency.

2   High-Order MRF Registration Model

2.1   General Form of the MRF Model

Many problems in image processing can be expressed as label problems in MRF. In image registration, consider the N-dimensional target image $I: \Omega \rightarrow \mathbb{R}^N$ and N-dimensional moving image $J: \Omega \rightarrow \mathbb{R}^N$, for each point $p \in \Omega$ (with spatial location x_p), there is a set of labels **L**, which corresponds to N-dimensional discrete displacements $\mathbf{d} = (d_1, d_2, d_3 \ldots d_N) \in \mathbf{L}$. Therefore, an energy function in image registration can be expressed as the following form

$$E = \sum_{p \in \Omega} |I(x_p) - J(x_p + \mathbf{D}(x_p))| + \omega \cdot R(\mathbf{D}) \tag{1}$$

where **D** is the deformation field composed of **d** and $\mathbf{D} = \sum_{p \in \Omega} \mathbf{d}_p$, $I(x_p)$ represents the intensity of the target image at point p, $J(x_p + \mathbf{D}(x_p))$ represents the intensity of the point p on the moving image after the deformation field is applied, R is a function of the deformation field **D**, and ω is a weight parameter, which determines the influence of the constraint term on the whole energy function.

In MRF, the constraint R is usually determined by the three elements of the neighborhood system $\mathcal{N}$, the cliques $\mathcal{C}$ and the potential functions $\mathcal{V}$ in the random field. In a 2D image, a common neighborhood system is a 4-neighborhood system as shown in Fig. 1(a) or 8-neighborhood system as shown in Fig. 1(b). Accordingly, in a 3D image, there are 6-neighborhood system, 18-neighborhood system and 26-neighborhood system. On this basis, we can define clique as a subset of points according to the set of pixel points in the neighborhood. According to the size of clique (the number of points p included in the clique), a one-element clique, a two-element clique and a three-element clique can be defined as: $\mathcal{C}_1 = \{p | p \in \Omega\}$, $\mathcal{C}_2 = \{(p, q) | p \in \Omega \; q \in \mathcal{N}_p\}$, $\mathcal{C}_3 = \{(p, q, r) | p \in \Omega, q, r \in \mathcal{N}_p\}$, where $\mathcal{N}_p$ (light blue dots in Fig. 1) represents the neighborhood of the point p (dark blue dots in Fig. 1) and Fig. 1(c), (d) and (e) show the structure of each clique in the 8-neighborhood system of the 2D image respectively.

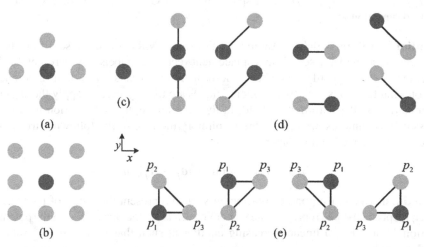

Fig. 1. Neighborhood system and clique categories in 2D image: (a) 4-neighborhood system (b) 8-neighborhood system (c) One-element cliques (d) Two-element cliques (e) Three-element cliques. See also the orientation of the coordinate system $\{x, y\}$. (Color figure online)

In MRF, the potential functions $\mathcal{V}$ are the core factor that determines the constraints. According to the Hammersley-Clifford theorem, the energy function E can be represented by the potential functions $\mathcal{V}$. Usually, a potential function is generally defined as a function of cliques, and its purpose is to calculate the cost of the corresponding potential function according to different order cliques. In image registration, the potential functions with one-element cliques represent a data term, the potential functions with two-element cliques and above are used to constrain the deformation field. The role of various potential functions with different cliques in image registration will be described in detail below.

2.2   Potential Functions with Different Cliques

Potential Functions with One-Element Cliques. In MRF, a potential function with one-element cliques (also called data term) is usually used to measure the similarity of control points in two images, that is, the data term in the energy function. In the CT image registration of the lung, due to the compression of the lungs and the local intensity inhomogeneity of the lung 4D CT images, in this paper, the Local Correlation (LCC) metric proposed in [14] is mainly used to construct the following potential function with one-element cliques

$$\mathcal{V}_1(x_p) = 1 - \frac{\langle I_p, J_p \rangle}{\sigma(I_p)\sigma(J_p)} \tag{2}$$

where $\bar{I}_p = H_w \otimes I_p$, $\bar{J}_p = H_w \otimes J_p$, $\sigma^2(I_p) = \overline{I_p^2} - \bar{I}_p^2$, $\sigma^2(I_p) = \overline{I_p^2} - \bar{I}_p^2$, $\langle I_p, J_p \rangle = \overline{I_p J_p} - \bar{I}_p \bar{J}_p$, H_w is a spatial-invariant Gaussian weighting kernel with standard deviation w.

Potential Functions with Two-Element Cliques. In MRF, since the set of labels for each point is defined as a set of the displacement vectors, so labels can represent the displacement vectors. In order to make the deformation field satisfy a certain smoothness, the potential functions with two-element cliques are usually used to apply the smoothing constraint to the deformation field. In the lung image, the deformation field is not necessarily continuous because of the respiratory motion, so the following truncation model in [15] is used in this paper

$$\mathcal{V}_2(\mathbf{d}_p, \mathbf{d}_q) = \min(r \cdot \|\mathbf{d}_p - \mathbf{d}_q\|, n) \tag{3}$$

where $\mathbf{d}_p, \mathbf{d}_q$ (the correspond displacement vector) represent the labels of point p and point q respectively. $\|\bullet\|$ represent the vector norm, n is the maximum cost of the potential function $\mathcal{V}_2$ and r is a linear increasing coefficient such that $r \cdot \|\mathbf{d}_p - \mathbf{d}_q\|$ satisfies $r \cdot \|\mathbf{d}_p - \mathbf{d}_q\| \ll n$.

Potential Functions with High-Order Cliques. Since the high-order cliques can describe the complex relationship between the labels, it can not only constrain the smoothness, but also maintain the topology of the deformation field, so that the deformation can be better constrained. In 2D images, the topology preservation term is built according to three-elements cliques, as shown in Fig. 1(e). Note that the four three-element cliques defined in Fig. 1(e) are analogous to the four corners from which the Jacobian is computed in [16], and we can get the Jacobian discriminants according to the different forms of the three-element cliques in Fig. 1(e)

$$J^{ff}\left(\mathbf{d}_{p1},\mathbf{d}_{p2},\mathbf{d}_{p3}\right) = \left(1+\mathbf{d}_{p2x}-\mathbf{d}_{p1x}\right)\left(1+\mathbf{d}_{p3y}-\mathbf{d}_{p1y}\right) - \left(\mathbf{d}_{p3x}-\mathbf{d}_{p1x}\right)\left(\mathbf{d}_{p2y}-\mathbf{d}_{p1y}\right)$$

$$J^{bf}\left(\mathbf{d}_{p1},\mathbf{d}_{p2},\mathbf{d}_{p3}\right) = \left(1+\mathbf{d}_{p1x}-\mathbf{d}_{p2x}\right)\left(1+\mathbf{d}_{p3y}-\mathbf{d}_{p1y}\right) - \left(\mathbf{d}_{p3x}-\mathbf{d}_{p1x}\right)\left(\mathbf{d}_{p1y}-\mathbf{d}_{p2y}\right)$$

$$J^{fb}\left(\mathbf{d}_{p1},\mathbf{d}_{p2},\mathbf{d}_{p3}\right) = \left(1+\mathbf{d}_{p2x}-\mathbf{d}_{p1x}\right)\left(1+\mathbf{d}_{p1y}-\mathbf{d}_{p3y}\right) - \left(\mathbf{d}_{p1x}-\mathbf{d}_{p3x}\right)\left(\mathbf{d}_{p2y}-\mathbf{d}_{p1y}\right)$$

$$J^{bb}\left(\mathbf{d}_{p1},\mathbf{d}_{p2},\mathbf{d}_{p3}\right) = \left(1+\mathbf{d}_{p1x}-\mathbf{d}_{p2x}\right)\left(1+\mathbf{d}_{p1y}-\mathbf{d}_{p3y}\right) - \left(\mathbf{d}_{p1x}-\mathbf{d}_{p3x}\right)\left(\mathbf{d}_{p1y}-\mathbf{d}_{p2y}\right) \quad (4)$$

where, in the isosceles right-angled triangle (the length of the right angle is the unit distance 1) formed by the clique elements p_1, p_2, p_3 (as shown in Fig. 1(e)), $\mathbf{d}_{p1},\mathbf{d}_{p2},\mathbf{d}_{p3} \in \mathbf{D}$ represents the displacement vector of the right-angled vertex, the displacement vector of the vertical right-angled vertex, and the displacement vector of the horizontal right-angled vertex, Eq. (4) is the Jacobian discriminant of the invariance of the constrained topology we need. According to the characteristics of the MCMC algorithm, in order to avoid the excessive cost value of the topology preservation terms composed of high-order cliques, we propose to use the logarithm function form to impose penalties on different deformations. Therefore, for 2D image, the topology preservation term can be expressed as follows:

$$\mathcal{V}_3\left(\mathbf{d}_{p1},\mathbf{d}_{p2},\mathbf{d}_{p3}\right) = \begin{cases} s \cdot \log(J\left(\mathbf{d}_{p1},\mathbf{d}_{p2},\mathbf{d}_{p3}\right)+1), & J\left(\mathbf{d}_{p1},\mathbf{d}_{p2},\mathbf{d}_{p3}\right) \geq 0, \\ m, otherwise. \end{cases} \quad (5)$$

where s is a linear increasing coefficient, m is the maximum cost value when the topology is not preserved and satisfies $\max(s \cdot \log(J\left(\mathbf{d}_{p1},\mathbf{d}_{p2},\mathbf{d}_{p3}\right)+1)) \ll m < 1$.

For 3D images, the corresponding topological constraint is mainly composed of an eight-element clique shown in Fig. 2(a). Calculating the eight-element clique constraint requires at least 64 Jacobian determinants, which imposes a huge computation burden in practice. Therefore, according to the literature [16], we constrain the topology by using 8 Jacobian matrices corresponding to the eight four-element cliques as shown in Fig. 2(b). In this way, not only the topology in the 3D image can be effectively maintained, but also the amount of calculation can be effectively reduced. Therefore, we can express a potential function with four-element cliques as follows

$$\mathcal{V}_4\left(\mathbf{d}_{p1},\mathbf{d}_{p2},\mathbf{d}_{p3},\mathbf{d}_{p4}\right) = \begin{cases} s \cdot \log(J\left(\mathbf{d}_{p1},\mathbf{d}_{p2},\mathbf{d}_{p3},\mathbf{d}_{p4}\right)+1), & J\left(\mathbf{d}_{p1},\mathbf{d}_{p2},\mathbf{d}_{p3},\mathbf{d}_{p4}\right) \geq 0, \\ m, otherwise. \end{cases} \quad (6)$$

2.3 High-Order MRF Registration Model

In the high-order MRF registration model, the energy function is usually composed of three-element cliques or higher-order cliques. It can be seen from the above potential

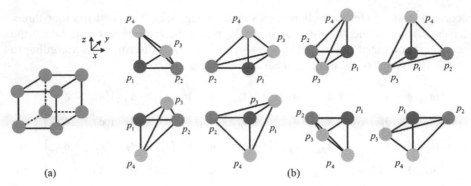

Fig. 2. Four-element cliques and eight-element clique in 3D image (a) Eight-element clique (b) Four-element cliques.

function analysis that in order to ensure that the deformation field can maintain the topology while having a certain smoothness, in the energy function, it is necessary to add not only a smoothing constraint term corresponding to two-element clique potential functions, but also it is necessary to add a topology preservation term with three-element clique potential functions or the four-element clique potential functions. Therefore, for 2D image registration, an energy function with the highest clique is a three-element clique can be expressed as follows:

$$E = \sum_{p \in C_1} V_1(x_p) + \sum_{p,q \in C_2} V_2(d_p, d_q) + \sum_{p,q,r \in C_3} V_3(d_p, d_q, d_r) \tag{7}$$

In the above equation, C_i represents a set of all i-element cliques, V_i is the potential function corresponding to i-element cliques. Correspondingly, in the 3D image, an energy function with the highest clique is a four-element clique can be expressed as follows:

$$E = \sum_{p \in C_1} V_1(x_p) + \sum_{p,q \in C_2} V_2(d_p, d_q) + \sum_{p,q,r,s \in C_4} V_4(d_p, d_q, d_r, d_s) \tag{8}$$

3 Model Analysis

In this section, we use the following images to discuss the role of the potential functions with one-element cliques, two-element cliques and three-element cliques corresponding to the data term, smooth term and topology preservation term in image registration. The main concern here is the effect of the potential functions with high-order cliques (three-element cliques) on the topological structure of the image deformation field.

In the experiments, we select Fig. 3(a) as the moving image, Fig. 3(b) is the target image. The size of both images is 128 × 128. It can be known from the MCMC algorithm that the data term consisting of one element cliques is the main deformation driving factor, the regular term with two-element cliques and other high-order cliques are the secondary deformation driving factor. Therefore, in the parameter selection, we should choose the parameters to control the cost value of the regular term does not exceed the cost value of the data term. According to this, it is more appropriate to select 0.3 for the smoothing parameter n and select 0.5 for the topology preservation parameter m. Under this criterion, selecting other parameters has little effect on the result.

First, we only rely on the similarity metric (data term) to register the moving and target image. It can be seen from Fig. 3(c) that the deformation field is distorted and folded. Figure 3(d) is the deformation field after registration by using the data term and the smoothing term, the smoothness of the deformation field is significantly improved compared to the deformation field after registration using only the data term. However, it can be seen from the partial enlarged image (Fig. 3(e)) of Fig. 3(d) that the deformation field of the image is still folded, and the topology cannot be maintained. Figure 3(f) is the deformation field after adding the topology preservation term in addition to the data term and the smoothing term in the energy function. Comparing Fig. 3(e) with Fig. 3(g), it can be seen that the folding phenomenon of the deformation field after the application of the potential functions with three-element cliques disappears obviously, which proves that the potential functions with three-element cliques are introduced into the energy function as the topology preservation term can maintain the topology of the deformation effectively.

4 Results

Since the lung 4D-CT image has high resolution and large range of local anatomical features, so it is difficult to directly register on the original resolution. To solve this problem, we use the multi-level processing strategy [17] to reduce the computational complexity. In the multi-level processing strategy, firstly, the original moving image and the original target image are respectively divided according to the preset multi-level number and corresponding grids of different spacing, and a corresponding number of multi-level control point grid maps are obtained. Secondly, the average intensity values of all the points in the corresponding grid map of the target image and the moving image of the first level are respectively calculated in the first level processing, and then use them as the intensity values of the control point of the grid map for the target image and the moving image in the level respectively. After applying the smoothing term constraint and the topology preservation term constraint in the energy function of the first level, the MCMC optimization algorithm is used to solve the optimization problem of the energy function and obtain the deformation field of the grid map. Finally, the deformation field grid map obtained by up-sampling the deformation field of the first level grid map according to the grid size of the next-level is used as the initial value of the next-level registration constraint.

The multi-level processing is performed until the preset levels, and the final level grid map deformation field is obtained, and then is up-sampled according to the original image size to obtain a final deformation field. The registered image is obtained by applying the final deformation field to the original moving image.

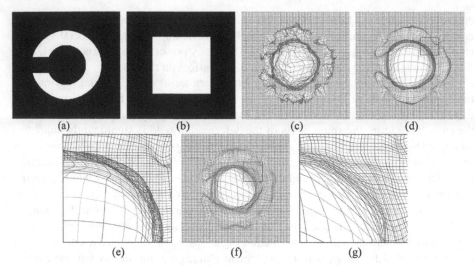

Fig. 3. Experimental results of the topology preservation for 2D deformation field. (a) Moving image. (b) Target image. (c) The deformation field after registration only based on data term. (d) The deformation field after registration based on data term and smoothing term. (e) Partial enlarged image of the red box area in Fig. 3(d). (f) The deformation field after registration based on data term, smoothing term and topology preservation term. (g) Partial enlarged image of the red box area in Fig. 3(f). (Color figure online)

On this basis, we evaluated our proposed HO-MRF method using DIR-lab dataset[1] that represents three-dimensional abdominal time series in respiratory motion. The DIR-lab dataset consists of 10 different sequences labeled 4D CT1-4D CT10, each of which is further divided into ten respiratory phase sequences from T00 to T90, where the maximum inspiratory phase (T00) and maximum the expiratory phase (T50) provides 300 expert landmarks. The average voxel resolution of the dataset is $1 \times 1 \times 2.5 \, mm^3$. Therefore, we clipped the image intensities between 50 and 1200 HU, and then select the image corresponding to the maximum inspiratory phase (T00) of each case as the moving image, and the image corresponding to the maximum expiratory phase (T50) as the target image for registration.

Figure 4 shows the lung CT coronal overlay image of the proposed HO-MRF method and the isopTV method for the fourth case in the 4D CT dataset. The overlay image is obtained by superimposing the T00 phase image or the registered T00 phase image and the T50 phase image. In the overlay image, the red portion and the green portion respectively indicate image under registration and image over registration. If the difference between the two images is small, the overlay image is dominated by gray. Otherwise, more red or green portions appear in the image. It can be seen from Fig. 4(a) that the T00 phase image and the T50 phase image have a large difference, in particular, there is a large displacement in the lower part of the lung, and there are many red regions in the figure, indicating severe under registration. Compared with Fig. 4(a), the overlay images of Fig. 4(b) and (c) obtained by HO-MRF method and isopTV method respectively are

---

[1] https://www.dir-lab.com/index.html.

obviously improved, but there are still over registration and under registration. Comparing the HO-MRF method with the isoPTV method, it can be seen that, in general, the under-registered red region and the over-registered green region in Fig. 4(c) are significantly less than those in Fig. 4(b) respectively. For obvious areas, see the corresponding red frame area marked in the Fig. 4.

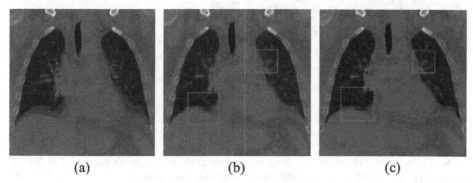

(a) (b) (c)

Fig. 4. Coronal overlay images for case 4 of 4D CT dataset. (a) Non-registered overlay image. (b) isopTV overlay image. (c) HO-MRF overlay image. (Color figure online)

In order to show the validity of the registration more intuitively, we take case 4 in the DIR-lab dataset as an example, and give the schematic diagrams of registration vector displacement error of the isopTV method (Fig. 5(a)) and the proposed HO-MRF method (Fig. 5(b)) respectively. The blue dots in the figure indicate the landmarks in the T50 phase, the red dots represent the landmarks in the T00 phase, and the lines connecting the registered points and the blue points are used as the displacement vectors of the landmarks. The range of the registration displacement vector error is marked by different colors. It can be seen from the Fig. 5 that for these two methods, the deformation of the registered landmarks basically maintains a smooth consistency, and the displacement vectors of the landmarks in the red squares marked in the two images are compared, it can be seen that the proposed HO-MRF method has better registration accuracy in the right lung than the isopTV method.

We also use the TRE (Target Registration Error) indicator to further evaluate the effect of image registration. The TRE is defined as the Euclidean distance between the coordinate of the expert landmarks and the coordinate of the landmarks registered by the deformation field transformation. The smaller the distance, the better the registration effect of the method. Table 1 provides the average TRE for various methods in the DIR-lab dataset. Among them, the MRF-based method (deeds [18]) can be used as a baseline to measure the validity of the registration. The NGF [19] method is currently best masked method (0.94 mm), which estimates the displacement field only inside the lungs. The isoPTV [12] method is currently the best among all mask-free methods in the DIR-Lab dataset. By comparison, it can be found that the registration results of the proposed HO-MRF method is better than the deeds method except that case1 is slightly higher than the deeds algorithm by 0.04 mm. Compared with the isoPTV method, the proposed HO-MRF method has one case equivalent and four cases superior to the isoPTV method.

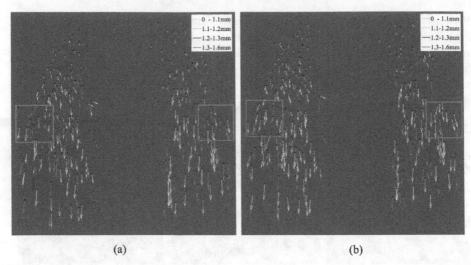

(a)   (b)

Fig. 5. The schematic diagrams of registration vector displacement errors for case 4 of 4D CT dataset. (a) isopTV method. (b) HO-MRF method. (Color figure online)

Table 1. Average TRE for various methods in DIR-lab dataset

Methods	TRE for 4D CT cases (mm)										Mean
	Case 1	Case 2	Case 3	Case 4	Case 5	Case 6	Case 7	Case 8	Case 9	Case 10	
No regis	4.01	4.65	6.73	9.42	7.10	11.10	11.59	15.16	7.82	7.63	8.52
cEPE [19]	0.80	0.77	0.92	1.22	1.21	0.90	0.98	1.16	1.00	0.99	1.00
NLR [20]	0.77	0.78	0.93	1.27	1.11	0.91	0.86	1.03	0.97	0.87	0.95
LMP [21]	**0.74**	0.78	0.91	1.24	1.17	0.90	0.87	1.04	0.98	0.89	0.95
SGM3D [22]	0.76	**0.72**	0.94	1.24	1.15	0.90	0.89	1.13	**0.91**	0.83	0.95
NGF [18]	0.78	0.79	0.93	1.27	**1.07**	0.90	0.85	1.03	0.94	0.86	0.94
aTV [23]	0.76	0.78	**0.82**	1.31	1.25	1.11	0.97	1.28	1.04	0.99	1.03
deeds [17]	0.80	0.86	1.14	1.17	1.77	1.88	2.21	2.78	1.35	1.50	1.60
isoPTV [11]	0.76	0.77	0.90	1.24	1.12	0.85	0.80	1.34	1.92	**0.82**	0.95
HO-MRF	0.80	0.79	1.01	**1.03**	1.25	**0.81**	**0.80**	**0.98**	0.95	0.84	**0.93**

In comparison with NGF method, the proposed method has one case equivalent and five cases better than NGF method. In addition, in the average TRE of all methods for 10 cases, the average TRE of the HO-MRF method is the best (0.93 mm).

5   Conclusions

The purpose of the high-order MRF-based image registration method proposed in this paper is to maintain the topological structure of the deformation field by introducing high-order clique constraints in the energy function. By comparing and analyzing the topological structure of the 2D image model deformation fields, we can find that compared with the low-order MRF registration algorithm with only smooth constraint in the energy function, although the energy function designed in this paper is more complicated, its ability to maintain the topological structure of the deformation field is also stronger, indicating that the design of the energy function with high-order groups is worthy of further study. The lung image registration experiments in 4DCT dataset show that the proposed method has obvious advantages in registration accuracy compared with the current optimal methods such as isoPTV and NGF.

Acknowledgements. This paper is supported by the Fundamental Research Funds for the Central Universities (China), the National Natural Science Foundation of China (Nos. 81671848 and 81371635) and Shandong Key Research and Development Program (2019GGX101022).

References

1. Kappes, J.H., et al.: A comparative study of modern inference techniques for structured discrete energy minimization problems. In: IEEE Conference on Computer Vision and Pattern Recognition, pp. 155–184. IEEE (2015)
2. Li, S.Z.: Markov Random Field Modeling in Image Analysis. Springer-Verlag, Tokyo (2001). https://doi.org/10.1007/978-4-431-67044-5
3. Wainwright, M.J., Jaakkola, T.S., Willsky, A.S.: Map estimation via agreement on trees: message-passing and linear programming. IEEE Trans. Inform. Theory **51**(11), 3697–3717 (2005)
4. Kolmogorov, V.: Convergent tree-reweighted message passing for energy minimization. IEEE Trans. Pattern Anal. Mach. Intell. **28**(10), 1568–1583 (2006)
5. Komodakis, N., Tziritas, G.: Approximate labeling via graph cuts based on linear programming. IEEE Trans. Pattern Anal. Mach. Intell. **29**(8), 1436–1453 (2007)
6. Rother, C., Kolmogorov, V., Lempitsky, V., Szummer, M.: Optimizing binary MRFs via extended roof duality. In: IEEE Conference on Computer Vision and Pattern Recognition, pp. 1784–1791. IEEE (2007)
7. Cordero-Grande, L., Vegas-Sanchez-Ferrero, G., Casaseca-de-la-Higuera, P., Alberola-Lopez, C.: A Markov random field approach for topology-preserving registration: application to object-based tomographic image interpolation. IEEE Trans. Image Process. **21**(4), 2047–2061 (2012)
8. Han, X.: Feature-constrained nonlinear registration of lung CT images. In: Medical Image Analysis for the Clinic, pp. 137–146 (2010)
9. Lugo, W., Seguel, J.: A fast and accurate parallel algorithm for genome mapping assembly aimed at massively parallel sequencers. In: 6th ACM Conference on Bioinformatics, Computational Biology, and Health Informatics, pp. 574–581. Association for Computing Machinery, Inc, Atlanta (2015)
10. Heinrich, M.P., et al.: MIND: modality independent neighborhood descriptor for multi-modal deformable registration. Med. Image Anal. **16**(7), 1423–1435 (2012)
11. Vishnevskiy, V., Gass, T., Szekely, G., Tanner, C., Goksel, O.: Isotropic total variation regularization of displacements in parametric image registration. IEEE Trans. Med. Imaging **36**(2), 385–395 (2017)

12. Ruehaak, J., et al.: Estimation of large motion in lung CT by integrating regularized keypoint correspondences into dense deformable registration. IEEE Trans. Med. Imaging **36**(8), 1746–1757 (2017)
13. Eppenhof, K.A.J., Pluim, J.P.W.: Pulmonary CT registration through supervised learning with convolutional neural networks. IEEE Trans. Med. Imaging **38**, 1097–1105 (2018)
14. Sun, X., Pitsianis, N.P., Bientinesi, P.: Fast computation of local correlation coefficients. In: Mathematics for Signal & Information Processing. International Society for Optics and Photonics (2008)
15. Glocker, B., Komodakis, N., Tziritas, G.: Dense image registration through MRFs and efficient linear programming. Med. Image Anal. **12**(6), 731–741 (2008)
16. Karacali, B., Davatzikos, C.: Estimating topology preserving and smooth displacement fields. IEEE Trans. Med. Imaging **23**(7), 868–880 (2004)
17. Heinrich, M.P., Jenkinson, M., Brady, M., Schnabel, J.A.: MRF-based deformable registration and ventilation estimation of lung CT. IEEE Trans. Med. Imaging **32**(7), 1239–1248 (2013)
18. Koenig, L., Ruehaak, J.: A fast and accurate parallel algorithm for non-linear image registration using normalized gradient fields. In: IEEE 11th International Symposium on Biomedical Imaging, pp. 580–583. IEEE (2014)
19. Hermann, S., Werner, R.: High accuracy optical flow for 3D medical image registration using the census cost function. In: Klette, R., Rivera, M., Satoh, S. (eds.) PSIVT 2013. LNCS, vol. 8333, pp. 23–35. Springer, Heidelberg (2014). https://doi.org/10.1007/978-3-642-53842-1_3
20. Ruehaak, J., Heldmann, S., Kipshagen, T., Fischer, B.: Highly accurate fast lung CT registration. In: Medical Imaging 2013: Image Processing, SPIE, Lake Buena Vista, pp. 86690Y-1–86690Y-9 (2013)
21. Polzin, T., et al.: Combining automatic landmark detection and variational methods for lung CT registration. In: International Workshop on Pulmonary Image Analysis (2013)
22. Hermann, S.: Evaluation of scan-line optimization for 3D medical image registration. In: IEEE Conference on Computer Vision and Pattern Recognition, pp. 3073–3080. IEEE (2014)
23. Vishnevskiy, V., Gass, T., Szekely, G., Goksel, O.: Total variation regularization of displacements in parametric image registration. In: Yoshida, H., Näppi, J., Saini, S. (eds.) ABD-MICCAI 2014. LNCS, vol. 8676, pp. 211–220. Springer, Cham (2014). https://doi.org/10.1007/978-3-319-13692-9_20

Image Segmentation

A Fully Automated Segmentation System of Positron Emission Tomography Studies

Albert Comelli[1,2]([✉]) [iD] and Alessandro Stefano[2] [iD]

[1] Fondazione Ri.MED, via Bandiera 11, 90133 Palermo, Italy
acomelli@fondazionerimed.com
[2] Institute of Molecular Bioimaging and Physiology,
National Research Council (IBFM-CNR), Cefalù, Italy

Abstract. In this paper, we present an automatic system for the brain metastasis delineation in Positron Emission Tomography images. The segmentation process is fully automatic, so that intervention from the user is never required making the entire process completely repeatable. Contouring is performed using an enhanced local active segmentation.

The proposed system is, at first instance, evaluated on four datasets of phantom experiments to assess the performance under different contrast ratio scenarios, and, successively, on ten clinical cases in radiotherapy environment.

Phantom studies show an excellent performance with a dice similarity coefficient rate greater than 92% for larger spheres. In clinical cases, automatically delineated tumors show high agreement with the gold standard with a dice similarity coefficient of $88.35 \pm 2.60\%$.

These results show that the proposed system can be successfully employed in Positron Emission Tomography images, and especially in radiotherapy treatment planning, to produce fully automatic segmentations of brain cancers.

Keywords: Active contour algorithm · Positron emission tomography imaging · Biological target volume · Segmentation

1 Introduction

Radiotherapy is often used to treat brain tumors otherwise inaccessible to conventional surgery. The classic and widespread approach to the identification of the volume to be treated is through the use of Magnetic Resonance Imaging (MRI). In particular, when soft-tissue contrast resolution needs to be high, as in brain malignancies, MRI is preferred over other approaches, e.g. computed tomography (CT). MRI has proved to be very efficient in reconstructing the anatomical properties of the investigated brain areas and for this reason, it has been almost the only diagnostic method employed for cancer delineation and treatment planning purposes [1–3]. Recently, the Positron Emission Tomography (PET) has been considered as a valuable source of information, in particular when the 11C-labeled Methionine (MET) radio-tracer is considered. MET-PET conveys complementary information to the anatomical information derived from

© Springer Nature Switzerland AG 2020
Y. Zheng et al. (Eds.): MIUA 2019, CCIS 1065, pp. 353–363, 2020.
https://doi.org/10.1007/978-3-030-39343-4_30

MRI or CT, and under favorable conditions, it may even deliver higher performance [4]. For these reasons, the integration of PET imaging in radiotherapy planning represents a desirable step forward in the treatment of gliomas and brain metastases.

Several PET delineation approaches have been proposed so far [5–8], and for a comprehensive review, the interested reader may refer to Foster *et al.* [9] and references therein. In general, segmentation algorithms can be categorized as semi-automatic or automatic. When the 18F-fluoro-2-deoxy-d-glucose (FDG) radio-tracer is used, for example, the lesion must be initially highlighted by the operator. Indeed, for some healthy structures a high FDG uptake is normal; the brain is a typical example. As a result, segmentation methods in FDG PET studies are exclusively semi-automatic. However, MET radio-tracer shows great sensitivity and specificity for the discrimination of healthy versus brain cancer tissues, making automatic approaches feasible. In the present study we used MET-PET to deploy a fully automatic method to delineate brain cancer and metastases.

Starting from our previous study [10] where we proposed a semi-automatic tool to segment general oncological lesions in PET studies, we obtained a fully automatic and operator independent system for MET PET studies on the brain. In the proposed application, the system performs all segmentation step automatically by individuating an optimal, operator independent, initial mask located on an automatically selected PET slice. Once the initial region of interest (ROI) has been identified, it is fed to an enhanced local active contour segmentation algorithm. The objective function was adapted to PET imaging and designed in such a way that its minimum corresponds to the best possible segmentation. To assess the performance of the system and to verify its suitability as medical decision tool in radio-treatments, four phantom experiments, and ten patient studies were considered.

2 Materials and Methods

2.1 Overview of the Proposed System

The main subject of the present study is a fully automatic and operator independent system for brain cancer segmentation, to be used in radiotherapy treatments. While the following subsections will illustrate different components of the system and their validation, in this section we present a brief overview of the system design. The data hereby discussed comprise a total of four phantom experiments and ten oncological patients (Sects. 2.2 and 2.3, respectively). Data from phantoms were used to assess the performances of the delineation algorithm. Concerning the practical use of the system on clinical cases, in order to normalize the voxel activity and to take into account the functional aspects of the disease, the PET images were pre-processed into SUV images (Sect. 2.5). The first step is the automatic identification of the optimal combination of starting ROI and slice containing the tumor. Then, this information is input to the subsequent components of the system (Sect. 2.6). Once the ROI is identified, the corresponding mask is fed into the next step of the system, where the segmentation is performed combining a Local region-based Active Contour (LAC) algorithm, appropriately adapted to handle PET images. The resulting segmentation is then propagated to the adjacent slices using a slice-by-slice marching approach. Each time convergence criteria are met for a specific slice, the corresponding optimal contour is propagated to the next, where the evolution is continued. Starting from the initial slice, the propagation is performed by contemporarily sweeping the data volume both upward and downward, until a suitable stopping condition, designed to detect a tumor-free slice, is met. Finally, the algorithm outputs a user independent Biological Tumor Volume (BTV). Detailed explanation of this task is provided in Sect. 2.7.

2.2 Phantom Studies

Four phantom experiments were used for preliminary assessment of the performance. The phantom is composed of an elliptical cylinder (minor axis = 24 cm, major axis = 30 cm, h = 21 cm) containing six different spheres (diameters: 10, 13, 17, 22, 28, and 37 mm) placed at 5.5 cm from the center of the phantom. The ratio between sphere and background radioactivity concentration ranged from 3:1 to 8:1. Performances were evaluated by grouping the results with respect to sphere diameters: small spheres, i.e. diameter smaller than 22 mm, and large spheres, with a diameter greater than 17 mm. This choice was motivated by the fact that large biases are introduced by the partial volume effect [11] in PET imaging.

2.3 Clinical Studies

Ten patients with brain metastases were retrospectively considered. These patients were referred to diagnostic PET/CT scan before Gamma Knife (Elekta, Stockholm, Sweden) treatment. Tumor segmentation was performed off-line without actually influencing the treatment protocol or the patient management. No sensitive patient data were accessed. As such, after all patients were properly informed and released their written consent, the institutional hospital medical ethics review board approved the present study protocol. Patients fasted 4 h before the PET examination, and successively were intravenously injected with MET. The PET/CT oncological protocol started 10 min after the injection.

2.4 PET/CT Acquisition Protocol

All acquisitions in this study were performed within the same hospital department and using the same equipment, a Discovery 690 PET/CT scanner (General Electric Medical Systems, Milwaukee, WI, USA). The PET protocol included a SCOUT scan at 40 mA, a CT scan at 140 keV and 150 mA (10 s), and 3D PET scans. The 3D ordered subset expectation maximization algorithm was used for the PET imaging. Each PET image consists of 256×256 voxels with a grid spacing of 1.17 mm^3 and thickness of 3.27 mm^3. Consequently, the size of each voxel is $1.17 \times 1.17 \times 3.27 \text{ mm}^3$. Thanks to the injected PET radio-tracer, tumor appears as hyper-intense region. The CT scan was performed contextually to the PET imaging and used for attenuation correction. Each CT image consists of 512×512 voxels with size $1.36 \times 1.36 \times 3.75 \text{ mm}^3$.

2.5 Pre-processing of PET Dataset

Pre-processing PET images is mandatory for inter-patient and follow-up comparisons. Among PET quantification parameters, body-weight SUV is the most widely used in clinical routine. For this reason, it was embedded in our system. SUV is the ratio of tissue radioactivity concentration (RC) and injected dose (ID) at the time of injection, divided by body weight. The RC is calculated as the ratio between the image intensity and the image scale factor. The ID is calculated as the product between actual activity and dose calibration factor.

2.6 Interesting Uptake Region Identification

In order to obtain a fully automatic BTV segmentation, an initial ROI enclosing the tumor must be produced, obviously without any intervention by the operator. Therefore, the system identifies the PET slice containing the maximum SUV (SUV_{max}) in the whole PET volume. By taking advantage of the great sensitivity and specificity of MET radio-tracers in discriminating between healthy and tumor tissues, we can confidently assume that such SUV_{max} resides inside the main lesion [12].

While this process takes place, an additional test is performed, in order to investigate the presence of isolated local maxima which may indicate metastases separated from the main lesion.

In the case that the presence of multiple (say "n") independent anomalies are recognized, each one is independently processed. A different local maximum (SUV_{max}^{j}, with $j = 1{:}n$) is identified for each lesion and, consequently, n regions are automatically identified. By design, the first identified lesion contains the global SUV_{max} (i.e., SUV_{max}^{1}).

Once the current slice with SUV_{max}^{j} has been identified, an automatic procedure to identify the corresponding ROI starts. The SUV_{max}^{j} voxel is used as target seed for a rough 2D segmentation based on the region growing (RG) method [13]. For each lesion, the obtained ROI represents the output of this preliminary step which is input to the next component of the system, where the actual delineation takes place. The latter is performed through an enhanced LAC segmentation algorithm. It is worth noting that the RG algorithm is used only to obtain a rough estimate of the tumor contour(s).

The same workflow is used to segment each metastasis independently, and the process is designed to carry on automatically. However, in case of multiple lesions the user will receive a warning message and if necessary, will be able to override the default behavior. In such a case, the algorithm can be paused, while the operator inspects the multiple metastases.

2.7 The Enhanced Local Active Contour Method

The model proposed by Lankton et al. [14] benefits of purely local edge based active contours and fully global region based active contours. At each point along a prominent intensity edge of the target, nearby points inside and outside the target will be modelled well by the mean intensities within the local neighborhoods on either side of the edge. The contour energy to be minimized is defined as:

$$E = \oint_C \left(\int_{R_{in}} \chi_l(x, s)(I(x) - u_l(s))^2 dx + \int_{R_{out}} \chi_l(x, s)(I(x) - v_l(s))^2 dx \right) ds \quad (1)$$

- R_{in} and R_{out} represent the regions inside and outside the curve C
- s represents the arc length parameter of C
- χ represents the characteristic function of the ball of radius l (local neighborhood) centered around a given curve point C(s)
- I represents the intensity function of the image to be segmented
- $u_l(s)$ and $v_l(s)$ denote the local mean image intensities within the portions of the local neighborhood $\chi_l(x, s)$ inside and outside the curve respectively (within R_{in} and R_{out}).

These neighborhoods are defined by the function χ, the radius parameter l, and the position of the curve C. Note that the function $\chi_l(x, s)$ evaluates to 1 in a local neighborhood around each contour point C(s) and 0 elsewhere. The contour C then divides each such local region into interior and exterior local pixels in accordance with the contour's rule to segment the domain of I.

Beyond the optimal identification of the starting slice containing the lesion, and, consequently, the identification of an initial operator independent mask for LAC segmentation (see Sect. 2.6) further improvements have been introduced in the LAC algorithm. In the following we summarize part of the method, described in [10]. To incorporate metabolic information, the intensity function I in (1) is replaced by the SUV, and $u_l(s)$ and $v_l(s)$ denote the local mean SUV intensities within the portions of the local neighborhood $\chi_l(x, s)$ inside and outside the curve. The shape of the contour C then divides each such local region into interior local points and exterior local points in accordance with the contour's segmentation of the SUV. The local means are specified as the ratios of $S_{I_l}(s)$, $S_{E_l}(s)$, $A_{I_l}(s)$, and $A_{E_l}(s)$ which represent the local sums of SUV intensities and the areas of their respective portions of the local neighborhood $\chi_l(x, s)$ inside and outside the curve. More precisely, the local interior region may be expressed as $R_{in} \cap \chi_l(x, s)$ and local exterior region as $R_{out} \cap \chi_l(x, s)$.

$$u_l(s) = \frac{S_{I_l}(s)}{A_{I_l}(s)}, \ v_l(s) = \frac{S_{E_l}(s)}{A_{E_l}(s)}$$

$$S_{I_l}(s) = \int_{R_{in}} \chi_l(x, s) SUV(x) \, dx, \quad S_{E_l}(s) = \int_{R_{out}} \chi_l(x, s) SUV(x) \, dx$$

$$A_{I_l}(s) = \int_{R_{in}} \chi_l(x, s) \, dx, \quad A_{E_l}(s) = \int_{R_{out}} \chi_l(x, s) \, dx$$

$$\chi_l(x, s) = \begin{cases} 1 \ when & x \in l - Ball(C(s)); \\ 0 & otherwise; \end{cases}$$

Once the ROI encircling the highest radio-tracer uptake area has been automatically identified (previous section), the resulting mask is used to initiate parallel segmentations on the neighboring slices above and below. Subsequently, for all the other slices in both directions, we similarly use the segmentation results of the previous slices as the initial mask inputs. The LAC method is inherently capable of locally widening or tightening where necessary when the contour is propagated from slice to slice. Since, this behavior is driven by the image properties rather than by an inherent knowledge of whether the cancer is present, a stopping criterion is necessary to prevent the LAC algorithm from misbehaving, or even diverging, when it reaches a slice where the cancer is absent (i.e. when there is nothing to be segmented).

Therefore, a fully automatic stopping condition is implemented. For the slice under consideration, at each point on the cancer edge, nearby points inside and outside the cancer must have a different local mean SUV. If the cancer is present, a positive difference between background and foreground intensity must occur, and consequently the algorithm can safely proceed with the next neighboring slice. When the system encounters a slice where the local mean $v_l(s)$ on R_{out} is greater or equal to the local mean SUV $u_l(s)$ on R_{in}, which is the opposite of what is expected, the slice is recognized as cancer-free and the slice-to-slice propagation is terminated in that direction. In this way, one slice at a time, the BTV is generated. Finally, the segmentation process is automatically stopped, thereby avoiding the need for any user intervention.

2.8 Framework for Performance Evaluation

Overlap-based and spatial distance-based metrics are considered to determine the accuracy achieved by the automatic segmentation system against the gold-standard [15]. In particular, the formulations of dice similarity coefficient (DSC), and Hausdorff distance (HD) are used.

DSC measures the spatial overlap between the reference volume and the segmentation system: a DSC value equal to 100% indicates a perfect match between two volumetric segmentations, while DSC = 0% indicates no overlap. Nevertheless, overlap-based metrics are not well suited for small anomalies. For this reason, distance-based metrics are preferable, especially when the boundary segmentation is critical, such as in BTV delineation for RTP. In particular, HD is used to measure the most mismatched boundary voxels between automatic and manual BTV: small HD means an accurate segmentation, while a large HD is synonymous of poor accuracy.

Finally, the performance of the proposed method is compared to other state of the art BTV segmentation methods: the original LAC method [17], the RG method [18], the enhanced RW method such as described in [19], and the FCM clustering method [20].

3 Results

3.1 Phantom Studies

Performance results from phantom experiments were divided considering small and large spheres, in four independent cases, each carried out with different signal ratios.

The accuracy improved for all spheres, regardless of their volume, when the signal ratio increased. In general, due to the partial volume effect, the separation of small targets from the background is very challenging, and the difficulty increases in condition of low signal contrast. The sphere volumes are underestimated with more false negatives than false positives. The dice similarity coefficient (DSC) rate is $77.51 \pm 3.46\%$ and the Hausdorff distance (HD) is 1.12 ± 0.15 voxels. For the spheres with a diameter greater than 17 mm, excellent performances are obtained with a DSC rate greater than 92% (HD $= 1.06 \pm 0.09$). The mean difference between segmented and actual volumes is positive (the sphere volumes are overestimated); larger margins can help to prevent the extension of tumor infiltration.

The performance of the system was compared to other state of the art PET image segmentation methods. In particular, the original LAC [17], the RG [18], the RW [19], and the FCM [20] methods have been used for comparison. Table 1 summarizes the results and shows that this automatic segmentation outperforms the methods tested for comparison for all the considered cases.

Table 1. DSC and HD values for the proposed method and other state of the art PET image segmentation methods.

	DSC	HD (voxels)
Our system	$84.79 \pm 8.00\%$	1.09 ± 0.12
Original LAC	$82.55 \pm 7.56\%$	1.44 ± 0.55
RW	$82.12 \pm 8.78\%$	1.22 ± 0.43
RG	$79.01 \pm 9.34\%$	1.67 ± 0.57
FCM	$77.13 \pm 8.79\%$	1.68 ± 0.49

3.2 Clinical Studies

The performance of the presented system is investigated considering ten metastases against the ground truth provided by three expert operators. In clinical cases, the histopathology analysis is unavailable after the gamma knife treatment. For this reason, the manual delineation performed by expert clinicians is a commonly accepted substitute for ground truth to assess the clinical effectiveness and feasibility of PET delineation methods. Consequently, manual segmentations performed by three experts are used to define a consolidated reference using STAPLE algorithm [16]. This simultaneous ground truth estimation tool combines a collection of segmentations into a single and consolidated ground truth segmentation. It computes a probabilistic estimate of the true segmentation estimating an optimal combination of the segmentations. This algorithm is formulated as an instance of the expectation maximization (EM).

Differently from the phantom studies, no discussion of the tumor volumes is provided here, mostly because all considered BTVs are greater than 2.5 ml (lesions with a sphere-equivalent diameter greater than 17 mm). In particular, tumor volumes ranged from 2.69 ml to 20.49 ml (mean $\pm$ std = 7.08 ± 5.81 ml). The ratio between lesion and background radioactivity concentration ranged from 2.76:1 to 7.40:1 (mean ± std = 3.88:1 ± 1.45:1). These values are included in the range of the phantom experiments used for preliminary performance testing. For this reason, although phantom studies don't replicate all the properties of real lesions, they represent a useful tool to assess performances across different segmentation methods.

Table 2 summarizes the comparison between this automatic segmentation and the original LAC and RW approaches. Since LAC and RW outperformed RG, and FCM methods on the phantom experiments, the latter algorithms were not considered in patient studies. The automatic algorithm performed better than LAC and RW methods minimizing the difference between references and automated BTVs.

Table 2. DSCs and HDs using the proposed system, original LAC and RW methods.

	DSC	HD
Our system	88.35 ± 2.60%	1.42 ± 0.57
Original LAC	83.77 ± 8.53%	2.97 ± 0.68
RW	87.01 ± 5.16%	1.38 ± 0.74

Figure 1 reports the comparison between the proposed segmentations and the gold-standards.

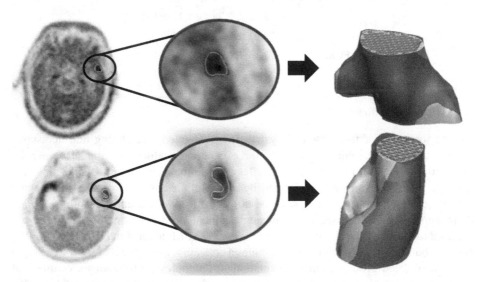

Fig. 1. Examples of automatic segmentations. The retrieved segmentations and the gold standards are shown in red and yellow, respectively. (Color figure online)

4 Discussion

In this study, a complex, semi-automatic system featuring an enhanced LAC algorithm purposely adapted to the PET imaging was further adapted to achieve the fully automatic BTV segmentations of brain cancers. The fully automatic approach leverages on the fact that MET-PET is capable of selectively highlight the ill regions of the brain, so avoiding false positives commonly encountered in other anatomic regions (e.g. as in FDG-PET studies). An automatic and operator-independent ROI is generated around the tumor(s) and used as input to an enhanced LAC algorithm. Then, the LAC performs the BTV delineation. The BTV is built by a slice-by-slice marching approach where the segmentation is performed on subsequent slices. In principle, segmentation through the evolution of a full 3D surface would be preferable. Indeed, while on the one hand we are currently investigating such a 3D approach, on the other hand, the present work moves an important step toward 3D data segmentation improving upon the model proposed by Lankton et al. [14] considering the issue of the PET slices thickness (3.27 mm^3) far greater than planar resolution (1.17 mm^3) which partially justifies the 2D approach. As a final remark, a fully automatic stop condition is provided. In this way, the proposed system produces segmentation results which are completely independent by the user.

Performance of the automatic system has been obtained by phantom studies consisting of hot spheres in a warm background. Nevertheless, phantom experiments cannot replicate all the aspects of a real clinical case but they represent a useful way to assess common performances across different algorithms. DSC greater than 92% for the larger spheres confirm better results in minimizing the difference between reference and automated BTVs than the other state-of-the-art algorithms. We would like to emphasize that original algorithms for both enhanced RW and original LAC methods [17, 19] were optimized for MET-PET brain metastases [10, 12]. Concerning RG and FCM methods, we used the source codes available on the web, and we adapted it to our PET dataset.

In patient studies, since radiotherapy treatment alters the cancer morphology over time, histopathology cannot provide a reliable ground truth. Consequently, manual delineation by experts, although it may differ between operators (for example, radiotherapy experts tend to draw larger boundaries than nuclear medicine physicians), is often used as surrogate gold-standard. In this study, we used manual delineations from three experts. To overcome the issue of differences in the manual delineations, a consolidated reference was built [16] and then used to assess the feasibility of the automatic segmentation. PET/CT data from ten patients before Gamma Knife treatment were considered. Results show that the proposed approach can be considered clinically feasible and could be used to extract PET parameters for therapy response evaluation purpose and to assist the BTV delineation during stereotactic radiosurgery treatment planning avoiding cancer recurrence. Finally, further investigations will be carried out to assess the usefulness to introduce in the segmentation a PET tissue classifier capable of influencing the local active contour toward what would be the segmentation performed by a human operator [21–25].

Acknowledgments. Authors would like to thank Prof. Anthony Yezzi, Dr. Samuel Bignardi, Dr. Giorgio Russo, MD. Maria Gabriella Sabini, and MD. Massimo Ippolito for their crucial support in the management of the proposed study.

References

1. Comelli, A., Bruno, A., Di Vittorio, M.L., et al.: Automatic Multi-seed Detection for MR Breast Image Segmentation, pp. 706–717. Springer, Cham (2017)
2. Chandarana, H., Wang, H., Tijssen, R.H.N., Das, I.J.: Emerging role of MRI in radiation therapy. J. Magn. Reson. Imaging **48**, 1468–1478 (2018). https://doi.org/10.1002/jmri.26271
3. Agnello, L., Comelli, A., Ardizzone, E., Vitabile, S.: Unsupervised tissue classification of brain MR images for voxel-based morphometry analysis. Int. J. Imaging Syst. Technol. (2016). https://doi.org/10.1002/ima.22168
4. Astner, S.T., Dobrei-Ciuchendea, M., Essler, M., et al.: Effect of 11C-methionine-positron emission tomography on gross tumor volume delineation in stereotactic radiotherapy of skull base meningiomas. Int. J. Radiat. Oncol. Biol. Phys. **72**, 1161–1167 (2008). https://doi.org/10.1016/j.ijrobp.2008.02.058
5. Comelli, A., Stefano, A., Bignardi, S., et al.: Active contour algorithm with discriminant analysis for delineating tumors in positron emission tomography. Artif. Intell. Med. **94**, 67–78 (2019). https://doi.org/10.1016/J.ARTMED.2019.01.002
6. Comelli, A., Stefano, A., Russo, G., et al.: K-nearest neighbor driving active contours to delineate biological tumor volumes. Eng. Appl. Artif. Intell. **81**, 133–144 (2019). https://doi.org/10.1016/j.engappai.2019.02.005
7. Hatt, M., Cheze Le Rest, C., Albarghach, N., et al.: PET functional volume delineation: a robustness and repeatability study. Eur. J. Nucl. Med. Mol. Imaging **38**, 663–672 (2011). https://doi.org/10.1007/s00259-010-1688-6
8. Berthon, B., Spezi, E., Galavis, P., et al.: Toward a standard for the evaluation of PET-Auto-Segmentation methods following the recommendations of AAPM task group No. 211: Requirements and implementation. Med Phys. (2017). https://doi.org/10.1002/mp.12312
9. Foster, B., Bagci, U., Mansoor, A., et al.: A review on segmentation of positron emission tomography images. Comput. Biol. Med. **50**, 76–96 (2014). https://doi.org/10.1016/j.compbiomed.2014.04.014
10. Comelli, A., Stefano, A., Russo, G., et al.: A smart and operator independent system to delineate tumours in Positron Emission Tomography scans. Comput. Biol. Med. (2018). https://doi.org/10.1016/J.COMPBIOMED.2018.09.002
11. Soret, M., Bacharach, S.L., Buvat, I.I.: Partial-volume effect in PET tumor imaging. J. Nucl. Med. **48**, 932–945 (2007). https://doi.org/10.2967/jnumed.106.035774
12. Stefano, A., Vitabile, S., Russo, G., et al.: A fully automatic method for biological target volume segmentation of brain metastases. Int. J. Imaging Syst. Technol. **26**, 29–37 (2016). https://doi.org/10.1002/ima.22154
13. Stefano, A., et al.: An automatic method for metabolic evaluation of gamma knife treatments. In: Murino, V., Puppo, E. (eds.) ICIAP 2015. LNCS, vol. 9279, pp. 579–589. Springer, Cham (2015). https://doi.org/10.1007/978-3-319-23231-7_52
14. Lankton, S., Nain, D., Yezzi, A., Tannenbaum, A.: Hybrid geodesic region-based curve evolutions for image segmentation. In: Proceedings of the SPIE 6510, Medical Imaging 2007: Physics of Medical Imaging, 16 March 2007, p. 65104U (2007). https://doi.org/10.1117/12.709700
15. Udupa, J.K., Leblanc, V.R., Zhuge, Y., et al.: A framework for evaluating image segmentation algorithms. Comput. Med. Imaging Graph. **30**, 75–87 (2006). https://doi.org/10.1016/j.compmedimag.2005.12.001
16. Warfield, S.K., Zou, K.H., Wells, W.M.: Simultaneous truth and performance level estimation (STAPLE): an algorithm for the validation of image segmentation. IEEE Trans. Med. Imaging **23**, 903–921 (2004). https://doi.org/10.1109/TMI.2004.828354

17. Lankton, S., Nain, D., Yezzi, A., Tannenbaum, A.: Hybrid geodesic region-based curve evolutions for image segmentation. 65104U (2007). https://doi.org/10.1117/12.709700
18. Day, E., Betler, J., Parda, D., et al.: A region growing method for tumor volume segmentation on PET images for rectal and anal cancer patients. Med. Phys. **36**, 4349–4358 (2009). https://doi.org/10.1118/1.3213099
19. Stefano, A., Vitabile, S., Russo, G., et al.: An enhanced random walk algorithm for delineation of head and neck cancers in PET studies. Med. Biol. Eng. Comput. **55**, 897–908 (2017). https://doi.org/10.1007/s11517-016-1571-0
20. Belhassen, S., Zaidi, H.: A novel fuzzy C-means algorithm for unsupervised heterogeneous tumor quantification in PET. Med. Phys. **37**, 1309–1324 (2010). https://doi.org/10.1118/1.3301610
21. Licari, L., et al.: Use of the KSVM-based system for the definition, validation and identification of the incisional hernia recurrence risk factors. Il Giornale di chirurgia **40**(1), 32–38 (2019)
22. Agnello, L., Comelli, A., Vitabile, S.: Feature dimensionality reduction for mammographic report classification. In: Pop, F., Kołodziej, J., Di Martino, B. (eds.) Resource Management for Big Data Platforms. CCN, pp. 311–337. Springer, Cham (2016). https://doi.org/10.1007/978-3-319-44881-7_15
23. Comelli, A., et al.: A kernel support vector machine based technique for Crohn's disease classification in human patients. In: Barolli, L., Terzo, O. (eds.) CISIS 2017. AISC, vol. 611, pp. 262–273. Springer, Cham (2018). https://doi.org/10.1007/978-3-319-61566-0_25
24. Comelli, A., Stefano, A., Benfante, V., Russo, G.: Normal and abnormal tissue classification in pet oncological studies. Pattern Recogn. Image Anal. **28**, 121–128 (2018). https://doi.org/10.1134/S1054661818010054
25. Comelli, A., Agnello, L., Vitabile, S.: An ontology-based retrieval system for mammographic reports. In: Proceedings of IEEE Symposium Computers and Communication (2016). https://doi.org/10.1109/ISCC.2015.7405644

A General Framework for Localizing and Locally Segmenting Correlated Objects: A Case Study on Intervertebral Discs in Multi-modality MR Images

Alexander O. Mader[1,2,3]([✉]), Cristian Lorenz[3], and Carsten Meyer[1,2,3]

[1] Institute of Computer Science, Kiel University of Applied Sciences, Kiel, Germany
alexander.o.mader@fh-kiel.de
[2] Faculty of Engineering, Department of Computer Science, Kiel University, Kiel, Germany
[3] Department of Digital Imaging, Philips Research, Hamburg, Germany

Abstract. Low back pain is a leading cause of disability that has been associated with intervertebral disc (IVD) degeneration by various clinical studies. With MRT being the imaging technique of choice for IVDs due to its excellent soft tissue contrast, we propose a fully automatic approach for localizing and locally segmenting spatially correlated objects—tailored to cope with a limited set of training data while making very few domain assumptions—and apply it to lumbar IVDs in multi-modality MR images. Regression tree ensembles spatially regularized by a conditional random field are used to find the IVD centroids, which allows to cut fixed-size sections around each IVD to efficiently perform the segmentation on the subvolumes. Exploiting the similar imaging characteristics of IVD tissue, we build an IVD-agnostic V-Net to perform the segmentation and train it on all IVDs (instead of a specific one). In particular, we compare the usage of binary (i.e., pairwise) CRF potentials combined with a latent scaling variable to tackle spine size variability with scaling-invariant ternary potentials. Evaluating our approach on a public challenge data set consisting of 16 cases from 8 subjects with 4 modalities each, we achieve an average Dice coefficient of 0.904, an average absolute surface distance of 0.423 mm and an average center distance of 0.59 mm.

Keywords: Object localization · Segmentation · Intervertebral discs · Multi modality · MRI

1 Introduction

Low back pain (LBP) is still a dominant health problem in the population affecting general well-being and work ability. Furthermore, it is a major cause of disability. Various clinical studies (e.g., [7]) repeatedly reported a strong association between LBP and the degeneration of the intervertebral discs (IVDs). Although all major medical imaging modalities were used to evaluate IVD degeneration,

© Springer Nature Switzerland AG 2020
Y. Zheng et al. (Eds.): MIUA 2019, CCIS 1065, pp. 364–376, 2020.
https://doi.org/10.1007/978-3-030-39343-4_31

magnetic resonance imaging (MRI) is still the imaging technique of choice. Its excellent soft tissue contrast without using ionizing radiation renders it ideal for the assessment of lumbar IVD anomalies. This created a major interest in methods for the automatic analysis and quantification of intervertebral discs in MR images of the lumbar spine [19].

Various approaches [1,4,5,9,18,20] have already been proposed to tackle the problem of segmenting intervertebral discs. A number of methods have been compared in a MICCAI grand challenge [18] focusing on T2-weighted MR images. In general, the approaches cover different modalities, e.g., T2-weighted MR images [3,18], CT images [17] and X-ray images [15], and different solution strategies, e.g., specifically crafted neural networks like the IVD-Net [4], graph-based methods [1] and approaches using mathematical morphology [2]. However, most of these approaches have in common that they either are not (or not easily) transferable to other applications, make strong assumptions about the application domain, generate sub-optimal results or are specifically crafted for the modality and/or the target object.

In this paper, we combine two existing generic methods for the localization of key points and the segmentation of objects to create a pipeline that requires little training data and makes very little domain assumptions. This yields an easily transferable framework to perform the localization and segmentation (or another local image analysis task) of arbitrary spatially correlated objects in arbitrary modalities. Here, we apply our framework to the segmentation of repetitive structures, namely lumbar intervertebral discs, in multi-modality MR images. First, regression tree ensembles regularized by a conditional random field (CRF) [11] are used to perform the labeled localization of the spatially correlated IVD centroids. We specifically compare binary (i.e., pairwise) CRF energy potentials— which are not scaling invariant—enhanced with a latent scaling variable, with ternary potentials which are designed to be scaling invariant, to tackle the spine size variability (caused, e.g., by growth or tissue degradation of the IVDs). Second, the well-known fully convolutional neural network V-Net [12] is used to perform an IVD-agnostic segmentation of the soft tissue in a volume-of-interest around each IVD centroid. To this end, we cut fixed-size reoriented sections around each IVD's centroid, effectively reducing the data to be segmented to a very small portion containing just the object. Moreover, the repetitive nature of the IVDs allows an effective V-Net training on only a few training cases. We evaluate the performance of our approach on the public data of the "MICCAI 2018 Challenge on Automatic IVD Localization and Segmentation from 3D Multi-modality MR (M3) Images" called "IVDM3Seg" [19].

2   Method

The task is to segment seven well-defined intervertebral discs (IVDs) in multi modality MR images (1.5T Dixon protocol in this case). For this we propose an algorithm consisting of four steps as illustrated in Fig. 1 and explained in detail in the following subsections. The key idea is to localize and label the

IVD centroids prior to performing the segmentation locally in small sections cut around each IVD centroid. Note, this is a general and fully automatic approach that requires very little domain assumptions, like the target object size to cut out fixed-size volumes of interest, and works in theory in 2D and 3D and with all major imaging modalities.

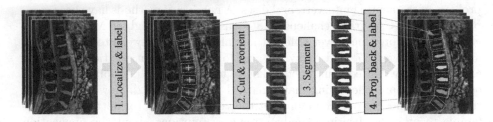

Fig. 1. Illustration of the four steps to predict labeled IVD segmentations.

2.1   Step 1: Localizing and Labeling IVD Centroids

The task of the first step is to localize the centroid of each individual IVD. This is achieved by utilizing the key point detection and localization framework proposed in [11]. This framework uses ensembles of regression trees—one for each key point—to regress voxel-wise pseudo probabilities by evaluating randomized intensity difference patterns of very few voxels taken from a patch (the size of which is derived from the target object size) centered around the regressed voxel. Alternatively, deep neural networks can be used for this purpose [10]. To prevent confusions between ambiguous key points, a conditional random field, modelling the global shape, is used to perform spatial regularization. This framework has been successfully applied to different detection and localization tasks in 2D and 3D images of different modalities, i.e., X-ray and CT, but not yet to multi-modality MR images. Here, we make the natural extension to multi channel volumes (e.g., aligned multi-modality MR images) by performing the intensity difference computations for each channel (4 in this case) and concatenating the individual intensity difference vectors.

Formally, we define an energy function

$$E(\mathbf{X} \mid \Lambda) = \sum_{l=1}^{L} \lambda_l \cdot \phi_l(\mathbf{X}_{\phi_l}) \tag{1}$$

that is parameterized by L weighted ($\Lambda = \{\lambda_l\}$) potential functions $\phi_l(\cdot)$. The ensemble pseudo probabilities (see above) are represented as unary potentials in this setup, one for each key point. Together with binary and/or ternary potentials, Eq. (1) computes the energy for a set of positions $\mathbf{X} = \{\mathbf{x}_1, \ldots, \mathbf{x}_N\}$ (one for each key point to localize) with $\mathbf{X}_{\phi_l}$ being the subset of positions whose key points are in ϕ_l's clique. To find the best position for each key point, given by

$$\hat{\mathbf{X}} = \arg \min_{\mathbf{X}} E(\mathbf{X}), \tag{2}$$

various inference strategies [16] can be used, potentially exploiting the graph structure for better runtime behavior.

Binary Potentials with Latent Scaling Variable. A very commonly used potential to manifest the spatial structure is a *binary (i.e., pairwise) Gaussian vector potential.* The idea is to model the vector spanned between two key points using a multivariate Gaussian distribution. Assuming that we estimated the distribution's mean $\boldsymbol{\mu}_{i,j}$ and covariance $\boldsymbol{\Sigma}_{i,j}$ on training data, we can define the potential as

$$\phi_{i,j}^{\text{vec}}(\mathbf{x}_i, \mathbf{x}_j) = -\log\left[f(s(\mathbf{x}_i - \mathbf{x}_j) \mid \boldsymbol{\mu}_{i,j}, \boldsymbol{\Sigma}_{i,j}) \right], \tag{3}$$

with $f(\cdot)$ being the probability density function (PDF) of a multivariate Gaussian distribution and s being a *latent scaling variable.*

Using such a potential allows to perform polynomial time inference on tree-structured graphs while providing scaling invariance, which—as we will see—is important for good performance on a data set showing for example the spine. Using a chain topology for the 7 IVD centroids and cliques formed between direct neighbors along the spinal chain, we obtain a graph with 6 pairwise potentials that we refer to as "binary w/latent variable" in the following. We apply belief propagation for CRF inference, which for a chain topology generates exact results in polynomial time.

Inherently Scaling Invariant Ternary Potentials. Alternatively, ternary potentials can be used to achieve scaling invariance without the need for a latent variable. The disadvantage is that the inference on graphs containing ternary potentials is more expensive. Here, we define two types of ternary potentials capturing the same information as the binary vector potential. The first one captures the relative distances between three key points by using a Gaussian distribution to model the ratio between two distances. With $g(\cdot)$ being the PDF of a Gaussian distribution and its mean $\mu_{i,j,k}^{\text{dist}}$ and variance $\sigma_{i,j,k}$ (of distance ratios) being estimated on training data, we define this potential as

$$\phi_{i,j,k}^{\text{dist}}(\mathbf{x}_i, \mathbf{x}_j, \mathbf{x}_k) = -\log\left[g\left(\|\mathbf{x}_i - \mathbf{x}_j\|_2 / \|\mathbf{x}_j - \mathbf{x}_k\|_2 \mid \mu_{i,j,k}^{\text{dist}}, \sigma_{i,j,k} \right) \right]. \tag{4}$$

The second ternary potential assesses the relative angle between two vectors projected to one plane by modeling this angle using a von Mises distribution. With $h(\cdot)$ being the PDF of such a distribution—its mean direction $\mu_{i,j,k}^{\text{ang}}$ and concentration $\kappa_{i,j,k}$ again being estimated on training data—and $\alpha(\cdot, \cdot)$ being a function to compute the minimal angle between two vectors projected onto a given plane, we can define our second ternary potential as

$$\phi_{i,j,k}^{\text{ang}}(\mathbf{x}_i, \mathbf{x}_j, \mathbf{x}_k) = -\log\left[h(\alpha(\mathbf{x}_i - \mathbf{x}_j, \mathbf{x}_j - \mathbf{x}_k) \mid \mu_{i,j,k}^{\text{ang}}, \kappa_{i,j,k}) \right]. \tag{5}$$

Note that the latent variable of the binary potentials allows to explicitly model the scale s of the spatial model to match the patient, which is not the case with the ternary potentials Eqs. (4) and (5). Again, using cliques of triplets formed along the spinal chain, we obtain a graph with $5 + 2 \cdot 5$ potentials (one distance and two rotation potentials for the sagittal and coronal plane per clique) that we refer to as "ternary" in the following. For efficient and exact inference we apply an A* inference algorithm (we can not use belief propagation since it is not a tree structure anymore).

Recursive Non-maximum Suppression to Generate Localization Hypotheses. Efficient runtime behavior is achieved by reducing the search domain of each key point in the CRF from the whole image domain to a set of localization hypotheses. These localization hypotheses are generated by applying non-maximum suppression to the regressed pseudo-probability maps of each key point. The quality of the selected localization hypotheses creates a natural upper bound on the CRF in terms of localization performance. Here, we propose a *recursive non-maximum suppression strategy* to find local maxima in the generated pseudo-probability maps. In contrast to standard non-maximum suppression (NMS), which uses a fixed distance D between maxima to be generated, we recursively apply NMS with different distances D_1 and D_2: First, we look for n_1 strong maxima using a rather large minimal distance D_1 between them. For each maximum found, we re-apply NMS in a local neighborhood to look for n_2 maxima using a much smaller minimal distance $D_2 \ll D_1$; this generates $n = n_1 n_2$ localization hypotheses in total. This procedure can be repeated multiple times depending on the structure of the search space. An example is illustrated in Fig. 2. Although both methods generate the same amount of localization hypotheses in this example, it is clearly visible that the recursive approach better samples the local neighborhood of large probability regions in the pseudo probability maps. The such sampled positions are closer to the IVD

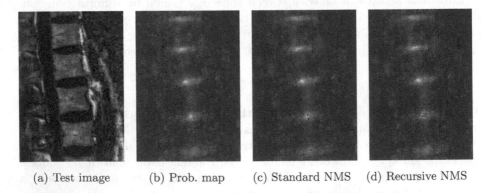

(a) Test image   (b) Prob. map   (c) Standard NMS   (d) Recursive NMS

Fig. 2. Recursive non-maximum suppression (d) compared to standard non-maximum suppression (c) applied to a pseudo probability map (b) generated by a regression tree ensemble on a cropped test image (a). A sagittal projection is shown with the white cross indicating the true position and red points being the derived local maxima. Best viewed in color with zooming. (Color figure online)

centroids and thus more meaningful. In our experiments we use the minimal per-axis distances $\mathbf{D}_1 = (2.4\ 1.875\ 1.875)$ cm and $\mathbf{D}_2 = (0.2\ 0.125\ 0.125)$ cm to find $n_1 = 10$ strong maxima and sample $n_2 = 10$ points around each one.

Training. The regression tree ensembles are trained as described in [11] using a patch size of roughly $7 \times 8.9 \times 8.9$ cm to match the target object, IVDs in this case. To estimate the weights Λ of the CRF and also optimize the graph topology, we minimize a max-margin hinge loss

$$L(\Lambda) = \frac{1}{K} \sum_{k=1}^{K} \max(0, m + E(\mathbf{X}_k^+ \mid \Lambda) - E(\mathbf{X}_k^- \mid \Lambda)) \tag{6}$$

on the K training samples using stochastic gradient descent in form of the Adam [8] optimizer. The key idea is to increase the difference in energy between the correct configuration of positions $\mathbf{X}_k^+$ for training example k and the *currently best rated incorrect* set of positions $\mathbf{X}_k^-$, determined by inference in each iteration, until the energy difference exceeds a margin $m = 1$. We carry out the optimization for 100 iterations using a learning rate of 0.01. Note that we train the regression tree ensembles and potentials on 60% of the training set and estimate the weights Λ on the remaining, exclusive 40% of the training set for better generalization.

2.2   Step 2: Sampling Reoriented IVD Sections

Given the IVD locations predicted by the previous step, we sample small reoriented fixed-size sections around each prediction. The size ($7.4 \times 7.2 \times 3.1$ cm), statistically estimated on training data, is chosen such that the IVD is fully contained plus some safety margin to compensate for slight localization errors. The sections are reoriented such that the IVDs are level inside the sections w.r.t. the transversal plane (see second step in Fig. 1). PCA was applied to the training segmentations to find the standard orientation of each IVD.

2.3   Step 3: Segmenting IVD Tissue Using a V-Net

As third step, we perform the actual segmentation of the disc tissue using the fully convolutional network V-Net [12]. To tackle the problem of few training data, we already reduced the original input image size to fixed size sections around each IVD. Additionally, we use an IVD-agnostic model to segment all seven IVD sections, effectively using the network to discriminate disc tissue from non-disc tissue (2-class problem instead of 8-class problem).

Training. We use the training setup as proposed by the V-Net authors Milletari et al. in [12] with a mini-batch size of 7 in combination with the generalized Dice loss optimized by the Adam optimizer [8]. The training data was created by cutting sections around the IVD centroids true positions. We also performed data augmentation in form of slight rotations (≤ 10 deg) and translations (≤ 5 mm) to create 24 additional augmented sections per original training section, effectively increasing the training set size by a factor of 25. The optimization was carried out for 5 epochs (1750 iterations) with a learning rate of 1e-4.

2.4   Step 4: Projecting Segmentations into Original Label Space

Finally, the resulting segmentations are transformed back into the original label space and relabeled according to the label predicted by the CRF in the first step.

3   Results and Discussion

3.1   IVDM3Seg Data

We evaluate our approach on the data used in the MICCAI 2018 IVDM3seg challenge [19]. It consists of 24 3D multi-modality MRI data sets, each showing *at least* the 7 well-defined IVDs, collected from 12 subjects in two time-spaced sessions to investigate potential IVD degeneration. The data sets were generated using a Dixon protocol with a 1.5T Siemens MRI scanner, thus each 3D multi-modality MRI data set is made of four aligned volumes (i.e., in-phase, opposed-phase, fat and water). Each volume has an anisotropic resolution of $2 \times 1.25 \times 1.25$ mm/px and comes with an unlabeled ground truth segmentation for all seven IVDs. We additionally labeled these segmentations and derived IVD centroids in form of center of mass positions.

Of the 24 MRI data sets, 16 (both data sets from 8 patients) were made publicly available for model training and validation; this data set is named "released (IVDM3Seg) data" in the following. The remaining 8 data sets remain non-disclosed as test data—named "non-disclosed (IVDM3Seg) data" in the following—for the organizers to perform a fair evaluation and comparison of methods (i.e., on these data, the algorithms are sent to and evaluated by the challenge organizers only). Note that we used histogram matching [13] prior to any processing to perform data normalization for each modality.

3.2   Metrics

We report the *segmentation performance* of our algorithm (calculated for the released data and reported as obtained from the challenge organizers for the non-released data) in terms of the evaluation metrics used by the challenge organizers [19]: The *Dice coefficient*, the *average absolute surface distance* (ASD) in mm and the *center distance* in mm, the latter being the distance between the center of mass position calculated for the predicted (step 4) and the true segmentation. These metrics are averaged over the individual IVDs and the test cases. Note that the IVD labels are ignored in assessing the segmentation performance reflecting the challenge design.

To assess the *localization performance* of step 1 of our algorithm (regression tree ensembles plus CRF) on the released data, we use the *mean localization error* (i.e., the Euclidean distance between the true IVD centroid and the location predicted by step 1 in mm). We also compute the *localization rate* measuring the number of correctly localized IVD centroids; a predicted centroid position is considered correct if the position is inside the IVD's true segmentation *and* if its localization error is less than 10 mm.

3.3 Evaluation on Released IVDM3Seg Data

We evaluated our method in detail on the released part of the data, i.e., on 16 data sets from 8 subjects. Since no subject identifiers were made available, we cannot perform a standard 8-fold cross-validation (CV) based on subject splits (such that data from a given subject appears only in the training *or* in the test data). Instead, we decided to apply our algorithm (training and evaluation) to every (out of $\binom{16}{2} = 120$) possible CV configuration. These 120 configurations contain the 8 true subject-splits (testing on one subject, training on the 7 remaining subjects) and 112 configurations where the second image from the two test subjects is contained in the training set. This renders our results on the released part of the data slightly over-optimistic. Nevertheless, we feel that this is a fair way to avoid a bias either towards a single test subject (with probability $8/120 = 6.67\%$) or still using a non-correct subject split (with probability 93.3%) when using a single, randomly selected CV fold. The goal of the evaluations on the released part of the data is to analyze each component of our algorithm in detail, which cannot be performed on the non-disclosed part of the data.

Localization Performance. We compare the localization performance of step 1 of our algorithm (regression tree ensemble plus CRF) for the IVD's centroids in the two previously defined CRF setups (Sect. 2.1). The average results over all 120 CV configurations are depicted in Fig. 3 for different numbers of scaling

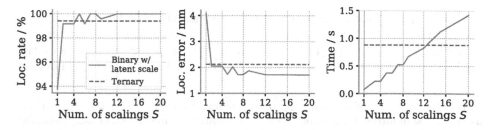

Fig. 3. Comparison of two different CRF setups used in step 1 of our approach in terms of localization rate (in %, see Sect. 3.2; first graph), mean error (in mm; second graph) and CRF computation plus inference time (in sec; third graph) as a function of a different number S of values for the latent scaling variable s in the pairwise CRF potentials (the ternary potentials are independent of s).

values, i.e., $s \in \{1\} \cup \{0.8 + 0.4 \cdot i/s \mid i = 1..S\}$. Looking at the graphs we can see that scaling invariance has to be considered to achieve optimal performance. Furthermore, we see that the CRF setup that utilizes binary potentials with a latent scaling variable performs better than the setup using ternary potentials, evaluated over all metrics. It correctly localized the 7 IVDs in all test images in all 120 CV folds with an average localization error of 1.72 mm for $S \geq 12$. In contrast, the CRF setup using ternary potentials failed in three of the 120 configurations, mis-localizing 10 of the 1680 IVDs, resulting in a localization rate of 99.4% and an average error of 2.12 mm. Looking at the runtime, we can see that the pairwise CRF potentials are faster to evaluate as long as $S \leq 12$ and provide better accuracy as the ternary CRF potentials if $S \geq 5$. In general, the increased amount of energy values to compute per potential ($n^3 \gg n^2$) plus the slower runtime of the A* inference algorithm make the ternary setup slower while not providing a better localization performance compared to pairwise CRF potentials as long as scaling invariance is accounted for in the latter setup. Note that we used a naïve Python implementation which was not optimized for runtime, but still allowed us to compare both setups. We set $S = 100$ in the following.

Segmentation Performance Given Perfect Localizations. Prior to running the full approach, we also assessed the optimal segmentation performance when excluding errors of the previous localization step 1. To this end, we used the ground truth IVD centroids to initialize step 2 of our algorithm (see Fig. 1) in a "cheating" experiment. Averaged over all 120 CV splits, we achieved a Dice coefficient of 0.903±0.027, an ASD of 0.431 mm and a center distance of 0.64 mm.

Further analysis revealed that by projecting the re-oriented, segmented IVDs back to the original image space, the Dice coefficient is slightly degraded, i.e., from 0.922 in re-oriented space to 0.903 in original space. This is likely caused by the small size of the volumes in relation to the surface of the IVDs and a voxel-based none-interpolating evaluation of the overlap, which results in an imprecise evaluation in the surface area.

Full Pipeline. Finally, we evaluated the segmentation performance running the full pipeline as depicted in Fig. 1. Again, we compare results for the two CRF setups to analyze the influence of different CRF parameterizations on the final segmentation output, the results of which are listed in Table 1. Looking

Table 1. Evaluation of our (full) approach comparing the two CRF parameterizations as mean ± standard deviations of the challenge evaluation metrics (Sect. 3.2), averaged over all 120 CV splits of the released IVD3MSeg data.

CRF setup	Dice	ASD/mm	Center dist./mm
Binary w/latent scale	**0.904 ± 0.027**	**0.423 ± 0.085**	**0.59 ± 0.394**
Ternary	0.902 ± 0.056	0.535 ± 2.238	0.715 ± 2.525

at the results, we can see that the better performance of the binary potentials in combination with a latent scaling variable manifested also in a better final segmentation result with lower mean ASD and center distance and much smaller standard deviation for all evaluation metrics. Also, we see that the small mean localization errors of step 1 (1.72 mm and 2.12 mm for the binary and ternary CRF parameterizations, respectively) barely have an influence on the segmentation performance.

3.4   Evaluation on Non-disclosed IVDM3Seg Data

For our approach to be evaluated on the non-disclosed data, we prepared a Docker container of our method and sent it to the challenge organizers to produce the results on the non-disclosed data. For this Docker container, we randomly selected 8 of the 120 models trained in the CV setup, requiring balanced training (i.e., each of the 16 training images was used in exactly 7 CV training runs), to form an ensemble of experts [14]. The final ensemble output is a voxel-wise majority vote on the (binary) segmentation outputs of the 8 ensemble members. The idea is that the deficiencies of each model from the low number of training subjects cancel out. Again, we compare our two CRF parameterizations w.r.t. the final segmentation output.

For the two cases 05 and 06 of the non-disclosed data, the entire IVD chain was shifted towards the head in both CRF configurations. An example is shown in Fig. 4 with the white arrow indicating the true start of the chain. Note that the IVD tissue was still accurately and precisely segmented. It is very surprising that the 8 localization models, the parts of which were trained on exclusive random training subsets (i.e., the decision tree ensemble was trained on a different subset of the training images than the CRF) and produced excellent CV results, majorly

Table 2. Comparison of our algorithm to the currently best performing method on the non-disclosed IVDM3Seg challenge data, again illustrating the mean ± the standard deviation of the challenge metrics (Sect. 3.2). We ignore (/) values where an artificial penalty was introduced by the challenge organizers for mis-localized IVDs, which does not allow for valid interpretation.

Method	Dice	ASD/mm	Center dist./mm
All 56 IVDs			
smartsoftV2 [6]	**0.911 ± 0.024**	**0.599 ± 0.200**	**0.756 ± 0.404**
Ours (ternary)	0.871 ± 0.177	/	/
Ours (binary w/lat. scale)	0.870 ± 0.177	/	/
52 matching out of totally 56 IVDs			
Ours (ternary)	**0.912 ± 0.021**	**0.57 ± 0.153**	**0.737 ± 0.375**
Ours (binary w/lat. scale)	**0.912 ± 0.021**	0.572 ± 0.153	0.741 ± 0.377
smartsoftV2 [6]	0.911 ± 0.025	0.602 ± 0.206	0.771 ± 0.408

agreed that the chain should be shifted. However, since the test data are not disclosed and only the final segmentation output was produced, the reason for the mis-localization cannot be determined. This mis-localization reduced the overall Dice coefficient to 0.870 and 0.871 for the binary w/latent variable and ternary configuration, respectively. By selectively looking at the data we can get a better view on the actual segmentation performance, i.e., differentiating between localization and segmentation errors. We do so by excluding the 4 none-matching (i.e., not overlapping) of the total $8 \cdot 7 = 56$ IVDs. We compare these numbers to the best reported numbers so far by Georgiev and Asenov [6] called "smartsoftV2", computed over all 56 IVDs and the 56 minus 4 IVDs for equal comparison. Looking at the results listed in Table 2, we can see that our approach is on par (favorable) with state of the art in terms of segmentation performance. Interestingly, the superiority of the binary setup in contrast to the ternary one suggested by the previous experiments is inverted (if only slightly) on the non-disclosed data, which might be caused by the ensemble setup and is statistically insignificant. Comparing the segmentation performance on the released data with the one on the non-disclosed data, we see an improvement of the Dice ($0.912 > 0.904$) that is likely caused by the ensemble setup.

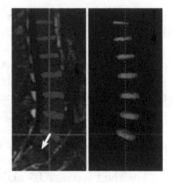

Fig. 4. Test image with an incorrect shift of all segmentations towards the head by one IVD (left: sagittal view; right: 3D rendering).

The full evaluation results and the used Docker containers are publicly available[1].

4   Conclusions and Future Work

In this paper, we combined a general localization method (regression tree ensembles spatially regularized by a conditional random field) with a general segmentation method (a fully convolutional neural network) to create an approach that can segment arbitrary spatially correlated objects in various modalities requiring very few training data and very few domain assumptions (i.e., target object size). The key idea is that localizing the object's centroids and exploiting the spatial information provided by the conditional random field one can cut object-specific sections to perform the segmentation on a much smaller sub-volume. This greatly reduces the amount of required training data, increases segmentation performance and reduces the runtime.

Here, we applied the approach to the task of segmenting seven well-defined intervertebral discs in multi-modality MR images using the data set provided by the MICCAI 2018 IVDM3Seg challenge [19]. We evaluated each part of our approach in detail and compared binary (i.e., pairwise) CRF potentials utilizing a latent scaling variable with ternary potentials to tackle the required scaling invariance in the spatial model. Averaged over all 120 cross-validation splits on

---

[1] https://github.com/fhkiel-mlaip/ivdm3seg.

the released data, the first step correctly localized all 1680 IVDs with a mean localization error of 1.72 mm. Following this first step, the final segmentation achieved an average Dice of 0.904, an absolute surface distance of 0.423 mm and a center distance of 0.59 mm. Evaluating the approach on the non-disclosed data showed that in two cases the whole spinal structure was shifted by one, i.e., the localization failed. However, evaluating the actual segmentation performance on the 52 (out of the totally 56) matching IVDs showed that our approach performs on par with the state-of-the-art, i.e., resulting in a Dice coefficient of 0.912, an ASD of 0.57 mm and a center distance of 0.737 mm.

For future work, we are currently tackling the common problem of group shifts in CRF setups on repetitive structures. Furthermore, we are actively looking for other data sets featuring spatially correlated objects, few data and challenging small objects.

Acknowledgements. This work has been financially supported by the Federal Ministry of Education and Research under the grant 03FH013IX5. The liability for the content of this work lies with the authors. Additionally, we would like to thank the challenge organizers Guoyan Zheng and Guodong Zeng for their efforts in maintaining the challenge and running evaluations for different model configurations.

References

1. Ben Ayed, I., Punithakumar, K., Garvin, G., Romano, W., Li, S.: Graph cuts with invariant object-interaction priors: application to intervertebral disc segmentation. In: Székely, G., Hahn, H.K. (eds.) IPMI 2011. LNCS, vol. 6801, pp. 221–232. Springer, Heidelberg (2011). https://doi.org/10.1007/978-3-642-22092-0_19
2. Carlinet, E., Géraud, T.: Intervertebral disc segmentation using mathematical morphology—a CNN-free approach. In: Zheng, G., Belavy, D., Cai, Y., Li, S. (eds.) CSI 2018. LNCS, vol. 11397, pp. 105–118. Springer, Cham (2019). https://doi.org/10.1007/978-3-030-13736-6_9
3. Chen, C., et al.: Localization and segmentation of 3D intervertebral discs in MR images by data driven estimation. IEEE TMI **34**(8), 1719–1729 (2015)
4. Dolz, J., Desrosiers, C., Ayed, I.B.: IVD-Net: intervertebral disc localization and segmentation in MRI with a multi-modal UNet. arXiv:1811.08305 (2018)
5. Dong, X., Zheng, G.: Automated 3D lumbar intervertebral disc segmentation from MRI data sets. In: Zheng, G., Li, S. (eds.) Computational Radiology for Orthopaedic Interventions. LNCVB, vol. 23, pp. 25–40. Springer, Cham (2016). https://doi.org/10.1007/978-3-319-23482-3_2
6. Georgiev, N., Asenov, A.: Automatic segmentation of lumbar spine 3D MRI using ensemble of 2D algorithms (smartsoftV2) (2018). https://ivdm3seg.weebly.com/smartsoftv2.html. Accessed 10 Feb 2019
7. Hebelka, H., Torén, L., Lagerstrand, K., Brisby, H.: Axial loading during MRI reveals deviant characteristics within posterior IVD regions between low back pain patients and controls. Eur. Spine J. **27**(11), 2840–2846 (2018)
8. Kingma, D.P., Ba, J.: Adam: a method for stochastic optimization. arXiv preprint arXiv:1412.6980 (2014)
9. Li, X., et al.: 3D multi-scale FCN with random modality voxel dropout learning for Intervertebral Disc Localization and Segmentation from Multi-modality MR Images. Med. Image Anal. **45**, 41–54 (2018)

10. Mader, A.O., Berg, J., Lorenz, C., Meyer, C.: A novel approach to handle inference in discrete markov networks with large label sets. In: PGM, pp. 249–259 (2018)
11. Mader, A.O., et al.: Detection and localization of spatially correlated point landmarks in medical images using an automatically learned conditional random field. CVIU **176**, 45–53 (2018)
12. Milletari, F., Navab, N., Ahmadi, S.A.: V-Net: fully convolutional neural networks for volumetric medical image segmentation. In: 3DV, pp. 565–571. IEEE (2016)
13. Nyúl, L.G., Udupa, J.K., Zhang, X.: New variants of a method of MRI scale standardization. IEEE TMI **19**(2), 143–150 (2000)
14. Perrone, M.P.: Improving regression estimation: averaging methods for variance reduction with extensions to convex measure optimization. Ph.D. thesis (1993)
15. Sa, R., et al.: Intervertebral disc detection in X-ray images using faster R-CNN. In: IEEE EMBC, pp. 564–567 (2017)
16. Wang, C., Komodakis, N., Paragios, N.: Markov random field modeling, inference & learning in computer vision & image understanding: a survey. CVIU **117**(11), 1610–1627 (2013)
17. Wong, A., Mishra, A., Yates, J., Fieguth, P., Clausi, D.A., Callaghan, J.P.: Intervertebral disc segmentation and volumetric reconstruction from peripheral quantitative computed tomography imaging. IEEE TBE **56**(11), 2748–2751 (2009)
18. Zheng, G., et al.: Evaluation and comparison of 3D intervertebral disc localization and segmentation methods for 3D T2 MR data: a grand challenge. Med. Image Anal. **35**, 327–344 (2017)
19. Zheng, G., Li, S., Belavy, D.: IVDM3Seg - MICCAI 2018 Challenge Intervertebral Disc Localization and Segmentation from 3D Multi-modality MR (M3) Images (2018). https://ivdm3seg.weebly.com/. Accessed 10 Feb 2019
20. Zhu, X., et al.: A method of localization and segmentation of intervertebral discs in spine MRI based on Gabor filter bank. Biomed. Eng. Online **15**(1), 32 (2016)

Polyp Segmentation with Fully Convolutional Deep Dilation Neural Network
Evaluation Study

Yunbo Guo$^{(\boxtimes)}$ and Bogdan J. Matuszewski

Computer Vision and Machine Learning (CVML) Research Group, School of Engineering,
University of Central Lancashire, Preston, UK
{ybguo1,bmatuszewski1}@uclan.co.uk

Abstract. Analyses of polyp images play an important role in an early detection of colorectal cancer. An automated polyp segmentation is seen as one of the methods that could improve the accuracy of the colonoscopic examination. The paper describes evaluation study of a segmentation method developed for the Endoscopic Vision Gastrointestinal Image ANAlysis – (GIANA) polyp segmentation sub-challenges. The proposed polyp segmentation algorithm is based on a fully convolutional network (FCN) model. The paper describes cross-validation results on the training GIANA dataset. Various tests have been evaluated, including network configuration, effects of data augmentation, and performance of the method as a function of polyp characteristics. The proposed method delivers state-of-the-art results. It secured the first place for the image segmentation tasks at the 2017 GIANA challenge and the second place for the SD images at the 2018 GIANA challenge.

Keywords: Fully convolutional dilation neural networks · Polyp segmentation · Data augmentation · Cross-Validation · Ablation tests

1 Introduction

Colorectal cancer is one of the leading causes of cancer deaths worldwide. To decrease mortality, an assessment of polyp malignancy is performed during colonoscopy examination, so it can be removed at an early stage. Currently, during colonoscopy, polyps are usually examine visually by a trained clinician. To automate analysis of colonoscopy images, machine learning methods have been utilised and shown to support polyp detectability and segmentation objectivity.

Polyp segmentation is a challenging task due to inherent variability of polyp morphology and colonoscopy image appearance. The size, shape and appearance of a polyp are different at different stages. In an early stage, colorectal polyps are typically small, may not have a distinct appearance, and could be easily confused with other intestinal structures. In the later stages, the polyp morphology changes and the size begin to increase. Illumination in colon screening is also variable, producing local overexposure highlights and specular reflections. Some polyps may look very differently from different

© Springer Nature Switzerland AG 2020
Y. Zheng et al. (Eds.): MIUA 2019, CCIS 1065, pp. 377–388, 2020.
https://doi.org/10.1007/978-3-030-39343-4_32

camera positions, do not have a visible transition between the polyp and its surrounding tissue, be affected by intestinal content and luminal regions (Fig. 1), inevitably leading to segmentation errors.

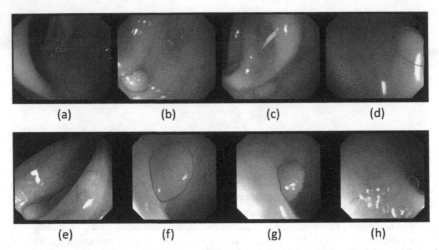

Fig. 1. Typical polyps in the GIANA SD training dataset: (a, h) Small size; (b) Blur; (c) Intestinal content; (d) Specular highlights/defocused; (e) Occlusion; (f) Large size; (g) Overexposed areas; (a, e, h) Luminal region.

The research reported here has been motivated by the limitations of previously proposed methods. This paper evaluates a novel fully convolutional neural network designed to accomplish this challenging segmentation task. The developed FCN method outputs polyp occurrence confidence maps. The final polyp delineation is either obtained by a simple thresholding of these maps or the hybrid level-set [1, 2] is used to smooth the polyp contour and eliminate small noisy network responses. The proposed method has been introduced in [3]. This paper aims to provide more in depth analysis of the method characteristics, focusing on the selection of the design parameters, adopted data augmentation scheme as well as overall validation of the proposed method. This analysis has not been published before.

2   Related Work

In literature on colonoscopy image analysis, various terms have been used to describe similar objectives. For example, some of the reported polyp detection and localisation methods provide heat maps and/or different levels of polyp boundary approximations, which could be interpreted as segmentation. On the other hand segmentation tools could be also seen as providing polyp detection and localisation functionality. Most of the reported techniques relevant to polyp segmentation can be divided into two main approaches based on either apparent shape or texture, with the methods using machine learning gradually gaining popularity. Some of the early approaches attempted to fit a predefined polyp shape models. Hwang et al. [4] used ellipse fitting techniques based on image curvature, edge distance and intensity values for polyp detection. Gross et al. [5] used Canny edge detector to process a prior-filtered images, identifying the relevant edges using a template matching technique for polyp segmentation. Breier et al. [6, 7] investigated applications of active contours for finding polyp outline. Although these methods perform well for typical polyps, they require manual contour initialisation.

The above mentioned techniques rely heavily on a presence of complete polyp contours. To improve the robustness, further research was focused on the development of robust edge detectors. Bernal et al. [8] presented a "depth of valley" concept to detect more general polyp shapes, then segment the polyp through evaluating the relationship between the pixels and detected contour. Further improvements of this technique are described in [9–11]. In the subsequent work, Tajbakhsh et al. [12] put forward a series of polyp segmentation method based on edge classification, utilising the random forest classifier and a voting scheme producing polyp localisation heat maps. In the follow-up work [13, 14] that approach was refined via use of several sub-classifiers.

Another class of polyp segmentation methods is based on texture descriptors, typically operating on a sliding window. Karkanis et al. [15] combined Grey-Level Co-occurrence Matrix and wavelets. Using the same database and classifier, Iakovidis et al. [16] proposed a method, which provided the best results in terms of area under the curve metric.

More recently, with the advances of deep learning, a hand-crafted feature descriptors are gradually being replaced by convolutional neural networks (CNN) [17, 18]. Ribeiro et al. [19] compared CNN with the state-of-art hand-crafted features on polyp classification problem, and found that CNN has superior performance. That method is based on a sliding window approach. The general problem with a sliding widow technique is that it is difficult to use image contextual information and approach is very inefficient. This has been addressed by the so called fully convolutional networks (FCN), with two key architectures proposed in [20, 21]. These methods can be trained end-to-end and output complete segmentation results, without a need for any post-processing. Vázquez et al. [22] directly segmented the polyp images using an off-the-shelf FCN architecture. Zhang et al. [23] use the same FCN, but they add a random forest to decrease the false positive. The U-net [21] is one of the most popular architectures for biomedical image segmentation. It has been also used for polyp segmentation. Li et al. [24] designed a U-net architecture for polyp segmentation to encourage smooth contours.

In recent years, it has been noticed that there is a relationship between size of CNN receptive field and the quality of segmentation results. A new layer, called dilation convolution, has been proposed [25] to control the CNN receptive field in a more efficient way. Chen et al. [26] utilised dilation convolution and developed architecture called atrous spatial pyramid pooling (ASPP) to learn the multi-scale features. The ASPP module consists of multiple parallel convolutional layers with different dilations.

In summary, colonoscopy image analysis (including polyp segmentation) is becoming more and more automated and integrated. Deep feature learning and end-to-end architectures are gradually replacing the hand-crafted and deep features operating on a sliding window. Polyp segmentation can be seen as a semantic instance segmentation problem and therefore, a large number of techniques developed in computer vision for generic semantic segmentation could be possibly adopted, providing effective and more accurate methods for polyp segmentation.

3 Method

The full processing pipeline of the proposed methodology is described in [3]. This section provides only the key information necessary for understating of the method evaluation described in the subsequent sections.

The proposed Dilated ResFCN polyp segmentation network is shown in Fig. 2. This architecture is inspired by [20, 26], and the Global Convolutional Network [27]. The proposed FCN consists of three sub-networks preforming specific tasks: feature extraction, multi-resolution classification, and fusion (deconvolution). The feature extraction sub-network is based on the ResNet-50 model [28]. The ResNet-50 has been selected, as for the polyp segmentation problem it has showed to provide a reasonable balance between network capacity and required resources. The multi-resolution classification sub-network consists of four parallel paths connected to the outputs from Res2 - Res5. Each such parallel path includes a dilation convolutional layer, which is used to increase the receptive field without increasing computational complexity. The larger receptive fields are needed to access contextual information about polyp neighborhood areas. The dilation rate is determined by the statistics of polyp size in the database used for training. For the lowest resolution path (the bottom path in Fig. 2) the 3×3 kernel can only represent a part of most polyps and the 7×7 kernel is too large. Therefore, 5×5 kernel, corresponds to dilation rate of 2, has been experimentally selected, as it can adequately represent 91% of all polyps in the training dataset. The regions of dilation convolutions should be overlapping and therefore the dilation rates increase with resolution. The dilation rates for sub-nets connected to Res5-Res2 are 2, 4, 8, 16 and the corresponding kernel sizes are 5, 9, 17, and 33. The fusion sub-network, corresponds to the deconvolution layers of FCN model. The segmentation results from each classification sub-network are up-sampled and fused by a bilinear interpolation.

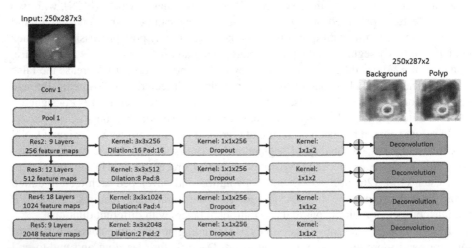

Fig. 2. Architecture of the proposed Dilated ResFCN network: feature extraction sub-network shown in blue, multiresolution feature classification sub-network shown in yellow, and fusion sub-network shown in green. (Color figure online)

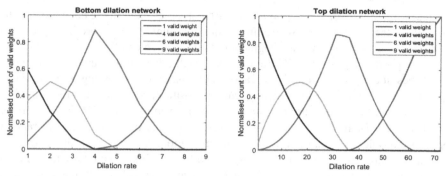

Fig. 3. The number of valid weights in bottom and top dilation networks in Fig. 2. (Color figure online)

Following the methodology described in [29], the number of active kernel weights at the top and bottom paths of the classification subnetwork are shown in Fig. 3. It can be seen, that with the dilation rate too high, the 3×3 kernel is effectively being reduced to a 1×1 kernel. On the other hand too small dilation rate leads to a small receptive field negatively affecting performance of the network. The selected dilation rates of 2 and 16 respectively for the "bottom" and "top" networks provide compromise with a sufficient number of kernels having 4–9 valid weights.

4 Implementation

4.1 Dataset

The proposed polyp segmentation method has been developed and evaluated on the data from the 2017 Endoscopic Vision GIANA Polyp Segmentation Challenge [30]. That data consist of Standard Definition (SD) and High Definition (HD) colonoscopy databases. The SD database has two datasets: training dataset, consisting of 300 low resolution, 500-by-574 pixel RGB images with the corresponding ground truth binary images. The images in that training dataset were obtained from 15 video sequences showing different polyps. The SD test dataset consists of 612 images with 288-by-384 pixels resolution. The HD database is composed of independent high-resolution RGB images of 1080-by-1920 pixels. The HD database includes 56 training images (with the corresponding ground truth) and 108 images used for testing. The results reported in this paper are based on a cross-validation approach using the training datasets only. Selected results obtained on the SD test dataset were reported in [3].

4.2 Data Augmentation

For the purpose of the method validation the SD and HD training datasets have been combined giving in total 355 training images. The performance of the CNN-based methods relies heavily on the size of training data used. Clearly, a set of 355 training images is very limited, at least from the perspective of a typical training set used in a context of deep learning. Moreover, some polyp types are not represented in the database, and for some others there are just a few exemplar images available. Therefore, it is necessary to enlarge the training set via data augmentation. Data augmentation is designed to provide more polyp images for CNN training. Although this method cannot generate new polyp types, it can provide additional data samples based on modelling different image acquisition conditions, e.g. illumination, camera position, and colon deformations.

All HD and SD images are rescaled to a common image size (250-by-287 pixels) in such a way that image aspect ratio is preserved. This operation includes random cropping equivalent to image translation augmentation. Subsequently, all images are augmented using four transformations. Specifically, each image is: (i) rotated with the ration angle randomly selected from [0°–360°) range, (ii) scaled with the scale factor randomly selected between 0.8 and 1.2, (iii) deformed using a thin plate spline (TPS) model with a fixed 10 × 10 grid and a random displacement of the each grid point with the maximum displacement of 4 pixels, (iv) colour adjusted, using colour jitter, with the Hue, Saturation, Value randomly changed, with the new values drawn from the distributions derived from the original training images [31]. In total after augmentation the training dataset consists of 19,170 images (Fig. 4).

| Original | Rotation | Deformation | Colour jitter | Scale |

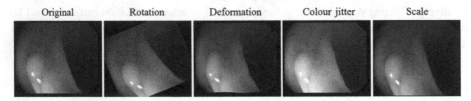

Fig. 4. A sample of augmented images using rotation, local deformation, colour jitter, and scale.

4.3   Evaluation Metrics

For a single segmented polyp, Dice coefficient (also known as F1 score), Precision, Recall, and Hausdorff distance are used to compare the similarity between the binary segmentation results and the ground truth. Precision and Recall are standard measures used in a context of binary classification:

$$Precision = \frac{TP}{TP+FP} \quad Recall = \frac{TP}{TP+FN} \tag{1}$$

where: TP, FN, and FP denotes respectively true positive, false negative and false positive. Precision and Recall could be used as indicators of over- and under- segmentation. Dice coefficient is often used in a context of image segmentation and is defined as:

$$Dice = \frac{2 \times TP}{2 \times TP + FN + FP} \tag{2}$$

Hausdorff Distance is the measure used to determine similarities between the boundaries G and S of two objectives. It is defined as:

$$H(G, S) = max\{sup_{x \in G} inf_{y \in S} d(x, y), sup_{x \in S} inf_{y \in G} d(x, y)\} \tag{3}$$

where: $d(x, y)$ denotes the distance between points x and y. The best result of this measure is 0, which means that the shapes of two objectives are completely overlapping.

4.4 Cross-Validation Data

For the purpose of validation original training images are divided into four V1-V4 cross-validation subsets with 56, 96, 97 and 106 images respectively. Following augmentation corresponding sets have 4784, 4832, 4821 and 4733 images for training. Following standard 4-fold validation scheme any three of these subsets are used for training (after image augmentation) and the remaining subset (without augmentation) for validation. Frames extracted from the same video are always in the same validation sub-set, i.e. they are not used for training and validation at same time.

5 Results

5.1 Comparison with Benchmark Methods

Two reference network architectures FCN8s [20] and ResFCN have been selected as benchmarks for evaluation of the proposed method. Whereas FCN8s is a well known fully convolutional network, the ResFCN is a simplified version of the network from Fig. 2 with the dilation kernels removed from the parallel classification paths. Table 1 lists the results (mean and standard deviation) for all three tested methods and all four-evaluation metrics. As it can be seen from that table the Dilated ResFCN achieves the best mean results for all the four metrics (the highest value for dice, precision, recall and the smallest value for the Hausdorff distance), as well as the smallest standard deviations for all the metrics, demonstrating the stability of the proposed method.

Table 1. Mean values and standard deviation obtained for different metrics on 4-fold validation data using FCN8s, ResFCN and Dilated ResFCN.

	Dice		Precision		Recall		Hausdorff	
	Mean	Std	Mean	Std	Mean	Std	Mean	Std
FCN8s	0.63	0.11	0.68	0.1	0.65	0.12	193	76
ResFCN	0.71	0.08	0.75	0.07	0.74	0.09	201	110
Dilated ResFCN	**0.79**	0.08	**0.81**	0.07	**0.81**	0.09	**54**	21

Figure 5 demonstrates results' statistics for all the methods and all the metrics using box-plot, with median represented by the central red line, the 25th and 75th percentiles represented by the bottom and top of each box and the outliers shown as red points. It can be concluded that the proposed method achieves better results than the benchmark methods. For all the metrics the true medians for the proposed method are better, with the 95% confidence, than for the other methods.

Significantly smaller Hausdorff distance measure obtained for the Dilated ResFCN results indicates a better stability of the proposed method with boundaries of segmented polyps better fitting to the ground truth data.

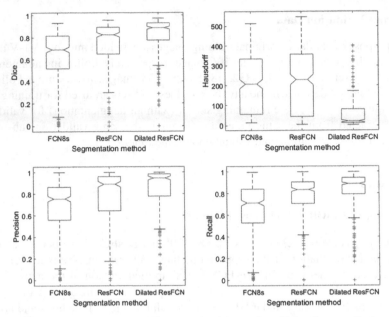

Fig. 5. The box-plot for different evaluation metrics. (Color figure online)

5.2 Data Augmentation Ablation Tests

As mentioned above, due to a very small training dataset, the data augmentation is an important step required for a suitable network training. In this section various data augmentations are investigated with the proposed Dilated ResFCN architecture. The result obtained after combining all the augmentations is also presented. Table 2 shows the mean Dice index obtained on each cross-validation subset along the overall mean dice index averaged across the four subsets. It is clear that the rotation seems to be the most informative augmentation method, followed by local deformations and colour jitter. It is also evident that the combination of different augmentation methods improves overall performance. It should be noted that for the "combined" augmentation, the same number of augmented images are used as for any other augmentation method tested.

Table 2. Mean Dice index obtained on 4-fold validation data using Dilated ResFCN network

Network	V1	V2	V3	V4	Mean
Combination	0.7583	**0.7420**	0.6086	**0.8518**	**0.7402**
Rotation	**0.7602**	0.7146	**0.6145**	0.8361	0.7314
Deformation	0.6772	0.7058	0.5917	0.7483	0.6807
Color jitter	0.6241	0.6957	0.5696	0.8019	0.6728
Scale	0.6536	0.6368	0.4742	0.7817	0.6366

The box-plots of the augmentation ablation tests are shown in Fig. 6. This confirms the conclusions drawn from the Table 2. Furthermore, it also demonstrates that the combined augmentation significantly improves the segmentation results when compared to any other standalone augmentation, with the real combined method median, being better than any other individual augmentation median with the 95% confidence level. Figure 6 shows also the distribution of the results as a function of the cross-validation folds. It can be seen that the results obtained on the fourth and third folds are respectively the best and worst. A closer examination of these folds reveals that images in the fourth fold are mostly showing larger polyps, whereas images in third fold are mostly depicting small polyps.

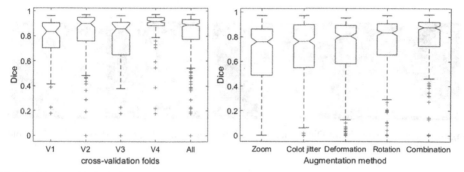

Fig. 6. Dice coefficient of Dilated ResFCN for cross-validation folds (left); and data augmentation ablation tests (right).

To further investigate the performance of the proposed method as a function of the polyp size, Fig. 7 shows the box-plot showing Dice index as a function of the polyp size. The "Small" and "Large" polyps are defined as having size smaller than the 25th and larger than 75th percentile of the polyp sizes in the training dataset. The remaining polyps as denoted as "Normal". The results demonstrate that the small polyps are hardest to segment. However it should be said that the metrics used are biased towards a larger polyps as a relatively small (absolute) over and under segmentation for a small polyps would led to more significant deterioration of the metrics. To combat this effect the authors proposed a secondary network, so called SE-Unet, designed specifically to segment small polyps [3]. The description of that method is though beyond the scope of this paper.

A typical segmentation results obtained using the Dilated ResNet network are shown in Fig. 8 with the blue and red contour representing respectively ground truth and the segmentation results[1].

---

[1] It is envisaged that supplementary results of the ongoing research on polyp segmentation will be gradually made available at: https://github.com/ybguo1/Polyp-Segmentation.

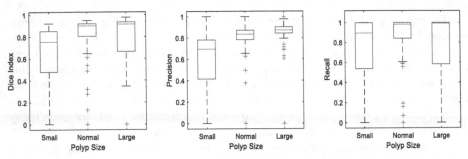

Fig. 7. Validation results obtained for the Dilated ResFCN network grouped as function of polyp size.

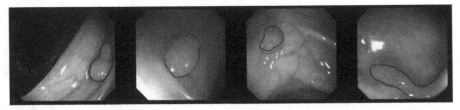

Fig. 8. Typical results, with Dice index (from left to right) of 0.97, 0.96, 0.71, and 0.69. (Color figure online)

6   Conclusion

The paper describes a validation framework for evaluation of the newly proposed Dilated ResFCN network architecture, specifically designed for segmentation of polyps in colonoscopy images. The method has been compared against two benchmark methods: FCN8s and ResFCN. It has been shown that suitably selected dilation kernels can improve performance of polyp segmentation on multiple evaluation metrics. In particular it has been shown that the proposed method matches well the shape of the polyp with the smallest and most consistent value of the Hausdorff distance. Due to a small number of training images, the data augmentation is the key for improving segmentation results. It has been shown that in that case the rotation is the strongest augmentation technique followed by local image deformation and colour jitter. Overall combination of different augmentation techniques has a significant effect on the results. The performance of the method as a function of the polyp size has been also analysed. Although some improvement on segmentation of small polyps have been achieved using architecture not reported in this paper a further improvement is still required, possibly through further optimisation of the dilation spatial pooling. The proposed method has been tested against state-of-the-art at the MICCAI's Endoscopic Vision GIANA Challenges, securing the first place for the SD and HD image segmentation tasks at the 2017 challenge and the second place for the SD images at the 2018 challenge.

Acknowledgements. The authors would like to acknowledge the organizers of the Gastrointestinal Image ANAlysis – (GIANA) challenges for providing video colonoscopy polyp images.

References

1. Zhang, Y., Matuszewski, B.J., Shark, L.K., Moore, C.J.: Medical image segmentation using new hybrid level-set method. In: 15th International Conference BioMedical Visualization: Information Visualization in Medical and Biomedical Informatics, pp. 71–76. IEEE (2008)
2. Zhang, Y., Matuszewski, B.J.: Multiphase active contour segmentation constrained by evolving medial axes. In: 16th IEEE International Conference on Image Processing, pp. 2993–2996. IEEE (2009)
3. Guo, Y., Matuszewski, B.J.: GIANA polyp segmentation with fully convolutional dilation neural networks. In: Proceedings of the 14th International Joint Conference on Computer Vision, Imaging and Computer Graphics Theory and Applications - Volume 4: GIANA, ISBN 978-989-758-354-4, pp. 632-641 (2019). https://doi.org/10.5220/0007698806320641
4. Hwang, S., Oh, J., Tavanapong, W., Wong, J., De Groen, P.C.: Polyp detection in colonoscopy video using elliptical shape feature. In: IEEE International Conference on Image Processing, vol. 2, pp. II–465. IEEE (2007)
5. Gross, S., et al.: Polyp segmentation in NBI colonoscopy. In: Meinzer, H.P., Deserno, T.M., Handels, H., Tolxdorff, T. (eds.) Bildverarbeitung für die Medizin 2009, pp. 252–256. Springer, Heidelberg (2009). https://doi.org/10.1007/978-3-540-93860-6_51
6. Breier, M., Gross, S., Behrens, A., Stehle, T., Aach, T.: Active contours for localizing polyps in colonoscopic NBI image data. In: Medical Imaging 2011: Computer- Aided Diagnosis, vol. 7963, p. 79632 M. International Society for Optics and Photonics (2011)
7. Breier, M., Gross, S., Behrens, A.: Chan-Vese segmentation of polyps in colono-scopic image data. In: Proceedings of the 15th International Student Conference on Electrical Engineering POSTER, vol. 2011 (2011)
8. Bernal, J., Sánchez, J., Vilarino, F.: Towards automatic polyp detection with a polyp appearance model. Pattern Recogn. **45**(9), 3166–3182 (2012)
9. Bernal, J., Sánchez, J., Vilarino, F.: Impact of image preprocessing methods on polyp localization in colonoscopy frames. In: 35th Annual International Conference of the IEEE Engineering in Medicine and Biology Society, pp. 7350–7354. IEEE (2013)
10. Bernal, J., Sánchez, F.J., Fernández-esparrach, G., Gil, D., Rodríguez, C., Vi-larino, F.: WM-DOVA maps for accurate polyp highlighting in colonoscopy: validation vs. saliency maps from physicians. Comput. Med. Imaging Graph. **43**, 99–111 (2015)
11. Bernal, J., et al.: Comparative validation of polyp detection methods in video colonoscopy: results from the MICCAI 2015 endoscopic vision challenge. IEEE Trans. Med. Imaging **36**(6), 1231–1249 (2017)
12. Tajbakhsh, N., Gurudu, S.R., Liang, J.: A Classification-Enhanced Vote Accumulation Scheme for Detecting Colonic Polyps. In: Yoshida, H., Warfield, S., Vannier, M.W. (eds.) ABD-MICCAI 2013. LNCS, vol. 8198, pp. 53–62. Springer, Heidelberg (2013). https://doi.org/10.1007/978-3-642-41083-3_7
13. Tajbakhsh, N., Chi, C., Gurudu, S.R., Liang, J.: Automatic polyp detection from learned boundaries. In: IEEE 11th International Symposium on Biomedical Imaging, pp. 97–100. IEEE (2014)
14. Tajbakhsh, N., Gurudu, Suryakanth R., Liang, J.: Automatic Polyp Detection Using Global Geometric Constraints and Local Intensity Variation Patterns. In: Golland, P., Hata, N., Barillot, C., Hornegger, J., Howe, R. (eds.) MICCAI 2014. LNCS, vol. 8674, pp. 179–187. Springer, Cham (2014). https://doi.org/10.1007/978-3-319-10470-6_23

15. Karkanis, S.A., Iakovidis, D.K., Maroulis, D.E., Karras, D.A., Tzivras, M.: Computer-aided tumor detection in endoscopic video using color wavelet features. IEEE Trans. Inf Technol. Biomed. **7**(3), 141–152 (2003)

16. Iakovidis, D.K., Maroulis, D.E., Karkanis, S.A., Brokos, A.: A comparative study of texture features for the discrimination of gastric polyps in endoscopic video. In: 18th IEEE Symposium on Computer-Based Medical Systems (CBMS) (2005). https://doi.org/10.1109/cbms.2005.6

17. LeCun, Y., Bottou, L., Bengio, Y., Haffner, P., et al.: Gradient-based learning applied to document recognition. Proc. IEEE **86**(11), 2278–2324 (1998)

18. Krizhevsky, A., Sutskever, I., Hinton, G.E.: Imagenet classification with deep convolutional neural networks. In: Advances in Neural Information Processing Systems, pp. 1097–1105 (2012)

19. Ribeiro, E., Uhl, A., Hafner, M.: Colonic polyp classification with convolutional neural networks. In: IEEE 29th International Symposium on Computer-Based Medical Systems, pp. 253–258. IEEE (2016)

20. Long, J., Shelhamer, E., Darrell, T.: Fully convolutional networks for semantic segmentation. In: Proceedings of the IEEE Conference on Computer Vision and Pattern Recognition, pp. 3431–3440 (2015)

21. Ronneberger, O., Fischer, P., Brox, T.: U-Net: Convolutional Networks for Biomedical Image Segmentation. In: Navab, N., Hornegger, J., Wells, W.M., Frangi, Alejandro F. (eds.) MICCAI 2015. LNCS, vol. 9351, pp. 234–241. Springer, Cham (2015). https://doi.org/10.1007/978-3-319-24574-4_28

22. Vázquez, D., et al.: A benchmark for endoluminal scene segmentation of colonoscopy images. J. Healthc. Eng. **2017** (2017)

23. Zhang, L., Dolwani, S., Ye, X.: Automated Polyp Segmentation in Colonoscopy Frames Using Fully Convolutional Neural Network and Textons. In: Valdés Hernández, M., González-Castro, V. (eds.) MIUA 2017. CCIS, vol. 723, pp. 707–717. Springer, Cham (2017). https://doi.org/10.1007/978-3-319-60964-5_62

24. Li, Q., et al.: Colorectal polyp segmentation using a fully convolutional neural network. In: 2017 10th International Congress on Image and Signal Processing, BioMedical Engineering and Informatics, pp. 1–5. IEEE (2017)

25. Yu, F., Koltun, V.: Multi-scale context aggregation by dilated convolutions. arXiv preprint arXiv:1511.07122 (2015)

26. Chen, L.C., Papandreou, G., Kokkinos, I., Murphy, K., Yuille, A.L.: Deeplab: Semantic image segmentation with deep convolutional nets, atrous convolution, and fully connected CRFs. IEEE Trans. Pattern Anal. Mach. Intell. **40**(4), 834–848 (2018)

27. Peng, C., Zhang, X., Yu, G., Luo, G., Sun, J.: Large kernel matters–improve semantic segmentation by global convolutional network. In: Proceedings of the IEEE Conference on Computer Vision and Pattern Recognition, pp. 4353–4361 (2017)

28. He, K., Zhang, X., Ren, S., Sun, J.: Deep residual learning for image recognition. In: Proceedings of the IEEE Conference on Computer Vision and Pattern Recognition, pp. 770–778 (2016)

29. Chen, L.C., Papandreou, G., Schroff, F. and Adam, H.: Rethinking atrous convolution for semantic image segmentation. arXiv preprint arXiv:1706.05587 (2017)

30. GIANA Challenge page, https://endovissub2017-giana.grand-challenge.org/polypsegmentation/. Accessed 02 Mar 2019

31. Must Know Tips/Tricks in Deep Neural Networks, http://lamda.nju.edu.cn/weixs/project/CNNTricks/CNNTricks.html. Accessed 02 Mar 2019

Ophthalmic Imaging

Convolutional Attention on Images for Locating Macular Edema

Maximilian Bryan[1](✉), Gerhard Heyer[1], Nathanael Philipp[1], Matus Rehak[2], and Peter Wiedemann[2]

[1] Universität Leipzig, Augustusplatz 10, 04109 Leipzig, Germany
`bryan@informatik.uni-leipzig.de`
[2] Klinik und Poliklinik für Augenheilkunde, Universitätsklinikum Leipzig, Liebigstraße 10, 04103 Leipzig, Germany

Abstract. Neural networks have become a standard for classifying images. However, by their very nature, their internal data representation remains opaque. To solve this dilemma, attention mechanisms have recently been introduced. They help to highlight regions in input data that have been used for a network's classification decision. This article presents two attention architectures for the classification of medical images. Firstly, we are explaining a simple architecture which creates one attention map that is used for all classes. Secondly, we introduce an architecture that creates an attention map for each class. This is done by creating two U-nets - one for attention and one for classification - and then multiplying these two maps together. We show that our architectures well meet the baseline of standard convolutional classifications while at the same time increasing their explainability.

Keywords: Neural networks · Convolutional neural networks · Attention

1 Introduction

When images of patients' eyes are created using OCT, a doctor has to look at them in order to check whether something pathologic can be identified. In order to support the doctor, we would like to create a model that gives an assessment of an eye's condition and tell whether an image shows a healthy eye or signs of a macular edema. To be of real help, the model should not just give a prediction which will be unquestionably used, but rather aid a doctor in his or her decision. To do this, along its prediction the model should also give some reasoning *why* it has come to its decision. In the visual context this should be a highlight of regions of the image which have been most helpful for that decision. The technique used for this visual highlighting is called *attention*.

Neural networks are designed as black boxes. When training, their task is to lower the training loss on a given training data set. During the course of training its layers and weights try to find abstract representations of the given data.

© Springer Nature Switzerland AG 2020
Y. Zheng et al. (Eds.): MIUA 2019, CCIS 1065, pp. 391–398, 2020.
https://doi.org/10.1007/978-3-030-39343-4_33

Here, with each added layer, the data representation becomes more and more abstract [21]. Finally, the representation has become that abstract that the network can classify complex images [6,10] and for some use cases even outperforms humans [4].

When using convolutional neural networks, it is possible to visualize its filters.[1,2] But as more and more layers are added to the network, the filters' learned features become more abstract and, in effect, less understandable. To tackle that problem, attention mechanisms have been created which all try to create a special connection between the input data and the network's output. Visualizations of this connection can then be used to explain the network's decision. It is an architectural trick to force the network not to use all of its given input data but only parts of it, and then do its classification task on that subset. One of the first architectures of visual attention exactly was created in order to create a link between images and their textual description [19], so that the authors were able to show which word of an image's description was linked to which region of the input image. In the medical field, visual attention has been used to segment images and extract objects [8,11,13]. Attention mechanisms have also been popular recently when working with text, especially when working with language tasks such as machine translation [1,9,12,16] or speech recognition [2,3].

When working with visual data, an *attention map* is created and functions as a bottle neck: The network is forced to create a feature map which is multiplied with the input data. The feature map is constrained by having the value 1 as a sum, which is done using softmax activation. Since the network has to do the classification task using the attention map, it has to decide which part of the input data it wants to keep and which part should be removed. A downside of this architecture is that the attention map has to keep information for all the classes: When the network has to decide between several classes, the attention map has to contain the information for all classes. But then it is not possible to trace which part of the input data has been used for which class.

In this paper, we are presenting two different attention architectures. The first architecture creates an attention map and classifies the images using that map. That way it can be visualized which part of the input image was used for the classification decision. Secondly, we introduce a modified architecture which is able to create an attention map for each class. We then use these architectures on a dataset of images of eyes, which are either healthy or show signs of a macular edema. We visualize our architectures' attention and are able to show that the classification results are better compared to a basic convolutional network, while at the same time being more expressive.

2   Visual Attention

Attention architectures enable users to understand which region of the input data has lead to a network's classification decision, but when having multiple

---

[1] https://ai.googleblog.com/2015/06/inceptionism-going-deeper-into-neural.html.

[2] https://blog.keras.io/how-convolutional-neural-networks-see-the-world.html.

classes the attention map is used for all classes. Thus, it cannot be seen which part of the input is used for which class. We first introduce a basic architecture for visual attention when classifying data, and then introduce a modification to have an attention map for each class.

2.1 Basic Visual Attention

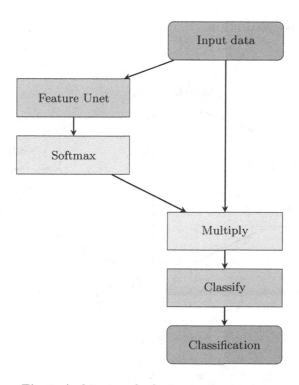

Fig. 1. Architecture for basic visual attention.

Starting point shall be a basic classifying neural network. When doing image classification, these networks often consist of several multiple layers [5,15] with multiple convolutions that take an input image and map that image onto several classes (Fig. 1).

When extending the network using attention, the input has to be multiplied by a feature map before being classified (similar to the approach presented in [17]). For creating such a feature map, a U-net [14] is a well suited choice: It contains an encoder and a decoder and shall be called *Feature U-net*. The encoder maps the input to a high dimensional vector using several convolutional layers and pooling steps, so that the encoded data contains all context information. When decoding, after each upsampling the values are concatenated with the values from the corresponding encoding step. At the last decoding step, the

U-net's input will be combined with all the contextual information from the encoding-decoding layers. Given the fact that we need to multiply the U-net's output with the input data, we need its output to be the same size as the input and to have only one feature dimension. That feature dimension will be activated using softmax, so that its sum equals 1. The effect of using softmax is that the network has to give high values to regions of the image that are needed for the classification, unimportant regions will get a low value.

2.2   Multi-class Attention

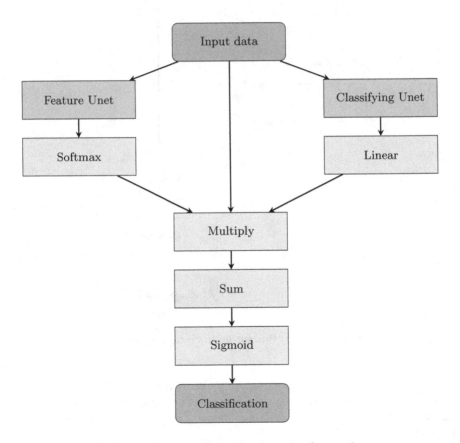

Fig. 2. Architecture for multi-class visual attention.

For the basic visual attention, one attention map was used to force the network to visualize the input image's regions that lead to its classification decision. When multiple classes exist, we want to have an attention map for each class. However, by doing this, multiple attention maps cannot be fed into our classifier since then no connection could be learned between each attention map and its corresponding class (Fig. 2).

Thus, we are adding a *Classifying U-net (CU)* of the same structure as the *Feature U-net* except for having a linear output. That output is being multiplied with the attention map, then summed and activated using sigmoid. The result is the network's confidence of a class being present in a given input image.

The effect is as follows: The *CU* has to learn for each class where that class can be seen in a given input image. In areas where it can be seen, the values have to be high so that when being multiplied with the attention map, the resulting value will also be high as well. If a class cannot be seen in a given image, the *CU* has to output small values to the feature map.

3   Experiments

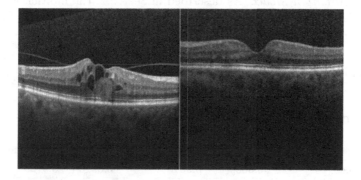

Fig. 3. Two examplary images from the dataset. Both images show the macula, whereas in the left image a macula edema can be seen, the right one is healthy.

The described architectures are being used on a dataset with OCT images of the macula of eyes. The eyes are either healthy or show signs of a macula edema (see Fig. 3). The dataset consists of 2821 images with signs of macular edema and 1161 images with no signs, which have been used for training. Additionally for testing, we have also used 722 images with signs of macular edema and 275 images with no sign. The images have been scaled to 128×128 pixels and are greyscale. First, we have been using a neural network without attention as base line. Then, we have been implementing the described attention architectures in Sect. 2. The models have all been trained using mean squared error for 100 epochs, one epoch consisting of 100 batches and each batch containing 16 training images.

3.1   Architectures

Simple Classifying Model. The basic convolutional neural network consists of 7 blocks, whereas each block contains 3 convolutional layers with batch normalization [7] and sigmoid activation. We are also using skip connections as seen in [18,20] in each block. Lastly, we are adding a so-called fully-connected layer before the final output layer. The output layer has two nodes, which represent either *healthy* or *edema*.

Simple Attention Model. The simple attention model implements the architecture described in Sect. 2.1. The encoding part of the U-net, the decoding part of the U-net as well as the classifying part of the model all consist of 7 blocks with exactly the same properties as the previously explained simple classifying model.

Multi-class Attention Model. As described in Sect. 2.2, the multiclass-attention model consists of two U-nets, which have the same properties as the previously described U-net.

3.2   Results

Table 1. Test results of the three architectures to investigate.

	Train loss	Train accuracy	Test loss	Test accuracy
Baseline	0.0669	0.9613	0.0916	0.9729
Simple attention	0.0514	0.9054	0.0383	0.9379
Multi-class attention	**0.0117**	**0.9752**	**0.0116**	**0.9825**

As can be seen in Table 1, the architecture with the basic attention is performing comparably well as our baseline classifying model. The multi-class attention with the two U-nets is performing significantly better. The reason for this may be that the basic attention model has to create an attention map which needs to contain interesting regions for both possible classes. When the classifying model is given the attention map, it has to identify patterns in that map that result in a class definition. Due to the softmax activation used for the attention maps, the information in these attention maps is very limited. The multi-class attention model on the other hand has dedicated attention maps which can concentrate on their own classes. Thus, whenever a region in these maps is highlighted, it is easy for the feature U-net to highlight the same region and create a high class confidence.

In Fig. 4, the multi-class attention can be seen. One can see that the attention highlights specific parts of the images. If there is no edema to be seen, the center

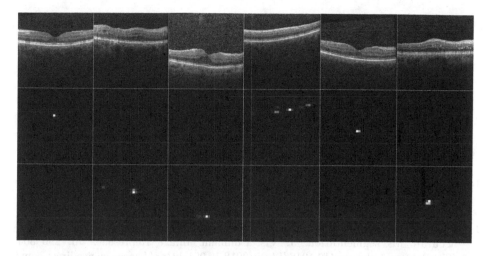

Fig. 4. Top row: input images. middle row: healthy attention, bottom row: macula edema attention.

of the image is highlighted. This can be interpreted that this area of the image is the most important one for checking for signs of edema. On the other hand, for the images with signs of macular edema, the corresponding area is always highlighted correctly.

4   Conclusion

We have introduced two architectures for highlighting parts of input images that result in a better understanding of why a network has come to its decision. The multi-class attention architecture presented has shown a better performance than the basic attention architecture and even a better performance than the baseline model that has been used just to classify the given data.

References

1. Bahdanau, D., Cho, K., Bengio, Y.: Neural machine translation by jointly learning to align and translate. CoRR abs/1409.0473 (2014). http://arxiv.org/abs/1409. 0473
2. Chan, W., Jaitly, N., Le, Q.V., Vinyals, O.: Listen, attend and spell. CoRR abs/1508.01211 (2015). http://arxiv.org/abs/1508.01211
3. Chorowski, J., Bahdanau, D., Serdyuk, D., Cho, K., Bengio, Y.: Attention-based models for speech recognition. CoRR abs/1506.07503 (2015). http://arxiv.org/abs/1506.07503
4. Ciresan, D.C., Meier, U., Schmidhuber, J.: Multi-column deep neural networks for image classification. CoRR abs/1202.2745 (2012). http://arxiv.org/abs/1202.2745
5. He, K., Zhang, X., Ren, S., Sun, J.: Deep residual learning for image recognition. CoRR abs/1512.03385 (2015). http://arxiv.org/abs/1512.03385

6. He, K., Zhang, X., Ren, S., Sun, J.: Delving deep into rectifiers: surpassing human-level performance on imagenet classification. In: Proceedings of the IEEE international conference on computer vision, pp. 1026–1034 (2015)

7. Ioffe, S., Szegedy, C.: Batch normalization: Accelerating deep network training by reducing internal covariate shift. CoRR abs/1502.03167 (2015). http://arxiv.org/abs/1502.03167

8. Jin, Q., Meng, Z., Sun, C., Wei, L., Su, R.: Ra-unet: a hybrid deep attention-aware network to extract liver and tumor in CT scans. CoRR abs/1811.01328 (2018). http://arxiv.org/abs/1811.01328

9. Kim, Y., Denton, C., Hoang, L., Rush, A.M.: Structured attention networks. CoRR abs/1702.00887 (2017). http://arxiv.org/abs/1702.00887

10. Krizhevsky, A., Sutskever, I., Hinton, G.E.: Imagenet classification with deep convolutional neural networks. In: Advances in neural information processing systems, pp. 1097–1105 (2012)

11. Lian, S., Luo, Z., Zhong, Z., Lin, X., Su, S., Li, S.: Attention guided U-net for accurate iris segmentation. J. Vis. Commun. Image Represent. **56**, 296–304 (2018). https://doi.org/10.1016/j.jvcir.2018.10.001. http://www.sciencedirect.com/science/article/pii/S1047320318302372

12. Luong, M., Pham, H., Manning, C.D.: Effective approaches to attention-based neural machine translation. CoRR abs/1508.04025 (2015). http://arxiv.org/abs/1508.04025

13. Oktay, O., et al.: Attention U-Net: Learning Where to Look for the Pancreas. arXiv e-prints arXiv:1804.03999 (Apr 2018)

14. Ronneberger, O., Fischer, P., Brox, T.: U-net: Convolutional networks for biomedical image segmentation. CoRR abs/1505.04597 (2015). http://arxiv.org/abs/1505.04597

15. Simonyan, K., Zisserman, A.: Very deep convolutional networks for large-scale image recognition. CoRR abs/1409.1556 (2014). http://arxiv.org/abs/1409.1556

16. Vaswani, A., et al.: Attention is all you need. CoRR abs/1706.03762 (2017). http://arxiv.org/abs/1706.03762

17. Wang, F., et al.: Residual attention network for image classification. CoRR abs/1704.06904 (2017). http://arxiv.org/abs/1704.06904

18. Wang, L., Yin, B., Guo, A., Ma, H., Cao, J.: Skip-connection convolutional neural network for still image crowd counting. Appl. Intell. **48**(10), 3360–3371 (2018). https://doi.org/10.1007/s10489-018-1150-1

19. Xu, K., et al.: Show, attend and tell: Neural image caption generation with visual attention. CoRR abs/1502.03044 (2015). http://arxiv.org/abs/1502.03044

20. Yamashita, T., Yamashita, J., Furukawa, H., Yamauchi, Y.: Multiple skip connections and dilated convolutions for semantic segmentation (2017)

21. Zeiler, M.D., Fergus, R.: Visualizing and Understanding Convolutional Networks. ArXiv e-prints (Nov 2013). https://arxiv.org/abs/1311.2901v3

Optic Disc and Fovea Localisation in Ultra-widefield Scanning Laser Ophthalmoscope Images Captured in Multiple Modalities

Peter R. Wakeford[1,2(✉)], Enrico Pellegrini[2], Gavin Robertson[2], Michael Verhoek[2], Alan D. Fleming[2], Jano van Hemert[2], and Ik Siong Heng[1]

[1] University of Glasgow, Glasgow G12 8QQ, Scotland
p.wakeford.1@research.gla.ac.uk
[2] Optos (Nikon), Carnegie Campus, Dunfermline KY11 8GR, Scotland

Abstract. We propose a convolutional neural network for localising the centres of the optic disc (OD) and fovea in ultra-wide field of view scanning laser ophthalmoscope (UWFoV-SLO) images of the retina. Images captured in both reflectance and autofluorescence (AF) modes, and central pole and eyesteered gazes, were used. The method achieved an OD localisation accuracy of 99.4% within one OD radius, and fovea localisation accuracy of 99.1% within one OD radius on a test set comprising of 1790 images. The performance of fovea localisation in AF images was comparable to the variation between human annotators at this task. The laterality of the image (whether the image is of the left or right eye) was inferred from the OD and fovea coordinates with an accuracy of 99.9%.

Keywords: Optic disc detection · Fovea detection · Laterality determination · Retinal images · Convolutional neural networks

1 Introduction

The optic disc (OD) and fovea are two anatomical landmarks found in the retina. The OD is where vasculature and nervous connections are made to the rest of body, and appears as a bright disc or oval in retinal images. The fovea, the centre of vision, is the area with the highest density of cone receptor cells, and is found on-axis to the lens of the eye.

Automating the location of these landmarks in retinal images is an important first step in the computer-aided diagnosis of many retinal diseases. For example, glaucoma can be graded by measuring morphology parameters of the OD [25], and automation of this measurement requires the OD to be located [8]. In the detection of diabetic retinopathy, it is often necessary to identify the OD before detection of exudates [5].

P. R. Wakeford—Supported by the EPSRC Centre for Doctoral Training in Applied Photonics.

© Springer Nature Switzerland AG 2020
Y. Zheng et al. (Eds.): MIUA 2019, CCIS 1065, pp. 399–410, 2020.
https://doi.org/10.1007/978-3-030-39343-4_34

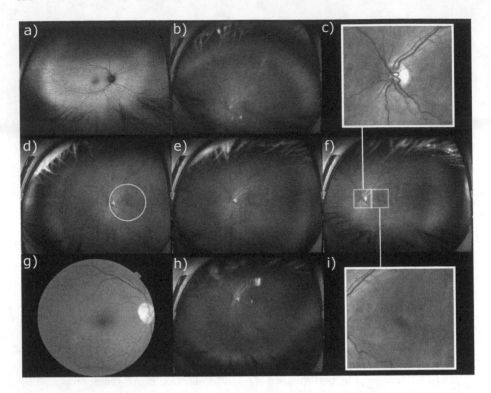

Fig. 1. Example UWFoV-SLO and fundus camera (FC) images; (a) Right eye autofluorescence (AF) central pole gaze (CP), (b) Left eye red-green (RG) reflectance eyesteered gaze (ES) Superior (c) Left eye optic disc (d) Left eye RG ES Nasal (approximate FoV of FC superimposed) (e) Left eye RG CP (f) Left eye RG ES Temporal (g) Right eye FC [16] (h) Left eye RG ES Inferior (i) Left eye Fovea. Images (b), (d): (f) and (h) make a complete set of CP and ES images of a left eye. (Color figure online)

The automatic classification of the *laterality* of the image, i.e., whether the image is of the right or left eye, is of clinical interest as it allows the image to be labeled without manual input from an operator. This can save valuable time in the clinical environment, with the benefit of classification accuracy similar to human grading [11], as operators are prone to error due to the repetitive nature of this task. The laterality of an image can be inferred from the coordinates of the OD and fovea; in images of the left eye, the OD is to the left of the fovea. The opposite is true for the right eye (see Fig. 1).

A fundus camera captures white-light photographs of the retina, whereas an scanning laser ophthalmoscope (SLO) uses directed laser beams to measure reflectance at each point on the retina to build a reflectance image. This technique allows imaging of a wider field of view—for instance, Optos ultra-wide field of view scanning laser ophthalmoscopes (UWFoV-SLOs) used in this study attain 200° field of view (FoV), covering 82% of the retina. Fundus camera images cover a smaller FoV (typically 45°), resulting in features such as the OD appearing proportionally larger with respect to the imaged area in fundus camera images compared to UWFoV-SLOs images. See Fig. 1(d) and (g) for a FoV comparison between fundus camera images and UWFoV-SLO images.

In addition to red-green (RG) reflectance images, UWFoV-SLO systems are capable of capturing autofluorescence (AF) images, in which a laser is used to stimulate emission at longer wavelengths. This allows cell respiration to be visualised, and aids the detection and diagnosis of retinal diseases [9]. AF images have higher intensity levels than reflectance images, and exhibit higher levels of noise. Figure 1 shows examples of RG and AF images for comparison.

The work presented here is novel for a number of reasons. Firstly, to the best of our knowledge this is the first time an OD and fovea localisation method has been reported on UWFoV-SLO images. Secondly, we are the first to investigate images captured in the AF modality. Thirdly, we show that the proposed convolutional neural network (CNN) architecture can be trained and tested with retinal images captured in multiple modalities. Lastly, we show that the accuracy of the laterality inferred from the OD and fovea coordinates predicted by the proposed method is higher than that of a second CNN which was specifically trained to perform laterality classification.

2   Related Work

Sinthanayothin et al. [22] were the first to propose an automated method to estimate the locations of the OD and fovea in digital fundus camera photographs. The OD was located by finding the region of highest intensity variation to the surrounding pixels, and the fovea located by finding the darkest region near the OD. Hoover et al. [10] showed that the direction of the retinal vasculature could be used to locate the OD, as vasculature extends from there.

Foracchia et al. [7] used the technique of fitting parametric models to the blood vessels to locate the OD after extracting the vasculature. Fleming et al. [6] showed that the retinal vasculature could be modeled as an elliptical form and analysed with the Hough transform, yielding approximate locations of the OD and fovea. These locations were refined with edge and intensity measurements. Tobin et al. [26] took the segmented retinal vasculature map, and spatially measured parameters such as vessel density and thickness. This was used with a geometric model of the retina for OD and fovea location estimation.

Niemeijer et al. [18] was first to treat retinal landmark localisation as a regression problem, in which the landmark coordinates were to be estimated. They showed that parameters estimated from the vasculature could be used to train a kNN regressor to return the coordinates of both the OD and fovea. Morphological operations on the blood vessels have also been used to determine the location of the OD, for example the work of Marin et al. [13].

With the advent of deep learning, methods employing these have been developed to locate retinal landmarks in fundus camera images. Calimeri et al. [3] employed transfer-learning of a CNN trained for face-detection for OD detection. Niu et al. [19] showed the region in the image containing the OD could be determined by generating saliency maps from the image (maps which indicate 'interesting' features, such as high spatial frequency). From these maps, a number of candidate OD regions were selected. These were then processed with a CNN, which was trained to classify each region as containing OD or not. Similarly, Mitra et al. [17] used a CNN to return a region-of-interest containing the OD, and Meng et al. [14] used a CNN trained on images where the blue channel is replaced with a vasculature map, to locate the OD.

Simultaneous OD and fovea localisation in fundus camera images with CNNs has been shown by Al-Bander et al. [1], whereby a single CNN returned four outputs, the x and y coordinates of the OD and fovea. This was further developed [2], in which two networks were utilised—the first gave an approximate location of the coordinates of both features, which were used to extract two image patches centred around the OD and fovea respectively. These image patches were then passed to the second network for refined feature centre estimation.

Recently, Meyer et al. [15] showed that a U-net architecture [20] could be used to predict the distance of each pixel from the nearest landmark. A predicted distance map was calculated, and image processing techniques were employed to identify the two landmarks in the map. At this stage, it was not known which landmark was the OD and which the fovea. A further step, based on assumptions of retinal brightness, was used to label the landmarks.

Automated laterality determination using fundus camera images has also been studied. For example Tan et al. [24] extracted blood vessels and the OD which were passed to a support vector machine for laterality classification. Roy et al. [21] deployed transfer-learning with a CNN, which also contained auxiliary inputs for extracted image data, such as blood vessel density and orientation. More recently, Jang et al. [11] demonstrated that CNNs could be put to this task, achieving 99.0% classification accuracy.

3   Materials

Images captured in both RG and AF modes were used in this experiment. The majority of images were captured in the central-pole (CP) gaze, in which the scan is centred on the fovea. Images were also captured with the eyesteered (ES) gazes, in which the scan was centred on the superior, inferior, nasal or temporal fields of the retina (Fig. 1). The private data set contained 4732 left eye images and 4786 right eye images. Images are not necessarily captured as left/right eye pairs, leading to the discrepancy in the number of left and right images. The data set was doubled in size by generating a horizontally-flipped version of each image, and the associated laterality class was swapped. Whilst this did not add new information, it allowed extra verification of the method when applied to laterality classification.

Table 1 shows the number of images and subjects from the RG and AF modes, split by gaze (CP or ES), including the horizontally flipped version of each original image. Some subjects were imaged in CP and ES or RG and AF so subject totals are not the sum of rows and columns.

Table 1. Number of images, including horizontally flipped versions, split by modality and gaze. Number of subjects indicated in parentheses.

Mode	Gaze		
	CP	ES	Total
RG	11132 (1153)	1818 (87)	12950 (1157)
AF	5792 (987)	294 (77)	6086 (993)
Total	16924 (2139)	2112 (164)	**19036 (2149)**

Annotations of all original images in the data set were obtained from a trained observer, who was required to annotate the location of the OD and fovea, and the image laterality and gaze. Three additional graders were required to annotate 100 RG and 100 AF images, so that inter-grader variation could be assessed.

4   Method

A CNN was devised, which accepted an UWFoV-SLO retinal image as input, and predicted four parameters; the OD and fovea x and y coordinates.

4.1   Preprocessing

To avoid overfitting and obtain a fair evaluation of results, images were split into training, validation and test sets on the constraints that (1) all images from any one subject were not put into more than one of the training, validation or test sets (2) The training, validation or test sets contained approximately 70%, 20% and 10% of the data set, respectively.

Images were downsampled from 3072×3900 to 768×975 pixels to reduce computational complexity. Only the green reflectance channel was used, so a single network could be devised to accept both reflectance and AF images, which are composed of one channel. The green reflectance channel was selected because the vasculature (the topography of which can be used for inferring laterality) contrast was higher. Images were not cropped, as landmarks can be located in the extremes of the capture area in ES images (Fig. 1). Image intensity values were scaled to have zero mean and unit standard deviation.

OD and fovea coordinates were normalised by dividing both x and y coordinates by the image width, W, so all coordinates were in the range [0, 1], and the loss function was not biased in favour of the y coordinate. Predicted values were mapped back to image coordinates at the inference stage. In the case of horizontally flipped images, the flipped x coordinates of the OD and fovea, x', were calculated by $x' = W - x$.

4.2   Architecture

The proposed architecture was selected based on performance of candidate architectures on the validation set. In the proposed network, five blocks of four convolutional layers were stacked, and convolutional layers in the same block had the same number of kernels. Blocks 1 and 2 had 32 kernels, 3 and 4 had 64 kernels, 5 had 128 kernels. The initial convolutional layer of each block had a stride of 2, all subsequent convolutional layers in each block had stride 1. The output from the final convolutional block was flattened, and two fully connected layers of 512 nodes followed, each preceded by dropout layers of dropout probability $p = 0.3$ [23]. The output layer was a fully connected layer of four nodes, one for each x and y coordinate of OD and fovea (Fig. 2).

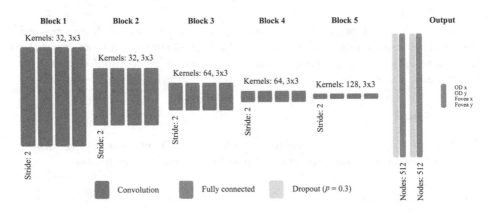

Fig. 2. Illustration of the proposed architecture for landmark localisation.

The network was optimised using the Adam optimiser [12] with learning rate of 0.0001 and decay rate parameters $\beta_1 = 0.9$ and $\beta_2 = 0.999$. The norm of the gradients was clipped at 1.0 to avoid instability of the loss during training. The mean squared error loss function between the ground truth landmark coordinates and the predicted coordinates was used.

The network (containing 49,882,660 trainable parameters) was trained using Keras with TensorFlow backend, and was performed on a mixture of RG and AF images presented in batches of 16 images. 13,550 training images from 1,587 subjects were used.

4.3   Landmark Localisation Evaluation

To allow direct comparison of results, the evaluation measure used by Meyer et al [15] was employed. The OD and fovea localisation accuracy was defined as the percentage of images for which the Euclidean distance between the predicted landmark location and the ground truth was below n OD radii (r).

To determine the OD radius, a modification to the OD diameter approximation proposed by Al-Bander et al. was used [2]. The OD radius r for each image was estimated from the Euclidean distance between the ground truth x and y

coordinates of the OD and fovea (Eq. 1) as

$$r = \frac{\sqrt{(x_{OD} - x_{fovea})^2 + (y_{OD} - y_{fovea})^2}}{5} \qquad (1)$$

The OD and fovea annotations of three additional graders for 100 RG and 100 AF images were compared to the ground-truth labels. The mean Euclidean distance between the annotations and the ground-truth labels, and the standard deviation, was calculated.

An open-source implementation [16] of the method for OD and fovea localisation in fundus camera images proposed by Meyer et al. [15] was tested on RG and AF UWFoV-SLO images. Given the constraints on the input size of the pretrained network, RG and AF images were resized to 384 × 512 pixels, and padded with zeros to 512 × 512 pixels. Distance probability maps were analysed with the image processing pipeline presented [16] to predict the OD and fovea coordinates.

4.4 Laterality Classification Evaluation

Laterality accuracy was calculated as the percentage of images which were classified as belonging to the correct class, from left or right eye. The class sizes were balanced as a result of the flipping operation, thus the accuracy did not require weighting.

The laterality of the input image can be inferred from the OD and fovea x coordinates, as shown in Eq. 2.

$$\text{Laterality} = \begin{cases} L, & \text{if } x_{OD} < x_{fovea}. \\ R, & \text{if } x_{OD} > x_{fovea}. \end{cases} \qquad (2)$$

A modified version of the proposed architecture was used to implement a laterality classifier. The output layer was replaced with a fully connected layer with a single, Sigmoid activated node to predict the laterality class of the input image. This allowed the laterality inferred from the landmark coordinates predicted by the proposed method to be compared with a classifier baseline. The classification network was trained in the same fashion as the landmark localisation network, but using binary cross-entropy loss between the ground truth class and predicted class labels.

5 Results

Results for landmark localisation accuracy and laterality classification accuracy for 1790 test images are shown in Table 2. Results of the proposed method are split by modality and gaze. Other methods demonstrated on fundus camera images are presented for comparison. Plots of OD and fovea accuracy as a function of distance are shown in Figs. 3 and 4. The accuracy of laterality inferred from the x coordinates of the OD and fovea is shown in Table 2. The laterality classification network accuracy is also shown.

Table 2. Accuracy for landmark localisation (expressed in OD radii, r), and accuracy of laterality classification inferred (Inf.) from the landmark x coordinates, and classified by classifier network (Class.). Results are split by mode and gaze. Accuracies are expressed as percentages. Fundus camera (FC) methods shown for comparison.

Method	Mode	Gaze	OD Accuracy			Fovea accuracy			Lat. Acc.		Images
			$0.25r$	$0.5r$	$1r$	$0.25r$	$0.5r$	$1r$	Inf.	Class.	
Jang [11]	FC		x	x	x	x	x	x	x	99.0	5180
Niemeijer [18]	FC		x	x	x	93.3	96.0	97.4	x	x	800
Marin [13]	FC		97.8	99.5	99.8	x	x	x	x	x	1200
Al-Bander [2]	FC		83.6	95.0	97.0	66.8	91.4	96.6	x	x	1200
Meyer [15]	FC		93.6	97.1	98.9	94.0	97.7	99.7	x	x	1136
Proposed	RG	All	67.6	93.7	99.5	56.0	88.2	99.1	99.8	99.0	1144
Proposed	RG	CP	71.3	95.1	99.5	58.7	90.4	99.0	99.8	98.9	1006
Proposed	RG	ES	40.6	83.3	99.3	36.2	72.5	100.0	100.0	100.0	138
Proposed	AF	All	73.4	96.3	99.2	48.6	85.3	99.1	100.0	100.0	646
Proposed	AF	CP	74.9	97.0	99.3	49.3	85.9	99.3	100.0	100.0	610
Proposed	AF	ES	47.2	83.3	97.2	36.1	75.0	94.4	100.0	100.0	36
Proposed	**All**	**All**	**69.7**	**94.6**	**99.4**	**53.4**	**87.2**	**99.1**	**99.9**	**99.4**	**1790**
Proposed	All	CP	72.6	95.9	99.4	55.2	88.7	99.1	99.9	99.3	1616
Proposed	All	ES	42.0	83.3	98.9	36.2	73.0	98.9	100.0	100.0	174

The mean Euclidean distance μ and standard deviation σ, normalised by r, between three additional graders and the ground-truth landmark coordinates is shown in Table 3. 100 RG and 100 AF images were used. The proposed method, tested on 1790 test images, is shown for comparison.

The pixel landmark distance prediction method proposed by Meyer [15] achieved 0% accuracy for both OD and fovea localisation within $1r$ on 1790 UWFoV-SLO images. The inferred laterality classification accuracy was 52.7%. These results could be expected given the model was trained on fundus camera images, and highlight the difference between the two image capturing systems.

Table 3. Mean Euclidean distance μ and standard deviation σ between three graders and the ground-truth landmark coordinates for 100 RG and 100 AF images. Distances normalised by r. The mean of grader scores, as well as the proposed method when tested on 1790 test images, are shown for comparison.

Grader	RG				AF			
	OD μ	OD σ	Fovea μ	Fovea σ	OD μ	OD σ	Fovea μ	Fovea σ
1	0.127	0.074	0.205	0.159	0.117	0.073	0.323	0.222
2	0.126	0.082	0.167	0.112	0.139	0.078	0.318	0.238
3	0.164	0.089	0.198	0.120	0.139	0.093	0.331	0.239
Mean	0.139	0.082	0.190	0.130	0.132	0.081	0.324	0.233
Proposed	**0.226**	**0.174**	**0.278**	**0.219**	**0.207**	**0.160**	**0.302**	**0.205**

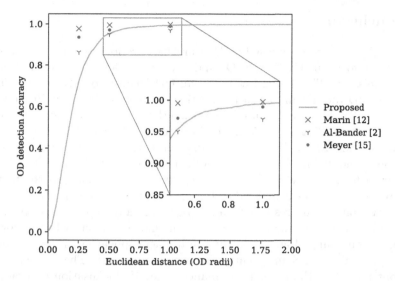

Fig. 3. Optic disc localisation accuracy for the proposed method as a function of Euclidean distance, expressed in optic disc radii. Results of other methods, achieved on fundus camera datasets, are shown for comparison.

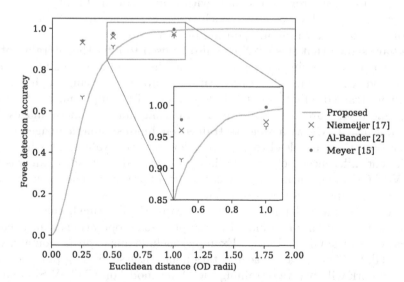

Fig. 4. Fovea localisation accuracy for the proposed method as a function of Euclidean distance, expressed in optic disc radii. Results of other methods, achieved on fundus camera datasets, are shown for comparison.

6　Conclusion

In this paper, we have proposed a CNN architecture for the localisation of the OD and the fovea in UWFoV-SLO images of two modalities—RG reflectance and AF emission. The method achieved an OD localisation accuracy of 99.4% within one OD radius, and fovea localisation accuracy of 99.1% within one OD radius (Table 2). These values are higher than those from other methods applied to fundus camera images, found in the literature, at $0.5r$ and $1r$ distances, but not at $0.25r$. However, a fair, direct comparison between these results cannot be made due to the differences between the imaging techniques. The UWFoV-SLO has a larger field of view (FoV), which results in features such as the OD and fovea appearing smaller compared to fundus camera images (Fig. 1).

Accuracy on CP images was found to be marginally higher than ES images. This may be due to imaging artefacts which are more prevalent when performing ES scans. For example, Fig. 1 shows bright artefacts appearing in the inferior and superior gaze images. ES images may also contain more lash. These artefacts may have contributed to the lower performance. The OD localisation accuracy was also found to be higher than the fovea localisation. This is expected, as the OD shows as a bright oval, whereas the centre of the fovea is less clearly-defined. This was reflected in the human annotation comparison results (Table 3), where inter-grader agreement was higher for the OD location than the fovea. When comparing performance with human graders, the proposed method achieved AF fovea localisation distance errors comparable with human graders (Table 3). However, for the remaining landmarks, humans achieved a higher accuracy. This may have been due to image downsampling during preprocessing.

We have shown that this CNN can also be used to infer the laterality of the input image by considering the predicted x coordinates of the retinal landmarks. This method achieves laterality classification accuracy values higher than a similar network trained only to predict the laterality of the input image (99.9% vs 99.4%), and results are in line with literature using fundus camera images [11].

Testing of a CNN with weights trained on fundus camera images [15,16] on UWFoV-SLO images lead to extremely low accuracy values. This highlights the inherent differences between the images captured by fundus cameras and UWFoV-SLOs, and confirms that direct transfer of pre-trained networks between the two image types is not possible.

In the cases where landmarks are not correctly identified, this may have repercussions on further automated image processing operations which require the locations of the OD and fovea. Examples include image projection and visualisation [4], biological feature segmentation and disease grading pipelines [25].

Future work will involve retraining the Meyer model on UWFoV-SLO images to compare performance. However, modification to the image processing technique for determining which landmark is the OD and fovea will be required, as the existing method assumes that the OD is brighter than the fovea, which does not hold in the case of AF images. Future work will also include training and testing our method on fundus camera data sets, ideally of subjects also imaged with UWFoV-SLOs, so direct comparison can be made.

References

1. Al-Bander, B., Al-Nuaimy, W., Al-Taee, M.A., Al-Ataby, A., Zheng, Y.: Automatic feature learning method for detection of retinal landmarks. In: Proceedings - 2016 9th International Conference on Developments in eSystems Engineering, pp. 13–18 (2016). https://doi.org/10.1109/DeSE.2016.4
2. Al-Bander, B., Al-Nuaimy, W., Williams, B.M., Zheng, Y.: Multiscale sequential convolutional neural networks for simultaneous detection of fovea and optic disc. Biomed. Signal Process. Control **40**, 91–101 (2018). https://doi.org/10.1016/j.bspc.2017.09.008
3. Calimeri, F., Marzullo, A., Stamile, C., Terracina, G.: Optic disc detection using fine tuned convolutional neural networks. In: Proceedings - 12th International Conference on Signal Image Technology and Internet-Based Systems, pp. 69–75 (2016). https://doi.org/10.1109/SITIS.2016.20
4. Croft, D.E., van Hemert, J., Wykoff, C.C., Clifton, D., Verhoek, M., Fleming, A., Brown, D.M.: Precise montaging and metric quantification of retinal surface area from ultra-widefield fundus photography and fluorescein angiography. Ophthalmic Surg. Lasers Imaging Retina **45**(4), 312–317 (2014)
5. Faust, O., Acharya, R., Ng, E.Y., Ng, K.H., Suri, J.S.: Algorithms for the automated detection of diabetic retinopathy using digital fundus images: a review. J. Med. Syst. **36**(1), 145–157 (2012). https://doi.org/10.1007/s10916-010-9454-7
6. Fleming, A.D., Goatman, K.A., Philip, S., Olson, J.A., Sharp, P.F.: Automatic detection of retinal anatomy to assist diabetic retinopathy screening. Phys. Med. Biol. **52**(2), 331–345 (2007). https://doi.org/10.1088/0031-9155/52/2/002
7. Foracchia, M., Grisan, E., Ruggeri, A.: Detection of optic disc in retinal images by means of a geometrical model of vessel structure. IEEE Trans. Med. Imaging **23**(10), 1189–1195 (2004). https://doi.org/10.1109/TMI.2004.829331
8. Haleem, M.S., Han, L., van Hemert, J., Li, B.: Automatic extraction of retinal features from colour retinal images for glaucoma diagnosis: a review. Comput. Med. Imaging Graph. **37**(7–8), 581–596 (2013). https://doi.org/10.1016/j.compmedimag.2013.09.005
9. Holz, F.G., Spaide, R.F., Schmitz-Valckenberg, S., Bird, A.C. (eds.): Atlas of Fundus Autofluorscence Imaging. Springer, Heidelberg (2007). https://doi.org/10.1007/978-3-540-71994-6
10. Hoover, A., Goldbaum, M.: Locating the optic nerve in a retinal image using the fuzzy convergence of the blood vessels. IEEE Trans. Med. Imaging **22**(8), 951–958 (2003). https://doi.org/10.1109/TMI.2003.815900
11. Jang, Y., Son, J., Park, K.H., Park, S.J., Jung, K.H.: Laterality classification of fundus images using interpretable deep neural network. J. Digit. Imaging 1–6 (2018). https://doi.org/10.1007/s10278-018-0099-2
12. Kingma, D.P., Ba, J.L.: Adam: A Method for Stochastic Optimization. CoRR abs/1412.6, 1–15 (2014). https://doi.org/10.1016/j.nano.2011.03.005. http://arxiv.org/abs/1412.6980
13. Marin, D., Gegundez-Arias, M.E., Suero, A., Bravo, J.M.: Obtaining optic disc center and pixel region by automatic thresholding methods on morphologically processed fundus images. Comput. Methods Programs Biomed. **118**(2), 173–185 (2015). https://doi.org/10.1016/j.cmpb.2014.11.003
14. Meng, X., Xi, X., Yang, L., Zhang, G., Yin, Y., Chen, X.: Fast and effective optic disk localization based on convolutional neural network. Neurocomputing **312**, 285–295 (2018). https://doi.org/10.1016/j.neucom.2018.05.114

15. Meyer, M.I., Galdran, A., Mendonça, A.M., Campilho, A.: A pixel-wise distance regression approach for joint retinal optical disc and fovea detection. In: Frangi, A.F., Schnabel, J.A., Davatzikos, C., Alberola-López, C., Fichtinger, G. (eds.) MICCAI 2018. LNCS, vol. 11071, pp. 39–47. Springer, Cham (2018). https://doi. org/10.1007/978-3-030-00934-2_5

16. Meyer, M.I., Galdran, A., Mendonca, A.M., Campilho, A.: Joint Retinal Optical Disc and Fovea Detection (2018). https://github.com/minesmeyer/od-fovea-regression

17. Mitra, A., Banerjee, P.S., Roy, S., Roy, S., Setua, S.K.: The region of interest localization for glaucoma analysis from retinal fundus image using deep learning. Comput. Methods Programs Biomed. **165**, 25–35 (2018). https://doi.org/10.1016/ j.cmpb.2018.08.003

18. Niemeijer, M., Abràmoff, M.D., van Ginnekena, B.: Fast detection of the optic disc and fovea in color fundus photographs. Med. Image Anal. **13**(6), 859–870 (2009). https://doi.org/10.1016/j.media.2009.08.003

19. Niu, D., Xu, P., Wan, C., Cheng, J., Liu, J.: Automatic localization of optic disc based on deep learning in fundus images. In: 2017 IEEE 2nd International Conference on Signal and Image Processing, pp. 208–212 (2017). https://doi.org/10. 1109/SIPROCESS.2017.8124534

20. Ronneberger, O., Fischer, P., Brox, T.: U-Net: convolutional networks for biomedical image segmentation. In: Navab, N., Hornegger, J., Wells, W.M., Frangi, A.F. (eds.) MICCAI 2015. LNCS, vol. 9351, pp. 234–241. Springer, Cham (2015). https://doi.org/10.1007/978-3-319-24574-4_28

21. Roy, P.K., Chakravorty, R., Sedai, S., Mahapatra, D., Garnavi, R.: Automatic eye type detection in retinal fundus image using fusion of transfer learning and anatomical features. In: 2016 International Conference on Digital Image Computing: Techniques and Applications, pp. 538–544 (2016). https://doi.org/10.1109/ DICTA.2016.7797012

22. Sinthanayothin, C., Boyce, J.F., Cook, H.L., Williamson, T.H.: Automated localisation of the optic disc, fovea, and retinal blood vessels from digital colour fundus images. Br. J. Ophthalmol. **83**(8), 902–910 (1999). https://doi.org/10.1136/bjo. 83.8.902

23. Srivastava, N., Hinton, G., Krizhevsky, A., Sutskever, I., Salakhutdinov, R.: Dropout : a simple way to prevent neural networks from overfitting. J. Mach. Learn. Res. **15**, 1929–1958 (2014)

24. Tan, N.M., et al.: Classification of left and right eye retinal images. In: Proceedings of SPIE, vol. 7624 (2010). https://doi.org/10.1117/12.844638. http://proceedings. spiedigitallibrary.org/proceeding.aspx?doi=10.1117/12.844638

25. Tangelder, G.J., Reus, N.J., Lemij, H.G.: Estimating the clinical usefulness of optic disc biometry for detecting glaucomatous change over time. Eye **20**(7), 755–763 (2006). https://doi.org/10.1038/sj.eye.6701993

26. Tobin, K.W., Chaum, E., Govindasamy, V.P., Karnowski, T.P.: Detection of anatomic structures in human retinal imagery. IEEE Trans. Med. Imaging **26**(12), 1729–1739 (2007)

Deploying Deep Learning into Practice: A Case Study on Fundus Segmentation

Sören Klemm, Robin D. Ortkemper, and Xiaoyi Jiang[✉]

Faculty of Mathematics and Computer Science,
University of Münster, Münster, Germany
{klemms,xjiang}@uni-muenster.de, robin.ortkemper@gmail.com

Abstract. We study the deployment of a deep learning medical image segmentation pipeline, which sees new input data, not contained in the training and evaluation database. Like in application, although this data shows the same properties, it may stem from a slightly different distribution than the training set because of differences in the hardware setup or environmental conditions. We show that, although cross-validation results suggest high generalization, segmentation score drops significantly with pre-processed data from a new database. The positive effects of a short fine-tuning phase after deployment, which seems to be necessary under such conditions, can be observed. To enable this study, we develop a segmentation pipeline comprising pre-processing steps to homogenize the data contained in 4 databases (DRIONS, DRISHTI-GS, RIM-ONE, REFUGE) and an artificial neural network (NN) segmenting optic disc and cup. This NN can be trained using exactly the same hyperparameters on all 4 databases, while achieving performance close to state-of-the-art methods specifically designed for the individual databases. Furthermore we deduct a hierarchy of the 4 databases with respect to complexity and broadness of contained samples.

Keywords: Image segmentation · Deep learning · Biomedical imaging

1 Introduction

Due to the great success of Machine Learning (ML) especially in the form of Deep Learning (DL) in the last decade, Neural Networks (NNs) became the state-of-the-art method for biomedical image segmentation and understanding [14]. That has led experts to arguably assume that practitioners will be replaced by DL algorithms [12]. While their application is not limited to the domain of biomedical image analysis, the consequences are particularly discussed in this field, as they have a great impact on society [6].

The question remains to what extent the results achieved with training data sets can be transferred to a real application in everyday clinical or laboratory business. This topic, also called generalizability, has been discussed mainly with a focus on overfitting a training data set. In this paper we extend this view to the deployment of trained algorithms to everyday application by practitioners.

© Springer Nature Switzerland AG 2020
Y. Zheng et al. (Eds.): MIUA 2019, CCIS 1065, pp. 411–422, 2020.
https://doi.org/10.1007/978-3-030-39343-4_35

Our findings are based on an exemplary problem: Automatic segmentation of two areas of the retina, optic disc (OD) and optic cup (OC). The diameter ratio of OD and OC, called cup to disc ratio (CDR), is predictive to glaucoma, which is a widely spread disease affecting more than 64 million individuals worldwide by 2013 [21]. Ophthalmoscopy or funduscopy is the visual analysis of the fundus of the eye to determine, amongst others, the diameter of the OC and OD and is the most common diagnostic method for glaucoma. While not curable, early treatment of glaucoma can reduce the risk of blindness by 50%. Automatic segmentation will help to find signs of glaucoma earlier in many patients.

The need for such methods caused a wave of research projects that tackle the automatic detection of glaucoma by calculating the CDR, many of which use ML approaches [15,19,22,24].

We show that a shrinked version of the famous U-Net architecture [17] can be trained to perform comparably to state-of-the-art methods on 4 different training data sets. By evaluating training and testing results, we estimate a hierarchy of the available databases with respect to complexity and self similarity of the contained data sets. As our common pre-processing pipeline allows inference on all databases using the same NN, we can furthermore gain insights about generalization of NNs not only from training to test sets but also to unseen databases. The latter can be seen as a simulated deployment to productive use, where different hardware or changes in environmental conditions might lead to shifts in the distribution of input data.

The remainder of the paper is structured as follows: In the next section, we give a brief overview of published research on generalization of DL methods, as well as disc and cup segmentation. In Sect. 3, we describe our training and testing data sets as well as the pre- and post-processing steps and the DL architecture trained to perform the segmentation task. The results of our study, especially the effect of fine-tuning and the limited significance of cross-validation results for generalization after deployment, are shown subsequently. In the last section, we discuss our results against the background of deployment of ML models in clinical applications.

2   Related Work

2.1   Generalization of ML Algorithms

In the last years, research on generalization of ML methods, especially NNs, increased with the growing popularity of DL. At the beginning of the broad application of DL, specific methods to improve generalization, like dropout, were proposed. Zhang et al. [23] published extensive theoretical research on the generalization properties of NNs focusing on the imbalance between trainable parameters and data samples. Furthermore, they claimed that common regularizers (data augmentation, dropout, weight decay) have less impact than changes to the model architecture.

Kawaguchi et al. suggest that architectures with increased trainability might be prone to less generalization [13]. While they focused on synthetic manipulations of the labels of the well known CIFAR10 and MNIST datasets, Geirhos et al. took a different approach and used artificially degraded input images from the ImageNet data set [9]. They concluded that NNs can be trained to be robust against noisy input. This robustness, however, does only apply to the specific degradations shown during training and has little to no effect on other types of noise. The same approach was taken by Dodge and Karam who showed that NNs outperforming humans on a subset of ImageNet rapidly loose their advantage as soon as noise and distortions are applied [4].

Only recently Recht et al. showed that quantitative results of state-of-the-art NNs are not matched if unseen data, which was not part of the original data set, is shown for inference [16].

2.2 Disc/Cup Segmentation

One key issue of ophthalmological image analysis is the detection of important image structures such as retinal blood vessels [2], retinal layers [5], disc and cup. The latter two are the focus of this work. Mahapatra et al. [15] published an approach for cup and disc segmentation in a pipeline of Field of Experts for feature selection, Random Forests to estimate probability maps for OD and OC and post-processing with an elliptical Hough transform and Graph Cuts to achieve smooth regions. Zilly et al. [24] employed a Convolutional Neural Network, trained with a boosting approach instead of gradient descent. They furthermore utilize entropy sampling to find image patches for the training procedure. Although their NN is rather small and shallow, with only 2 layers and 34 filter kernels, they achieve highly competitive results on both segmentation tasks. In the same year, Sevastopolski used the U-Net architecture [17,18] and minimal pre-processing to apply the DL philosophy to the CDR problem, achieving results comparable with Zilly et al. at the expense of a more than 300-fold larger parameter space to be learned. In 2018 Sevastopolsky et al. [19] proposed the stacked U-Net an application of the residual network approach [11] with U-Nets as building blocks. Their 15-fold stacked U-Net increases the number of trainable parameters again, while achieving only small improvements on some benchmarks. Wang et al. [22] participated at the REFUGE challenge [3] with a cascade of three NNs, consecutively applied to first pre-segment a region of interest (ROI) around the OD and later performing fine-segmentation of OC and OD. To increase the available training data, they use a method called source adaption. Source adaption employs the generative adversarial network (GAN) approach to learn a mapping from the feature space of the DRISHTI-GS database to the REFUGE feature space.

3 Methods

As we want to study the deployment of a trained machine learning model, we need a NN trainable on different databases. In addition, the available databases differ significantly in terms of resolution and data acquisition. Hence, we apply a pre-processing pipeline to all data sets that homogenizes the input data of the NN.

3.1 Databases

While many of the previous works perform training and validation on samples from the same database, we investigate the performance of trained models on samples from a different database, taken under similar conditions. We conduct our study on 4 databases. Figure 1 shows examples from all databases after pre-processing, which is described in the next section.

The DRISHTI-GS database [20] consists of a total of 101 images in PNG format. The images of the data set were taken from patients at the Aravind Eye Clinic in Madurai, India. The images were first taken with centered papilla and a resolution of 2896×1944 pixels and then tailored to the region with the structures of the retina, which causes slight variations in the resolutions of the images. Ground truth data was created with the help of four experts. The majority vote procedure is used to determine whether at least three of the four experts agree on the segmentation as ground truth. This corresponds to the procedure proposed by the editor of the database, which was also used in related publications.

Images of the RIM-ONE v.3 database [8] were taken from patients from three Spanish hospitals: Hospital Universitario de Canarias, Hospital Clinico San Carlos and Hospital Universitario Miguel Servet. The third version of the database, published in 2015, contains 159 stereo recordings in JPEG format. Since the presented algorithm does not deal with stereo recordings, only the left view is used. The resolution of a single image is 1072×1424 pixels. In contrast to DRISHTI-GS, the images have already been manually cropped to the approximate area of the OD. The database contains masks for the disc and cup segmentation of two experts, as well as an average of the two segmentations, which was used as ground truth in the presented procedure.

DRIONS-DB [1] comprises 110 images of patients from the Miguel Servet Hospital in Zaragoza, Spain and was released in 2008. DRIONS-DB has two significant differences compared to the other records used: First, images were taken with an analog fundus camera and later digitized with a scanner. Second, the database contains only ground truth data for the segmentation of the disc. The images have a resolution of 600×400 pixels and were taken with an approximately centered disc. Two experts labeled the contour of the disc with 36 points. The convex hull of all points was used as ground truth segmentation.

The data set of the REFUGE Challenge contains 1200 recordings, which are divided into training, test and validation sets of identical size. Since we only had access to the training set, the 400 recordings of this set are used for training and evaluation. The images are in JPEG format and have a resolution of 2124×2056 pixels. The database contains ground truth data in the form of a segmentation mask for disc and cup. To create these masks, seven experts from the Zhongshan Ophthalmic Center at Sun Yat-Sen University in China first created segmentations, which were then merged into a single mask by another expert [3].

3.2 Pre- & Post-processing

DRIONS   DRISHTI   REFUGE   RIM-ONE

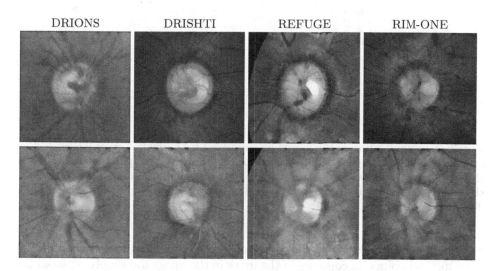

Fig. 1. Examples of the images after pre-processing. All images have been cropped and scaled to 512 × 512 pixels. Larger vessels were partially removed using in-painting and contrast limited adaptive histogram equalization was applied. Every column shows two images of the same database.

First, all input images are cropped to a quadratic ROI around the OD and then scaled to 512 × 512 pixels, using bilinear interpolation.

Larger blood vessels in the area of the disc, which are known to have a detrimental effect on segmentation performance [10], are removed. The green color channel is used for segmenting the blood vessels as follows.

The image is inverted and a Gaussian filter with a standard deviation of 0.45 is applied. This serves as a basis for further processing by a morphological top-hat transformation with a circular structuring element of 18 pixels in size. A binary mask is created, with a threshold value selected such that at least 75% of the gray values lie beneath it. This is because about 12.5% of the pixels of a fundus camera image belong to blood vessels [7]. However, since the area of the disc selected as the region of interest in the first step of the pre-processing contains significantly more and wider blood vessels than the rest of the retina, we use the doubled value. To remove possible artifacts from the created binary image, a smoothed version of the mask is created using a 15 × 15 pixel median filter. The new mask is the union of all pixels in the original mask and the result of smoothing by the median filter. The result is dilated with a 5-pixel circular structuring element to close any gaps and smooth the contours of the segmented blood vessels. The segmentation is concluded with an analysis of the 8-connected components. In the final mask, only those components are retained that are responsible for at least five percent of the overall pixels on the mask after the dilation. With the

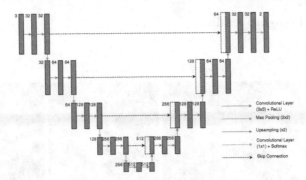

Fig. 2. The architecture of the neural network used for evaluation is very close to the original U-Net as proposed in [17]. To reduce the number of free parameters, we halve the number of channels in every step of the contraction and expansion path. Following the same philosophy, the up-sampling is realized using bilinear interpolation instead of deconvolution.

created mask the segmented blood vessels are removed from the image by in-painting. All points on contours of the mask in the image are simply replaced by the arithmetic mean of all points from a 9×9 neighborhood, while points that are part of the mask are not considered. If the neighborhood exceeds the image dimensions, zero padding is used. In the last step of the pre-processing, contrast limited adaptive histogram equalization is performed.

After the NN has inferred a segmentation, the continuous probabilities are binarized using a threshold of 0.5. Afterwards, only the largest connected component is retained in order to remove small artifacts. Because it is known that OD and OC are convex objects, the convex hull of this largest connected component is taken as predicted segmentation.

3.3   Deep Learning

The architecture used for image segmentation is a shrinked version of the U-Net [17]. While the neural network is trained individually for segmentation of disc and cup, the architecture and hyperparameters are identical in both cases and across all databases. Figure 2 shows a design overview of this architecture. Each blue rectangle represents a blob of data being passed from one layer to the next. The numbers in the upper left corner display the number of channels. The sizes of the rectangles give an impression of the spatial dimensions of each blob. For the main part, our implementation is very close to the original implementation. We use 3×3 convolutions, 2×2 max-pooling and ReLU activations for the contract-ing path. In the expansive path we use bilinear interpolation for the upsampling step instead of deconvolution. Throughout the whole network, the number of chan-nels in the feature maps are halved compared to the original implementation. This reduces the number of free parameters and increases trainability of the network architecture.

For the training of this architecture, we use the multinomial logistic loss:

$$E = -\frac{1}{N} \sum_{n=1}^{N} \log\left(\boldsymbol{y}_n \cdot \boldsymbol{l}_n\right), \tag{1}$$

where N is the total number of training pixels, $\boldsymbol{y}_n$ is the output of the last network layer at pixel n, $\boldsymbol{l}_n$ is the corresponding one-hot-encoded label vector and $\cdot$ is the scalar product. We furthermore use an Adam optimizer with standard parameters ($\beta_1 = 0.9$, $\beta_2 = 0.999$, $\epsilon = 10^{-8}$) and a base learning rate $\eta = 5 \cdot 10^{-4}$. To prevent overfitting, L_2 weight regularization with a weight decay of $5 \cdot 10^{-4}$ is used. The training is run for 300 epochs. As it is known that dropout is a powerful regularization method, we also train a variant of our network with 20% dropout in the input layer.

All networks are trained using exactly this hyperparameter set and NN architecture.

4 Results

We evaluate the performance of our trained models using the F_1 score:

$$F_1 = 2 \cdot \frac{PR \cdot RC}{PR + RC} = \frac{2 \cdot TP}{2 \cdot TP + FN + FP} \tag{2}$$

with PR and RC as precision and recall, respectively. In other terms, TP, FN and FP are true positives, false negatives and false positives. All measures are taken pixel-wise across the whole test data set. Afterwards, the F_1 score is calculated on the cumulative values.

As we perform five-fold cross-validation, we generate five models per training database. For every model we calculate the F_1 score on every test data set, which is either the left-out set from cross-validation or another database. If the test data stems from a different database, we test the models on all data sets of this database. Our reported values are the mean and standard deviation across all five trained models.

Table 1. Mean F_1-Score of state-of-the-art methods compared to the best results achieved with our evaluated architecture. While the scores achieved by our model are the results of training on different data sets, the model architecture and training parameters were not changed throughout all training runs.

Method	DRIONS disc	DRISHTI disc	DRISHTI cup	REFUGE disc	REFUGE cup	RIM-ONE disc	RIM-ONE cup
Zilly et al. [24]	-	**0.97**	0.87	-	-	0.94	0.82
Sevastopolsky et al. [18]	**0.96**	**0.97**	**0.89**	-	-	0.95	0.84
Haleem et al. [10]	-	0.95	0.81	-	-	0.91	**0.89**
Wang et al. [22]	-	0.96	0.86	**0.95**	**0.89**	-	-
shrinked U-Net	**0.96**	0.93	0.85	**0.95**	0.87	**0.96**	0.87

Table 2. Mean F_1 scores and standard deviation (small numbers) across all five models trained per database in the cross-validation. All models were evaluated against unseen data sets. These are either part of a different database or part of the left-out cross-validation set. Results of the latter case are shaded gray.

| tested on→ | DRIONS | DRISHTI | | REFUGE | | RIM-ONE | | MEAN | |
↓trained on	disc	disc	cup	disc	cup	disc	cup	disc	cup
DRIONS	0.87.05	0.90.02	-	0.86.01	-	0.73.03	-	0.84	-
w. dropout	0.67.13	0.49.09	-	0.73.06	-	0.14.07	-	0.51	-
DRISHTI	0.76.06	0.93.04	0.85.02	0.89.02	0.73.01	0.79.02	0.60.03	0.84	0.73
w. dropout	0.82.02	0.92.04	0.79.09	0.89.01	0.72.01	0.64.05	0.52.06	0.82	0.68
REFUGE	0.73.06	0.87.04	0.50.04	0.95.00	0.84.02	0.66.03	0.47.02	0.80	0.60
w. dropout	0.56.14	0.67.19	0.47.10	0.94.03	0.87.04	0.37.12	0.43.08	0.64	0.59
RIM-ONE	0.67.07	0.72.05	0.57.08	0.77.02	0.38.14	0.88.02	0.70.05	0.76	0.55
w. dropout	0.70.06	0.80.08	0.44.06	0.78.05	0.59.05	0.96.03	0.75.10	0.81	0.59
MEAN	0.76	0.86	0.64	0.87	0.65	0.77	0.59		
w. dropout	0.69	0.72	0.58	0.84	0.72	0.54	0.57		

Results of our best-performing models are shown in Table 1. It becomes clear that, although it is not the goal of our study to achieve best segmentation performance, the performance of our model including the pre- and post-processing steps is comparable to other methods. More importantly, our shrinked U-Net can be trained with the same hyperparameters across all databases and achieves comparable performance to state-of-the-art methods, tuned specifically for one database.

At the end of our first evaluation step, we have trained specialized models that have seen data from only one database and produce a probability map for either disc or cup. As five-fold cross-validation is applied, we train five models per database and segmentation goal. In our deployment study we apply these modes to different databases. The complete results of this deployment study are shown in Table 2. Large numbers show the mean F_1 score across all five cross-validation models. Smaller numbers show the standard deviation. Diagonal entries show the standard evaluation protocol, where every model is evaluated on the kept-out data sets of cross-validation. All off-diagonal entries represent models evaluated on a different database than what they were trained on. In those cases, the whole database was used for evaluation. Although pre-processing is limiting the influence of database specifics, the scores drop significantly if models are evaluated on a different database. The only exception to that observation are models trained with DRIONS, which perform comparably on REFUGE and even better on DRISHTI if no dropout is used. When observing results of the models using dropout, the effects are less clear. While dropout seems to have a stabilizing effect on DRISHTI models, it has an opposite effect for models trained on the REFUGE database.

Table 3. F_1 scores after transfer learning a model for 10 and 50 epochs after pre-training on DRISHTI for 300 epochs. For reference, the first two rows repeat the scores of the pre-trained model from Table 2. Evaluations on the database used for fine tuning are shaded.

tested on→ ↓trained on	DRIONS disc	DRISHTI disc	cup	REFUGE disc	cup	RIM-ONE disc	cup	MEAN disc	cup
baseline	$0.76_{.06}$	$0.93_{.04}$	$0.85_{.02}$	$0.89_{.02}$	$0.73_{.01}$	$0.79_{.02}$	$0.60_{.03}$	0.84	0.73
w. dropout	$0.82_{.02}$	$0.92_{.04}$	$0.79_{.09}$	$0.89_{.01}$	$0.72_{.01}$	$0.64_{.05}$	$0.52_{.06}$	0.82	0.68
DRIONS									
10 epochs	$0.93_{.04}$	$0.92_{.00}$	-	$0.89_{.00}$	-	$0.75_{.02}$	-	0.87	-
w. dropout	$0.75_{.08}$	$0.71_{.08}$	-	$0.84_{.02}$	-	$0.35_{.12}$	-	0.66	-
50 epochs	$0.96_{.02}$	$0.93_{.01}$	-	$0.89_{.01}$	-	$0.75_{.01}$	-	0.88	-
w. dropout	$0.74_{.08}$	$0.71_{.07}$	-	$0.85_{.02}$	-	$0.32_{.09}$	-	0.66	-
RIM-ONE									
10 epochs	$0.72_{.02}$	$0.84_{.01}$	$0.56_{.09}$	$0.78_{.02}$	$0.65_{.04}$	$0.94_{.01}$	$0.82_{.01}$	0.82	0.68
w. dropout	$0.73_{.03}$	$0.86_{.01}$	$0.60_{.09}$	$0.70_{.02}$	$0.70_{.04}$	$0.91_{.02}$	$0.81_{.04}$	0.80	0.70
50 epochs	$0.70_{.03}$	$0.80_{.03}$	$0.49_{.09}$	$0.73_{.03}$	$0.67_{.05}$	$0.94_{.04}$	$0.87_{.04}$	0.79	0.68
w. dropout	$0.73_{.01}$	$0.84_{.01}$	$0.55_{.09}$	$0.71_{.02}$	$0.66_{.04}$	$0.93_{.01}$	$0.86_{.03}$	0.80	0.69

As models trained on the DRISHTI database show the best overall performance, we do another evaluation step, which involves transfer learning of these models. Out of the five models trained on DRISHTI for each task, we chose the model reaching the median score. We now have four models, each trained for disc or cup segmentation, with or without dropout on DRISHTI respectively. These are now fine-tuned on the DRIONS and RIM-ONE databases, using the same cross-validation folds as before. We train for 10 and 50 more epochs, simulating translation into clinical use, where some more labeled images are shown to the models before deployment. The results of this transfer learning approach can be seen in Table 3. As before, we make sure that evaluation takes place on previously unseen data only, i.e. evaluation is performed on the left-out cross-validation set for DRISHTI and DRIONS or RIM-ONE respectively. The results clearly show that a short transfer learning phase after the simulated deployment increases the performance on the new database by a large margin for models trained without dropout. Transfer learning on models trained with dropout, however, shows no benefits.

Figure 3 illustrates how fine tuning affects individual training samples. The scatter plots show segmentation quality before and after the fine-tuning phase of 50 epochs without dropout for the disc segmentation task. On the left, it is clear that DRISHTI samples performing sub-par before the fine-tuning, profit from the additional training. Training on RIM-ONE causes higher improvement than training on DRIONS. For samples performing well before, fine-tuning generally has an adverse effect, especially for RIM-ONE. The following plots show changes for samples from the databases used for fine-tuning. As expected, almost all samples from the database used for fine-tuning profit from the additional training.

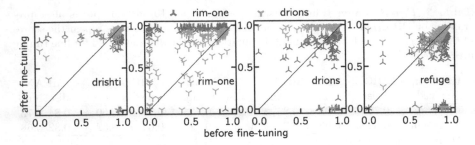

Fig. 3. Change of disc segmentation F_1 score for all individual samples from the different databases after fine tuning for 50 epochs without dropout. The diagonal in every graph shows the improvement boundary. Samples above this diagonal are segmented better after fine-tuning, samples below suffered from degradation. The additional tick on the x-axis marks the average score across all databases after pre-training ($F_1 = 0.84$).

The comparison of these two plots also shows that negative effects on samples from DRIONS are more severe when fine-tuned on RIM-ONE than the other way around. This is supported by the overall statistics shown in Table 3. The reasoning from that table is also supported in the right plot, where fine-tuning on RIM-ONE has more negative effects on the samples from the unseen REFUGE database. Three of the four plots show a tendency of well performing samples degrading drastically. Clusters in the bottom right of the respective sub-plots indicate this. The causes for that are worth a deeper investigation in the future.

5   Discussion and Conclusion

We have shown that after deployment of a DL model, a short period of transfer learning should be applied in order to adopt to slight differences in the input data. In case of the DRIONS database, the model trained with transfer learning produced much better results than the model trained solely on this database. Furthermore, we could not observe any positive effects on generalization when using dropout in the input layer. In some cases, especially when training on the RIM-ONE database, training seems to benefit slightly from dropout. There are also instances, where the use of dropout causes large drops in the overall performance.

We can also draw the conclusion that the REFUGE database, which is by far the largest available to us, is relatively easy to segment and seems to be very homogeneous. First, the mean score across all training databases is the highest for segmentation of OC and OD, which can be seen in the bottom rows of Table 2. Second, models trained with REFUGE only perform relatively bad on all other databases. RIM-ONE, which constantly produces the worst results when models trained on it are evaluated on other databases, is also hard to segment for models trained on other databases. This suggests that many of the data sets included in RIM-ONE differ from the data included in all other databases. On the other hand, DRIONS is also hard to segment, but produces good evaluation results with other databases. This implies coverage of a larger feature space that includes many of the other data sets.

This assumption is also seconded by the results after transfer learning, where the DRISHTI model tuned on DRIONS outperforms the model tuned on RIM-ONE across all databases by a large margin. Tuning on RIM-ONE actually decreases the scores on all databases but itself. This also seconds the idea of a small overlap of the samples contained in RIM-ONE and the others.

In summary, we have shown that great care has to be taken when deploying a ML algorithm into practice. Although cross-validation results might suggest a high generalizability of the trained method, slight differences in the input data might cause a drastic performance drop. We have also shown that fine-tuning the already trained network with examples from the deployment domain quickly leads to good results. Hence, this approach seems not only advisable but necessary. Nevertheless, in our study, fine-tuning only lead to a performance increase for the new data but did not lead to an improvement for any other database. This training stage should hence only be performed on data samples taken on site with the actual hardware setup. Furthermore, we have shown that the relatively new REFUGE database might - despite its size - not be a good choice for a training database of a general OC and OD segmentation pipeline, whereas the DRISHTI database seems to provide the best training baseline.

We can not yet draw conclusions on how to select training samples or databases to achieve a high generalization. A deeper analysis of individual samples and their influence on the performance has to be carried out in the future. We currently do not see why this should not be applicable to other machine learning tasks, but hope for further insights in the future, when those factors are taken into account and are investigated further.

References

1. Carmona, E.J., Rincón, M., García-Feijoó, J., Martínez-de-la-Casa, J.M.: Identification of the optic nerve head with genetic algorithms. Artif. Intell. Med. **43**(3), 243–259 (2008)
2. Chakraborti, T., Jha, D.K., Chowdhury, A.S., Jiang, X.: A self-adaptive matched filter for retinal blood vessel detection. Mach. Vis. Appl. **26**(1), 55–68 (2015)
3. Consortium for Open Medical Image Computing: Refuge - Retinal Fundus Glaucoma Challenge. https://refuge.grand-challenge.org. Accessed 02 Mar 2019
4. Dodge, S.F., Karam, L.J.: A study and comparison of human and deep learning recognition performance under visual distortions. In: Proceedings of International Conference on Computer Communication and Networks, pp. 1–7 (2017)
5. Duan, J., Tench, C.R., Gottlob, I., Proudlock, F., Bai, L.: Automated segmentation of retinal layers from optical coherence tomography images using geodesic distance. Pattern Recogn. **72**, 158–175 (2017)
6. Fischer, J.R., Chamoff, L.: AI: A tool or a replacement for radiologists? (2017). https://www.dotmed.com/news/story/39033. Accessed 02 Mar 2019
7. Fraz, M.M., et al.: Blood vessel segmentation methodologies in retinal images - a survey. Comput. Methods Programs Biomed. **108**(1), 407–433 (2012)

8. Fumero, F., Alayón, S., Sánchez, J.L., Sigut, J.F., González-Hernández, M.: RIM-ONE: an open retinal image database for optic nerve evaluation. In: Proceedings of IEEE International Symposium on Computer-Based Medical Systems, pp. 1–6 (2011)

9. Geirhos, R., Temme, C.R.M., Rauber, J., Schütt, H.H., Bethge, M., Wichmann, F.A.: Generalisation in humans and deep neural networks. In: Proceedings of Advances in Neural Information Processing Systems, pp. 7549–7561 (2018)

10. Haleem, M.S., et al.: A novel adaptive deformable model for automated optic disc and cup segmentation to aid glaucoma diagnosis. J. Med. Syst. **42**(1), 1–20 (2018)

11. He, K., Zhang, X., Ren, S., Sun, J.: Deep residual learning for image recognition. In: Proceedings of IEEE Conference on Computer Vision and Pattern Recognition, pp. 770–778 (2016)

12. Hinton, G.E.: Comment on radiology and deep learning at the 2016 Machine Learning and Market for Intelligence Conference in Toronto (2016). https://www.youtube.com/watch?v=2HMPRXstSvQ. Accessed 02 Mar 2019

13. Kawaguchi, K., Kaelbling, L.P., Bengio, Y.: Generalization in deep learning. CoRR abs/1710.05468 (2017)

14. Litjens, G.J.S., et al.: A survey on deep learning in medical image analysis. Med. Image Anal. **42**, 60–88 (2017)

15. Mahapatra, D., Buhmann, J.M.: A field of experts model for optic cup and disc segmentation from retinal fundus images. In: Proceedings of IEEE International Symposium on Biomedical Imaging, pp. 218–221 (2015)

16. Recht, B., Roelofs, R., Schmidt, L., Shankar, V.: Do CIFAR-10 Classifiers Generalize to CIFAR-10? CoRR abs/1806.00451 (2018)

17. Ronneberger, O., Fischer, P., Brox, T.: U-Net: convolutional networks for biomedical image segmentation. In: Navab, N., Hornegger, J., Wells, W.M., Frangi, A.F. (eds.) MICCAI 2015. LNCS, vol. 9351, pp. 234–241. Springer, Cham (2015). https://doi.org/10.1007/978-3-319-24574-4_28

18. Sevastopolsky, A.: Optic disc and cup segmentation methods for glaucoma detection with modification of U-Net convolutional neural network. Pattern Recogn. Image Anal. **27**(3), 618–624 (2017)

19. Sevastopolsky, A., Drapak, S., Kiselev, K., Snyder, B.M., Georgievskaya, A.: Stack-U-Net: refinement network for image segmentation on the example of optic disc and cup. CoRR abs/1804.11294 (2018)

20. Sivaswamy, J., Krishnadas, S.R., Joshi, G.D., Jain, M., Tabish, A.U.S.: Drishti-GS: retinal image dataset for optic nerve head (ONH) segmentation. In: Proceedings of IEEE International Symposium on Biomedical Imaging, pp. 53–56 (2014)

21. Tham, Y.C., Li, X., Wong, T.Y., Quigley, H.A., Aung, T., Cheng, C.Y.: Global prevalence of glaucoma and projections of glaucoma burden through 2040: a systematic review and meta-analysis. Ophthalmology **121**(11), 2081–2090 (2014)

22. Wang, S., Yu, L., Heng, P.A.: Optic disc and cup segmentation with output space domain adaptation. https://refuge.grand-challenge.org/media/REFUGE/public_html/Proceedings/REFUGE-CUHKMED.pdf. Accessed 02 Mar 2019

23. Zhang, C., Bengio, S., Hardt, M., Recht, B., Vinyals, O.: Understanding deep learning requires rethinking generalization. CoRR abs/1611.03530 (2016)

24. Zilly, J.G., Buhmann, J.M., Mahapatra, D.: Glaucoma detection using entropy sampling and ensemble learning for automatic optic cup and disc segmentation. Comput. Med. Imaging Graph. **55**, 28–41 (2017)

Joint Destriping and Segmentation of OCTA Images

Xiyin Wu[1,2,3(✉)], Dongxu Gao[3], Bryan M. Williams[3], Amira Stylianides[4], Yalin Zheng[3,4], and Zhong Jin[1,2]

[1] School of Computer Science and Engineering,
Nanjing University of Science and Technology, Nanjing 210094, China
xiyinwu1990@gmail.com, zhongjin@njust.edu.cn
[2] Key Laboratory of Intelligent Perception and System for High-Dimensional Information of Ministry of Education, Nanjing University of Science and Technology, Nanjing 210094, China
[3] Department of Eye and Vision Science, Institute of Ageing and Chronic Disease, University of Liverpool, Liverpool L7 8TX, UK
yalin.zheng@liverpool.ac.uk
[4] St. Paul's Eye Unit, Royal Liverpool University Hospital, Liverpool L7 8XP, UK

Abstract. As an innovative retinal imaging technology, optical coherence tomography angiography (OCTA) can resolve and provide important information of fine retinal vessels in a non-invasive and non-contact way. The effective analysis of retinal blood vessels is valuable for the investigation and diagnosis of vascular and vascular-related diseases, for which accurate segmentation is a vital first step. OCTA images are always affected by some stripe noises artifacts, which will impede correct segmentation and should be removed. To address this issue, we present a two-stage strategy for stripe noise removal by image decomposition and segmentation by an active contours approach. We then refine this into a new joint model, which improves the speed of the algorithm while retaining the quality of the segmentation and destriping. We present experimental results on both simulated and real retinal imaging data, demonstrating the effective performance of our new joint model for segmenting vessels from the OCTA images corrupted by stripe noise.

Keywords: Vessels segmentation · Destriping · OCTA

1 Introduction

Retinal blood vessels health is integral to high quality human vision. Changes in the retinal vasculature have a close relationship with many ophthalmological and cardiovascular diseases, such as diabetic retinopathy (DR) and age-related macular degeneration (AMD) [13]. As an important computer-aided image analysis technique, retinal vessel segmentation is the first and most important step to detect the retinal vasculature. The segmentation of retinal vessels is a valuable precursor for further processing and analysis, such as retinal image registration,

© Springer Nature Switzerland AG 2020
Y. Zheng et al. (Eds.): MIUA 2019, CCIS 1065, pp. 423–435, 2020.
https://doi.org/10.1007/978-3-030-39343-4_36

feature extraction and localization of retinal structures, such as the fovea and optic disk [18].

To acquire imagery of fine retinal vessels, the commonly used techniques are fluorescein angiography (FA) and indocyanine green angiography (ICGA) [8]. Both FA and ICGA require intravenous dye injections, which can have adverse side effects and still only provide information relating to superficial blood vessels. Recently, an innovative technology called optical coherence tomography angiography (OCTA) has emerged as an effective way of visualizing the retinal vessels up to capillary level [14]. Retinal vessels at different depths of the retina revealed by OCTA are illustrated in Fig. 1. OCTA generates structural images of the retina based on laser light reflectance on the surface of moving red blood cells. Unlike FA and ICGA, OCTA is a non-invasive, fast, depth-resolved approach. OCTA can fully visualize choroidal neovascularization in AMD and small retinal neovascularization in DR, which are difficult to identify in FA. Therefore, the usage of OCTA in the diagnosis of vascular diseases is expected to increase significantly in the near future. Our work concerns the automated segmentation of retinal blood vessels in OCTA images.

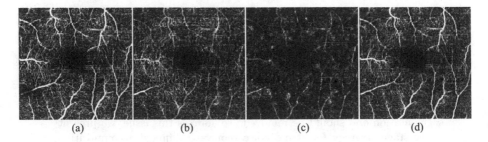

(a)   (b)   (c)   (d)

Fig. 1. OCTA images taken from different retinal layers. (a) Superficial vascular plexus (SVP): from internal limiting membrane to inner plexiform layer. (b) Deep vascular plexus (DVP): from inner plexiform layer to outer plexiform layer. (c) Avascular layer (AL): from outer plexiform layer to Bruch's membrane. (d) Whole retina (WR): from internal limiting membrane to Bruch's membrane.

As a relatively new modality, few studies exist and analyze the retinal vessels in OCTA and work in the OCTA vessel segmentation is still at an early stage in its development. Eladawi et al. [9] presented a joint Markov-Gibbs random field method to segment blood vessels from OCTA scans. The authors further estimated three local features from the segmented vessels to distinguish the status of DR patients [10]. Compared with OCTA, research in retinal vessel segmentation with color fundus images has a longer history. Many methods have been proposed over the last two decades, such as active contour models, wavelets methods, Gaussian mixture models, Adaboost and support vector machine, to name a few. One of the most commonly used segmentation approaches is active contours, such as the Chan-Vese model [5]. Chan-Vese model applied global statistics to the extraction of objects and was useful for objects with homogeneous intensity. To handle the non-uniform situation of the Chan-Vese model, Sum et al. [15] developed a modified version by combining the local image contrast into a level set based active contour. Bashir et al. [1] grew a Ribbon of Twins active contour model to locate vessel edges under different conditions. Zhao et al. [17] segmented retinal vessels by developing an infinite perimeter active counter model with hybrid region information.

There are two limitations in the mentioned literatures in contributing to clinical diagnostics. Firstly, previous work on OCTA images have neglected the unavoidable noise problem. Additional image noise can arise from the OCTA image acquisition, eye motion and image pre-processing strategies. One of the most common types of noise is white horizontal stripe noise (see Fig. 1), which results from patient eye motion [12]. This problem has to be tackled before segmenting the retinal vessels to achieve effective, good-quality segmentation. Secondly, color fundus images can not provide depth information, limiting the analysis of choroidal neovascularization and small retinal neovascularization.

To overcome the above limitations, we first present a two-stage strategy for stripe noise removal and retinal vessel segmentation in OCTA images. This strategy removes stripe noise and Guassian noise from the image before segmenting the denoised image. Then we propose a joint model which achieves the two tasks simultaneously. Although there is no literature about removing stripe noise in OCTA images, remote sensing images suffer from similar noise and lots of related works have been explored. Image decomposition based methods have been considered for remote sensing image destriping in recent years and achieve good results [7]. Thus, image decomposition theory is chosen for the OCTA destriping problem. For vessel segmentation, a good choice is active contour models, which can not only provide smooth and closed contours but also achieve subpixel accuracy on vessel boundaries.

2   Methodology

2.1   Two-Stage Strategy for Segmenting OCTA Images Corrupted by Stripe Noise

A two-stage strategy is designed for segmenting OCTA images affected by stripe noise. In this strategy, we first remove the stripe noise and subsequently segment the retinal vessels.

Stripe Noise Removal. Considering the stripe acquired along with the image content, we proceed with an image decomposition strategy, i.e. the given noisy image is decomposed to the desired clean image, the stripe noise and Guassian white noise. The model of the observed image can be formulated as:

$$Y = I + S + N \tag{1}$$

where Y is the given noisy image; I denotes the desired clean image; S is the additive stripe component; and N represents the linear assumption error and Gaussian white noise. Stripe noise removal aims to estimate both I and S simultaneously from Y.

As shown in Fig. 1, the stripe noise has a salient structural characteristic, showing only in the horizontal direction. This is characterized by small rank for both periodical and nonperiodical stripes based on the analysis of [7]. Thus, the low-rank constraint is used for the stripe component. For the clean image, the anisotropic total variation (TV) regularization is used to achieve sharper boundaries in the reconstructed image, which will be important for segmenting vessels accurately. The image decomposition model is given as follows:

$$\min_{I,S} \frac{1}{2}\|I + S - Y\|_F^2 + \tau\|I\|_{\text{TV}} + \lambda\text{rank}(S) \tag{2}$$

where $\|I\|_{\text{TV}} = \int_\Omega |\nabla I| dx$, τ and λ are two positive regularization parameters balancing the three terms. Ω is an open set representing the image domain, and $\nabla = (\nabla_x; \nabla_y)$ is the horizontal and vertical derivative operators of I, respectively. Due to the non-convexity of the rank constraint, the nuclear norm is introduced to replace the low-rank constraint as its convex substituted function [11]. The nuclear norm based image decomposition model is given by

$$\min_{I,S} \frac{1}{2}\|I + S - Y\|_F^2 + \tau\|I\|_{\text{TV}} + \lambda\|S\|_*. \tag{3}$$

Segmentation of Vessels from Stripe Denoised Images. Active contour models have demonstrated excellent performance in dealing with challenging segmentation problems including vessel segmentation [1,17]. The Global Minimization of the Active Contour/Snake model (GMAC) [3] is introduced here for vessel segmentation. This model provides a convex solution for the well-known Chan-Vese (CV) model [5] and incorporates information from an edge detector. The GMAC model can be formulated as the energy minimization problem below:

$$\min_{u,c_1,c_2} \int_\Omega g(x,y)|\nabla u(x,y)|dxdy + \beta \int_\Omega u(x,y)(I(x,y) - c_1)^2 dxdy$$
$$+ \beta \int_\Omega (1 - u(x,y))(I(x,y) - c_2)^2 dxdy \tag{4}$$

where u is a characteristic function of a closed set Ω_C, C represents the boundaries of Ω, c_1 and c_2 are the average of $I(x,y)$ inside and outside of the segmented region respectively, and β is a small, positive balancing parameter. The first term is the TV-norm of u with weighted edge indicator function $g(x,y)$.

Thus, the vessel segmentation from stripe removed images is achieved by solving the problem (3) to obtain the stripe removed and denoised image I, followed segmenting I by solving the problem (4). The solvers of these two problems will be described later.

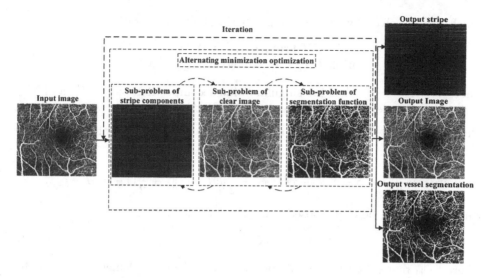

Fig. 2. Illustration of our joint model for segmentation images with stripe noise.

2.2 A New Joint Model for Segmenting Images Corrupted by Stripe Noise

In this section, we develop a joint model for simultaneously removing stripe noise and segmenting images. The model tackles the problems of image segmentation and considers the possible presence of stripe noise in a single model. The new joint model is formed by replacing the function I in the segmentation model (4) with the stripe-removed and denoised image term. Furthermore, the constraints of the decomposition problem (3) should be included in the joint model. The illustration of our proposed joint model is shown in Fig. 2. Building the constraints into the segmentation model, the new joint minimisation model is presented as follows:

$$\min_{I,S,u,c_1,c_2} \frac{1}{2}\|I + S - Y\|_F^2 + \tau\|I\|_{\mathrm{TV}} + \lambda\|S\|_* + \alpha\|u\|_{\mathrm{TV},g} + \frac{\beta}{2}u\|I - c_1\|_F^2$$
$$+ \frac{\beta}{2}(1 - u)\|I - c_2\|_F^2 \tag{5}$$

where $\|\cdot\|_{\mathrm{TV},g}$ is the TV-norm with weighted function, α, β are small, positive balancing parameters. If α and β equal to 0, then the model (5) becomes the destriping model (3). The segmentation terms (fourth, fifth and sixth terms of (5)) obtain the intensity and spatial information from the stripe removed image I. To solve the joint model, each of the arguments is solved in turn. The detailed optimization process is presented as follows.

Stripe Component S. Fixing other arguments, S is evaluated by minimising the following function:

$$\hat{S} = \arg\min_S \frac{1}{2}\|I + S - Y\|_F^2 + \lambda\|S\|_* \tag{6}$$

which can be solved by the following soft-thresholding operation:

$$\begin{cases} S^{k+1} = U(\text{shrink\_L}_*(\sum, \lambda))V^{\mathrm{T}} \\ \text{shrink\_L}_*(\sum, \lambda) = \text{diag}\{\max(\sum_{ii} - \lambda, 0)\}_i \end{cases} \quad (7)$$

where $Y - I = U \sum V^{\mathrm{T}}$ is the singular value decomposition of $Y - I^k$, and $\sum_{ii}$ is the diagonal element of the singular value matrix $\sum$.

Clean Image I. Keeping other arguments fixed and minimising with respect to I, we have

$$\hat{I} = \arg\min_I \frac{1}{2}\|I + S - Y\|_F^2 + \tau\|I\|_{\mathrm{TV}} + \frac{\beta}{2}u\|I - c_1\|_F^2 + \frac{\beta}{2}(1-u)\|I - c_2\|_F^2 \quad (8)$$

where $\|I\|_{\mathrm{TV}}$ can be decomposed along the directions x and y:

$$\|I\|_{\mathrm{TV}} = \tau_x\|\nabla_x I\|_1 + \tau_y\|\nabla_y I\|_1 \quad (9)$$

where $\|\cdot\|_1$ represents the sum of absolute value of matrix elements. The weight τ along directions x and y is different due to the directional characteristics of the stripe component. The alternative direction multiplier method (ADMM) is introduced to solve the problem (8) efficiently. We convert this problem into two sub-problems. Let $D_x = \nabla_x I$, $D_y = \nabla_y I$, $D = [D_x, D_y]^{\mathrm{T}}$, $\nabla = [\nabla_x, \nabla_y]^{\mathrm{T}}$, $\tau = [\tau_x, \tau_y]^{\mathrm{T}}$, (8) equals to the following problem:

$$\begin{aligned} \{\hat{I}, \hat{D}\} = \arg\min_{I, D_x, D_y} &\frac{1}{2}\|I + S - Y\|_F^2 + \tau\|D\|_1 + \frac{\beta}{2}u\|I - c_1\|_F^2 \\ &+ \frac{\beta}{2}(1-u)\|I - c_2\|_F^2 + \frac{\gamma}{2}\|D - \nabla I - \frac{J}{\gamma}\|_F^2. \end{aligned} \quad (10)$$

The I-related sub-problem is followed by

$$\begin{aligned} \hat{I} = \arg\min_{I, D_x, D_y} &\frac{1}{2}\|I + S - Y\|_F^2 + \frac{\beta}{2}u\|I - c_1\|_F^2 \\ &+ \frac{\beta}{2}(1-u)\|I - c_2\|_F^2 + \frac{\gamma}{2}\|D - \nabla I - \frac{J}{\gamma}\|_F^2. \end{aligned} \quad (11)$$

We can obtain a closed form of (11) via fast 2-D Fourier transform (FFT):

$$I^{k+1} = \mathcal{F}^{-1}\left(\frac{\mathcal{F}(Y - S^{k+1} + \beta u^k c_1^k + \beta(1 - u^k)c_2^k) + \nabla^{\mathrm{T}}(\gamma^k D^k - J^k)}{1 + \beta + \gamma^k(\mathcal{F}(\nabla))^2}\right) \quad (12)$$

The D-related sub-problem is followed by

$$\hat{D} = \arg\min_D \tau\|D\|_1 + \frac{\gamma}{2}\|D - \nabla I - \frac{J}{\gamma}\|_F^2 \quad (13)$$

which can be solved by a soft shrinkage operator

$$\begin{cases} D^{k+1} = \text{shrink\_L}_1(\nabla I^{k+1} + \frac{J^k}{\gamma^k}, \frac{\tau}{\gamma^k}) \\ \text{shrink\_L}_1(r, \xi) = \frac{r}{|r|} * \max(r - \xi, 0). \end{cases} \quad (14)$$

Region Average Intensity Values c_1, c_2. Keeping other arguments fixed and minimising with respected c_1 and c_2, respectively. We have the equations for c_1 and c_2

$$c_1 = \frac{\int_\Omega u(x,y)I(x,y)dxdy}{\int_\Omega u(x,y)dxdy}, c_2 = \frac{\int_\Omega (1-u(x,y))I(x,y)dxdy}{\int_\Omega (1-u(x,y))dxdy} \qquad (15)$$

which can be evaluated directly by giving the average intensities inside and outside of the segmentation contour. c_1 and c_2 in GMAC can be solved in a same way.

Segmentation Indicator Function u. Minimising model (5) with respecting to u and fixing the other arguments, we have

$$\hat{u} = \arg\min_u \alpha\|u\|_{TV,g} + \frac{\beta}{2}u\|I - c_1\|_F^2 + \frac{\beta}{2}(1-u)\|I - c_2\|_F^2. \qquad (16)$$

A convex regularization variational model is used according to [2]:

$$\{\hat{u},\hat{v}\} = \arg\min_{u,v} \alpha\|u\|_{TV,g} + \frac{1}{2\theta}\|u + v - I\|_F^2$$
$$+ \frac{\beta}{2}(I-v)\|I - c_1\|_F^2 + \frac{\beta}{2}(1 - I + v)\|I - c_2\|_F^2 \qquad (17)$$

where the parameter θ is small so that we can approximate $I = u + v$. The function u denotes the geometric information, while v represents the texture information in the clean image I. Problem (17) can be solved by minimizing u and v iteratively.

The u-related sub-problem is followed by

$$\hat{u} = \arg\min_u \alpha\|u\|_{TV,g} + \frac{1}{2\theta}\|u + v - I\|_F^2. \qquad (18)$$

The solution of (18) can be achieved efficiently by a fast dual projection algorithm. The derived solution is:

$$u^{k+1} = v^k - \alpha\theta\mathrm{div}p^{k+1} \qquad (19)$$

where p is given by a fixed point method as follows:

$$p^{k+1} = \frac{p^k + \delta t\nabla(\mathrm{div}(p^k) - v/(\alpha\theta))}{1 + \frac{\delta t}{g(x,y)}|\nabla(\mathrm{div}(p^k) - v/(\alpha\theta))|}. \qquad (20)$$

where $\delta t \leq 1/8$ is the temporal step.

The v-related sub-problem is followed by

$$\hat{v} = \arg\min_v \frac{1}{2\theta}\|u + v - I\|_F^2 + \frac{\beta}{2}(I-v)\|I - c_1\|_F^2 + \frac{\beta}{2}(1 - I + v)\|I - c_2\|_F^2. \qquad (21)$$

The v-minimization can be achieved through the following update:

$$v^{k+1} = \min\{\max\{u(x,y) - \frac{\beta\theta}{2}[(I(x,y) - c_2)^2 - (I(x,y) - c_1)^2], 0\}, 1\}. \qquad (22)$$

u in GMAC can be solved in a similar way. The overall algorithm is presented in Algorithm 1.

Algorithm 1. Segmenting Images Corrupted by Stripe Noises by the Joint Model

---

Require: Image Y with stripe noise, parameters $\tau, \lambda, \alpha, \beta, \gamma, \theta, \rho$.
Ensure: Clean image I, stripe component S and segmentation indicator funcion u.
1: **Initialize:** set $J^1 = 0, I^1 = Y, u^1 = 0, v^1 = 0$
2: **for** $k = 1 : N$ **do**
3:     Compute S^{k+1} by solving (7)
4:     Compute I^{k+1} by solving (12)
5:     Compute D^{k+1} by solving (14)
6:     Compute c_1^{k+1} and c_2^{k+1} by solving (15)
7:     Compute u^{k+1} by solving (19)
8:     Compute v^{k+1} by solving (22)
9:     Update Lagrangian multipliers $J^{k+1} = J^k + \gamma^k(\nabla I^{k+1} - S^{k+1})$
10:    Update penalization parameters $\gamma^{k+1} = \gamma^k \cdot \rho$
11: **end for**

---

3   Results and Discussion

The proposed algorithms are evaluated in two parts: the effectiveness of destriping and the effectiveness of segmentation. All experiments are run in MATLAB (R2018a) (Mathworks, MA) on a desktop with 8GB RAM, Intel (R) Core (TM) i5-7500 CPU @ 3.40 GHz. We denote the models to be compared and tested as follows:

D1: The weighted median filter denoising method [4].
D2: The wavelet denoising method [6].
N1: The two-stage strategy presented in this paper by destriping model (3) followed by segmentation model (4).
N2: The proposed joint model (5) for destriping and segmentation simultaneously.

3.1   Effectiveness of Destriping

In the experiment of stripe noise removal, simulated data and real data are employed to compare the performance of our methods (N1 and N2) with two denoising methods (D1 and D2). Two evaluation metrics are utilized. i.e. the peak signal-to-noise ratio (PSNR) and the structural similarity (SSIM) [16]. The two metrics are calculated by using Eqs. (23) and (24), respectively.

$$PSNR = 10 \cdot \log_{10}\left(\frac{MAX_I^2}{MSE}\right) \tag{23}$$

$$SSIM(x, y) = \frac{(2\mu_x\mu_y + c_1)(2\sigma_{xy} + c_2)}{(\mu_x^2 + \mu_y^2 + c_1)(\sigma_x^2 + \sigma_y^2 + c_2)} \tag{24}$$

where MAX_I is the maximum pixel value of the noise-free image and MSE is the mean squared error between noise-free image and noisy image. x and y are two windows of an image. $\mu_x, \mu_y, \sigma_x^2, \sigma_y^2$ are the average of x, y and the variance of x, y, respectively. σ_{xy} is the covariance of x and y. c_1, c_2 are two variables.

Simulated Experiments. In simulated experiments, stripe noise is added into an FA image. As shown in Fig. 3(b), the destriping results of four methods are shown in Fig. 3 (c)–(f). Both D1 and D2 fail to remove the stripe noise. The proposed N1 and N2 can remove the stripes and retain stripe-free areas well.

The highest PSNR and SSIM values in each intensity scenario of simulated experiments are marked in bold in Table 1. Five different levels of stripes are added to the original image to test the robustness of our methods. The intensity denotes the mean absolute value of the stripe lines. Our joint model N2 achieves the best performance according to Table 1. As the stripe level increase, the advantage of our methods N1 and N2 over other methods becomes clearer.

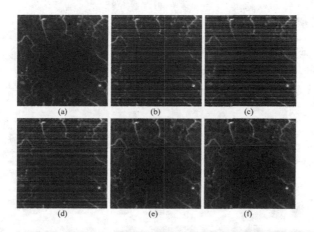

Fig. 3. Illustration of simulated destriping results. (a) Original image. (b) Image with added stripe noise. (c) Result of D1. (d) Result of D2. (e) Result of N1. (f) Result of N2.

Table 1. Quantitative results (PSNR (dB) and SSIM values) on simulated data.

Method	Stripe Noise									
	Intensity=10		Intensity=20		Intensity=30		Intensity=40		Intensity=50	
	PSNR	SSIM	PSNR	SSIM	PSNR	SSIM	PSNR	SSIM	PSNR	SSIM
Degraded	26.65	0.636	21.24	0.360	18.29	0.230	16.33	0.161	14.90	0.120
D1	26.78	0.624	21.27	0.377	18.15	0.254	16.09	0.187	14.60	0.147
D2	29.81	0.763	22.99	0.447	19.35	0.266	16.99	0.177	15.42	0.128
N1	34.99	0.945	33.81	**0.942**	31.31	0.923	28.61	**0.896**	26.03	0.858
N2	**35.00**	**0.947**	**33.82**	**0.942**	**31.94**	**0.928**	**28.66**	**0.896**	**26.11**	**0.859**

Real Experiments. The real OCTA stripe images are tested in this subsection. The proposed method is evaluated on 30 images collected from the Royal Liverpool University Hospital. Each image is taken in a 3 mm * 3 mm field of view centered on the fovea from the internal limiting membrane (ILM) to the inner plexiform layer (IPL). We choose a representative image (see Fig. 4) and zoom in two regions of this image, i.e. R1 and R2. It is shown that our proposed method N1 and N2 can remove most stripe noise, while D1 and D2 can just remove a little stripe noise.

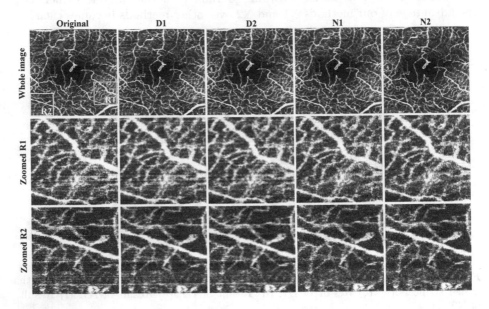

Fig. 4. Illustration of real OCTA destriping results. The first column shows the original OCTA image. The destriping results of D1, D2, N1 and N2 are shown from the second to fifth columns, respectively. The second and third rows show the zoomed regions R1 and R2 of the whole image, respectively.

3.2   Effectiveness of Segmentation

All images of the OCTA dataset mentioned in the real experiments (Sect. 3.1) are evaluated. The quantitative evaluation metric is accuracy, which is calculated as follows [17]:

$$Accuracy = \frac{TP + TN}{TP + FP + TN + FN} \tag{25}$$

where TP, FP, TN, FN indicate the true positive, false positive, true negative and false negative, respectively.

The computation of accuracy is based on the center line of the segmentation results and the center line annotation (see Fig. 5(d)). The average accuracy values and CPU times of N1 and N2 are listed in Table 2. Compared to the two-stage strategy N1, our proposed joint model N2 can achieve higher accuracy in less time. We also give two segmentation examples in Fig. 5. N2 can generate competitive segmentation results to N1 efficiently.

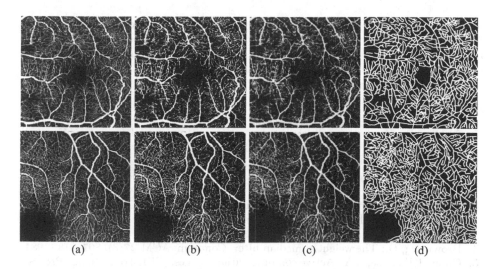

    (a)   (b)   (c)   (d)

Fig. 5. Illustration of OCTA segmentation results. (a) Original image. (b) Segmentation result of N1. (c) Segmentation result of N2. (d) Center line annotation.

Table 2. Average accuracy values and cpu times of OCTA images segmentation.

Method	Accuracy	Times (s)
N1	0.9204	97.63
N2	0.9356	67.45

4   Conclusion

This paper presented a two-stage strategy and proposed a new joint model for segmenting retinal blood vessels in OCTA images corrupted by stripe noise. The two-stage strategy removed the stripe noise by a low-rank representation model, and then segmented the retinal blood vessels by an active contour model called GMAC. The joint model combined the models of stripe noise removal and vessel segmentation. These models solved the two problems efficiently, with the joint model solving the problems simultaneously, faster and more accurately.

We tested the performance of our methods on an OCTA dataset from the Royal Liverpool University Hospital. The quantitative and efficiency comparison results showed that our proposed joint model provided excellent vessel segmentation results. It is believed that our work can inspire a way to consider the inclusion of stripe removal and clinical vessel analysis models.

Acknowledgments. This work is partially supported by National Natural Science Foundation of China under Grant nos. 61872188, U1713208, 61602244, 61672287, 61702262, 61773215, and Postgraduate Research & Practice Innovation Program of Jiangsu Province under grant no. KYCX18_0426.

References

1. Al-Diri, B., Hunter, A., Steel, D.: An active contour model for segmenting and measuring retinal vessels. IEEE Trans. Med. Imaging **28**(9), 1488–1497 (2009)
2. Aujol, J.F., Gilboa, G., Chan, T., Osher, S.: Structure-texture image decomposition-modeling, algorithms, and parameter selection. Int. J. Comput. Vision **67**(1), 111–136 (2006)
3. Bresson, X., Esedoḡlu, S., Vandergheynst, P., Thiran, J.P., Osher, S.: Fast global minimization of the active contour/snake model. J. Math. Imaging Vis. **28**(2), 151–167 (2007)
4. Brownrigg, D.: The weighted median filter. Commun. ACM **27**(8), 807–818 (1984)
5. Chan, T.F., Vese, L.A.: Active contours without edges. IEEE Trans. Image Process. **10**(2), 266–277 (2001)
6. Chang, S.G., Yu, B., Vetterli, M.: Adaptive wavelet thresholding for image denoising and compression. IEEE Trans. Image Process. **9**(9), 1532–1546 (2000)
7. Chang, Y., Yan, L., Wu, T., Zhong, S.: Remote sensing image stripe noise removal: From image decomposition perspective. IEEE Trans. Geosci. Remote Sens. **54**(12), 7018–7031 (2016)
8. De Carlo, T.E., Romano, A., Waheed, N.K., Duker, J.S.: A review of optical coherence tomography angiography (octa). Int. J. Retina Vitreous **1**(1), 5 (2015)
9. Eladawi, N., Elmogy, M., Helmy, O., Aboelfetouh, A., Riad, A., Sandhu, H., Schaal, S., El-Baz, A.: Automatic blood vessels segmentation based on different retinal maps from octa scans. Comput. Biol. Med. **89**, 150–161 (2017)
10. Eladawi, N., et al.: Early diabetic retinopathy diagnosis based on local retinal blood vessel analysis in optical coherence tomography angiography (octa) images. Med. Phys. **45**(10), 4582–4599 (2018)
11. Fazel, M.: Matrix rank minimization with applications. Ph.D. thesis, Stanford University (2002)
12. Moult, E., et al.: Ultrahigh-speed swept-source oct angiography in exudative amd. Ophthalmic Surg. Lasers Imaging Retina **45**(6), 496–505 (2014)
13. Sheng, B., et al.: Retinal vessel segmentation using minimum spanning superpixel tree detector. IEEE Trans. Cybern. **99**, 1–13 (2018)
14. Spaide, R.F., Fujimoto, J.G., Waheed, N.K., Sadda, S.R., Staurenghi, G.: Optical coherence tomography angiography. Progress Retinal Eye Res. **64**, 1–55 (2018)
15. Sum, K., Cheung, P.Y.: Vessel extraction under non-uniform illumination: a level set approach. IEEE Trans. Biomed. Eng. **55**(1), 358–360 (2008)
16. Wang, Z., Bovik, A.C., Sheikh, H.R., Simoncelli, E.P., et al.: Image quality assessment: from error visibility to structural similarity. IEEE Trans. Image Process. **13**(4), 600–612 (2004)

17. Zhao, Y., Rada, L., Chen, K., Harding, S.P., Zheng, Y.: Automated vessel segmentation using infinite perimeter active contour model with hybrid region information with application to retinal images. IEEE Trans. Med. Imaging **34**(9), 1797–1807 (2015)
18. Zhu, C., et al.: Retinal vessel segmentation in colour fundus images using extreme learning machine. Comput. Med. Imaging Graph. **55**, 68–77 (2017)

Posters

AI-Based Method for Detecting Retinal Haemorrhage in Eyes with Malarial Retinopathy

Xu Chen[1(✉)], Melissa Leak[2], Simon P. Harding[1,3], and Yalin Zheng[1,3]

[1] Department of Eye and Vision Science, Institute of Ageing and Chronic Disease,
University of Liverpool, Liverpool L7 8TX, UK
{xuchen,yzheng}@liverpool.ac.uk
[2] School of Life Sciences, University of Liverpool, Liverpool L69 7ZB, UK
[3] St Paul's Eye Unit, Royal Liverpool University Hospital, Liverpool L7 8XP, UK
http://www.liv-cria.co.uk

Abstract. Cerebral Malaria (CM) as one of the most common and severe diseases in sub-Saharan Africa, claimed the lives of more than 435,000 people each year. Because Malarial Retinopathy (MR) is as one of the best clinical diagnostic indicators of CM, it may be essential to analysing MR in fundus images for assisting the CM diagnosis as an applicable solution in developing countries. Image segmentation is an essential topic in medical imaging analysis and is widely developed and improved for clinic study. In this paper, we aim to develop an automatic and fast approach to detect/segment MR haemorrhages in colour fundus images. We introduce a deep learning-based haemorrhages detection of MR inspired by Dense-Net based network called one-hundred-layers tiramisu for the segmentation tasks. We evaluate our approach on one MR dataset of 259 annotated colour fundus images. For keeping the originality of raw MR colour fundus images, 6,098 sub-images are extracted and split into a training set (70%), a validation set (10%) and a testing set (20%). After implementation, our experimental results testing on 1,669 annotated sub-images, show that the proposed method outperforms commonly mainstream network architecture U-Net.

Keywords: Malarial Retinopathy · Retinal haemorrhage · Fundus imaging · Deep learning · Image segmentation

1 Introduction

Cerebral Malaria (CM), as a form of severe malaria, is one of the most common diseases in sub-Saharan Africa, which is transmitted by the Anopheles mosquito when it feeds on previously infected humans. According to the World Malaria Report 2018 [20], malaria affected over 219 million cases of malaria in 2017 and claimed the lives of more than 435,000 people per year. 266,000 (61%) malaria deaths were children aged under 5 years. Retinopathy is clinically significant in

© Springer Nature Switzerland AG 2020
Y. Zheng et al. (Eds.): MIUA 2019, CCIS 1065, pp. 439–449, 2020.
https://doi.org/10.1007/978-3-030-39343-4_37

CM because the similarity of eye and brain embryonic origins have led to shared features of the relevant microvascular systems [18]. Lewallen *et al.* [15] coined the term malarial retinopathy (MR), which is one of the best clinical diagnostic indicators of CM. MR is characterized by retinal whitening, papilledema, capillary non-perfusion (CNP) and varying haemorrhage presence in the ocular fundus [2]. Therefore, screening MR may assist in improving the accuracy of diagnosis of CM by using binocular indirect ophthalmoscopy (BIO) or analysing fundus images by ophthalmic expert team [4,19]. In developing countries, however, there is a barrier to use BIO widely because of expensive equipment and technical expertise required. A fully automated analysing fundus images for MR may provide fast diagnostic assistance. MR detection methods for CM diagnosis using retinal colour images were proposed [1,8,9].

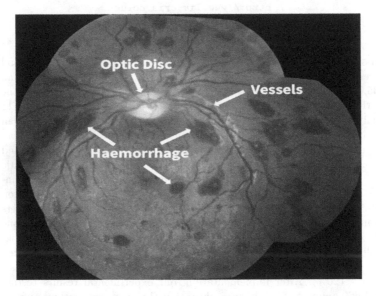

Fig. 1. An example of a montaged colour fundus image showing the presence of haemorrhages

As retinal haemorrhages are one of the visible signs on colour fundus images of MR as well as diabetic retinopathy (DR), retinal haemorrhages detection is of high importance in both MR and DR detection steps for further automated diagnosis [17]. Figure 1 shows an example of a colour fundus image showing the presence of haemorrhages of MR. In the previous works of retinal haemorrhages detection for DR diagnosis, approaches based on splat feature classification [23] and mathematical morphology [10] were proposed. For MR haemorrhages detection, Joshi *et al.* [9] developed an automated segmentation method based on splat classification for the detection of retinal haemorrhages to characterize MR. However, this work has a few limitations. the performance and running time of haemorrhages segmentation are sensitive to the size of splats so this study may not be robust for other various retinal haemorrhages case. A new deep learning network based on Dense-Net called one-hundred-layers tiramisu (Tiramisu) was proposed for medical image segmentation by *Jégou et al.* [7]. Tiramisu model overcomes this limitation of U-Net in various medical image applications [7,16].

To the best knowledge of the authors, this is the first work that applied Deep Learning-based methods on MR. We use Tiramisu method to segment haemorrhages within colour fundus images. In this paper, we make the following contributions: (i) We develop a deep learning-based framework for segmentation in colour fundus photography. (ii) Based on this, we implement the Tiramisu model for automatic MR related haemorrhages detection and compare with general U-Net. (iii) We evaluate our segmentation performance on a supervised biomedical segmentation setting. The remaining of the paper is organised as follows. Section 2 describes the proposed method. Section 3 provides a description of the experimental settings. In Section 4, experimental results are presented and compared to existing methods. Section 5 & Section 6 gives discussion, future directions and conclusions.

2   Methodology

For many challenging tasks in computer vision, Convolutional Neural Networks (CNNs), as a branch of deep neural networks, show remarkable performance, which is able to work by extracting hierarchical features for learning. Decades later, since the power of graphical computing in the hardware field was increased through the improvements in graphical processor units (GPUs), CNNs-based methods are viable for more complex computer vision tasks, *e.g*, object segmentation, scene reconstruction, motion estimation, image restoration etc. [13]. There are many different CNNs architectures, such as LeNet-5 [14] proposed in 1995, Alex-Net [12], VGG-Net [22], ResNet [5] and Dense-Net [6] have been developed and introduced into various computer vision tasks. In terms of image segmentation, since CNNs-based models achieved state-of-the-art results, the image segmentation problem was transformed and solved as a problem of pixelwise classification. In which, each pixel from images will be treated as single independent objects for classification [3]. On the other hand, such as U-Net proposed by Ronneberger *et al.* [21] treats each image as the input and the output of neural network model as an end-to-end fashion and the performance is better when compared to pixel-wised approaches [7,16,24]. However, there are a few limitations in this image-wise method. Small objects would be segmented wrongly by U-Net framework around target annotations, because of the lack of consideration on the outside region of the object. In order to tackle this limitation, Tiramisu model makes each layer to connect with others in a feedforward fashion for encouraging extracted features to be reused efficiently as well as strengthening feature propagation for ignoring gradually the influence from outside of targets.

2.1 CNNs Architectures

In this subsection, the architectures of U-Net and Dense-Net based Tiramisu will be described respectively.

U-Net: U-Net was proposed and widely for semantic segmentation with high precise results. There are two essential paths: the down-sampling path and up-sampling path composing U-Net main architecture. In the down-sampling path of U-Net, the architecture is as similar as a typical CNNs in which each layer consists of two convolution layers with kernel size 3×3, rectified linear unit (ReLU) and pooling layers. In the other path, each layer consists of a 2×2 up-convolution layer, one concatenation operation with related feature map by skipped connections and two convolution layers with kernel size 3×3. As one of the important components in the U-Net architecture, skipped connections are designed for forwarding feature maps from down-sampling path to the up-sampling path to localize high resolution features for generating the segmentation result as final output. At the end of the last layer, a softmax classifier is attached to provide the probability distribution for each pixel. In total, the architecture of U-Net has 23 layers.

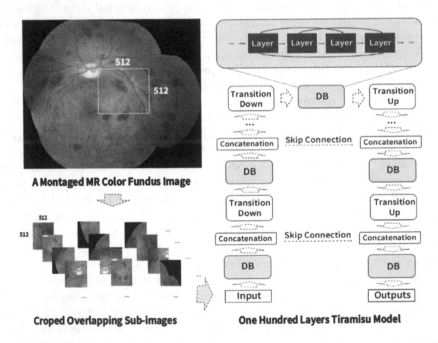

Fig. 2. Overview of our proposed method which takes the object's area and length of the boundaries into account during training.

Tiramisu: Compared to U-Net architecture, the most difference in the Dense-Net based Tiramisu architecture are: (1) Dense Block (DB) including Batch Normalization (BN) layer, ReLU, convolution layer with kernel size 3 × 3 and dropout layer with probability $p = 0.2$; (2) Transition down composed of BN, ReLU, 3 × 3 convolution and dropout with $p = 0.2$; (3) Transition up including one transposed convolution with kernel size 3 × 3. In each DB, every layer is connected densely with other layers. The benefits of introducing DB is for preserving the feed-forward nature and reusing extracted features more efficiently and effectively. 5 Transition Down and 5 Transition Up are introduced following DB layers. Skipped connections and concatenation operations are used as well. At the end of the last layer in Tiramisu, softmax is attached as a classification layer for generating the final probability distribution. Overall, there are 38 layers in the down-sampling path and up-sampling path, respectively. The bottleneck path consists of 15 layers. Tiramisu network has 103 layers totally including Transition Down and Transition Up blocks. In Fig. 2, Tiramisu architecture is presented.

2.2 Loss Functions

In order to train a CNNs-based model for achieving prediction with high accuracy, loss function could play an essential role in the machine learning field. Loss function is a measure of prediction which can be back-propagated to the previous layers for updating and optimising the weights of CNNs. As one of the most commonly used loss functions, Cross-Entropy (CE) Loss is a pixel-wise measure function. CE loss is expressed in the following Eq. 1:

$$loss_{CE}(T, P) = -\sum_{i=1}^{\Omega} [T_i \cdot \log(P_i) + (1 - T_i) \cdot \log(1 - P_i)] \tag{1}$$

where, ground truth image and the prediction are denoted as T, $P \in [0, 1]$ respectively; i indexes each pixel value in image spatial space Ω.

3 Experiments

We used Tiramisu as our MR haemorrhages segmentation framework with loss function CE. The performance will be compared with U-Net framework.

3.1 Dataset

We demonstrate our model on a sample of 259 montaged MR colour fundus images from a sample of children suffering from CM that had been admitted to the Paediatric Research Ward, Queen Elizabeth Central Hospital Malawi.

Various patients who had met the WHO criteria for CM were involved, that were sampled for haemorrhage analysis. The original fundus images were captured with the use of a Nikon-E1 digital camera and produced a field view of 50°. These images were derived from two groups of patients labelled died vs survived, and consist of one image per patient. After approval by the local ethics committees at the University of Malawi College of Medicine and the University of Liverpool, Nick Beare collected the representative images for the further CM analysis. The mentioned image montages were formed by stitching together multiple original images of the same eye to visualise a larger area of the retina as shown in Fig. 1. The corresponding ground truth label maps (haemorrhage regions and background) are annotated by Melissa Leak.

3.2   Data Preprocessing

In order to keep the originality of raw colour fundus image as well as increase the number of MR fundus images for training and validation, all 259 fundus images and the corresponding annotation are cropped into about 23 overlapped sub-images/patches for each image with relatively smaller sizes: 512 × 512 as a new dataset with 6098 sub-images. Example images and their corresponding ground truth are shown in the first left columns of Fig. 3. In our experiment, the new dataset is partitioned into three subsets: training (3,543), validation (886) and testing (1,669). And then, all of the 1,669 sub-images will be recombined back to the original image size for a better demonstration.

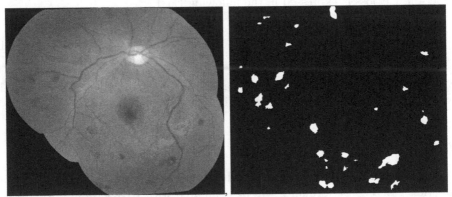

(a) An Original Image (2924 × 2623 × 3 pixels)   (b) MR Related Haemorrhage Annotation

Fig. 3. An example of MR colour fundus images and the corresponding annotation of haemorrhage

3.3   Implementation

We implemented our networks using Keras-gpu 2.0.5 and Tensorflow-gpu 1.7 as backend. We trained our models with Adam optimizer [11] by learning rate of 10^{-5}. All the experiments were performed by using an NVIDIA K40 GPU and 32 GB memory.

3.4   Evaluation Criteria

Dice Coefficient (DC) Score: DC score is widely used for evaluating the performance of segmentation. DC score measures the size of overlapping regions from the ground truth reference and segmentation results. Higher DC score means better segmentation performance. DC score can be expressed as:

$$DC(T, P) = 2 \cdot \frac{\sum_{i=1}^{\Omega}(T_i \times P_i)}{\sum_{i=1}^{\Omega}(T_i + P_i)} \tag{2}$$

where, T, P are represented as the ground truth and the prediction of segmentation, respectively; i is an index of each pixel value within an image spatial space Ω.

Area Under Receiving Operator Characteristics (ROC) Curve (AUC): And also, AUC is calculated as another evaluation metric. AUC is able to reflect the trade-offs between the true-positive rate (sensitivity) and the false-positive rate (specificity) at various threshold settings. Higher AUC means better performance.

4   Results

We test our MR haemorrhages segmentation framework with 1669 sub-images with manual annotation. As shown in Fig. 4, four examples from testing dataset are randomly selected to present the segmentation performance compared with one of general segmentation solutions: U-Net framework. From left to right, the example fundus image, manual annotation, segmentation results by U-Net and Tiramisu are shown respectively. DC score is introduced to quantify the segmentation performance. For MR haemorrhages segmentation, the DC scores of U-Net (0.9062) are lower than the DC scores of Tiramisu which is 0.9950. As shown in Fig. 5, the segmentation framework of Dense-Net based Tiramisu model achieves better AUC results (0.9358) than U-Net segmentation framework (0.6307) under the same segmentation task.

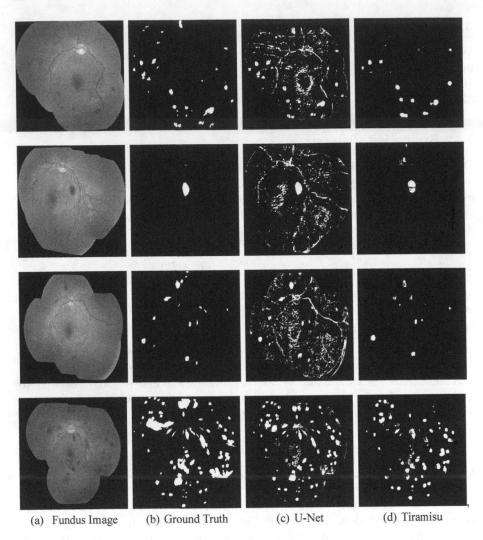

(a) Fundus Image   (b) Ground Truth   (c) U-Net   (d) Tiramisu

Fig. 4. Haemorrhage segmentation results of four random examples images. From left to right, the original MR fundus image, ground truth, segmentation results by U-Net and Tiramisu are shown respectively.

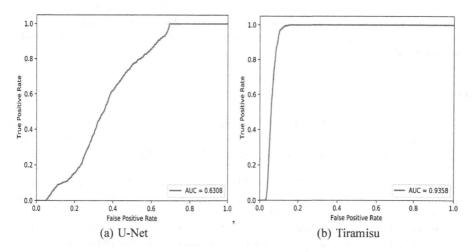

Fig. 5. Performance of different segmentation methods, in terms of area under the curve (AUC) for MR haemorrhage detection with 1,669 annotated fundus sub-images

5   Discussion

Our method based on the idea of Tiramisu is able to detect the location of MR haemorrhage. We also present one general segmentation models U-Net as the basic model to prove that our proposed method is better and more robust than U-Net in the same segmentation tasks, because of the MR haemorrhage feature reuse enabled by dense connections. However, in order to extract sub-images with the size of 512 × 512 for preserving the originality of raw colour fundus images, the size of training and validation datasets are increased so that it needs more memory for loading and is time-consuming also. In future work, we will investigate new strategies for optimising memory during the training process or preserving details of a dataset. And also, because in clinical practice, the number of MR haemorrhage is concerned, the function of automatically counting haemorrhage will be developed for the benefits of the clinic.

6   Conclusions

In this paper, we introduced an AI-based haemorrhages detection of MR in colour fundus images inspired by Dense-Net for the segmentation tasks. After implementation, the results showed that the proposed approach outperforms state-of-the-art U-Net approaches. It is believed that this new development will be applied to other challenging segmentation tasks. The experiment results have proved that the performance is better than commonly mainstream architectures when we test on a 1669 annotated fundus images dataset.

Acknowledgments. Xu Chen would like to acknowledge the PhD funding from Institute of Ageing and Chronic Disease at University of Liverpool and Liverpool Vascular Research Fund.

References

1. Ashraf, A., Akram, M.U., Sheikh, S.A., Abbas, S.: Detection of macular whitening and retinal hemorrhages for diagnosis of malarial retinopathy. In: 2015 IEEE International Conference on Imaging Systems and Techniques (IST), pp. 1–5. IEEE (2015)
2. Beare, N.A., Taylor, T.E., Harding, S.P., Lewallen, S., Molyneux, M.E.: Malarial retinopathy: a newly established diagnostic sign in severe malaria. Am. J. Trop. Med. Hyg. **75**(5), 790–797 (2006)
3. Coupé, P., Manjón, J.V., Fonov, V., Pruessner, J., Robles, M., Collins, D.L.: Patch-based segmentation using expert priors: application to hippocampus and ventricle segmentation. NeuroImage **54**(2), 940–954 (2011)
4. Essuman, V.A., et al.: Retinopathy in severe malaria in Ghanaian children-overlap between fundus changes in cerebral and non-cerebral malaria. Malaria J. **9**(1), 232 (2010)
5. He, K., Zhang, X., Ren, S., Sun, J.: Deep residual learning for image recognition. In: Proceedings of the IEEE Conference on Computer Vision and Pattern Recognition, pp. 770–778 (2016)
6. Huang, G., Liu, Z., Van Der Maaten, L., Weinberger, K.Q.: Densely connected convolutional networks. In: IEEE Conference on Computer Vision and Pattern Recognition, vol. 1, p. 3 (2017)
7. Jégou, S., Drozdzal, M., Vazquez, D., Romero, A., Bengio, Y.: The one hundred layers tiramisu: fully convolutional densenets for semantic segmentation. In: 2017 IEEE Conference on Computer Vision and Pattern Recognition Workshops (CVPRW), pp. 1175–1183. IEEE (2017)
8. Joshi, V., et al.: Automated detection of malarial retinopathy in digital fundus images for improved diagnosis in malawian children with clinically defined cerebral malaria. Sci. Rep. **7**, 42703 (2017)
9. Joshi, V.S., et al.: Automated detection of malarial retinopathy-associated retinal hemorrhages. Invest. Ophthalmol. Vis. Sci. **53**(10), 6582–6588 (2012)
10. Kande, G.B., Savithri, T.S., Subbaiah, P.V.: Automatic detection of microaneurysms and hemorrhages in digital fundus images. J. Digit. Imaging **23**(4), 430–437 (2010)
11. Kingma, D.P., Ba, J.: Adam: a method for stochastic optimization. arXiv:1412.6980 (2014)
12. Krizhevsky, A., Sutskever, I., Hinton, G.E.: Imagenet classification with deep convolutional neural networks. In: Advances in Neural Information Processing Systems, pp. 1097–1105 (2012)
13. LeCun, Y., Bengio, Y., Hinton, G.: Deep learning. Nature **521**(7553), 436 (2015)
14. LeCun, Y., et al.: Comparison of learning algorithms for handwritten digit recognition. In: International Conference on Artificial Neural Networks, Perth, Australia, vol. 60, pp. 53–60 (1995)
15. Lewallen, S., Taylor, T.E., Molyneux, M.E., Wills, B.A., Courtright, P.: Ocular fundus findings in malawian children with cerebral malaria. Ophthalmology **100**(6), 857–861 (1993)

16. Li, X., Chen, H., Qi, X., Dou, Q., Fu, C.W., Heng, P.A.: H-DenseUNet: Hybrid densely connected UNet for liver and liver tumor segmentation from CT volumes. arXiv:1709.07330 (2017)
17. Looareesuwan, S., et al.: Retinal hemorrhage, a common sign of prognostic significance in cerebral malaria. Am. J. Trop. Med. Hyg. **32**(5), 911–915 (1983)
18. Maude, R.J., Dondorp, A.M., Sayeed, A.A., Day, N.P., White, N.J., Beare, N.A.: The eye in cerebral malaria: what can it teach us? Trans. R. Soc. Trop. Med. Hyg. **103**(7), 661–664 (2009)
19. Maude, R.J., Hassan, M.U., Beare, N.A.: Severe retinal whitening in an adult with cerebral malaria. Am. J. Trop. Med. Hyg. **80**(6), 881 (2009)
20. Organization, W.H.: World malaria report 2018. World HealthOrganization (2018)
21. Ronneberger, O., Fischer, P., Brox, T.: U-Net: convolutional networks for biomedical image segmentation. In: Navab, N., Hornegger, J., Wells, W.M., Frangi, A.F. (eds.) MICCAI 2015. LNCS, vol. 9351, pp. 234–241. Springer, Cham (2015). https://doi.org/10.1007/978-3-319-24574-4_28
22. Simonyan, K., Zisserman, A.: Very deep convolutional networks for large-scale image recognition. arXiv:1409.1556 (2014)
23. Tang, L., Niemeijer, M., Abramoff, M.D.: Splat feature classification: detection of the presence of large retinal hemorrhages. In: 2011 IEEE International Symposium on Biomedical Imaging: From Nano to Macro, pp. 681–684. IEEE (2011)
24. Tran, P.V.: A fully convolutional neural network for cardiac segmentation in short-axis MRI. arXiv:1604.00494 (2016)

An Ensemble of Deep Neural Networks for Segmentation of Lung and Clavicle on Chest Radiographs

Jingyi Zhang[1,2], Yong Xia[1,2(✉)], and Yanning Zhang[2]

[1] Research & Development Institute of Northwestern Polytechnical University in Shenzhen,
Shenzhen 518057, China
yxia@nwpu.edu.cn

[2] National Engineering Laboratory for Integrated Aero-Space-Ground-Ocean Big Data
Application Technology, School of Computer Science and Engineering,
Northwestern Polytechnical University, Xi'an 710072, China

Abstract. Accurate segmentation of lungs and clavicles on chest radiographs plays a pivotal role in screening, diagnosis, treatment planning, and prognosis of many chest diseases. Although a number of solutions have been proposed, both segmentation tasks remain challenging. In this paper, we propose an ensemble of deep segmentation models (enDeepSeg) that combines the U-Net and DeepLabv3+ to address this challenge. We first extract image patches to train the U-Net and DeepLabv3+ model, respectively, and then use the weighted sum of the segmentation probability maps produced by both models to determine the label of each pixel. The weight of each model is adaptively estimated according to its error rate on the validation set. We evaluated the proposed enDeepSeg model on the Japanese Society of Radiological Technology (JSRT) database and achieved an average Jaccard similarity coefficient (JSC) of 0.961 and 0.883 in the segmentation of lungs and clavicles, respectively, which are higher than those obtained by ten lung segmentation and six clavicle segmentation algorithms. Our results suggest that the enDeepSeg model is able to segment lungs and clavicles on chest radiographs with the state-of-the-art accuracy.

Keywords: Lung segmentation · Clavicle segmentation · Chest radiograph · Deep learning

1 Introduction

Chest diseases pose serious health threats to human beings. For instance, pneumonia alone affects approximately 450 million people globally per year (i.e. 7% of the world population), resulting in about 4 million deaths, mostly in third-world countries, and lung cancer accounts for about 27% of all cancer deaths, which is considered as the leading cause of cancer-associated death worldwide [1]. Among various medical imaging modalities, the chest radiography, colloquially called chest X-ray, is one of the most commonly accessible radiological examinations, due to its cost-effectiveness

© Springer Nature Switzerland AG 2020
Y. Zheng et al. (Eds.): MIUA 2019, CCIS 1065, pp. 450–458, 2020.
https://doi.org/10.1007/978-3-030-39343-4_38

and dose-effectiveness. It is treated as a basic and important radiological tool for screening, diagnosis, treatment planning, and prognosis of many chest diseases, in which the segmentation of critical regions, such as lungs and clavicles, is an essential step. Such segmentation, however, is a challenging task, since the complex anatomical structures of tissues and organs may mutually overlap on a radiograph.

A number of algorithms for the segmentation of lungs and clavicles on chest radiographs have been proposed in the literature. Many of them incorporate the prior domain knowledge into the segmentation process. Seghers et al. [2] incorporated object-specific gray-level appearance and shape characteristics into a supervised scheme for model-based segmentation of medical images. Shao et al. [3] proposed a hierarchical deformable framework that integrates the scale-dependent shape and appearance information to segment lungs on chest radiographs. Candemir et al. [4] presented a nonrigid registration-driven robust lung segmentation algorithm, which consists of a content-based image retrieval approach for identifying training images, creating the initial patient-specific anatomical model of lung shape using SIFT-flow for deformable registration, and a graph cut based refinement of lung boundaries. Ibragimov et al. [5] presented an approach to landmark detection for landmark-based lung field segmentation, in which the appearance of individual landmarks is characterized by Haar-like features and a random forest classifier, and the spatial relationships among landmarks are modeled by Gaussian distributions augmented by shape-based random forest classifiers. Ginneken et al. [6] compared several methods for lung and clavicle segmentation, including the active shape model (ASM), active appearance model (AAM), pixel classification method, and combinations thereof.

Recently, deep convolutional neural networks (DCNNs) have led to major breakthroughs on many image segmentation tasks, since they provide an 'end-to-end' framework to learn image representation and classification and thus free users from the troublesome handcrafted feature extraction. Such breakthroughs have prompted many researchers to apply DCNNs to the segmentation of lungs and clavicles on chest radiographs. Novikov et al. [7] proposed fully convolutional architectures, which combine delayed subsampling, exponential linear units, highly restrictive regularization, and a large number of high-resolution low-level abstract features, to segment lungs, clavicles, and the heart on chest radiographs. Dai et al. [8] proposed the structure correcting adversarial network (SCAN) to segment lung fields and the heart on chest radiographs, which incorporates a critic network to impose structural regularities on the convolutional segmentation network and guide the segmentation model to achieve realistic outcomes. The state-of-art methods have a Jaccard similarity coefficient (JSC) of 0.954 for lung segmentation [4] and 0.868 for clavicle segmentation [7], which still have much room for improvement.

The U-Net [9] is a widely used convolutional network architectures for fast and precise segmentation of images and is known to work with less number of images. The DeepLabv3+ model [10] is one of the latest semantic image segmentation model, which is based on an inverted residual structure where the shortcut connections are between the thin bottle-neck layers. In this paper, we propose an ensemble of deep segmentation models (enDeepSeg) that combines the U-Net and DeepLabv3+ for the segmentation of lungs and clavicles on chest radiographs. We evaluated the proposed model against several existing solutions on the Japanese Society of Radiological Technology (JSRT) database [11] and achieved the state-of-the-art performance.

2   Dataset

For this study, we used the publicly available JSRT database [11], which consists of 247 chest radiographs collected from 13 institutions in Japan and one in the United States. The images were scanned from films to a size of 2048 × 2048 pixels, a spatial resolution of .175 mm/pixel and 12 bit gray levels. The ground truth annotations of lungs and clavicles, which are available in the Segmentation in Chest Radiographs (SCR) database [6], were acquired by two trained observers independently, who were allowed to zoom and adjust brightness and image contrast, and could take unlimited time for segmentation. When in doubt, they reviewed cases with the same radiologist and the radiologist provided the segmentation he believed to be correct. Following other studies [2–8], we use the segmentations of the first observer, who is a medical student familiar with medical images and medical image analysis, as gold standard. However, the ground truths in the SCR database have a size of 1024 × 1024 pixels. For the ease of comparison, we resized each chest radiograph to 1024 × 1024 for this study.

3   Method

The proposed enDeepSeg model consists of three steps (1) extracting 512 × 512 image patches on each radiograph with a horizontal stride of 64 and a vertical stride of 512, (2) training the U-Net and DeepLabv3+ model, respectively, and (3) fusing the obtained segmentation results. A diagram that summarizes the proposed algorithm is shown in Fig. 1. We now delve into the second and third steps.

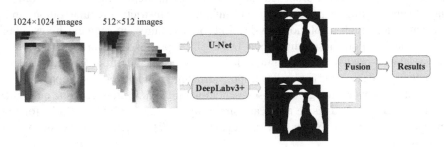

Fig. 1. Diagram of the proposed enDeepSeg model.

3.1   Training U-Net

The structure of the U-Net used for this study is shown in Fig. 2. It consists of an input layer that accepts 512 × 512 chest radiograph patches, nine blocks (namely Block-1 to

Block-9), and a Softmax output layer. Block-1 has two convolutional layers and a 2 × 2 max-pooling layer, where each convolutional layer consists of 64 kernels of size 3 × 3 and uses a padding of 1. Block-2 to Block-5 are similar to Block-1, except for the number of kernels. In addition, Block-5 has an extra up-sampling layer. Block-6 accepts the combined output of Block-4 and Block-5. Block-7 to Block-9 are similar to Block-6. Exceptionally, Block-9 has a 1 × 1 convolutional layer. We use the cross entropy as the loss function.

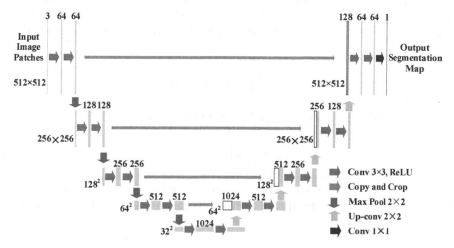

Fig. 2. Architecture of the U-Net used for this study.

We employed a pre-trained VGG16 model [12] to initialize the parameters of the U-Net. To fine tune this model, we adopted the online data augmentation technique, in which the augmentation operations include rotation of any angle, blurring, horizontal flip, vertical flip, and zooming, to increase the number of training patches to five times. We set the maximum iteration number to 200 and the batch size to 4, and chose the Adam optimizer with an initial learning rate of 1×10^{-4} to update parameters. The learning rate was reduced by a default factor of 0.1 after 100 epochs.

3.2 Training DeepLabv3+ Model

The DeepLabv3+ model [10] combines the spatial pyramid pooling module and encoder-decoder structure. This model has an encoder-decoder architecture, which is able to capture sharper object boundaries by gradually recovering the spatial information. The encoder block mainly applies several parallel atrous convolution with different rates, called the atrous spatial pyramid pooling (ASPP), which can encode multi-scale contextual information by probing the input with multiple fields-of-view. The output of the encoder block is upsampled bilinearly by a factor of 4, followed by the concatenation with the low-level features in the decoder block. The concatenated features are first fed into a 3 × 3 convolutional layer, and then bilinearly upsampled again to produce the output segmentation maps.

We adopted a ResNet-101 model [13] that has been converged on the ImageNet dataset to initialize the DeepLabv3+ model. We chose the momentum optimizer to minimize the cross entropy loss. We fixed the maximum iteration number to 150, set the batch size to 4, the initial learning rate to 7×10^{-3}, and the end learning rate to 10^{-6}.

3.3 Fusing Segmentation Maps

Let the segmentation probability maps obtained by feeding an image patch to both models be denoted by M_U and M_D, respectively. To combine the advantages from both models, we define the segmentation map produced by the ensemble model as follows [14]

$$M_{en} = \frac{1}{\alpha_U + \alpha_D}(\alpha_U M_U + \alpha_D M_D) \tag{1}$$

where the weighting parameters α_U and α_D can be estimated as

$$\alpha_\# = \frac{1}{2}\ln\left(\frac{1 - \varepsilon_\#}{\varepsilon_\#}\right) \tag{2}$$

where $\varepsilon_\#$ is the error rate of the model $\#$ on the validation set. Finally, we applied the soft-max function to the probability of each pixel to determine if it belongs to the target (i.e. lungs or clavicles) or not.

3.4 Evaluation

The proposed enDeepSeg model was evaluated on the JSRT database using the three-fold cross-validation (3-Fold CR). In the training stage, we randomly choose 10% of training images to form a validation set, aiming to estimate the error rate of each sub-model. In the testing stage, each test image of size 1024×1024 was split into four non-overlapping patches of size 512×512. Those four patches were fed to the trained enDeepSeg model sequentially, and the obtained results were concatenated to form the image-level segmentation. The segmentation results of lungs and clavicles were quantitatively assessed by JSC and Dice similarity coefficient (DSC), which can be calculated as follows

$$JSC = \frac{|GT \cap SR|}{|GT \cup SR|} \tag{3}$$

$$DSC = \frac{2|GT \cap SR|}{|GT| + |SR|} \tag{4}$$

where GT represents the ground truth region, SR represents the segmented region, and $|R|$ denotes the number of pixels in the region R. The values of JSC and DSC range from 0 to 1, with a higher value representing a more accurate segmentation result.

4 Experiment and Results

Figure 3 shows four example test images, the corresponding results of lung and clavicle segmentation obtained by the proposed enDeepSeg model, and the ground truth. Since the lungs and clavicles are partly overlapped, the segmentation results and ground truth of them were given in different subfigures. It shows that the obtained segmentation results are highly similar to the ground truth.

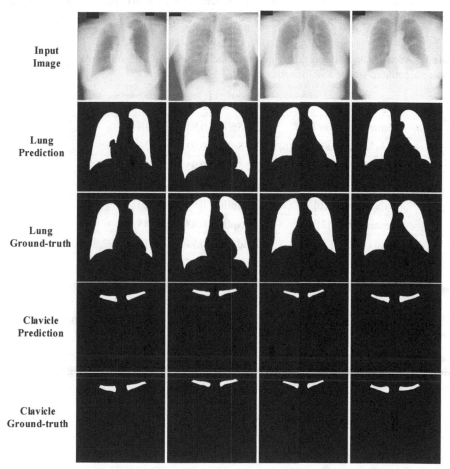

Fig. 3. Illustration of segmentation results: (1st row) four example test images, (2nd and 4th rows) the results of lung and clavicle segmentation, and (3rd and 5th rows) the ground truth.

Table 1 gives the average lung segmentation performance of the proposed enDeepSeg model and 10 existing algorithms on the JSRT database. To ensure a fair comparison, the performance of existing algorithms was adopted in the literature. It shows that the proposed model achieved an average JSC value of 0.961 and an average DSC value of 0.980, which are higher than the performance achieved by other algorithms and set the new state of the art on this database.

Table 2 gives the average clavicle segmentation performance of the proposed enDeepSeg model and six existing algorithms on the JSRT database. Similarly, the performance of existing algorithms was adopted in the literature. It shows that the proposed model achieved an average DSC value of 0.935, which is slightly higher than that obtained in [7], and an average JSC value of 0.883, which is substantially better than other algorithms except for human observer. And the advantage of human observer comes from the guidance of a professional radiologist. The results reported in Tables 1 and 2 reveal that our model is able to produce, to our knowledge, the most accurate segmentation of lungs and clavicles on chest radiographs.

Table 1. Average lung segmentation performance of the proposed enDeepSeg model and 10 existing algorithms on the JSRT database.

Algorithm	Validation	JSC	DSC
Human observer [6]	SCR Split	0.946 ± 0.018	/
Hybrid AAM and PC [6]	SCR Split	0.933 ± 0.026	/
Hybrid ASM and PC [6]	SCR Split	0.934 ± 0.037	/
PC [6]	SCR Split	0.938 ± 0.027	/
Hybrid voting [6]	SCR Split	0.949 ± 0.020	/
Seghers et al. [2]	SCR Split	0.930 ± 0.045	/
Shao et al. [3]	2-Fold CR	0.946 ± 0.019	0.972 ± 0.010
Ibragimov et al. [5]	Not Clear	0.953 ± 0.020	/
Novikov et al. [7]	3-Fold CR	0.950	0.974
Candemir et al. [4]	Not Clear	0.954 ± 0.015	0.967 ± 0.008
Proposed	**3-Fold CR**	**0.961 ± 0.004**	**0.980 ± 0.002**

Table 2. Average clavicle segmentation performance of the proposed enDeepSeg model and six existing algorithms on the JSRT database.

Algorithm	Validation	JSC	DSC
Human observer [6]	SCR Split	0.896 ± 0.037	/
Hybrid AAM and PC [6]	SCR Split	0.613 ± 0.206	/
Hybrid ASM and PC [6]	SCR Split	0.663 ± 0.157	/
ASM tuned [6]	SCR Split	0.734 ± 0.137	/
Hybrid voting [6]	SCR Split	0.736 ± 0.106	/
Novikov et al. [7]	3-Fold CR	0.868	0.929
Proposed	**3-Fold CR**	**0.883 ± 0.013**	**0.935 ± 0.010**

5   Discussion

5.1   Ablation Study

The proposed enDeepSeg model is an ensemble of U-Net and the DeepLabv3+ model, each of which has its merit. To demonstrate that the ensemble of both models results in performance gain, we compared the performance of the proposed model to that of U-Net and the DeepLabv3+ model in Table 3. It shows that the DeepLabv3+ model outperforms U-Net on both segmentation tasks, and combining U-Net with the DeepLabv3+ model leads to further performance improvement, which is slight in lung segmentation, but substantial in clavicle segmentation. The reason may lies in the fact that, since lung segmentation is much easier than clavicle segmentation, it is hard to further improve the already-high accuracy of lung segmentation.

Table 3. Accuracy comparison of lung between U-Net and DeepLabv3+

Organ	JSC			DSC		
	U-Net	DeepLabv3+	Proposed	U-Net	DeepLabv3+	Proposed
Lungs	0.953	0.959	**0.961**	0.975	0.979	**0.980**
Clavicles	0.873	0.870	**0.883**	0.928	0.928	**0.935**

5.2   Time Complexity

It took about 13 h to train the U-Net and about 19 h to train the DeepLabv3+ model (Intel Xeon E5-2640 V4 CPU, NVIDIA Titan Xp GPU, 512 GB Memory). However, applying the proposed enDeepSeg model to the segmentation of lungs and clavicles on each chest radiograph costs less than one second. The fast online testing suggests that this model could be used in a routine clinical workflow.

6   Conclusion

This paper proposes the enDeepSeg model for the segmentation of lungs and clavicles on chest radiographs, which is an ensemble of U-Net and the DeepLabv3+ model. Our results show that the proposed model outperforms ten existing lung segmentation algorithms and six clavicle segmentation algorithms on the JSRT database. In our future work, we plan to use synthetic data generated by generative adversarial networks to improve the accuracy of clavicle segmentation.

Acknowledgement. This work was supported in part by the Science and Technology Innovation Committee of Shenzhen Municipality, China, under Grants JCYJ20180306171334997, in part by the National Natural Science Foundation of China under Grants 61771397, in part by Synergy Innovation Foundation of the University and Enterprise for Graduate Students in Northwestern Polytechnical University under Grants XQ201911, and in part by the Project for Graduate Innovation team of Northwestern Polytechnical University. We appreciate the efforts devoted to collect and share the JSRT database and SCR database for comparing interactive and (semi)-automatic algorithms for the segmentation of lungs and clavicles on chest radiographs.

References

1. Ruuskanen, O., Lahti, E., Jennings, L.C., et al.: Viral pneumonia. Lancet **377**, 1264–1275 (2011)
2. Seghers, D., Loeckx, D., Maes, F., et al.: Minimal shape and intensity cost path segmentation. IEEE Trans. Med. Imaging **26**, 1115–1129 (2007)
3. Shao, Y., Gao, Y., Guo, Y., et al.: Hierarchical lung field segmentation with joint shape and appearance sparse learning. IEEE Trans. Med. Imaging **33**, 1761–1780 (2014)
4. Candemir, S., Jaeger, S., Palaniappan, K., et al.: Lung segmentation in chest radiographs using anatomical atlases with nonrigid registration. IEEE Trans. Med. Imaging **33**, 577–590 (2014)
5. Ibragimov, B., Likar, B., Pernu, F., et al.: Accurate landmark-based segmentation by incorporating landmark misdetections. In: IEEE International Symposium on Biomedical Imaging (ISBI 2016), pp. 1072–1075 (2016)
6. Ginneken, B.V., Stegmann, M.B., Loog, M.: Segmentation of anatomical structures in chest radiographs using supervised methods: a comparative study on a public database. Med. Image Anal. **10**, 19–40 (2006)
7. Novikov, A.A., Lenis, D., Major, D., et al.: Fully convolutional architectures for multi-class segmentation in chest radiographs. IEEE Trans. Med. Imaging **37**, 1865–1876 (2018)
8. Dai, W., Dong, N., Wang, Z., Liang, X., Zhang, H., Xing, E.P.: SCAN: structure correcting adversarial network for organ segmentation in chest X-rays. In: Stoyanov, D., et al. (eds.) DLMIA/ML-CDS -2018. LNCS, vol. 11045, pp. 263–273. Springer, Cham (2018). https://doi.org/10.1007/978-3-030-00889-5_30
9. Ronneberger, O., Fischer, P., Brox, T.: U-Net: convolutional networks for biomedical image segmentation. In: Navab, N., Hornegger, J., Wells, William M., Frangi, Alejandro F. (eds.) MICCAI 2015. LNCS, vol. 9351, pp. 234–241. Springer, Cham (2015). https://doi.org/10.1007/978-3-319-24574-4_28
10. Chen, L.C., Zhu, Y., Papandreou, G., et al.: Encoderdecoder with atrous separable convolution for semantic image segmentation. arXiv preprint arXiv:1802.02611 (2018)
11. Shiraishi, J., Katsuragawa, S., Ikezoe, J., et al.: Development of a digital image database for chest radiographs with and without a lung nodule: receiver operating characteristic analysis of radiologists' detection of pulmonary nodules. Am. J. Roentgenol. **174**, 71–74 (2000)
12. Simonyan, K., Zisserman, A.: Very deep convolutional networks for large-scale image recognition. arXiv preprint arXiv:1409.1556 (2014)
13. He, K., Zhang, X., Ren, S., et al.: Deep residual learning for image recognition. In: Computer Vision and Pattern Recognition (CVPR), pp. 770–778 (2016)
14. Hastie, T., Rosset, S., Zhu, J., et al.: Multi-class adaboost. Stat. Interface **2**(3), 349–360 (2009)

Automated Corneal Nerve Segmentation Using Weighted Local Phase Tensor

Kun Zhao[1,2], Hui Zhang[1(✉)], Yitian Zhao[2,3(✉)], Jianyang Xie[2], Yalin Zheng[3,4], David Borroni[4,6], Hong Qi[5], and Jiang Liu[2,7]

[1] School of Information and Control Engineering, Shenyang Jianzhu University, Shenyang, China
zhanghui11.11@163.com
[2] Ningbo Institute of Industrial Technology, Chinese Academy of Sciences, Ningbo, China
yitian.zhao@nimte.ac.cn
[3] Department of Eye and Vision Science, University of Liverpool, Liverpool, UK
[4] St. Paul's Eye Unit, Royal Liverpool University Hospital, Liverpool, UK
[5] Department of Ophthalmology, Peking University Third Hospital, Beijing, China
[6] Department of Ophthalmology, Riga Stradins University, Riga, Latvia
[7] Department of Computer Science and Engineering, Southern University of Science and Technology, Shenzhen, China

Abstract. There has been increasing interest in the analysis of corneal nerve fibers to support examination and diagnosis of many diseases, and for this purpose, automated nerve fiber segmentation is a fundamental step. Existing methods of automated corneal nerve fiber detection continue to pose difficulties due to multiple factors, such as poor contrast and fragmented fibers caused by inaccurate focus. To address these problems, in this paper we propose a novel weighted local phase tensor-based curvilinear structure filtering method. This method not only takes into account local phase features using a quadrature filter to enhance edges and lines, but also utilizes the weighted geometric mean of the blurred and shifted responses to allow better tolerance of curvilinear structures with irregular appearances. To demonstrate its effectiveness, we apply this framework to 1578 corneal confocal microscopy images. The experimental results show that the proposed method outperforms existing state-of-the-art methods in applicability, effectiveness, and accuracy.

Keywords: Corneal nerve · Curvilinear structure · Segmentation · Local phase

1 Introduction

Over the last decade, several studies [1–3] have confirmed that numerous corneal nerve properties, such as nerve fiber branching, density, length, tortuosity, etc., are linked to both systemic diseases and conditions of the eye [4,5]. Early detection of these properties may help to reduce the incidence of vision loss and

© Springer Nature Switzerland AG 2020
Y. Zheng et al. (Eds.): MIUA 2019, CCIS 1065, pp. 459–469, 2020.
https://doi.org/10.1007/978-3-030-39343-4_39

blindness. For this to be possible, accurate detection and analysis of the nerve fiber is essential [6].

In vivo confocal microscopy (IVCM) is the common technique of choice for the imaging and inspection of corneal nerves: in particular, for the non-invasive acquisition of the subbasal nerve plexus [7]. The manual identification of nerve fiber by ophthalmologists is tedious, highly labour-intensive and subject to human error, while the available commercial software still relies heavily on manual refinement. Consequently, development of an automatic vascular tracing method is indispensable to overcome time constraints and avoid human error.

Extensive work has been carried out on automatic curvilinear structure segmentation (see [8,9] for extensive reviews). Although it bears a superficial similarity to other curvilinear structure segmentation tasks, segmenting corneal nerve fiber is more challenging because of poor contrast and fragmented and multiple scales of tortuous fiber in the image, as shown as Fig. 1. Moreover, many images contain potentially confusing non-target structures such as dendritic cells that can be easily mistaken for fiber given their similar appearance.

Fig. 1. Corneal nerves images with poor contrast (left), non-target structures (middle), and discontinuous and multiple spatial scales of fibers (right).

In this work, we introduce a new curvilinear feature enhancement metric: namely a weighted local phase tensor. This tensor is enabled by log-Gabor filter, Hessian transform, blurring and shifting functions, to resolve the weak response and discontinuities yielded by most filter-based methods.

1.1   Related Works

Numerous corneal nerve fiber segmentaion methods have been proposed during the last decade. Ruggeri et al. [10] and Scarpa et al. [11] adopted Gabor filtering to enhance nerve visibility: a tracking procedure was then implemented, starting from a set of automatically-defined seed points. Fuzzy c-mean clustering was then applied to classify the pixels as nerve or non-nerve pixels. Poletto [12] further extended the method of [10] to a dataset consisting of 30 epithelium corneal images, the nerves were extracted by connecting the seed points using their minimum cost paths. Dabbah et al. [13] further proposed a multi-scale adaptive dual-model detection algorithm, based on random forest and neural networks.

Ferreira et al. [14] enhanced IVCM images using phase-shift analysis, then identified the nerve structures by using a phase symmetry-based filter. Guimaraes et al. [15] removed illumination artefacts by applying top-hat filtering and a bank of log-Gabor filters. The hysteresis threshold approach was used to determine the candidate nerve segments, and true and false nerve segments were distinguished using support vector machines (SVM). Annunziata et al. [16] proposed a hand-crafted ridge detector, SCIRD, which utilized curved-support Gaussian models to compute the second order directional derivative in the gradient direction at each pixel. Al-Fahdawi et al. [17] used a coherence filter to improve the IVCM image quality: morphological operations were then applied to remove epithelial cells; the corneal nerves were extracted using an improved edge detection method which was able to bridge the nerve discontinuities. More recently, Colonna et al. [18] used a deep learning approach, the U-Net-based Convolutional Neural Network [19], to trace the corneal nerve. It consisted of a contracting path, which captured nerve descriptors, and a symmetric expanding path, which enabled precise nerve localization.

However, the above-mentioned corneal nerve tracing methods need further improvement, as confocal corneal images contain spurious illumination artefacts, and the whole image may appear dimmed due to focusing problems. Moreover, bright, elongated structures other than nerve segments (e.g., cells), are normally present, and these may cause false-positives. For these reasons, image and nerve enhancement are two essential steps in the process of discriminating the corneal nerve tree.

2   Method

A quadrature filter is a well-designed tool for distinguishing intrinsic features in the image that are invariant to changes in illumination. In this section, we propose the use of a weighted local phase tensor to enhance the fiber structures within the IVCM images.

2.1   Local Phase Tensor

For a one-dimensional (1D) problem, by using the so-called analytical function, the amplitude $A(x)$ and phase $\phi(x)$ of a given signal $f(x)$ is defined as $A(x) = \|f(x) - if_{\mathcal{H}}(x)\| = \sqrt{f^2(x) + f_{\mathcal{H}}^2(x)}$, and $\phi(x) = \arctan\left(\frac{f(x)}{f_{\mathcal{H}}(x)}\right)$, where $i = \sqrt{-1}$, and $f_{\mathcal{H}}(x)$ is the Hilbert transform of $f(x)$ [20].

In order to enhance spatial localization and to avoid the problems posed by the analytic signal for 2D or higher dimensions and the 2D Hilbert transform, the analysis of the signal must be take over a narrow range of frequencies at different locations in the 2D signal. Boukerroui et al. [20] suggested that local phase should be estimated by a quadrature filter with even-symmetric and odd-symmetric parts. In consequence, the definition of local phase in 2D application may be rewritten as:

$$\phi(x) = \arctan\left(\frac{f_e(x) * f(x)}{f_o(x) * f(x)}\right) = \arctan\left(\frac{E(x)}{O(x)}\right). \tag{1}$$

where $f_e(x) * f(x)$ is the even (symmetric) band-pass filter and denotes as $E(x)$, while $f_o(x) = \mathcal{H}(f_e(x))$ is the Hilbert transform of the even filter $f_e(x)$, and denotes as $O(x)$. In particular, the log-Gabor (log normal) filter is a commonly used quadrature filter, and $E(x)$ and $O(x)$ are the responses of even and odd quadrature pair filter to an image can be estimated by:

$$E(x) = \text{real}\{F^{-1}(\text{LG}(\omega) \times F(x))\}, \tag{2}$$

$$O(x) = \text{imag}\{F^{-1}(\text{LG}(\omega) \times F(x))\}, \tag{3}$$

where LG, F and F^{-1} indicate the log-Gabor filtering, the forward and inverse Fourier transforms, respectively.

Then the local phase tensor in symmetric and asymmetric aspects $\mathcal{T}_E$ and $\mathcal{T}_O$ are calculated as [21]:

$$\mathcal{T}_E = [\mathbf{H}(E(x))] \cdot [\mathbf{H}(E(x))]^T, \tag{4}$$

$$\mathcal{T}_O = -\frac{1}{2}([\nabla(O(x))] \cdot [\nabla\nabla^2(O(x))]^T \\ + [\nabla\nabla^2(O(x))] \cdot [\nabla(O(x))]^T), \tag{5}$$

where $\mathbf{H}$ denotes the Hessian operation, and ∇, ∇^2 indicates the Gradient and Laplacian operations, respectively. The superscript T denotes the transpose operation.

According to the monogenic signal analysis [22], the local phase tensor can be obtained by:

$$\mathcal{T} = \sum_{\theta=1}^{\Theta}\left\{\sqrt{(\mathcal{T}_E^\theta)^2 + (\mathcal{T}_O^\theta)^2}\right\} \cdot \cos(\varphi). \tag{6}$$

In practice, multiple orientations are needed to capture structures at different directions. Furthermore, in order to achieve a rationally invariant tensor, filters for all directions have to be combined. Θ indicates the set of directions under consideration: $\Theta = \{\frac{\pi}{16}, \frac{2\pi}{16}, \frac{3\pi}{16}, \cdots, \frac{15\pi}{16}, \pi\}$. The instantaneous phase φ presenting the local contrast independently of feature type (line and edge), and may be defined as

$$\varphi(x) = E^\theta(x) + |O^\theta(x)|i, \tag{7}$$

As observed from Eqs. (2) and (3), $E(x)$ reaches the maximal response at lines while $O(x)$ is almost 0, the filter response is purely real, and leads to a line-like signal. While for edges, the $E(x)$ is 0 and $O(x)$ has the maximal response, and the filter response is purely imaginary. This suggests that image edges align with the zero crossing of the real part of the phase map. Therefore, the real part of the response $E^\theta(x)$ and the absolute value of the imaginary part $O^\theta(x)$ are used, with a view to avoid confusion caused by changes in structural direction.

2.2   Weighted Local Phase Tensor

In real applications, due to poor contrast the extracted responses of the corneal nerves are represented as discontinuities. In consequence, in order to permit greater tolerance of the positions, deformations and scales of the respective contours, blurring and shifting operations are applied to the local phase tensor. The blurring operation is able to suppress the noise or background, and the shifting operation is used to enhance the response of the quadrature filter by maximizing all neighboring pixels in low-contrast in the dark and low contrast regions [23].

The blurring consists of a dilation of the filter response with a Gaussian function G_σ with standard deviation σ: $\sigma = \sigma_0 + \alpha\rho$. The σ_0 and α are the constants that regulate the tolerance to deformation of the concerned responses, and ρ is the radius parameter, representing a linear function of the distance from the centre of the quadrature filter. We then shift each blurred local phase tensor by a distance ρ_i in the direction opposite to ϕ_i. Formally, the i-th blurred and shifted responses $S_{\sigma_i,\rho_i,\phi_i}$ of the local phase tensor $\mathcal{T}$ can be calculated by:

$$
\begin{aligned}
&S_{\rho_i,\phi_i}(u,v) \\
&= \max_{u',v'}\{\mathcal{T}_{\sigma_i}(u - \Delta u_i - u', v - \Delta v_i - v')G_{\sigma'}(u',v')\}.
\end{aligned}
\tag{8}
$$

where $-3\sigma \leq u'$, $v' \leq -3\sigma$. The above configuration process represents a convolution of the weighting function with respect to the filter center u, v, and $S = \{(\rho_i, \phi_i)|i = 1,\ldots,n\}$, where n indicates the number of quadrature responses. $\Delta u_i = -\rho_i\cos\phi_i$ and $\Delta v_i = -\rho_i\sin\phi_i$ is the shift vector of i-th quadrature responses in Cartesian coordinates. This shift operation is able to assemble all the responses at the proposed filter centre. The parameter values of S can be automatically determined from the aforementioned filter settings of the standard deviation of the filter responses, kernel size, and orientations: $\rho_i \in \{0,2,4\}$ and $\phi_i \in \{0,0.5\pi,\pi,1.5\pi\}$.

Then the weighted local phase tensor $\hat{\mathcal{T}}$ for a given image is defined as a threshold ϵ of a multiplication of S_{ρ_i,ϕ_i}:

$$
\hat{\mathcal{T}} = \left|\prod_{i=1}^{|S|}(S_{\rho_i,\phi_i})\right|^\epsilon.
\tag{9}
$$

where ϵ denotes the control parameter that sets the threshold of the response at a fraction ϵ ($0 \leq \tau \leq 1$) of the maximum response.

In order to enhance curvilinear structures of different sizes present within a single image, multiple scales fusion is needed. In our study, the given image is uniformly down-sampled to $1/m$ of its original size, and the fused local phase tensor is defined as:

$$
\hat{\mathcal{T}} = \frac{\sum_{m=1}^{M} \mathcal{T}_m|\mathcal{T}_m|^\gamma}{\sum_{m=1}^{M} |\mathcal{T}_m|^\gamma}
\tag{10}
$$

where $m \in \{1, \cdots M\}$, and M denotes the number of scales ($M = 3$ in this work). γ is the order number of the power of the magnitude of the filter response at each scale. Figure 2 demonstrates that the proposed weighted local phase tensor acts as a general curvilinear structure indicator, providing an improved local phase tensor. It may be observed that the proposed weighted local phase tensor is able to better preserve poorly-imaged, low-contrast fibers, which appear as discontinuous to the unweighted tensor (Fig. 2(c) and (d): red arrows).

(a)   **(b)**   **(c)**   **(d)**

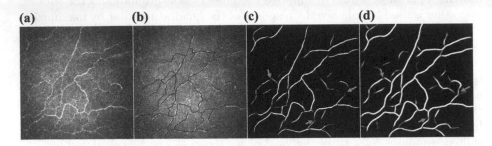

Fig. 2. Illustrative enhancement results using the local phase tensor and weight local phase tensor. (a) original image; (b) groundtruth; (c) response of local phase tensor; (d) response of weighted local phase tensor. (Color figure online)

3   Materials and Evaluation Metrics

3.1   Dataset

A total of 1578 images of corneal subbasal epithelium from 108 normal and pathological subjects were acquired using a Heidelberg Retina Tomograph equipped with a Rostock Cornea Module (HRT-III) microscope (Heidelberg Engineering Inc.). The 108 subjects included: 30 healthy subjects; 18 subjects with diabetes; and 60 subjects with dry eye disease. The image resolution was 384×384 pixels, and the field of view was $400 \times 400 \mu m^2$. The nerves appear as bright curvilinear structures lying over a darker background. The reference groundtruth was segmented manually by an ophthalmologist, who traced the centerlines of all visible nerves using an in-house program written in Matlab (Mathworks R2017, Natwick).

3.2   Evaluation Metrics

To compare the nerve fiber tracing performance of the proposed method with the corresponding groundtruth, we computed the *sensitivity* and *false discovery rate* (FDR) [15] between the predicted centerlines and groundtruth centerlines. Sensitivity is the fraction of the number of pixels in the correctly detected nerve segments (true positives) nerves over the total number of pixels in the groundtruth nerves. FDR is defined as the fraction of the total number of pixels incorrectly detected as nerve segments (false positives) nerves over the total number of pixels of the traced nerves in groundtruth. The use of *specificity*, defined as the number of pixels correctly rejected as non-nerve structures (true negatives), is not adequate for the evaluation of this tracing task, since the vast majority of pixels do not belong to corneal nerves.

It is worth noting that, as is customary in the evaluation of methods extracting one pixel-wide curves [15], a three-pixel tolerance region around the manually traced nerves is considered to be a true positive. In other words, a predicted centerline point is considered as true positive if is no more than three pixels distant from the nearest ground truth centerline point.

4 Experimental Results

We validated the effectiveness of the proposed weighted local phase tensor-based nerve fiber segmentation method against three other state-of-the-art curvilinear structure enhancement methods: Frangi's Vesselness filter (FVF) [24], multiscale hessian filter (MHF) [25], and combination of shifted filter responses (COS-FIRE) [23]. An infinite perimeter active contour with hybrid region (IPACHR) method [26] is used to segment the fibers from the filtered curvilinear structures. However, other sophisticated segmentation methods may work equally well. For a fair comparison, the parameters of these filters were optimized for best performance as follows. **FVF** scales: 1–8, scale ratio: 2; **MHF** scales: 1–3, spacing resolution: [3; 3]; **COSFIRE** scales: 1–4, orientation: $\theta \in \{\frac{\pi i}{8} | i = 0, \cdots 7\}$, threshold value: 0.35.

Figure 3 demonstrates the filtered curvilinear structures obtained by applying these different methods. Overall, all methods demonstrated similar performance on fibers with large diameters. It can be seen that FVF is able to enhance most larger fibers, but falsely enlarges background features where intensity inhomogeneities are present. MHF misses most fibers with small diameters, and also enhances some background regions, which leads to inaccurate identification of the fiber structures. As for the COSFIRE, the fiber edges are clearly enhanced, and this method achieves better results in distinguishing fiber from background. However, the COSFIRE also enhanced some surrounding non-target structures.

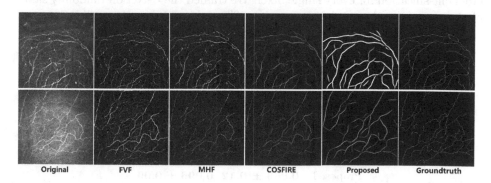

Fig. 3. Illustrative results of corneal nerve filtered by different curvilinear structure enhancement methods.

In contrast to these filters, we can see that the proposed method generally demonstrated superior performance in detecting nerve fiber regions (uniform responses at both high and low intensities), and provided relatively stronger responses to small fibers than other methods. In other words, the proposed method is not only able to enhance fiber regions so as to stand out more conspicuously from background, but also has the ability to reject non-fiber features. Such properties are due to the proposed filter retaining the intrinsic information of features that are invariant to changes in intensity, location and scale, which permits better detection of curvilinear structures under varying conditions. Table 1 shows this superior segmentation performance based on the proposed curvilinear structures enhancement method, demonstrating both higher sensitivity and lower FDR by significant margins.

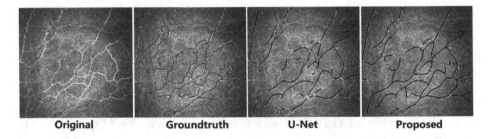

Original   **Groundtruth**   **U-Net**   **Proposed**

Fig. 4. Illustrative results of corneal nerves traced by U-Net and by the proposed method.

It is interesting to note that the visual results (Fig. 4) and evaluation metrics (Table 1) demonstrate that our method performs better than the deep learning-based approach: U-Net [19], which has recently attracted attention. We employed the U-Net-based convolutional neural network [18] for the fully automatic segmentation of corneal nerves from the IVCM image. This network comprises a contracting-encoder and an expanding-decoder, which allows the user to obtain a label classification for every single pixel. We trained the U-Net on randomly sampled images from the database, reserving 20% of this database as a validation set.

Table 1. Performance of five different segmentation methods.

Methods	Sensitivity	FDR
FVF	0.912 ± 0.25	0.181 ± 0.12
MHF	0.933 ± 0.19	0.142 ± 0.11
COSFIRE	0.950 ± 0.24	0.113 ± 0.09
U-Net	0.956 ± 0.17	0.105 ± 0.08
Proposed	$\mathbf{0.963 \pm 0.12}$	$\mathbf{0.096 \pm 0.09}$

5   Conclusions

Nerve fiber segmentation is the fundamental step in automated diagnosis of many nerve-related diseases, and it remains a challenging medical image analysis problem despite considerable effort in research. Many factors come together to make this problem difficult to address, such as uneven illumination and noise in the original image. In this paper, we have presented a weighted local phase tensor-based curvilinear structure filtering method, and have applied it successfully to corneal nerve fiber segmentation. The proposed filter exploits the advantages of a local phase tensor, making use of the geometric mean of blurred and shifted quadrature filter responses to allow more tolerance in the position of the respective contours. The evaluation results demonstrate the superiority of our model when compared with other state-of-the-art methods. It is our intention in our future work to measure the tortuosity, length, and density of the extracted nerves, so as to further evaluate the significance of changes in these morphological features and their association with nerve-related diseases, such as diabetic neuropathy.

Acknowledgment. This work was supported by National Natural Science Foundation of China (61601029, 61602322), Zhejiang Provincial and Ningbo Natural Science Foundation (LZ19F010001, 2018A610055).

References

1. Annunziata, R., Kheirkhah, A., Aggarwal, S., Hamrah, P., Trucco, E.: A fully automated tortuosity quantification system with application to corneal nerve fibres in confocal microscopy images. Med. Image Anal. **32**, 216–232 (2016)
2. Edwards, K., Pritchard, N., Vagenas, D., Russell, A.W., Malik, R.A., Efron, N.: Standardizing corneal nerve fibre length for nerve tortuosity increases its association with measures of diabetic neuropathy. Diabet. Med. J. Br. Diabet. Assoc. **31**(10), 1205–1209 (2014)
3. Kim, J., Markoulli, M.: Automatic analysis of corneal nerves imaged using in vivo confocal microscopy. Clin. Exp. Optom. **101**(2), 147–161 (2018)
4. Zhao, Y., et al.: Uniqueness-driven saliency analysis for automated lesion detection with applications to retinal diseases. In: Frangi, A.F., Schnabel, J.A., Davatzikos, C., Alberola-López, C., Fichtinger, G. (eds.) MICCAI 2018, Part II. LNCS, vol. 11071, pp. 109–118. Springer, Cham (2018). https://doi.org/10.1007/978-3-030-00934-2_13
5. Zhao, Y., et al.: Retinal artery and vein classification via dominant sets clustering-based vascular topology estimation. In: Frangi, A.F., Schnabel, J.A., Davatzikos, C., Alberola-López, C., Fichtinger, G. (eds.) MICCAI 2018, Part II. LNCS, vol. 11071, pp. 56–64. Springer, Cham (2018). https://doi.org/10.1007/978-3-030-00934-2_7
6. Zhao, Y., et al.: Saliency driven vasculature segmentation with infinite perimeter active contour model. Neurocomputing **259**, 201–209 (2017)
7. Scarpa, F., Zheng, X., Ohashi, Y., Ruggeri, A.: Automatic evaluation of corneal nerve tortuosity in images from in vivo confocal microscopy. Investig. Ophthalmol. Vis. Sci. **52**(9), 6404–6408 (2011)

8. Fraz, M., et al.: Blood vessel segmentation methodologies in retinal images - a survey. Comput. Methods Programs Biomed. **108**, 407–433 (2012)
9. Zhao, Y., Zheng, Y., Liu, Y., Na, T., Wang, Y., Liu, J.: Automatic 2D/3D vessel enhancement in multiple modality images using a weighted symmetry filter. IEEE Trans. Med. Imaging **37**(2), 438–450 (2018)
10. Ruggeri, A., Scarpa, F., Grisan, E.: Analysis of corneal images for the recognition of nerve structures. In: 2006 International Conference of the IEEE Engineering in Medicine and Biology Society, pp. 4739–4742 (2006)
11. Scarpa, F., Grisan, E., Ruggeri, A.: Automatic recognition of corneal nerve structures in images from confocal microscopy. Investig. Ophthalmol. Vis. Sci. **49**(11), 4801–4807 (2008)
12. Poletti, E., Ruggeri, A.: Automatic nerve tracking in confocal images of corneal subbasal epithelium. In: Proceedings of the 26th IEEE International Symposium on Computer-Based Medical Systems, pp. 119–124 (2013)
13. Dabbah, M.A., Graham, J., Petropoulos, I.N., Tavakoli, M., Malik, R.A.: Automatic analysis of diabetic peripheral neuropathy using multi-scale quantitative morphology of nerve fibres in corneal confocal microscopy imaging. Med. Image Anal. **15**(5), 738–747 (2011)
14. Ferreira, A., Morgado, A., Silva, J.: A method for corneal nerves automatic segmentation and morphometric analysis. Comput. Methods Programs Biomed. **107**(1), 53–60 (2012)
15. Guimarães, P., Wigdahl, J., Ruggeri, A.: A fast and efficient technique for the automatic tracing of corneal nerves in confocal microscopy. Transl. Vis. Sci. Technol. **5**, 7 (2016)
16. Annunziata, R., Kheirkhah, A., Hamrah, P., Trucco, E.: Boosting hand-crafted features for curvilinear structure segmentation by learning context filters. In: Navab, N., Hornegger, J., Wells, W.M., Frangi, A.F. (eds.) MICCAI 2015. LNCS, vol. 9351, pp. 596–603. Springer, Cham (2015). https://doi.org/10.1007/978-3-319-24574-4_71
17. Al-Fahdawi, S., et al.: A fully automatic nerve segmentation and morphometric parameter quantification system for early diagnosis of diabetic neuropathy in corneal images. Comput. Methods Programs Biomed. **135**, 151–166 (2016)
18. Colonna, A., Scarpa, F., Ruggeri, A.: Segmentation of corneal nerves using a U-Net-based convolutional neural network. In: Stoyanov, D., et al. (eds.) OMIA/COMPAY -2018. LNCS, vol. 11039, pp. 185–192. Springer, Cham (2018). https://doi.org/10.1007/978-3-030-00949-6_22
19. Ronneberger, O., Fischer, P., Brox, T.: U-Net: convolutional networks for biomedical image segmentation. In: Navab, N., Hornegger, J., Wells, W.M., Frangi, A.F. (eds.) MICCAI 2015. LNCS, vol. 9351, pp. 234–241. Springer, Cham (2015). https://doi.org/10.1007/978-3-319-24574-4_28
20. Boukerroui, D., Noble, J., Brady, M.: On the choice of band-pass quadrature filters. J. Math. Imaging Vis. **21**, 53–80 (2004)
21. Hacihaliloglu, I., Rasoulian, A., Abolmaesumi, P., Rohling, R.: Local phase tensor features for 3D ultrasound to statistical shape+pose spine model registration. IEEE Trans. Med. Imaging **33**, 2167–2179 (2014)
22. Felsberg, M., Sommer, G.: The monogenic signal. IEEE Trans. Signal Process. **49**, 3136–3144 (2001)
23. Azzopardi, G., Strisciuglio, N., Vento, M., Petkov, N.: Trainable COSFIRE filters for vessel delineation with application to retinal images. Med. Image Anal. **19**, 46–57 (2015)

24. Frangi, A.F., Niessen, W.J., Vincken, K.L., Viergever, M.A.: Multiscale vessel enhancement filtering. In: Wells, W.M., Colchester, A., Delp, S. (eds.) MICCAI 1998. LNCS, vol. 1496, pp. 130–137. Springer, Heidelberg (1998). https://doi.org/10.1007/BFb0056195
25. Jerman, T., Penus, F., Likar, B., Spiclin, Z.: Enhancement of vascular structures in 3D and 2D angiographic images. IEEE Trans. Med. Imaging **35**(9), 2107–2118 (2016)
26. Zhao, Y., Rada, L., Chen, K., Zheng, Y.: Automated vessel segmentation using infinite perimeter active contour model with hybrid region information with application to retinal images. IEEE Trans. Med. Imaging **34**(9), 1797–1807 (2015)

Comparison of Interactions Between Control and Mutant Macrophages

José A. Solís-Lemus[1]([⊠]), Besaid J. Sánchez-Sánchez[2], Stefania Marcotti[2], Mubarik Burki[2], Brian Stramer[2], and Constantino C. Reyes-Aldasoro[1]

[1] School of Mathematics, Computer Science and Engineering,
City, University of London, London, UK
`jose.solis-lemus.2@city.ac.uk`
[2] Randall Centre for Cell and Molecular Biophysics,
King's College London, London, UK

Abstract. This paper presents a preliminary study on macrophages migration in *Drosophila* embryos, comparing two types of cells. The study is carried out by a framework called `macrosight` which analyses the movement and interaction of migrating macrophages. The framework incorporates a segmentation and tracking algorithm into analysing motion characteristics of cells after contact. In this particular study, the interactions between cells is characterised in the case of control embryos and Shot3 *mutants*, where the cells have been altered to suppress a specific protein, looking to understand what drives the movement. Statistical significance between control and mutant cells was found when comparing the direction of motion after contact in specific conditions. Such discoveries provide insights for future developments in combining biological experiments to computational analysis.

Keywords: Segmentation · Cell tracking · Track analysis · Macrophages

1 Introduction

Cell migration is highly involved in development and adult life, in maintaining homeostasis with processes such as wound healing and immune response [1,2]. Moreover, pathological conditions such as cancer or autoimmune disease, are related to dysfunctional cell migration. At present, many aspects of cell migration are known, however no single model is able to integrate all the cues driving motion [3]. Macrophages are highly migratory cells of the immune system that have ranges of functions ranging from tissue repair to immune responses to foreign pathogens [4]. However, excessive migration can be related to autoimmune disease and cancer [5]. An ideal model system to study in vivo cell migration are the embryonic macrophages of *Drosophila megalonaster*, as they are amenable to high spatio-temporal resolution live imaging [6]. In [7], contact inhibition of locomotion was described in these cells, showing that these cell-cell interactions

© Springer Nature Switzerland AG 2020
Y. Zheng et al. (Eds.): MIUA 2019, CCIS 1065, pp. 470–477, 2020.
https://doi.org/10.1007/978-3-030-39343-4_40

are needed for functional migration. In the present study, macrophages from control embryos were compared to Shot3 embryos.

Cell tracking is defined as the linking between objects in a temporal context. In this work, tracking of cells is achieved by segmenting the cells to obtain positions and then linking between the same object in two positions in consecutive time frames. Segmentation and tracking of cells is a widely studied area [8,9]. However, few studies have been made on identifying patterns in the migration or providing a biological context to the tracks obtained. In previous work, the analysis of macrophages' movement has been studied in the context of the cell-shape evolution [10], as well as the comparison of movement patterns of interacting cells from non-interacting [11]. In [12], a framework was presented to analyse the tracks of migrating macrophages, analysing the movement related to the interactions.

In this work, a study on novel data is presented, where time sequences of control and mutant macrophages were acquired and an underlying difference in the motion is searched for. The main contribution in this work consists of the use of a software framework to provide quantitative measurements to provide comparative quantitative measurements of different conditions. Figure 1 represents the differences in movement patterns hypothesised in this work: to distinguish through image analysis cases of (a) control and (b) mutant cells.

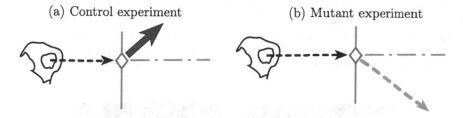

Fig. 1. Illustration of the hypothesis behind this paper. Different movement patterns from control to mutant experiments are represented by the different types of line and colours in the diagram. (Color figure online)

2   Materials

Fourteen time sequences of macrophages in *Drosophila* embryos were acquired following the protocol described in [6,7], with nuclei labelled in red and microtubules in green. Each image in the time-lapse sequences was obtained every ten seconds at a pixel density of $0.21\mu m$. The 14 experiments consist of images of size $(n_w, n_h, n_d) = (512, 672, 3)$.

The datasets are classified as control or mutant experiments. Four control and ten Shot3 mutant experiments were analysed, in which mutation affects the cytoskeletal crosslinking. The number of frames within the control experiments range between 137 and 272, while the mutant experiments range between 135 and 422 frames. Figure 2 shows a comparison between four frames of a control experiment against four frames of a mutant experiment.

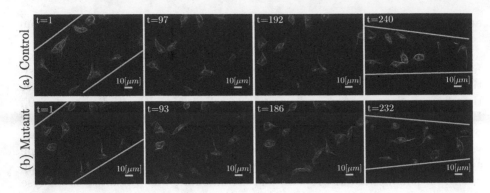

Fig. 2. Comparison between five frames of (a) control against five of (b) mutant experiments. The datasets were chosen because they had a similar number of frames and thus a similar spacing between the frames in both experiments could be shown ($\approx$95). Yellow lines have been manually added to the first and last frames on both experiments to showcase the apparent change of focus of the microscope as time evolves. (Color figure online)

Overlapping events, defined as *clumps* are relevant to the study of interactions caused by cell-cell contact, as presented in Fig. 3. Given certain circumstances, cells have been shown to align their microtubules and drastically change their direction of movement [7]. The contact observed in certain *clumps* suggest an alteration of the migration patterns of the cells involved. This type of interaction was measured first in [12], where cell-cell contact was shown to be influential in the movement of cells.

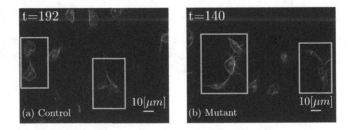

Fig. 3. Representation of *clumps* in control and mutant experiments. Both datasets present overlapping events, where *clumps* are formed. The detail of two frames from Fig. 2 is shown, highlighting *clumps* in both types of experiment.

3   Methods

Macrosight [12] is a framework for the analysis of moving macrophages capable of segmenting the two layers of fluorescence in the dataset presented previously, and apply the keyhole tracking algorithm inside the PhagoSight framework [13] on the centroids of the segmented nuclei.

Figure 4 shows an illustration of the flow of information in `macrosight`. Each track generated T_r contains information on the (i) position $\mathbf{x}_t$ at a given time frame t, (ii) track identifier r, (iii) velocity v_t, and whether the cell is part of a clump.

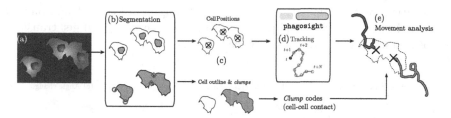

Fig. 4. Illustration of part of the `macrosight` framework. (a) Represents the original image sequences. The two levels of fluorescence are segmented in (b) based on a hysteresis threshold where the levels are selected by the [14] algorithm. The segmentation of the red channel provides the positions necessary to produce the tracks (d) of the cells using the keyhole tracking algorithm [13]. Finally, the tracks' information is combined with the clump information from the segmented green channel to allow analysis of movement based on contact events (e).

Each *clump* can be uniquely identified through a code $c(r, q)$, where $r > q$ indicate that at a certain time frame t, tracks T_r and T_q belong in the same clump. This allows for each interaction to be analysed. Several tracks can come together into a single clump, thus the *clump* codes evolve. Figure 5 represents the evolution of a given track T_2 and its involvement in two different clumps.

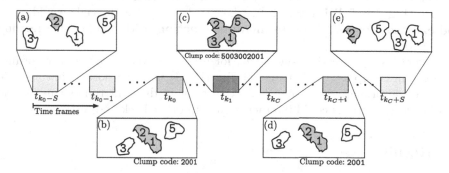

Fig. 5. Illustration of clump codes to the different time frames for a particular track T_2. The horizontal axis represents the time, and the detail of five frames is presented to illustrate the evolution of track T_2 as it interacts with other cells. In (a) and (e), track T_2 is not in contact with any other cell, thus no clump is present. (b) Represents the moment when T_2 and T_1 start interacting in clump 2001. Following in (c), tracks T_3 and T_5 become present in the clump, thus the *clump* code changes to 5003002001.

3.1   Movement Analysis Experiments

The events of interest in this paper consist of analysing cell-cell contact events of two cells. The change of direction $\theta_x \in (-\pi, \pi)$ is calculated by taking the positions of the tracks $\mathcal{T}_r$ and $\mathcal{T}_q$ up to S frames prior the first contact at time frame t_{k_0}, as well as the positions up to S frames after the last time frame of contact t_{k_c}. The time in clump $TC = t_{k_c} - t_{k_0}$ refers to the number of frames the two tracks interact in a given instance of the clump, and it is not taken into consideration for the calculation of angle θ_x. A diagram of the calculation of θ_x is provided in Fig. 6, where the positions on the image $\mathbf{x} = (x, y)$ get translated and rotated into new frame of reference (x', y').

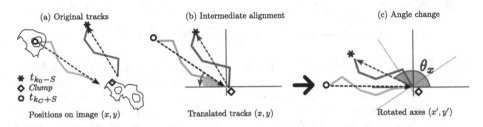

Fig. 6. Illustration of direction change (θ_x) measurement. Three markers represent different positions of a given track. The markers are (○) represents S frames before contact; (◇) represents the starting instant of the clump; and (∗) represents the position where the experiment is finalised. Notice the translation and rotation into the new frame of reference (x', y').

3.2   Selection of Experiments

All available datasets were segmented and tracked. The tracks' information was searched to find types of *clumps* which fulfil the criteria: (i) **only two cells interacting**; (ii) **full interaction**, where at least one of the cells would enter and exit the *clump*; and (iii) **immediate reaction**, with a value of S ranging from 3 to 5.

The changing direction angles, θ_x, for each case were calculated, recording the time in clump TC and change of direction. It is worth noting that a single clump could provide more than one experiment in different time spans, as the two interacting tracks could interact with each other back and forth.

4   Results

After the processes of segmentation, tracking, and selection of suitable experiments, twenty four control and thirty nine mutant cases were selected for analysis. Table 1 shows the number of cases per dataset selected, it is worth noting the different numbers of experiments fitting the criteria between datasets.

Table 1. Number of suitable experiments per dataset. Notice that not all datasets provided the same number of experiments for the analysis, as one or more of the selecting criteria would not be fulfilled. Also, the mutant datasets 01,02 and 09 did not provide any suitable experiments due to clumps always involving more than two cells.

Dataset ID	n experiments	Dataset ID	n experiments		
CONTROL01	14	MUTANT03	10	MUTANT07	3
CONTROL02	4	MUTANT04	2	MUTANT08	2
CONTROL04	4	MUTANT05	2	MUTANT10	4
CONTROL05	2	MUTANT06	9	MUTANT11	7
TOTAL	24			TOTAL	39

The resulting tracks representing changes of direction are shown in Fig. 7 for (a) control and (b) mutant. Differences can be observed in the displacement of the cells towards and from the centre, in the horizontal direction x'.

Boxplots showing the change of direction angle θ_x, time in clump TC and distances from the centre in the x' are presented in Fig. 8. Notice that in Fig. 8(a), the angle θ_x for mutant experiments appear to be distributed with more cases towards the lower angles, or a smaller change of direction after the contact. However, on its own, this measurement did not provide a statistically significant difference. The data points where $\theta_x < 90$ were chosen, as seen in Fig. 9. A t-test was calculated between the remaining angles showing statistical significance ($p = 0.03 < 0.05$).

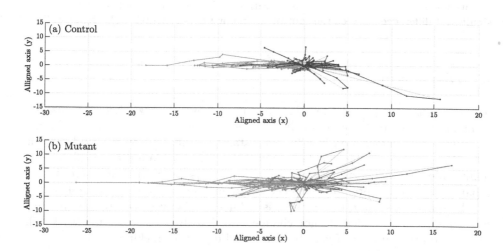

Fig. 7. Comparison of aligned tracks for (a) Control and (b) Mutants experiments. Refer to Fig. 6 on how to read this figure.

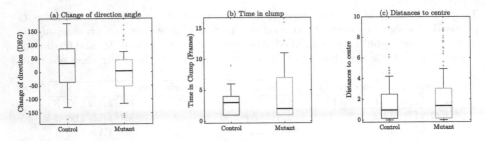

Fig. 8. Comparison of relevant variables between Control (blue) and Mutant (red) experiments. (a) Change of direction angle, θ_x, coming from Fig. 7. (b) Time in clump TC in frames. Finally, (c) shows the distances to the centre. (Color figure online)

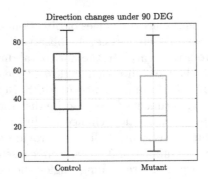

Fig. 9. Change of direction differences between Control (blue) and Mutant (red) experiments for angles under 90°. After observation of Fig. 7(a), the smaller angles show a significant difference between the control and mutant experiments. (Color figure online)

5   Discussion

The previous work presented in [12] presented a novel framework for the analysis of macrophages migration in a controlled environment. In this work, the framework was extensively used in different datasets comparing control and mutant cells. While some of the calculations still did not provide a statistically significant difference between control and mutant cells, some insights were found. Apart from the qualitative differences between the measurements presented it can be noted, as seen in Fig. 9, there is a difference between control and mutant experiments in cells that do not change direction drastically. Future work will improve on the number of variables collected from the tracking.

References

1. Martinez, F.O., Sica, A., Mantovani, A., Locati, M.: Macrophage activation and polarization. Front. Biosci. J. Virtual Libr. **13**, 453–461 (2008)
2. Wood, W., Martin, P.: Macrophage functions in tissue patterning and disease: new insights from the fly. Dev. Cell **40**(3), 221–233 (2017)
3. Petrie, R.J., Doyle, A.D., Yamada, K.M.: Random versus directionally persistent cell migration. Nat. Rev. Mol. Cell Biol. **10**(8), 538–549 (2009)
4. Chawla, A., Pollard, J.W., Wynn, T.A.: Macrophage biology in development, homeostasis and disease. Nature **496**(7446), 445 (2013)
5. Pocha, S.M., Montell, D.J.: Cellular and molecular mechanisms of single and collective cell migrations in Drosophila: themes and variations. Annu. Rev. Genet. **48**, 295–318 (2014)
6. Stramer, B., Wood, W.: Inflammation and wound healing in Drosophila. In: Jin, T., Hereld, D. (eds.) Chemotaxis. Methods in Molecular Biology™ (Methods and Protocols), vol. 571, pp. 137–149. Humana Press, Totowa (2009). https://doi.org/10.1007/978-1-60761-198-1_9
7. Stramer, B., et al.: Clasp-mediated microtubule bundling regulates persistent motility and contact repulsion in Drosophila macrophages in vivo. J. Cell Biol. **189**(4), 681–689 (2010)
8. Ulman, V., et al.: An objective comparison of cell-tracking algorithms. Nat. Methods **14**, 1141–1152 (2017)
9. Maška, M., et al.: A benchmark for comparison of cell tracking algorithms. Bioinformatics **30**(11), 1609–1617 (2014)
10. Solís-Lemus, J.A., Stramer, B., Slabaugh, G., Reyes-Aldasoro, C.C.: Shape analysis and tracking of migrating macrophages. In: IEEE 15th International Symposium on Biomedical Imaging (ISBI 2018), pp. 1006–1009. Washington, DC, April 2018
11. Solís-Lemus, J.A., Stramer, B., Slabaugh, G., Reyes-Aldasoro, C.C.: Analysis of the interactions of migrating macrophages. In: Nixon, M., Mahmoodi, S., Zwiggelaar, R. (eds.) MIUA 2018. CCIS, vol. 894, pp. 262–273. Springer, Cham (2018). https://doi.org/10.1007/978-3-319-95921-4_25
12. Solís-Lemus, J.A., Stramer, B., Slabaugh, G., Reyes-Aldasoro, C.C.: Macrosight: a novel framework to analyze the shape and movement of interacting macrophages using MATLAB®. J. Imaging **5**(1), 17 (2019)
13. Henry, K.M., Pase, L., Ramos-Lopez, C.F., Lieschke, G.J., Renshaw, S.A., Reyes-Aldasoro, C.C.: PhagoSight: an open-source MATLAB® package for the analysis of fluorescent neutrophil and macrophage migration in a Zebrafish model. PLoS ONE **8**(8), e72636 (2013)
14. Otsu, N.: A threshold selection method from gray-level histograms. IEEE Trans. Syst. Man, Cybern. **9**(1), 62–66 (1979)

Comparison of Multi-atlas Segmentation and U-Net Approaches for Automated 3D Liver Delineation in MRI

James Owler[✉], Ben Irving, Ged Ridgeway, Marta Wojciechowska, John McGonigle, and Sir Michael Brady

Perspectum Diagnostics Ltd., Oxford, UK
jamesowler97@gmail.com

Abstract. Segmentation of medical images is typically one of the first and most critical steps in medical image analysis. Manual segmentation of volumetric images is labour-intensive and prone to error. Automated segmentation of images mitigates such issues. Here, we compare the more conventional registration-based multi-atlas segmentation technique with recent deep-learning approaches. Previously, 2D U-Nets have commonly been thought of as more appealing than their 3D versions; however, recent advances in GPU processing power, memory, and availability have enabled deeper 3D networks with larger input sizes. We evaluate methods by comparing automated liver segmentations with gold standard manual annotations, in volumetric MRI images. Specifically, 20 expert-labelled ground truth liver labels were compared with their automated counterparts. The data used is from a liver cancer study, HepaT1ca, and as such, presents an opportunity to work with a varied and challenging dataset, consisting of subjects with large anatomical variations responding from different tumours and resections. Deep-learning methods (3D and 2D U-Nets) proved to be significantly more effective at obtaining an accurate delineation of the liver than the multi-atlas implementation. 3D U-Net was the most successful of the methods, achieving a median Dice score of 0.970. 2D U-Net and multi-atlas based segmentation achieved median Dice scores of 0.957 and 0.931, respectively. Multi-atlas segmentation tended to overestimate total liver volume when compared with the ground truth, while U-Net approaches tended to slightly underestimate the liver volume. Both U-Net approaches were also much quicker, taking around one minute, compared with close to one hour for the multi-atlas approach.

Keywords: Deep learning · Multi-atlas segmentation · Biomedical image segmentation

1 Introduction

Over the last few decades, the rapid development of non-invasive imaging technologies has given rise to large amounts of data; analysis of such large datasets

© Springer Nature Switzerland AG 2020
Y. Zheng et al. (Eds.): MIUA 2019, CCIS 1065, pp. 478–488, 2020.
https://doi.org/10.1007/978-3-030-39343-4_41

has become an increasingly complex task for clinicians. For example, in abdominal magnetic resonance imaging (MRI), image segmentation can be used for measuring and visualising internal structures, analysing changes, surgical planning, and extracting quantitative metrics. The high variability of location, size, and shape of abdominal organs makes segmentation a challenging problem. Segmentation of medical images is often one of the first and most critical steps in medical image analysis. Manual segmentation of volumetric medical images is a labour-intensive task that is prone to inter-rater and intra-rater variability. There is a need to automate the process to increase both efficiency and reproducibility and decrease subjectivity. Developing a robust automated segmentation method using deep learning has been an area of intense research in recent years [1].

Popular automated segmentation techniques include: using statistical models [2], image registration, classical machine learning algorithms [3] and, most recently, deep learning. Statistical models and classical machine learning algorithms do not generalise as well. Classical machine learning algorithms require careful feature engineering which is a time consuming and complex process. In this paper we compare 'feature engineering free' state-of-the-art multi-atlas based segmentation to recent deep-learning approaches.

Multi-atlas segmentation (MAS) was first introduced and popularised by Rohlfing et al. in 2004 [4]. Since then, substantial progress has been made [5]. MAS is a process that warps a number an expert-labelled atlas images (the moving images) into the coordinate space of a target image, via non-linear image registration. The iterative process of image registration involves optimising some similarity metric, such as cross-correlation, between the warped and target image by means of deformation of the 'moving' image [5]. Atlas labels are then propagated onto the target image and fused together in a way that can use the relationship between the target image and registered atlases, alongside the propagated labels [6]. MAS has the ability to capture anatomical variation much better than a model-based average or registration to one chosen case [4]. MAS is a computationally intensive process and has consequently grown in popularity due to an increase in computational resources [5].

Deep learning refers to neural networks with numerous layers that are capable of automatically extracting features [1]. This self-learning capability has a significant advantage over traditional machine learning algorithms, namely that features do not have to be hand-crafted. Deep learning has been applied to numerous fields [1]; image analysis tasks, such as object recognition, is one such. Convolutional neural networks (CNNs) were first introduced in 1989 [7] but have only recently become popular after a breakthrough in 2012 [9], along with rapid increases in GPU power, memory and accessibility. CNNs are currently by far the most popular approach for image analysis tasks [8].

Advancements in algorithms, and GPU technology, along with increased availability of training sets, has enabled larger, more robust networks. CNNs can automatically learn representations of image data with increasing levels of abstraction via convolutional layers [1]. These convolutional layers drastically reduce the number of parameters, when compared with traditional 'fully connected' neural networks, as weights are shared among convolutional layers [9].

U-Net was first introduced in 2015 by Ronneberger et al. as a deep CNN architecture geared towards biomedical image segmentation [10]. U-Net is a development of the fully convolutional network architecture [11]. A contracting encoder, which analyses the full image, is followed by an expanding decoder to provide the final segmentation; shallower layers in the network capture local information while deeper layers, whose retrospective field is much larger, capture global information [10]. The expanding decoder aims to recover a full-resolution pixel-to-pixel label map, from the different feature maps created in the contracting layers. Previously, 2D U-Net architectures have been thought of as more appealing than their 3D versions due to limitations in computational cost and GPU memory [12]. This said, current advancements in GPU memory and accessibility (cloud services such as Amazon Web Services) has enabled deeper 3D networks, with larger input sizes.

Here we compare the performance of a state-of-the-art multi-atlas segmentation approach with more recent 2D and 3D U-Net approaches. We evaluate the performance of each method by comparing manual ground truth liver labels with their automated counterparts, in challenging volumetric MRI images. Segmenting the liver is a process often used in surgical planning [13].

2   Materials

Data for the evaluation of methods was from 'HepaT1ca' a liver cancer study. T1-weighted 3D-SPGR images of the abdomen were collected according to the HepaT1ca study protocol [14]. HepaT1ca data implies working with a varied and challenging dataset, consisting of subjects with large anatomical variations responding from different tumours and resections. The different segmentation methods should ideally be robust to such dramatic changes between different images, and complicates both testing and training.

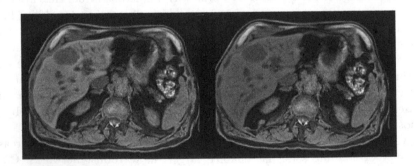

Fig. 1. Example 2D slice from a volumetric MRI image in the dataset with the corresponding liver label.

135 livers were labelled by a trained analyst. Subjects in the dataset have various tumours and resections of different shapes and sizes. The dataset was split into: 115 training cases and 20 test cases (Fig. 1).

3   Methods

3.1   Multi-atlas Segmentation Method

For the multi-atlas segmentation, 45 subjects were chosen at random from the 'training' set. A number of random atlases were used in order to capture anatomical variation within a population (one of the underlying principles of MAS). For each subject in the test set, the 45 random atlases were non-linearly registered to the test image.

The registration step was divided into two parts: affine registration (scaling, translation, rotation and shear mapping), followed by non-linear registration. We used the affine transform from ANTs (advanced normalisation tools) package [15]. We chose the DEEDS (dense displacement sampling) algorithm [16] for the non-linear part. DEEDS has been shown to yield the best registration performance in abdominal imaging [17], when compared with other common registration algorithms. Image grids are represented as a minimum spanning tree; a global optimum of the cost function is found using discrete optimisation. After all of the atlases (and their corresponding labels) had been registered to the target image, we used STEPS [6], as a template selection and label fusion algorithm, to produce the final liver segmentation.

3.2   Deep Learning Segmentation Methods

For the deep learning methods, we used slightly different model architectures for the 3D and 2D U-Net implementations. Figure 2 illustrates architecture of the 2D U-Net implementation; an expansion of the network previously used for quantitative liver segmentation [18].

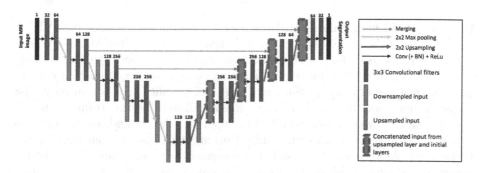

Fig. 2. Architecture of the extended 2D U-Net method used in the comparison

Like the original 2D U-Net, our implementation has a contracting and expanding path. The input size of the network is a 288 × 224 image with 1 channel (black and white image). Images were padded to ensure the output size of feature maps were the same size as the input image. We used batch normalisation (BN), after each convolution, to improve performance and stability of the network [19]. BN was followed by a rectified linear unit (ReLu) activation function. As suggested in [20], in the contracting path we doubled the number of channels, prior to max pooling, to avoid any bottlenecks in the network. We also applied this same principle in the expanding path. An addition of two max pooling (downsampling) layers, that further reduce the dimensionality of the image, resulted in better localisation, and as such, improve final segmentation performance of the network.

The 3D U-Net implementation was essentially identical to the 2D network, but with 3D operations instead, e.g. a 3 × 3 × 3 convolution instead of a 3 × 3 convolution. The input size to the network was 224 × 192 × 64 voxels. We also used one fewer max pooling layer; the lowest resolution/ highest dimensional representation of an image was 14 × 12 × 4. This allowed for better depth localisation. If we still had 2 additional max pooling layers there would only have been 2 layers in the depth dimension.

Each model was implemented using the Keras framework, with Tensorflow as the backend.

Pre-processing. Each image underwent some pre-processing before being fed into the network. First, we applied 3 rounds of N4 Bias Field Correction [21] to remove any image contrast variations due to magnetic field inhomogeneity. All intensity values were then normalised between a standard reference scale (between 0 and 100). We also winsorised images, by thresholding the maximum intensity value to the 98th percentile; a heuristic that gives a reasonable balance between the reduction of high signal artefacts and image contrast.

For the 2D U-Net network, volumes were split into their respective 2D slices in the axial plane. Slices could then be reassembled into their respective volumes after a liver segmentation had been predicted by the network.

Training and Data Augmentation. Both networks were trained on NVIDIA Tesla V100 GPUs for 100 epochs, with a learning rate of 0.00005. We used a batch size of 10 and 1 for the 2D and 3D networks, respectively. During training we employed strategies to prevent overfitting; an important process that ensures that true features of images are learnt, instead of specific features that only exist in the training set. In addition to batch normalisation, when training the 2D network, slices were randomly shuffled between all subjects and batches for each epoch. Batch order was also randomized during each epoch when training the 3D network. Anatomically plausible data augmentation was applied 'on-the-fly' to further reduce the risk of overfitting. We applied small affine transformations with 5 degrees of rotation, 10% scaling and 10% translation. Both networks used Adam optimisation [22] with binary cross-entropy as the loss function. Each network took around 6 h to train.

After each network was trained, liver masks were predicted for each volumetric image in the test set.

3.3 Evaluation

Methods were evaluated by comparing the automated liver segmentations, produced by the automated segmentation methods, with the expert-labelled ground truth image. The first comparison metric we used was the Dice overlap score.

$$Dice = \frac{(2 \times \sum(X \cap Y))}{(\sum X + \sum Y)} \tag{1}$$

Dice measures the number of voxels that overlap between the ground truth segmentation (X) and the automated segmentation (Y). A score of 1 represents a perfect overlap between two 3D segmentations, while 0 represents no overlap between segmentations. In addition to Dice, we also measured performance by calculating the percentage difference in volume between the ground truth and automated segmentations. This highlighted if methods tended to underestimate or overestimate total liver volume.

$$dV = \frac{V2 - V1}{V1} \times 100 \tag{2}$$

V1 represents the liver volume of the ground truth segmentation. V2 represents the liver volume of the automated segmentation. The time taken for each segmentation method to run was also recorded for each test image.

We then used a paired t-test, to evaluate the differences in mean and variance between Dice metrics and volume percentage differences, between the segmentations produced by each method.

4   Results

Figure 3 shows a boxplot of the Dice overlap scores between the automated liver segmentations and ground truth annotations, for all images in the test set.

The multi-atlas approach was found to perform significantly worse than both the 3D U-Net (t = 3.397, p = 0.003)[1] and 2D U-Net (t = 2.628, p = 0.017)[2] approaches, while between the U-Net approaches, the 3D version slightly outperforms the 2D version (t = 2.016, p = 0.051) (Fig. 4).

Figures 5 and 6 show examples of a single slice from different volumetric images, their corresponding automated liver segmentations and the ground truth liver segmentations. Figure 5 shows a more challenging case in the test set, whereby the subject has had a previous liver resection and is missing a substantial part of the liver. Figure 6 highlights an image with an exemplar liver.

---

[1] (t = 4.886, p = 0.0001) excluding the MAS outlier.
[2] (t = 3.499, p = 0.003) excluding the MAS outlier.

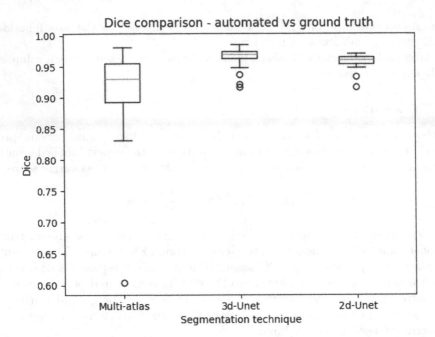

Fig. 3. Boxplot of dice scores for each segmentation method. The box extends from the lower to upper quartile values of the data, with a line at the median. The whiskers extend from the box to show the range of the data. Flier points are those past the end of the whiskers. The outlier seen in the multi-atlas dice scores is 3.8 standard deviations away from the mean.

	MULTI-ATLAS	**3D U-NET**	**2D U-NET**
Median Dice	0.931	0.970	0.957
Min Dice	0.604	0.916	0.916
Max Dice	0.981	0.985	0.970

Fig. 4. Numerical values from the dice scores

The multi-atlas approach tended to overestimate the overall volume of the liver. The percentage differences in volume, of the automated multi-atlas segmentations with the ground truth, were significantly different than differences in volume for the 2D U-Net ($t = 3.432$, $p = 0.003$)[3] and the 3D U-Net ($t = 3.812$, $p = 0.001$)[4]. The 3D U-Net tended to underestimate the volume

---

[3] ($t = 4.906$, $p = 0.0001$) excluding the MAS outlier.
[4] ($t = 5.381$, $p = 0.00004$) excluding the MAS outlier.

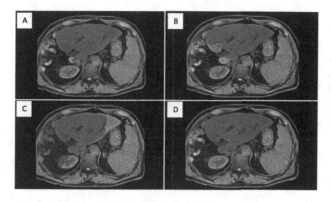

Fig. 5. A more challenging case for the automated techniques. (A) 3D U-Net, (B) 2D U-Net, (C) Multi-atlas, (D) Manual annotation.

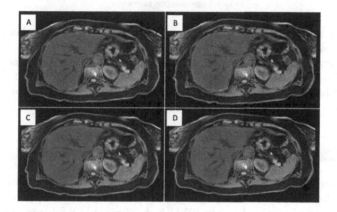

Fig. 6. A case with an exemplar liver. (A) 3D U-Net, (B) 2D U-Net, (C) Multi-atlas, (D) Manual annotation.

(median $= -3.01\%$) more the then 2D U-Net (median $= -0.15\%$). The distributions in volume differences between the 2D U-Net and 3D U-Net are different ($t = 3.824$, $p = 0.001$).

The multi-atlas based approach took close to 1 h to compute a liver segmentation. The deep-learning based approaches, once trained, took around 1 min to compute a liver segmentation (Fig. 7).

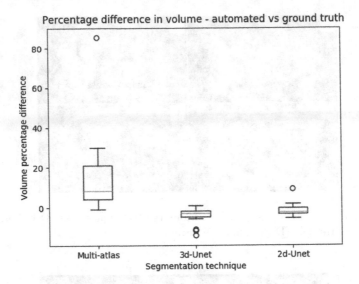

Fig. 7. Boxplot of the percentage differences in volumes for all test cases for each segmentation method. The box extends from the lower to upper quartile values of the data, with a line at the median. The whiskers extend from the box to show the range of the data. Flier points are those past the end of the whiskers. The outlier seen in the multi-atlas volume percentage difference scores is 3.8 standard deviations away from the mean.

5   Discussion

Both deep learning models significantly outperformed the multi-atlas based app-roach, with the 3D U-Net achieving slightly better performance, in terms of over-lap with the ground truth, than the 2D U-Net. This could be due to spatial encod-ing in the 3D U-Net; inputs to the 2D network are completely independent slices that have no information about where, within the volume, the slice was located.

A disadvantage of fully convolutional 3D networks is that they have much larger computational cost and GPU memory requirements. Previously, this has limited the depth of a network and the filters' field-of-view, two key factors for performance gains, resulting in better performance from 2D networks. More complex network architectures have been developed to avoid some of these draw-backs by using 2D networks with encoded spatial information [12]. However, recently state-of-the-art GPUs are now easily accessible on Cloud services, such as Amazon Web Services, and have increasingly larger amounts of GPU pro-cessing power and memory, allowing for deeper networks and larger inputs. We believe utilising these state-of-the-art GPUs was a contributing factor to the superior performance from the 3D network. We did not have to employ patch-based methods in order to effectively use a 3D network. Input images were only slightly downsampled (to around 90% of the original dimension) to fit the 3D model in GPU memory. This slight downsampling could be a reason why the 3D

U-Net underestimated liver volume more than the 2D U-Net. 2D U-Nets could be more useful for automated liver volumetry; however, this does not mean 2D U-Nets are the best for other applications such as surgical planning and extracting quantitative metrics.

The deep learning approaches were several orders of magnitude faster than the multi-atlas based approach. Although, in a clinical workflow for fully-automated segmentation, this may not be a limiting factor, faster segmentation time does provide significant advantages when analysing larger datasets. Inter-observer variability is a factor to consider when assessing the performance of an automated segmentation. However, here ground-truth delineations were provided by a single annotater which was appropriate given the tests were of how close the methods resembled the annotations they were trained on.

When using the multi-atlas segmentation method, we saw a much larger variation in segmentation accuracy when compared with the deep learning approaches (Fig. 3). The probabilistic multi-atlas approach did not generalize well when compared with the deep learning approaches, this could be due to insensitivity to biologically-relevant variance (as seen in Fig. 5). Variance could be more apparent in this dataset due to a larger variation in liver shapes and sizes between subjects, from tumours and previous liver resections. That being said, although not studied in depth here, the number of tumours within a liver did not seem to alter the segmentation performance of any of the automated techniques. There are more advance atlas selection techniques [23] which could reduce the variance; however, it does not alleviate the computational time drawback of a multi-atlas technique.

In conclusion, the U-Net approaches were much more effective at automated liver delineation (once trained), both in terms of time and accuracy, than the multi-atlas segmentation approach.

Acknowledgements. Portions of this work were funded by a Technology Strategy Board/Innovate UK grant (ref 102844). The funding body had no direct role in the design of the study and collection, analysis, and interpretation of data and in writing the manuscript.

References

1. Lecun, Y., Bengio, Y., Hinton, G.: Deep learning. Nature **521**, 436–444 (2015)
2. Heimann, T., Meinzer, H.P.: Statistical shape models for 3D medical image segmentation: a review. Med. Image Anal. **13**, 543–563 (2009)
3. Fritscher, K., Magna, S., Magna, S.: Machine-learning based image segmentation using Manifold Learning and Random Patch Forests. In: Imaging and Computer Assistance in Radiation Therapy (ICART) Workshop, MICCAI 2015, pp. 1–8 (2015)
4. Rohlfing, T., Russakoff, D.B., Maurer Jr., C.R.: An expectation maximization-like algorithm for multi-atlas multi-label segmentation. In: Proceedings of the Bildverarbeitung frdie Medizin, pp. 348–352 (2004)
5. Iglesias, J.E., Sabuncu, M.R.: Multi-atlas segmentation of biomedical images: a survey. Med. Image Anal. **24**, 205–219 (2015)

6. Jorge Cardoso, M., et al.: STEPS: similarity and truth estimation for propagated segmentations and its application to hippocampal segmentation and brain parcelation. Med. Image Anal. **17**, 671–684 (2013)

7. Lecun, Y., Jackel, L.D., Boser, B., Denker, J.S., Gral, H., Guyon, I.: Handwritten digit recognition. IEEE Commun. Mag. **27** (1989)

8. Zhao, Z.-Q., Zheng, P., Xu, S., Wu, X.: Object detection with deep learning: a review. IEEE Trans. Neural Netw. Learn. Syst. (2019)

9. Krizhevsky, A., Sutskever, I., Hinton, G.: ImageNet classification with deep convolutional neural networks. In: Advances in Neural Information Processing Systems, vol. 2, pp. 1097–1105 (2012)

10. Ronneberger, O., Fischer, P., Brox, T.: U-Net: convolutional networks for biomedical image segmentation. In: Navab, N., Hornegger, J., Wells, W.M., Frangi, A.F. (eds.) MICCAI 2015. LNCS, vol. 9351, pp. 234–241. Springer, Cham (2015). https://doi.org/10.1007/978-3-319-24574-4_28

11. Long, J., Shelhamer, E., Darrell, T.: Fully convolutional networks for semantic segmentation. In: Proceedings of the IEEE Conference on Computer Vision and Pattern Recognition, pp. 3431–3440 (2015)

12. Li, X., Chen, H., Qi, X., Dou, Q., Fu, C., Heng, P.: H-DenseUNet: hybrid densely connected UNet for liver and tumor segmentation from CT volumes. IEEE Trans. Med. Imaging **37**, 2663–2674 (2018)

13. Gotra, A., et al.: Liver segmentation: indications, techniques and future directions. Insights Imaging **8**, 377–392 (2017)

14. Mole, D.J., et al.: Study protocol: HepaT1ca, an observational clinical cohort study to quantify liver health in surgical candidates for liver malignancies. BMC Cancer **18**, 890 (2018)

15. Avants, B.B., Tustison, N.J., Song, G., Cook, P.A., Klein, A., Gee, J.C.: A reproducible evaluation of ANTs similarity metric performance in brain image registration. Neuroimage **54**(3), 2033–2044 (2010)

16. Heinrich, M.P., Jenkinson, M., Brady, S.M., Schnabel, J.A.: Globally optimal deformable registration on a minimum spanning tree using dense displacement sampling. In: Ayache, N., Delingette, H., Golland, P., Mori, K. (eds.) MICCAI 2012. LNCS, vol. 7512, pp. 115–122. Springer, Heidelberg (2012). https://doi.org/10.1007/978-3-642-33454-2_15

17. Xu, Z., et al.: Evaluation of six registration methods for the human abdomen on clinically acquired CT. IEEE Trans. Biomed. Eng. **63**, 1563–1572 (2016)

18. Irving, B., et al.: Deep quantitative liver segmentation and vessel exclusion to assist in liver assessment. In: Valdés Hernández, M., González-Castro, V. (eds.) MIUA 2017. CCIS, vol. 723, pp. 663–673. Springer, Cham (2017). https://doi.org/10.1007/978-3-319-60964-5_58

19. Ioffe, S., Szegedy, C.: Batch normalization: accelerating deep network training by reducing internal covariate shift (2015)

20. Szegedy, C., Vanhoucke, V., Ioffe, S., Shlens, J., Wojna, Z.: Rethinking the inception architecture for computer vision. In: Proceedings of the IEEE Conference on Computer Vision and Pattern Recognition, pp. 2818–2826 (2016)

21. Tustison, N.J., et al.: N4ITK: improved N3 bias correction. IEEE Trans. Med. Imaging **29**, 1310–1320 (2010)

22. Kingma, D.P., Ba, J.: Adam: a method for stochastic optimization. arXiv Preprint arXiv:1412.6980 (2014)

23. Antonelli, M., et al.: GAS: a genetic atlas selection strategy in multi-atlas segmentation framework. Med. Image Anal. **52**, 97–108 (2019)

DAISY Descriptors Combined with Deep Learning to Diagnose Retinal Disease from High Resolution 2D OCT Images

Joshua Bridge[✉], Simon P. Harding, and Yalin Zheng

Department of Eye and Vision Science, University of Liverpool,
Liverpool L7 8TX, UK
{jbridge,sharding,yzheng}@liverpool.ac.uk,
http://www.liv-cria.co.uk

Abstract. Optical Coherence Tomography (OCT) is commonly used to visualise tissue composition of the retina. Previously, deep learning has been used to analyse OCT images to automatically classify scans by the disease they display, however classification often requires downsampling to much lower dimensions. Downsampling often loses important features that may contain useful information. In this paper, a method is proposed which incorporates DAISY descriptors as 'intelligent downsampling'. By avoiding random downsampling, we are able to keep more of the useful information to achieve more accurate results. The proposed method is tested on a publicly available dataset of OCT images, from patients with diabetic macula edema, drusen, and choroidal neovascularisation, as well as healthy patients. The method achieves an accuracy of 76.6% and an AUC of 0.935, this is an improvement to a previously used method which uses InceptionV3 with an accuracy of 67.8% and AUC of 0.912. This shows that DAISY descriptors do provide good representations of the image and can be used as an alternative to downsampling.

Keywords: DAISY descriptors · Deep learning · Retina · OCT · GRU

1 Introduction

Optical Coherence Tomography (OCT) is an imaging method commonly used to analyse tissue composition [1]. It has recently been shown that deep learning has the ability to detect several retinal diseases from OCT images [6,10]. One challenge encountered when analysing OCT data is the high resolution of the scans, passing these images straight to a deep learning network often results in an out of memory error. Current methods often require downsampling to much lower resolutions to make computation practical. Conventional downsampling methods often lose important features [7]. Previous methods often focus on making the image appear similar to a human observer, who may rely on different features to those which a computer may recognise. The method proposed here incorporates DAISY image descriptors to greatly reduce image dimension, followed by

© Springer Nature Switzerland AG 2020
Y. Zheng et al. (Eds.): MIUA 2019, CCIS 1065, pp. 489–496, 2020.
https://doi.org/10.1007/978-3-030-39343-4_42

a three layer Gated Recurrent Unit (GRU) network [5] to classify the images according to disease (see Fig. 1). The use of image descriptors provides a data efficient alternative to downsampling and allows us to use more of the useful information contained in each image. The aim of this study is to demonstrate that DAISY descriptors can successfully represent images, acting as 'intelligent downsampling', we aim to demonstrate that the method is a viable alternative to downsampling.

We demonstrated the method on a publicly available dataset of OCT scans [10]. The OCT images are split into four groups; normal, drusen, choroidal neovascularisation (CNV), and diabetic macular edema (DME). The normal group have no visible disease from the OCT scan and are classed as being healthy. The drusen group have small lipid deposits in the macula, which are commonly found in older people. Patients with drusen are more likely to develop age related macular degeneration (AMD) in the future [12] AMD is a leading cause of vision loss in older patients and greatly affects daily activities. The CNV group have signs of CNV which is indicative of wet AMD [12], this is when new blood vessels begin to form in the choroid [9]. CNV can often lead to blindness, although there are some treatments options available [3]. The final group consists of patients with DME. DME is a common result of diabetic retinopathy and causes vision problems in patients with diabetes [2].

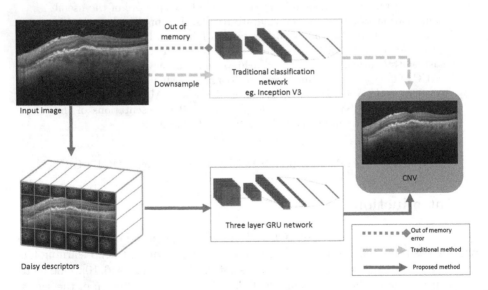

Fig. 1. The proposed framework aims to utilise DAISY descriptors as 'intelligent downsampling', followed by a GRU network to classify the images from the descriptors.

2 Methods

The method is extendible to many different types of images, here demonstrate the method using OCT, which has been likened to ultrasound, using light instead of sound to produce a cross-sectional view of tissue composition [1]. Often OCT images are combined to produce a 3D representation of the tissue. OCT is commonly used to image the back of the eye (fundus) to diagnose eye disease.

2.1 Dataset

The data consists of OCT images, collected by Shiley Eye Institute and made publicly available [10]. In this dataset, a single OCT B-scan made up each image. Examples of OCT images are shown (see Fig. 3), depicting the 4 ocular diseases contained in this dataset. The original dataset contains 108,314 training images from 4,686 patients and 1,000 validation images from a separate 663 patients, split into 4 groups according to the disease type they display. In this preliminary work, we used a subset of this data to save time, 20,020 images were used for training, 4,112 images for validation, and the original 1,000 images were used for testing (Fig. 2). Original images ranged in size from 512×496 to 1536×496. Images were first rescaled to 1500×1000 pixels, as all images must be the same size in this method.

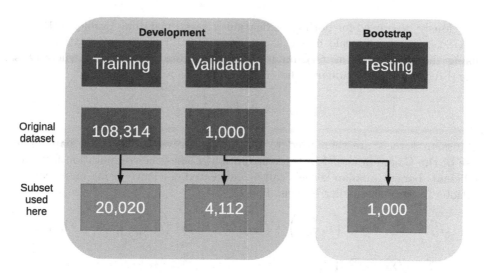

Fig. 2. Visual representation of the dataflow.

2.2 DAISY

DAISY is a method of image description which is similar to SIFT but is much faster to compute, due to it's implementation of Gaussian kernels. DAISY has

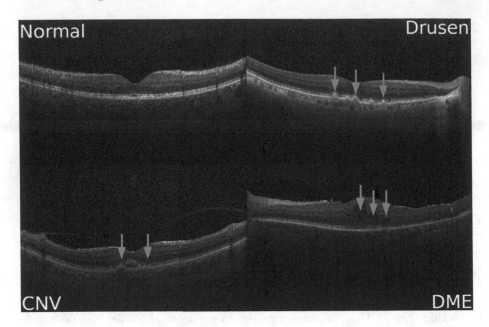

Fig. 3. Examples from each of the four groups included in the dataset. Arrows indicate areas highlighting the identified pathology.

previously been used for both classification [11] and matching problems [13,14]. DAISY takes a greyscale image as an input, I, and creates orientation maps using the gradient norm, G_O, for a specified number of directions, where O is the direction of the gradient. The orientation maps are calculated as:

$$G_O = max\left(0, \frac{\partial I}{\partial O}\right).$$

Gaussian kernels are then used to produce convolved orientation maps. The use of the Gaussian kernel in the convolutions makes the computation fast and efficient. Large Gaussian kernels can be calculated efficiently using many convolutions from much smaller kernels. For $\Sigma_1 < \Sigma_2$:

$$G_O^{\Sigma_2} = G_{\Sigma_2} * max\left(0, \frac{\partial I}{\partial O}\right) = G_\Sigma * G_{\Sigma_1} * max\left(0, \frac{\partial I}{\partial O}\right) = G_\Sigma * G_O^{\Sigma_1},$$

using $\Sigma = \sqrt{\Sigma_2^2 - \Sigma_1^2}$ [14].

At each pixel location, DAISY produces a vector of values obtained from the convolved orientation maps. The final output is a 3D tensor representation of the image, which is smaller than the original image. DAISY hyperparameters can be chosen to produce either sparse or dense descriptions of images, DAISY is mainly used for dense description as it is efficient compared to similar methods, such as SIFT and GLOH [14]. A visual representation of DAISY descriptors is displayed on an example image (Fig. 4).

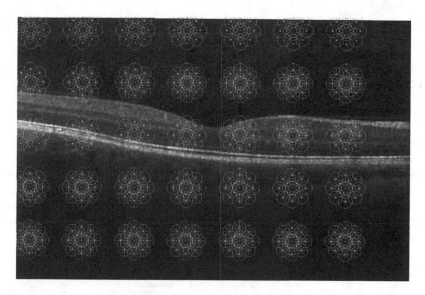

Fig. 4. Visual representation of DAISY descriptors on an example image. The DAISY hyperparameters were chosen to produce a sparse representation to clearly display the rings in DAISY. Each descriptor consists of a centre ring, 8 middle rings, and 8 outer rings, each with 8 directions. The output of the DAISY descriptors algorithm shown here is a $4 \times 6 \times 136$ tensor of descriptors. This example shows a sparse representation to highlight how DAISY describes the images.

2.3   Classification Network

The classification network consists of a 3 layer GRU network [4]. GRU networks are a recurrent neural network, which uses gated units to control the flow of information, update and reset gates. GRU has been shown to perform better than older recurrent units such as tanh and is at least as good as LSTM [5].

This network was trained using Keras 2.2.4 on an Ubuntu 18.04 machine with a Titan X 12 GB GPU and 32 GB of memory. Training was performed for 500 epochs of 200 steps each using the Nesterov Adam optimiser, batch size was set to 32, with categorical cross-entropy used as the loss function. Parametric Rectified Linear Unit (PReLU) was used as the activation function for the hidden GRU layers and Softmax was used in the output layer. Early stopping with a patience of 10 epochs was used to prevent overfitting, model checkpoints were used to select the best classification model, based on the loss in the validation dataset.

2.4   Model Performance

Model performance was assessed using loss, accuracy, and the Area Under the receiver operating Curve (AUC), with categorical cross-entropy used as the loss function. The dataset included a testing dataset with patients independent to those in the training and validation dataset [10]. The model was assessed on this testing dataset, with bootstrapping used to construct confidence intervals.

3 Results

3.1 Training and Validation

The output of the DAISY algorithm was a $16 \times 24 \times 200$ tensor which is a suitable size for computation. This took an average of 4.9 s per image to produce, training the deep learning model then took 29 s per epoch. The model was trained until convergence (see Fig. 5), at this point the best model based on validation accuracy was chosen. In the training dataset we achieved an overall accuracy of 93.7% and AUC of 0.9358, in the validation dataset accuracy was 76.5% and AUC was 0.9359.

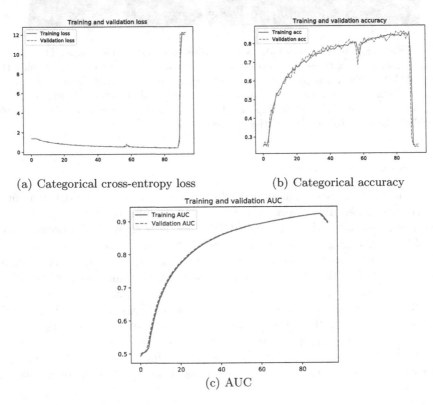

(a) Categorical cross-entropy loss   (b) Categorical accuracy

(c) AUC

Fig. 5. Model performance at each epoch in the testing and validation datasets, for loss 5(a), accuracy 5(b), and AUC 5(c). After around 90 epochs, model performance appears to have converged and no further epochs were required.

3.2 Testing Dataset Performance

Bootstrapping [8] was performed on 1,000 samples, the median values and 95% confidence intervals were 77.1% (74.3%, 79.7%) for accuracy and 0.928 (0.928, 0.929) for AUC. Bootstrapping required no further model training and was very quick to calculate confidence intervals.

3.3 Comparisons

To assess the usefulness of our method, we compared our results to results using Inception V3, which has previously been used on the full dataset [10]. To provide a fair comparison we use exactly the same conditions as before, only changing the network itself. The results are presented in Table 1.

Table 1. Our method shows improved performance over the previously used Inception V3 network. All performance measures show improved performance, with the exception of loss in the testing dataset which showed a non-statistically significant improvement. Confidence intervals were calculated using a 1000 iteration bootstrap procedure.

Method	Measure	Training	Validation	Testing (95% confidence interval)
Inception V3	Loss	0.901	0.830	0.729 (0.668, 0.790)
	Accuracy	62.4%	68.1%	67.8% (65.1, 70.9)
	AUC	0.705	0.709	0.912 (0.912, 0.914)
Our method	Loss	**0.433**	**0.347**	0.614 (0.559, 0.677)
	Accuracy	**83.7%**	**86.3%**	**76.6% (74.0, 79.1)**
	AUC	**0.918**	**0.918**	**0.935 (0.935, 0.936)**

4 Discussion and Conclusions

This paper has briefly outlined a two-stage method which combines DAISY descriptors with a deep learning network. This has provided a more data efficient alternative to downsampling. On an example dataset of OCT images, the two-stage method achieved an overall multi-class accuracy of 76.6%, an improvement over a previously used method. With large volumes of high dimensional OCT images, the proposed method has given a quick method of classifying disease type. Acting as a type of 'intelligent downsampling', DAISY descriptors have enabled our method to maintain more of the information contained within images, while only using a three layer recurrent neural network. In future this method can provide a better alternative to random downsampling.

The main limitation of this study is that DAISY descriptors give a 3D output, there are very few pretrained deep neural networks for 3D data such as those for 2D images. Model performance is expected to greatly increase with further exploration of the classification network.

References

1. Adhi, M., Duker, J.S.: Optical coherence tomography-current and future applications (2013)
2. Beck, R.W., et al.: Three-year follow-up of a randomized trial comparing focal/grid photocoagulation and intravitreal triamcinolone for diabetic macular edema. Arch. Ophthalmol. **127**(3), 245–251 (2009)
3. Chen, Y., et al.: Myopic choroidal neovascularisation: current concepts and update on clinical management. Br. J. Ophthal. **99**(3), 289–296 (2014)
4. Cho, K., et al.: Learning Phrase Representations using RNN Encoder-Decoder for Statistical Machine Translation (2014)
5. Chung, J., Gulcehre, C., Cho, K., Bengio, Y.: Empirical Evaluation of Gated Recurrent Neural Networks on Sequence Modeling (2014)
6. De Fauw, J., et al.: Clinically applicable deep learning for diagnosis and referral in retinal disease. Nat. Med. **24**(9), 1342–1350 (2018)
7. Díaz García, J., Brunet Crosa, P., Navazo Álvaro, I., Pérez, F., Vázquez Alcocer, P.P.: Feature-preserving downsampling for medical images. In: EuroVis 2015: The EG/VGTC Conference on Visualization: Posters Track (2015)
8. Efron, B.: Bootstrap methods: another look at the jackknife. Ann. Stat. **7**(1), 1–26 (1979)
9. Green, W.R., Wilson, D.J.: Choroidal neovascularization. Ophthalmology **93**(9), 1169–1176 (1986)
10. Kermany, D.S., et al.: Identifying medical diagnoses and treatable diseases by image-based deep learning. Cell **172**(5), 1122–1124.e9 (2018)
11. Lei, B., Thing, V.L., Chen, Y., Lim, W.Y.: Logo classification with Edge-based DAISY descriptor. In: Proceedings - 2012 IEEE International Symposium on Multimedia, ISM 2012, pp. 222–228. IEEE (2012)
12. Parodi, M.B., Evans, J.R., Virgili, G., Michelessi, M., Bacherini, D.: Laser treatment of drusen to prevent progression to advanced age-related macular degeneration. Cochrane Database of Systematic Reviews (2015)
13. Tola, E., Lepetit, V., Fua, P.: A fast local descriptor for dense matching. In: 26th IEEE Conference on Computer Vision and Pattern Recognition, CVPR, pp. 1–8. IEEE (2008)
14. Tola, E., Lepetit, V., Fua, P.: DAISY: an efficient dense descriptor applied to wide-baseline stereo. IEEE Transact. Pattern Anal. Mach. Intell. **32**(5), 815–830 (2010)

Segmentation of Left Ventricle in 2D Echocardiography Using Deep Learning

Neda Azarmehr[1,2]([⊠]), Xujiong Ye[1], Stefania Sacchi[3], James P. Howard[2],
Darrel P. Francis[2], and Massoud Zolgharni[2,4]

[1] School of Computer Science, University of Lincoln, Lincoln, UK
nAzarmehr@lincoln.ac.uk
[2] National Heart and Lung Institute, Imperial College London, London, UK
[3] Cardiovascular Rehabilitation Department, San Raffaele University Hospital, Milan, Italy
[4] School of Computing and Engineering, University of West London, London, UK

Abstract. The segmentation of Left Ventricle (LV) is currently carried out manually by the experts, and the automation of this process has proved challenging due to the presence of speckle noise and the inherently poor quality of the ultrasound images. This study aims to evaluate the performance of different state-of-the-art Convolutional Neural Network (CNN) segmentation models to segment the LV endocardium in echocardiography images automatically. Those adopted methods include U-Net, SegNet, and fully convolutional DenseNets (FC-DenseNet). The prediction outputs of the models are used to assess the performance of the CNN models by comparing the automated results against the expert annotations (as the gold standard). Results reveal that the U-Net model outperforms other models by achieving an average Dice coefficient of 0.93 ± 0.04, and Hausdorff distance of 4.52 ± 0.90.

Keywords: Deep learning · Segmentation · Echocardiography

1 Introduction

To evaluate the cardiac function in 2D ultrasound images, quantification of the LV shape and deformation is crucial, and this relies on the accurate segmentation of the LV contour in end-diastolic (ED) and end-systolic frames [1, 2]. Currently, the manual segmentation of the LV has the following problems such as, it needs to be performed only by an experienced clinician, the annotation suffers from inter-and intra-observer variability, and it should be repeated for each patient. Consequently, it is a tedious and time-costing task. Therefore, the automatic segmentation methods have been proposed to resolve this issue that can lead to increase patient throughput and can reduce the inter-user discrepancy.

There are many proposed methods for 2D LV segmentation. Recently deep CNN has shown very promising results for image segmentation [8, 9, 11].

This study aims to adapt and evaluate the performance of different state-of-the-art deep learning semantic segmentation methods to segment the LV border on 2D echocardiography images automatically. The rest of the paper is structured as follows. In Sect. 2,

© Springer Nature Switzerland AG 2020
Y. Zheng et al. (Eds.): MIUA 2019, CCIS 1065, pp. 497–504, 2020.
https://doi.org/10.1007/978-3-030-39343-4_43

the dataset and the several neural networks models are described. In Sect. 3, evaluation measures of the performance and accuracy of the neural network are addressed. Experimental results and discussion are presented in Sect. 4. Finally, conclusion and future work are provided in Sect. 5.

2 Methodology

2.1 Dataset

The study population consisted of 61 patients (30 males), with a mean age of 64 ± 11, who were recruited from patients who had undergone echocardiography with Imperial College Healthcare NHS Trust. Only patients in sinus rhythm were included. No other exclusion criteria were applied. The study was approved by the local ethics committee and written informed consent was obtained.

Each patient underwent standard Transthoracic echocardiography using a commercially available ultrasound machine (Philips iE33, Philips Healthcare, UK), and by experienced echocardiographers. Apical 4-chamber views were obtained in the left lateral decubitus position as per standard clinical guidelines [3].

All recordings were obtained with a constant image resolution of 480×640 pixels. The operators performing the exam were advised to optimise the images as would typically be done in clinical practice. The acquisition period was 10 s to make sure at least three cardiac cycles were present in all cine loops. To take into account, the potential influence of the probe placement (the angle of insonation) on the measurements, the entire process was conducted three times, with the probe removed from the chest and then placed back on the chest optimally between each recording. A total of three 10-s 2D cine loops was, therefore, acquired for each patient. The images were stored digitally for subsequent offline analysis.

To obtain the gold-standard (ground-truth) measurements, one accredited and experienced cardiology expert manually traced the LV borders. Where the operator judged a beat to be of extremely low quality, the beast was declared invalid, and no annotation was made. We developed a custom-made program which closely replicated the interface of echo hardware. The expert visually inspected the cine loops by controlled animation of the loops using arrow keys and manually traced the LV borders using a trackball for the end-diastolic and end-systolic frames. Three heartbeats (6 manual traces for end-diastolic and end-systolic frames) were measured within each cine loop. Out of 1098 available frames (6 patients $\times$ 3 positions $\times$ 3 heartbeats $\times$ 2 ED/ES frames), a total of 992 frames were annotated. To investigate the inter-observer variability, a second operator repeated the LV tracing on 992 frames, blinded to the judgment of the first operator. A typical 2D 4-chamber view is shown in Fig. 1, where the locations of manually segmented endocardium by the two operators are highlighted.

2.2 Neural Network for Semantic Segmentation

All images were resized to a smaller dimension of 320×240 pixels for feeding into the deep learning models. From the total of 992 images, 595 (60%) were randomly selected

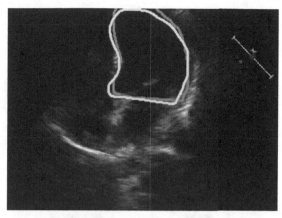

Fig. 1. An example 2D 4-chamber view. The blue and yellow curves represent the annotations by Operator-A and Operator-B, respectively. (Color figure online)

for training, 20% of total data used for validation, and the remaining 20% was used for testing.

Standard and well-established U-Net neural network architecture was firstly used since this architecture is applicable to multiple medical image segmentation problems [4]. The U-Net architecture comprises of three main steps such as down-sampling, up-sampling steps and cross-over connections. During the down-sampling stage, the number of features will increase gradually while during up-sampling stage the original image resolution will recover. Also, cross-over connection is used by concatenating equally size feature maps from down-sampling to the up-sampling to recover features that may be lost during the down-sampling process.

Each down-sampling and up-sampling has five levels, and each level has two convolutional layers with the same number of kernels ranging from 64 to 1024 from top to bottom correspondingly. All convolutions kernels have a size of (3 × 3). For down-sampling Max pooling with size (2 × 2) and equal strides was used.

In addition to the U-net, SegNet and FC-DenseNet models were also investigated. The SegNet model contains an encoder stage, a corresponding decoder stage followed by a pixel-wise classification layer. In SegNet model, to accomplish non-linear up-sampling, the decoder performs pooling indices computed in the max-pooling step of the corresponding encoder [5]. The number of kernels and kernel size was the same as the U-Net model.

FC-DenseNet model is a relatively more recent model which consists of a down-sampling and up-sampling path made of dense block. The down-sampling path is composed of two Transitions Down (TD) while an up-sampling path is containing two Transitions Up (TU). Before and after each dense block, there is concatenation and skip connections (see Fig. 2). The connectivity pattern in the up-sampling is different from the down-sampling path. In the down-sampling path, the input to a dense block is concatenated with its output, leading to linear growth of the number of feature maps, whereas in the up-sampling path, it is not [6].

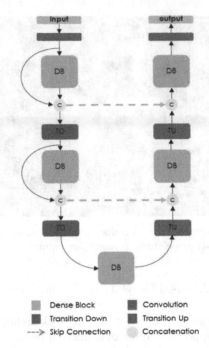

Fig. 2. Diagram of FC-DenseNet architecture for semantic segmentation [6].

All models produce the output with the same spatial size as the input image (i.e., 320 × 240). Pytorch was used for the implementations [10], where Adam optimiser with 250 epochs and learning rate of 0.00001 were used for training the models. The network weights are initialised randomly but differ in range depending on the size of the previous layer [7]. Negative log-likelihood loss is used as the network's objective function. All computations were carried using an Nvidia GeForce GTX 1080 Ti GPU.

All models were trained separately and indecently using the annotations provided by either of the operators, and following acronyms are used for the sake of simplicity: GT_{OA} and T_{OB} as ground-truth segmentations provided by Operator-A and Operator-B, respectively; P_{OA} and P_{OB} as Predicted LV borders by deep learning models trained using GT_{OA} and T_{OB}.

3   Evaluation Measures

The Dice Coefficient (DC), Hausdorff distance (HD), and intersection-over-union (IoU) also known as the Jaccard index were employed to evaluate the performance and accuracy of the CNN models in segmenting the LV region. The DC (1) was calculated to measure the overlapping regions of the Predicted segmentation (P) and the ground truth (GT). The range of DC is a value between 0 and 1, which 0 indicates there is not any overlap between two sets of binary segmentation results while 1, indicates complete overlap.

$$DC = \frac{2|P \cap GT|}{|P| + |GT|} \tag{1}$$

Also, the HD was calculated using the following formula for the contour of segmentation where, $d(j, GT, P)$ is the distance from contour point j in GT to the closest contour point in P. The number of pixels on the contour of GT and P specified with O and M respectively.

$$HD = \max\left(max_{j\in[0,O-1]}d(j, GT, P), max_{j\in[0,M-1]}d(j, P, GT)\right) \quad (2)$$

Moreover, the IoU was calculated image-by-image between the Predicted segmentation (I_P) and the ground truth (GT). For a binary image (one foreground class, one background class), IoU is defined for the ground truth and predicted segmentation GT and I_P as

$$IoU(GT, I_P) = \frac{|GT \cap I_P|}{|GT \cup I_P|} \quad (3)$$

4   Experiment Results and Discussion

Figure 3 shows example outputs from the three models when trained using annotation provided by Operator-A (i.e., GT_{OA}). The contour of the predicted segmentation was used to specify the LV endocardium border. The red, solid line represents the automated results, while the green line represents the manual annotation.

As can be seen, the U-Net model achieved higher DC (0.98), higher IoU (0.99), and lower HD (4.24) score. A visual inspection of the automatically detected LV border also confirms this. The LV border obtained from the SegNet and FC-DenseNet models seems to be less smooth compared to that in the U-Net model. However, all three models seem to perform with reasonable accuracy.

U-Net	SegNet	FC-DenseNet
DC = 0.98	0.96	0.91
HD = 4.24	6	6.78
IoU = 0.99	0.98	0.96

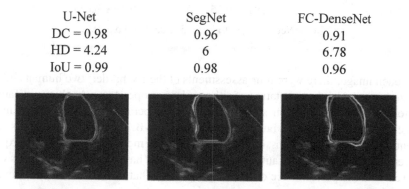

Fig. 3. Typical outputs from U-Net, SegNet, and FC-DenseNet models. (Color figure online)

Figure 4 illustrates the results for a sample failed case, for which all three models seem to struggle with the task of LV segmentation. By closer scrutiny of the echo images for such cases, it is evident that the image quality tends to be lower due to missing borders,

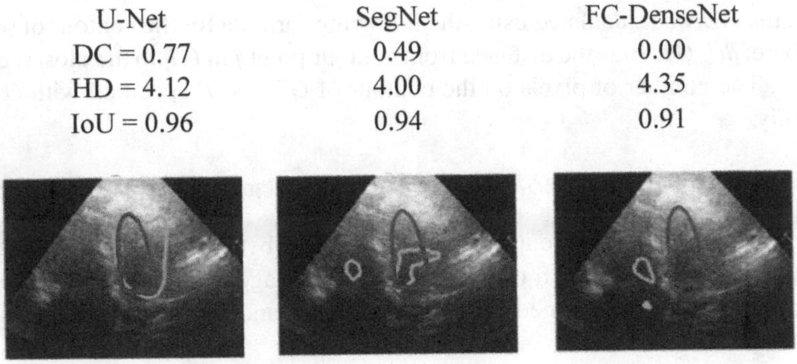

U-Net	SegNet	FC-DenseNet
DC = 0.77	0.49	0.00
HD = 4.12	4.00	4.35
IoU = 0.96	0.94	0.91

Fig. 4. Failed case example outputs from U-Net, SegNet, and FC-DenseNet models.

presence of speckle noise or artefacts, and poor contrast between the myocardium and the blood pool.

Table 1 provides the average Dice coefficient, Hausdorff distance, and Intersection-over-Union for the three models, across all testing images (199 images). The U-Net model, in comparison with the SegNet and FC-DenseNet models, achieved relatively better performance. The average Hausdorff distance, however, was higher for the FC-DenseNet, compared to the other two models.

Table 1. Comparison of evaluation measures of dice coefficient (DC), Hausdorff distance (HD), and intersection-over-union (IoU) between the three examine models, expressed as mean ± SD.

Model	DC	HD	IoU
U-Net	0.93 ± 0.04	4.52 ± 0.90	0.98 ± 0.01
SegNet	0.91 ± 0.06	4.65 ± 0.89	0.98 ± 0.01
FC-DenseNet	0.84 ± 0.11	5.05 ± 0.69	0.96 ± 0.02

For each image, there were four assessments of the LV border; two human and two automated (trained by the annotation of either of human operators). As shown in Table 2, the automated models perform similarly to human operators. The automated model disagrees with the Operator-A, but so does the Operator-B. Since different experts make different judgments, it is not possible for any automated model to agree with all experts. However, it is desirable for the automated models do not have larger discrepancies when compared with the performance of human judgments; that is, to behave approximately as well as human operators.

Table 2. Comparison of evaluation measures (Dice coefficient, Hausdorff distance, and intersection-over-union) for the U-Net model between five possible scenarios.

Compared scenarios	DC	HD	IoU
OA vs OB	**0.88 ± 0.06**	**4.50 ± 0.87**	**0.83 ± 0.03**
P_{OA} vs OA	0.93 ± 0.04	4.52 ± 0.90	0.98 ± 0.01
P_{OA} vs OB	**0.89 ± 0.04**	**4.76 ± 0.91**	**0.97 ± 0.01**
P_{OB} vs OB	0.91 ± 0.05	4.87 ± 0.85	0.98 ± 0.01
P_{OB} vs OA	**0.89 ± 0.06**	**4.82 ± 0.82**	**0.98 ± 0.01**

5 Conclusion and Future Work

The time-consuming and operator-dependent process of manual annotation of left ventricle border on a 2D echocardiographic recording could be assisted by the automated models that do not require human intervention. Our study investigated the feasibility of such automated models which perform no worse than human experts.

The automated models demonstrate larger discrepancies with the gold-standard annotations when encountered with the lower image qualities. This is potentially caused by the lack of balanced data in terms of different image quality levels. Since the patient data in our study was obtained by the expert echocardiographers, the distribution leans more towards higher average and higher quality images. This may result in the model forming a bias towards the more condensed quality-level images. Future investigations will examine the correlation between the performance of the deep learning model and the image qualities, as well as using more balanced datasets.

The patients were a convenience sample drawn from those attending a cardiology outpatient clinic. They, therefore, may not be representative of patients who enter trials with particular enrolment criteria or of inpatients or the general population. A further investigation will look at a wide range of subjects in any cardiovascular disease setting. The segmentation of other cardiac views, and using data acquired by various ultrasound vendors can also be considered for a comprehensive examination of the deep learning models in echocardiography.

Acknowledgements. N.A. was supported by the School of Computer Science PhD scholarship at the University of Lincoln.

References

1. Raynaud, C., et al.: Handcrafted features vs ConvNets in 2D echocardiographic images. In: 2017 IEEE 14th International Symposium on Biomedical Imaging (ISBI 2017), Melbourne, Australia, pp. 1116–1119. IEEE (2017)
2. Lang, R.M., et al.: Recommendations for cardiac chamber quantification by echocardiography in adults: an update from the American Society of Echocardiography and the European Association of Cardiovascular Imaging. Eur. Heart J. Cardiovasc. Imaging 16(3), 233–271 (2015)

3. Zhang, J., et al.: Fully automated echocardiogram interpretation in clinical practice: feasibility and diagnostic accuracy. Circulation **138**(16), 1623–1635 (2018)
4. Ronneberger, O., Fischer, P., Brox, T.: U-Net: convolutional networks for biomedical image segmentation. In: Navab, N., Hornegger, J., Wells, W.M., Frangi, A.F. (eds.) MICCAI 2015. LNCS, vol. 9351, pp. 234–241. Springer, Cham (2015). https://doi.org/10.1007/978-3-319-24574-4_28
5. Badrinarayanan, V., Kendall, A., Cipolla, R.: SegNet: a deep convolutional encoder-decoder architecture for image segmentation. IEEE Trans. Pattern Anal. Mach. Intell. **39**(12), 2481–2495 (2017)
6. Jégou, S., Drozdzal, M., Vazquez, D., Romero, A., Bengio, Y.: The one hundred layers tiramisu: fully convolutional DenseNets for semantic segmentation. In: 30th Proceedings of the IEEE Conference on Computer Vision and Pattern Recognition Workshops, Honolulu, Hawaii, pp. 11–19. IEEE (2017)
7. He, K., Zhang, X., Ren, S., Sun, J.: Delving deep into rectifiers: surpassing human-level performance on ImageNet classification. In: International Conference on Computer Vision, Santiago, Chile, pp. 1026–1034. IEEE (2015)
8. Smistad, E., Østvik, A.: 2D left ventricle segmentation using deep learning. In: 2017 IEEE International Ultrasonics Symposium (IUS), Washington DC, United States, pp. 1–4. IEEE (2017)
9. Jafari, M.H., et al.: A unified framework integrating recurrent fully-convolutional networks and optical flow for segmentation of the left ventricle in echocardiography data. In: Stoyanov, D., et al. (eds.) DLMIA/ML-CDS 2018. LNCS, vol. 11045, pp. 29–37. Springer, Cham (2018). https://doi.org/10.1007/978-3-030-00889-5_4
10. Paszke, A., et al.: Automatic differentiation in PyTorch. In: 31st Conference on Neural Information Processing Systems (NIPS 2017), Long Beach, CA, USA, pp. 1–4 (2017)
11. Goceri, E., Goceri, N.: Deep learning in medical image analysis: recent advances and future trends. In: International Conferences Computer Graphics, Visualization, Computer Vision and Image Processing, July 2017

Author Index

Printed in the United States
By Bookmasters